Intelligent Systems Reference Library

Volume 207

The aim of this series is to publish a Reference Library, including novel advances and developments in all aspects of Intelligent Systems in an easily accessible and well structured form. The series includes reference works, handbooks, compendia, textbooks, well-structured monographs, dictionaries, and encyclopedias. It contains well integrated knowledge and current information in the field of Intelligent Systems. The series covers the theory, applications, and design methods of Intelligent Systems. Virtually all disciplines such as engineering, computer science, avionics, business, e-commerce, environment, healthcare, physics and life science are included. The list of topics spans all the areas of modern intelligent systems such as: Ambient intelligence, Computational intelligence, Social intelligence, Computational neuroscience, Artificial life, Virtual society, Cognitive systems, DNA and immunity-based systems, e-Learning and teaching, Human-centred computing and Machine ethics, Intelligent control, Intelligent data analysis, Knowledge-based paradigms, Knowledge management, Intelligent agents, Intelligent decision making, Intelligent network security, Interactive entertainment, Learning paradigms, Recommender systems, Robotics and Mechatronics including human-machine teaming, Self-organizing and adaptive systems, Soft computing including Neural systems, Fuzzy systems, Evolutionary computing and the Fusion of these paradigms, Perception and Vision, Web intelligence and Multimedia.

Indexed by SCOPUS, DBLP, zbMATH, SCImago.

All books published in the series are submitted for consideration in Web of Science.

More information about this series at http://www.springer.com/series/8578

Md Atiqur Rahman Ahad ·
Atsushi Inoue
Editors

Vision, Sensing and Analytics: Integrative Approaches

Springer

Editors
Md Atiqur Rahman Ahad
Professor
Department of Electrical
and Electronic Engineering
University of Dhaka
Dhaka, Bangladesh

Specially Appointed Associate Professor
Department of Intelligent Media
Osaka University
Suita, Japan

Atsushi Inoue
Solutions Architect for Greenfield
and Startups
Amazon Web Services, USA

Visiting Professor
Graduate School of Regional Innovation
Mie University, Japan

ISSN 1868-4394 ISSN 1868-4408 (electronic)
Intelligent Systems Reference Library
ISBN 978-3-030-75489-1 ISBN 978-3-030-75490-7 (eBook)
https://doi.org/10.1007/978-3-030-75490-7

This Springer imprint is published by the registered company Springer Nature Switzerland AG
The registered company address is: Gewerbestrasse 11, 6330 Cham, Switzerland

Foreword

It is clear that analyzing large amounts of data, especially visual and other sensor data, has great utility and an extremely wide range of important applications and opportunities. Fundamental advances in sensing, in machine learning, in computer vision, and in data science will impact society in multiple ways, including the critical areas of medicine and healthcare. This volume, edited by Md Atiqur Rahman Ahad and Atsushi Inoue, provides multiple perspectives on several key topics of interest, including advances in deep learning, sensing and understanding people, advanced data analysis, contactless sensing and interaction, and biomedical and healthcare analyses.

The editors and chapter authors share their considerable expertise throughout the book.

The chapters describe recent research and informed perspectives in these topics, and they will be of significant interest to students, researchers, engineers, and managers seeking to gain broad insight and to understand state-of-the-art advances in vision, sensing, and analytics. From fundamental research to timely COVID-19 applications, the book focuses on integrative approaches in conceiving and applying these advances.

I expect that readers will enjoy the variety of topics and learn a great deal about trends in vision, sensing, and analytics. I hope that many readers will ask "What can be done next? What should be done?" and continue to advance the field in their own work.

Kudos to all of the contributors for a fine and timely contribution.

April 2021 Matthew Turk

Preface

Recent technology advancements in vision, sensing, and analytics have brought to us new trends and made significant impacts in our societies. Especially, the advancement of their tools and services opens the door to highly impactful innovations and applications as a result of effective and efficient use of them rather than making them from scratch. Such advancements have brought to scientists and engineers different skills and mind-sets. These include but not are necessarily limited to rapid development frameworks, a scalable architecting, and system design driven by applications, users, and their contexts. This demands a new guideline with different perspectives, especially for our young and new generations.

This book, *Vision, Sensing and Analytics: Integrative Approach*, aims at such a new guideline. We present a collection of carefully selective cases based on contributions from multiple IEEE technically co-sponsored international conferences related to vision, sensing, and analytics (http://cennser.org/ICIEV, http://cennser.org/IVPR). Furthermore, we focus on cases with significant added values as a result of integrations, e.g., multiple sensing, analytics with different data sources, and comprehensive monitoring with many different sensors. We also commit those contributions to be readable by young and new generations such as senior-level undergraduate and early-year graduate students to stimulate their innovations.

This book consists of introductory monographs, R&D cases, and future aspects of related matters that are contributed by top specialists with great diversity of gender, experience levels, and demography (i.e., from over 20 different universities and institutes from 8 countries). This is responsive to our social issues too. Majority of contests are related to biomedical and health care, and there are three contributions directly addressing COVID-19. Such biomedical and healthcare matters are so sophisticated as their nature and in strongly regulated contexts that the integrative approaches are mandatory for its rapid development and scalable architectures. We are confident that this book presents it quite effectively.

We are grateful to Prof. Matthew Turk (Fellow, IEEE; Fellow, IAPR; President, Toyota Technological Institute at Chicago, USA) for his enormous time to write the Foreword for the book. The journey of editing this book is extremely challenging, and we took a very lengthy process to finalize it. Our review process has been significantly more rigorous (e.g., 4–7 times of correspondence) than usual book editing and journal article reviews. This is extremely important in order to sustain our quality in this new guideline. We hope that readers find this book as a unique, irreplaceable guideline of the integrative approach.

Md Atiqur Rahman Ahad
Atsushi Inoue

Contents

1 **Deep Architectures in Visual Transfer Learning** 1
Walid Gomaa

2 **Deep Reinforcement Learning: A New Frontier in Computer Vision Research** 29
Sejuti Rahman, Sujan Sarker, A. K. M. Nadimul Haque, and Monisha Mushtary Uttsha

3 **Deep Learning for Data-Driven Predictive Maintenance** 71
Muhammad Sohaib, Shiza Mushtaq, and Jia Uddin

4 **Multi-criteria Fuzzy Goal Programming Under Multi Uncertainty** 97
Junzo Watada, Nureize Binti Arbaiy, and Qiuhong Chen

5 **Skeleton-Based Human Action Recognition on Large-Scale Datasets** 125
Tonmoy Hossain, Sujan Sarker, Sejuti Rahman, and Md Atiqur Rahman Ahad

6 **Sensor-Based Human Activity and Behavior Computing** 147
Anindya Das Antar, Masud Ahmed, and Md Atiqur Rahman Ahad

7 **Radar-Based Non-Contact Physiological Sensing** 177
Shekh Md Mahmudul Islam

8 **Biomedical Radar and Antenna Systems for Contactless Human Activity Analysis** 213
Aniqa Tabassum and Md Atiqur Rahman Ahad

9 **Contactless Monitoring for Healthcare Applications** 243
K. M. Talha Nahiyan and Md Atiqur Rahman Ahad

10 Personalized Patient Safety Management: Sensors and Real-Time Data Analysis 267
Md. Jasim Uddin and Monika Nasrin Munni

11 Electrical Impedance Tomography Based Lung Disease Monitoring 307
Aniqa Tabassum and Md Atiqur Rahman Ahad

12 Image Analysis with Machine Learning Algorithms to Assist Breast Cancer Treatment 327
Abu Asaduzzaman, Fadi N. Sibai, Shigehiko Kanaya, Md. Altaf-Ul-Amin, Md. Jashim Uddin, Kishore K. Chidella, and Parthib Mitra

13 Role-Framework of Artificial Intelligence in Combating the COVID-19 Pandemic 357
Mohammad Shorif Uddin, Sumaita Binte Shorif, and Aditi Sarker

14 Time Series Analysis for CoVID-19 Projection in Bangladesh 371
Kawser Ahammed and Mosabber Uddin Ahmed

15 Challenges Ahead in Healthcare Applications for Vision and Sensors 397
Manan Binth Taj Noor, Nusrat Zerin Zenia, and M. Shamim Kaiser

Chapter 1
Deep Architectures in Visual Transfer Learning

Walid Gomaa

Abstract Deep learning is a recent form of machine learning that depends on structural compositional models that can represent mapping functions that otherwise require an exponentially larger size flat models. The most successful realization of such learning paradigm is deep neural networks that have in recent years achieved state-of-the-art performance in tasks related, in particular, to visual data, audio data, and natural language processing. The main drawback of such methods is that, albeit their superhuman performance, that performance is achieved solely on very specific individual tasks. Intelligence must however, be broader and more general. So the next big research program is to build machines that are smart in a more general sense over multiple tasks and domains. One way to achieve that is through transfer learning. Transfer learning refers to a broad set of techniques, all aimed towards the reuse of knowledge gained from solving some problem towards the solution of some other problem. In this paper we study the effectiveness of known deep architectures in transfer learning in visual tasks. We consider two of the VGG family, namely, VGG16 and VGG19, the Xception architecture, DenseNet121, and finally the ResNet50 architecture. They are already pretrained on the ImageNet dataset, we tune and test their transfer performance on the Caltech-256 image set. Four of these architectures have shown good training/validation performance on the latter dataset. In addition, DenseNet121 and Xception have shown exceptional superior performance over the VGG variants and ResNet50, though thay are an order of magnitude less in size. However, they are an order of magnitude deeper confirming the lasting conjecture that deep compositional architectures are exponentially more representative and expressive than flatter networks.

Keywords Deep learning · ImageNet · VGG architectures · Xception architecture · Object recognition · Transfer learning · DenseNet121 · ResNet50

W. Gomaa (✉)
Cyber Physical Systems Lab., Egypt Japan University of Science and Technology, New Borg El-Arab City, Alexandria, Egypt
e-mail: walid.gomaa@ejust.edu.eg

Faculty of Engineering, Alexandria University, Alexandria, Egypt

M. A. R. Ahad and A. Inoue (eds.), *Vision, Sensing and Analytics: Integrative Approaches*, Intelligent Systems Reference Library 207, https://doi.org/10.1007/978-3-030-75490-7_1

1.1 Introduction

Object recognition refers to a plethora of automatic visual tasks involving the identification, classification, and localization objects in digital photos. *Image classification* is concerned with the prediction of the objects' classes in a given image. *Object localization* aims at identifying the spatial locations of one or more objects in a given image, such identification can be represented by drawing a rectangular bounding box around the spatial extent of the identified object. *Object detection* merges these latter tasks into the localization and classification of one or more objects in a given digital image. In order for all such tasks to be realized in the modern sense, we need large image datasets in addition to powerful enough deep network architectures in order to huge amounts of computational resources.

At the core of modern computer vision research is the existence of datasets that are large enough and highly qualitative [51]. These are needed at all stages of the (deep) learning pipeline in visual tasks from training to validation to testing. They should cover all recent relevant visual tasks including classification, detection, segmentation, and localization algorithms. The most notably of such datasets is ImageNet [17, 38]. This dataset has been the driving force behind all recent advances in machine learning in general, and neural computing in particular, specifically the resurgence of the latter after the last winter. The publicly available image datasets such as UIUC [7], Caltech 4 [22], Caltech 101 [43], and Caltech-256 [23] have contributed a major role in the recent upsurge in modern computer vision research, advancing the field by providing a common ground for algorithms development and evaluation.

Deep neural networks DNNs have dominated the landscape of statistical machine learning with the publication of the seminal work of Hinton and Salakhutdinov [34]. Since then there have been tremendous shift in the machine learning research and industrial communities, towards the analysis of deep learning methods, creation of new architectures and optimization methodologies, application in very diverse range of domains, and even innovating new applications not imaginable before the introduction of such new paradigm. This can be mainly attributed to the huge performance gains in predictive performance [39] constantly and continually defining state-of-the-art performance in almost all domains at an unprecedented rate. Deep methods have, in some sense, refined and justified our previous intuitions about neural computing. The most prominent of which is the notion of *hierarchical representation* as articulated in [14, 34], where (deep) neural networks are believed to learn and represent the data features in an incremental, *compositional* manner. This recursive compositionlity is very powerful in the sense that it allows for exponential splitting of the input space at a polynomial cost (in terms of the network depth). This hierarchical representation has been empirically validated in some works including [56] and [29] where the learned features and activation maps are visualized in image-based convolutional neural networks. A direct implication of this sort of *hierarchical learning* is that it enables the **sharing** or **re-use** of knowledge (at all levels of representation from lower to upper levels) gained from solving one problem in the solution of another problem. For example, the problem of recognizing two different objects shares similar requirements at a lower level of representation such as boundaries, edges, and/or constituent geometric components of the objects, up until some point where the visual characteristics are sufficiently different that allows for the discrimination between the two objects (e.g., cats and flowers have completely different appearances when considered in their entirety (the shape), however, they are both composed of lines and curves). Hence, use deep networks have become more and more feasible through the reuse of lower level knowledge in solving several problems, saving tremendous amount of computational resources and helping in democratizing deep learning, and machine learning in general. This is described as *transfer learning* [50] and is very commonly used in computer vision tasks [40, 49].

Transfer learning refers to the situation where what has been learned in one setting and/or task exploited to improve generalization in another setting and/or task [37]. From some perspective it can be considered as the improvement of learning in a new target task through the transfer of learned knowledge from a related base/source task [21]. Transfer learning, though boosted by the recent revolution in deep learning, is not particularized to the latter. The concept has popped up in some way or another in several historical contexts. For example, in the NIPS 1995 workshop Learning to Learn: knowledge fusion and transfer is believed to have provided the initial drive and motivation for research in this field The key motivations, especially in the context of deep learning is that, solving complex enough problems needs both huge computational power along side vast amount of labeled data. In most cases the latter is neither available nor feasible to manually annotate such data. A rather recent survey on the progress of transfer learning for classification, regression, and clustering problems can be found in [50].

Lisa Torrey and Jude Shavlik [21] described three possible benefits to gain when applying transfer learning on a target task using a pretrained model on a source task:

1. **Higher start:** The initialization of the target is higher than train-from-scratch configuration.
2. **Higher rate:** The rate at which the target task improves over time during fine-tuning of the pretrained source model is higher than the train-from-scratch configuration.
3. **Higher asymptote/limit:** The convergence of the pretrained source model on the target task is better than it otherwise would be.

These benefits are more concretely illustrated in Fig. 1.1 where it hows two performance curves. The green curve is the training performance using trained-from-scratch model on the target task whereas the red curve shows the performance of the fine-tuned source model. On a more foundational level, the work done in [55] empirically studies the quantitative contribution of each neuron in each layer of the deep convolution network in the specificity to the current task vs. the generality beyond the current task. The authors concluded from their results that transferability is inversely affected by two factors: (1) the specialization/customization of higher layer neurons to their original source task compromises their performance on the target task, and (2) hardness of optimization in regard to splitting the network between co-adapted neurons. They validated their hypothesis using an example network trained on ImageNet and showed that either of these two issues may dominate, depending on the network location where the features are transferred, that is, whether the features are transferred from the bottom, middle, or top of the network. They also concluded that, in addition to being self-evident, transferability of features degrades as the distance between the source base task and target task increases, but in any case transferability is better than random initialization (training-from-scratch). A final result that enforces the previous phrase is that initializing a network with transferred features from almost anywhere at any level of layers can produce a boost to generalization, though the quality, effectiveness, and rate of performance improvement depend on the particular layer at which the feature transfer occurs.

In this paper we examine the effectiveness of the transfer learning methodology using some state-of-the-art deep learning architectures. We pick five architectures. The first two are a variant of the VGG architecture, the third is the Xception network, the fourth and fifth are two attempts to have effective and efficient very deep networks which are DenseNet121 and ResNet50. These networks are already pretrained on the ImageNet dataset. We use these networks for transfer learning in the following way: We download the pretrained network excluding the top layers (these are the layers tailored to the supervised learning task), we then add a convolution layer and a final dense layer of softmax units for image classification, finally, we fine tune the network by training on some given dataset that defines that task at hand. We compare the effectiveness of transfer learning using these deep architectures over a rather hard task of image classification in the Caltech-250 dataset. The Caltech-256 [23] consists of 30,607 images comprising 257 categories. Our results confirm

Fig. 1.1 Different ways of learning improvement using transfer learning

the theoretical hypothesis that drive the empirical successes which is that deeper network architectures achieve better predictive results that at a reasonable number of computing nodes that would need an exponentially more nodes at wide shallow networks. This is manifested in the superior predictive performance achieved by the DenseNet121 and Xception networks over the VGGs at lesser number of training parameters.

The paper is organized as follows. Section 1.1 is an introduction. Section 1.2 gives background and some literature review about transfer learning. Section 1.3 introduces the main datasets used in this work, and the deep architectures used over these datasets are presented in Sect. 1.4. The experimental work, results, and discussions are given in Sect. 1.5. Finally, the paper concludes with brief discussion about this works and points to future research in Sect. 1.6.

1.2 Transfer Learning

Transfer learning refers to a broad set of techniques, all aimed towards the reuse of knowledge gained from solving some problem towards the solution of some other problem. The notion of skill transfer for purposes of performance enhancements on disparate tasks has been studied extensively in different contexts [54]. In [42], the authors propose the transfer of extracted features from one source dataset to another target dataset. A Deep Sparse Autoencoder (DSAE) is trained on raw pre-extracted feature vectors in order to act as a feature extractor. Then, three Support Vector Machine classifiers SVCs are trained: (1) one SVC is trained on the features extracted from source-domain samples using DSAE, (2) another SVC is trained on features extracted from target domain samples using DSAE and (3) the last SVC is trained on raw low-level features also extracted from target domain samples. The authors use fusion techniques in order to combine the outputs of the SVCs to yield a final decision. They as well compare their approach to two state-of-the-art transfer learning based classifiers and report better performance than both of them.

In [15], transfer learning scheme based on convolution neural network (CNN) is proposed in the domain of human activity recognition from inertial motion data. The authors pretrain a CNN classifier on a source/base domain, and then use the learned weights as initialization for anther CNN that is structured similarly, however, used for another target domain. They studied the efficacy and effectiveness of their method over three datasets spanning different users, device placements and sensor modalities. They report $F1$-scores up to 0.94 in some of the evaluated scenarios. The authors in [50] introduce a taxonomy for transfer learning approaches. Under this taxonomy, transfer learning in the manner adopted in deep learning may be considered to be *Inductive* since labeled data are available for the transfer/new task. Specifically, it involves the transfer of feature representations between tasks. Other inductive approaches include instance-based transfer (when data from one task are repurposed for use by another task) parameter-based (when parameters are shared between tasks) or relational (when knowledge of relationships from one task are reused in another task).

The authors in [35] repurpose the activity instances gathered in the performance of some task for the training of a classifier on another set of instances. They utilize two activity datasets publicly available on the web. They authors built a similarity map that transforms an instance of some activity in the source domain to an instance of some other activity in the target domain. Their empirical work showed the feasibility and effectiveness of their approach.

The authors in [41] have proposed a kind of transfer learning that is instance-based. In their work labelled samples from one set of activities (i.e., the source dataset) are re-use as samples for matching common activities in another activity set (the target dataset), based on a **transferability metric**. Afterwards, they used K-Means clustering to detect anomalous clusters of the unseen/uncommon activities in the target activity set. They train a classifier on samples from both the common and uncommon activities. The resultant models govern how new instances are classified. For instance, given a new sample a discriminatory classifier is used to place the sample as belonging to a common or uncommon activity. Accordingly, the new sample is matched either against a classifier based on the boosted samples or against the anomalous clusters. The authors experiment using three datasets achieving up to 85% recall in inter-dataset testing, where the target dataset contains previously unseen activities.

In [45] the authors use a pretrained deep convolutional neural network with transfer learning for the purpose of detecting several kinds of damage to old buildings in the cities of Fez and Meknes in Morocco. They validated the robustness of their approach on different architectures and a small set of images not used in the learning and validation phase. Another vision-based application of transfer learning is done in the transportation arena. The work done by [28] employs a neural network classifier pretrained on ImageNet for the automatic detection of pavement distress and cracks. The authors in [44] use a pretrained convolutional network to recognize the type of neonatal pain expression. They contend that the use of transfer learning alleviates the occurrence of over-fitting as well as accelerate the training procedure. In principal transfer learning is essential in such application. Learning-from-scratch would require the existence of large scale **labeled** datasets in order to achieve reasonable predictive performance. However, this kind of labeled data is not available for neonatal pain expression. They use several state-of-the-art pretrained networks such as AlexNet, VGG-16, Inception-V3, ResNet-50 and Xception. They used as an unsupervised feature extractors, and then fine-tuned using the small neonatal pain expression image dataset. The fine-tuned VGG-16 achieved the best recognition accuracy of 78.3%, indicating that fine-tuning on a small dataset can lead to effective results. Such medical application, among others, show much promise in the effective use of machine learning tools and technologies in clinical diagnosis. Another medical work is done by [13]. The authors apply transfer learning with pretrained deep convolutional networks to identify 11 different types of serous cells in effusion cytology. They used four different kinds of pretrained network including AlexNet, GoogleNet, ResNet and DenseNet. These are used as feature extractors and then fine-tuned on a serous cell dataset. The authors evaluated their approach on both original and augmented sets of serous cells. Among the four networks, ResNet and DenseNet obtained the highest accuracies of 93.44% and 92.90%.

The use of transfer learning for sequential or time-series data has been made possible through the use of recurrent neural networks and their variants, which are generally difficult to train and prone to overfitting. In [4] the authors developed an approach for transfer learning applicable to fixed and variable-length activity recognition time-series data. They train a convolutional neural network, use it (excluding the top layers)as a feature extractor. And finally, they train a feedforward neural network as a classifier over the extracted features for other datasets. The approach is evaluated on five activity (inertial time-series) datasets. The experiments showed the efficacy effectiveness of the approach. The results stayed within 5% of trained-from-scratch training whilst obtaining a huge speedup of 24–52× in the training time. The same authors have done a similar work in [3], using deep metric learning. They used a deep Triplet network in order to generate fixed-length descriptors from activity samples. These descriptors are used for activity classification. They evalauted their approach again on five activity datasets which radically different in their data collection process and the underlying activities. They achieved classification accuracies up to 96% in self-testing scenarios (training-from-scratch) and up to 91% in cross-dataset testing without retraining (transfer learning without even fine-tuning).

A very interesting work that carries the spirit of transfer learning, even though it is not strictly so is the work carried out in [5]. Collection and annotation of training samples in IMU-based activity recognition pose significant difficulties and can very much reduce the performance of models trained on such limited data. Hence, there is an urgency need to find techniques and approaches that can tackle this problem. One such scenario is explored in [5] where the authors investigate the feasibility of reusing inertial streams collected from different "source" body locations for activity recognition at different "target" body locations. This is done through the use of bidirectional recurrent neural networks for mapping source locations to target locations. They denote such mapping as "roaming". The authors studied the predictive performance of the transferred samples relative to the performance resulting from samples collected genuinely at the target body locations. The results indicate that such roaming models can permit the reuse of cross-body samples. More specifically, they validated their approach using the REALDISP dataset [12], which is an activity dataset (33 activities) collected from multiple sensors placed at different body locations. The results obtained using this dataset indicate the feasibility and effectiveness of roaming models between proximal body locations, and the highest sustained performance is obtained, in particular, from activities whose dynamics are sufficiently captured by the source location.

Almost all the previous discussion assume transfer learning via automatic feature extraction using the deep convolutional neural network paradigm. It is worth investigating transfer learning via feature engineering. Consider again human activity recognition using inertial motion data. The work done in [1] extracts two kinds of features: coefficients coming from the wavelet transform of the input time-series and the coefficients of vector autoregressive models of these time-series. These are used individually to train four classifiers including variants of random forests. Empirical work is done over three datasets each is used to train-from-scratch the four classifiers. It is then worth investigating the cross-testing performance, that is, train-from-scratch over a dataset and cross-test on the other ones. This can be said as well of the work done in [27] where the autocorrelation function of inertial signals is used as the feature vector, and in [8, 11] where LSTM variants are used, not over the original sequential signal, but over statistical and frequency domain features extracted sequentially from a sliding window over the inertial signals. Similar observations can be addressed for the work done in [2, 9, 24–26] for human activity recognition and as well in [6, 46] for gait analysis. In all of these it is worth investigating what the best features are for transferability. It is also worth investigating the transferability of hidden Markov models, especially when applied to human activity recognition and gait analysis as is done in [10].

On the other hand, very few works have been done for transfer learning over video streams. An example is the work done by the authors in [53] where their rationale is that networks trained-from-scratch on video datasets suffer from severe overfitting and accordingly they perform poorly on test sets. This particularly limits the usage of deep models on a wide range of computer vision problems where obtaining training data is difficult. To handle this problem the authors propose the use of transfer learning from images to videos and utilize the knowledge embedded in weakly labeled image corpus for video related tasks. This latter corpus is used to learn visual patterns that are overlooked by models trained only on the video corpus. Eventually, the resultant network has better generalizability and recognition rate. They show that through the use of transfer learning from image to video, a frame-based recognizer can be learned with only $4k$ videos. Transfer learning shows show a very promising arena for advancement in video content analytics and its wide variety of applications. For example, the work done in [47, 48] in crowd analysis use a set of geometrically overlapping LSTM to model the scene dynamics and then use such models for a multitude of purposes such as generating hypothetical trajectories in the scene, anomaly detection, etc. The question that arises is whether such models can be just fine-tuned on other crowd scenes (not experimented with in the given work) and used with effective performance on such alien scenes without training-from-scratch. The same applies for the work done in [31, 32] in crowd scene analysis. Also, of particular interest is the use of transfer learning in medical applications. We have mentioned one application above, however, there are tremendous opportunities in this area. For example, the work done in [18–20] for the quality assessment and abnormality detection in human action performance.

In the current work, the empirical study can then be considered to be inductive, feature-based transfer learning since we leverage the feature extraction abilities gained from one dataset on other, previously-unseen datasets.

1.3 Datasets

Two important image datasets are considered in the current work. The first one is ImageNet which has already been used to pretrain the three deep architectures with which we experiment. The second one is Caltech-256 which is used to fine tune the pretrained networks for its new task of classifying the images in this dataset.

1.3.1 ImageNet

ImageNet [17, 38] is an image dataset that consists of over 15 millions labeled high-resolution images. The images span a range of roughly 22, 000 categories. The images have been collected from the web and labeled by humans using Amazon's Mechanical Turk crowd-sourcing tool. Based on this dataset, an annual competition called the ImageNet Large-Scale Visual Recognition Challenge (ILSVRC) has been held since 2010. ILSVRC uses a subset of ImageNet with roughly 1000 images in each of 1000 categories. So roughly in this competition, there are 1.2 million training images, 50, 000 validation images, and 150, 000 testing images. The images in this dataset are not fixed in size and they have variable resolution. Therefore, they have been down-sampled to the fixed resolution of 256×256.

1.3.2 Caltech-256 Dataset

The Caltech-256 [23] image set is a sequel to its predecessor, the Caltech-101 dataset [43]. New features have been added including size increase, new and larger clutter categories, and overall increased difficulty. It is designed for training models for visual recognition. Within each category there is high variability, particularly, with regard to the object spatial location. Figure 1.2 shows some samples from this dataset and Table 1.1 shows its basic characteristics.

1.4 Deep Architectures

Deep learning in general, and deep neural computing in particular, has proven capable of continually achieving state-of-the-art performance in increasingly challenging tasks. They achieved such power through its intrinsic characteristics of distributed representation, automatic feature extraction, hierarchical feature representations, and high expressiveness of the input space using only polynomial resources. The most widespread and successful manifestation of deep learning are realized by deep neural networks. Such networks are extensions of the classical shallow networks by signifi-

(a) Bear (b) Frisbee (c) Cake (d) Iris

Fig. 1.2 Samples from caltech-256 dataset

Table 1.1 Caltech-256 basic characteristics

Dataset	Released	No. of categories	Total No. of images	Images per category			
				Min	Med	Mean	Max
Caltech-256	2006	257	30,607	80	100	119	827

Table 1.2 Deep neural architectures

Model	Size	No. parameters	Depth
VGG16	528 MB	138,357,544	23
VGG19	549 MB	143,667,240	26
Xception	88 MB	22,910,480	126
DenseNet121	33 MB	8,062,504	121
ResNet50	98 MB	25,636,712	50

cantly increasing the depth of the network whose training have been made possible by advancement of computational learning techniques, availability of computing power, and massive amount of data.

Several main deep architectures have been proposed over the years since 2012 progressively achieving state-of-the-art results, particularly, on the ImageNet, at an unprecedented rate and scale. These are essentially pretrained on ImageNet and the trained model, in a transfer learning fashion, can then be downloaded and fine tuned for other tasks. In the current work we investigate five architectures, two variants of the VGG network, the Xception network, DenseNet121 and finall ResNet50. Table 1.2 compares the basic architectures of the three chosen networks. Depth refers to the feedforward topological depth of the network. This includes activation layers, batch normalization layers, etc.

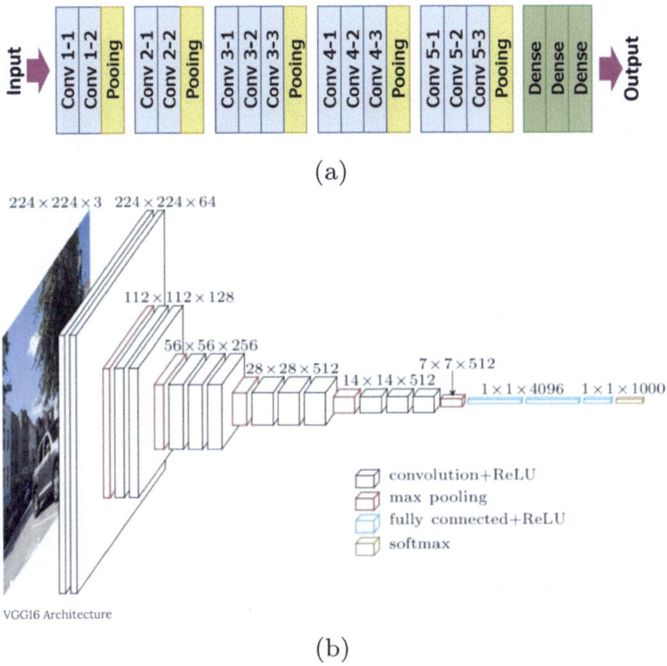

Fig. 1.3 The VGG16 architecture

1.4.1 VGG16

VGG16 is a convolutional neural network architecture that was proposed by K. Simonyan and A. Zisserman from the University of Oxford in 2014 [52]. This network top-5 test accuracy of 92.7% on a version of ImageNet that contained over 14 million images spanning 1000 classes. It was a sequel to the stardom AlexNet where the large kernels in the latter (11 and 5 in the first and second convolutional layer, respectively) have been replaced by a multitude of smaller 3 × 3 kernels VGG16 was trained over ImageNet for weeks using NVIDIA Titan Black GPU's. The architecture is shown in Fig. 1.3.

1.4.2 VGG19

VGG19 is a variant of VGG model which consists of 19 layers (16 convolution layers, 3 Fully connected layer, 5 MaxPooling layers and 1 SoftMax layer). VGG-19 is trained on a subset of ImageNet consisting of more than a million images [52]. This subset spans about 1000 object categories including natural objects, sports, athletics, plants, fungus, etc. Fig. 1.4 shows the architecture of VGG19.

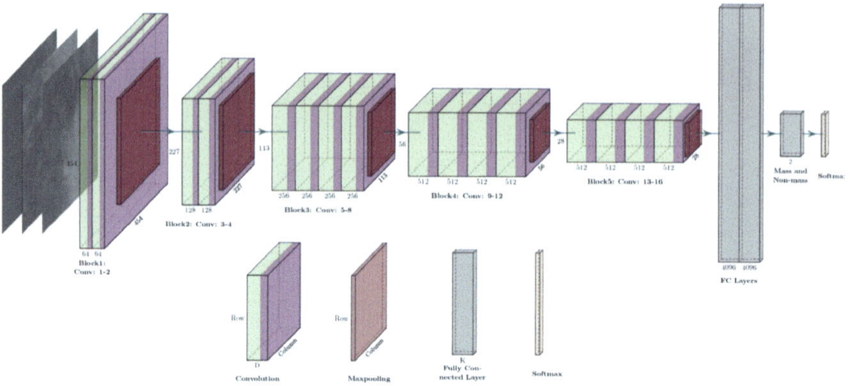

Fig. 1.4 The VGG19 architecture [30]

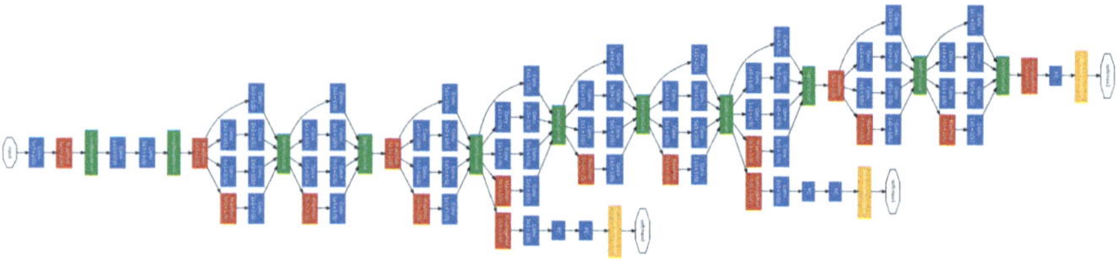

Fig. 1.5 The Xception architecture

1.4.3 Xception

Xception was proposed by François Chollet [16], the creator and chief maintainer of the Keras library, see Fig. 1.5. It is a CNN with depth 71. The publicly available downloadable version of the network is trained on a subset of ImageNet containing more than a million images. [38], and it can classify up to 1000 object categories. Hence, the network, viewed in the context of unsupervised feature extractor, has learned a rich diverse set of image feature representations. Xceptiont is an extension of the Inception architecture where the standard Inception modules are replaced with depthwise separable convolutions.

1.4.4 DenseNet121

Dense convolutional (DenseNet) networks were created to realize the paradoxical idea that: CNNs can go deeper, achieve better predictive performance, and at the same time are efficient to train. This is realized through the creations of shortcuts in the network: creating connections between layers close to the input and those close to the output[36]. Every layer in the network receives inputs from the feature maps of

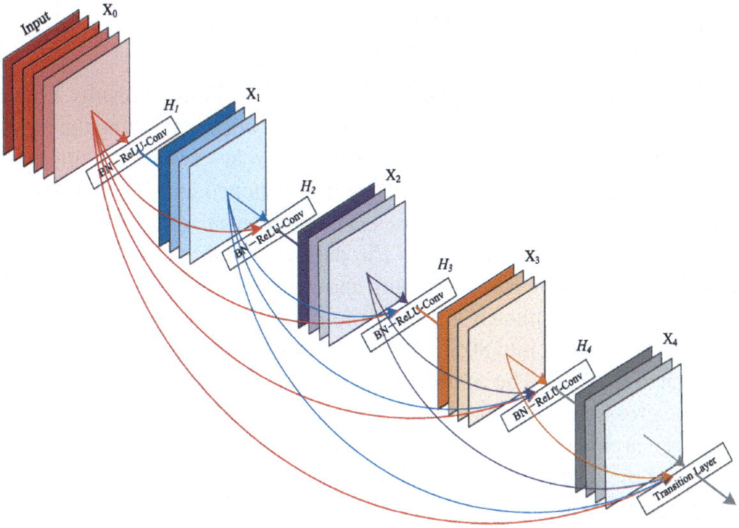

Fig. 1.6 The DenseNet architecture [36]

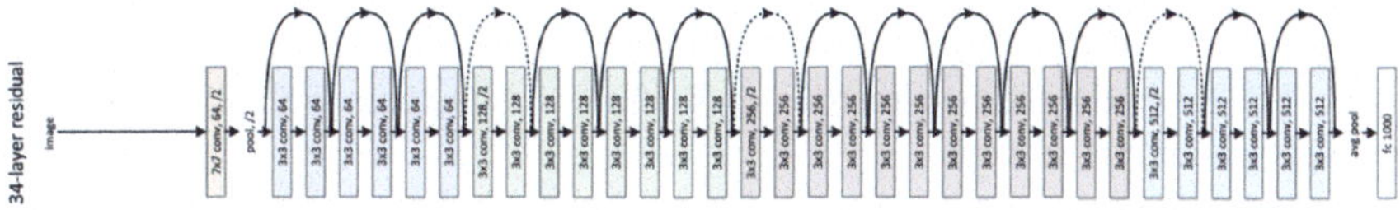

Fig. 1.7 The ResNet architecture [33]

all the preceding layers in addition to its own induced feature maps are fed as inputs to all subsequent layers. The authors [36] articulate the beneficial characteristics of DenseNets including the mitigation of the vanishing gradient problem, strengthening the propagation of features downstream the network, the encourage of feature reuse in addition to significantly reducing the number of parameters. Figure 1.6 shows a schematic diagram of the notion of dense convolutional networks. In the current paper we use DenseNet121 implementation that has 121 layers.

1.4.5 ResNet50

The philosophy of residual networks (ResNets) is to implement the concept of *residual learning* which aims at easing the training of networks that are substantially deep [33]. The authors redefined the purpose of layers as learning *residual functions* with reference to the layer inputs, instead of learning unreferenced functions [33]. Figure 1.7 shows a schematic diagram of the notion of residual convolutional networks. In the current paper we use ResNet50 implementation that has 50 layers.

1.5 Experiments and Results

In this section we describe our experimental work and the induced results. For each of the three architectures we download the pretrained network (excluding the top layers) and then we replace the top part by a new convolution layer with 256 filters each of size 3×3 with padding; the convolution layer is flattened and finally connected to a softmax layer for classification purposes. The number of softmax units is equal to the number of categories in the Caltech-256 dataset which is 257.

For each architecture we test its training and validation performance. The metrics we use for testing performance are both the classification accuracy and the loss function. We split the data into 80% : 20%, 80% for training and 20% for validation. We let a batch size of 32. The optimization method used is RMSProp and the loss function is the cross-entropy. We vary two hyper-parameter settings in the same way across the three architectures: the learning rate and the dropout ratio.

We first give the results for each architecture separately and then we do comparisons among the three architectures.

1.5.1 VGG16

Figure 1.8 shows the training metrics of VGG16 architectures with all variants of the hyper-parameter settings.

As expected the no dropout model (red curve) gives the best training performance. Dropout is a regularization technique that improves the generalization performance by lowering the overfitting effect at the expense of some level of degradation in the training performance. It is also evident that lowering the learning rate by one order of magnitude (the dashed green curve) greatly decreases the convergence rate of the model. We can also observe that after 10 epochs, all models (except the latter with

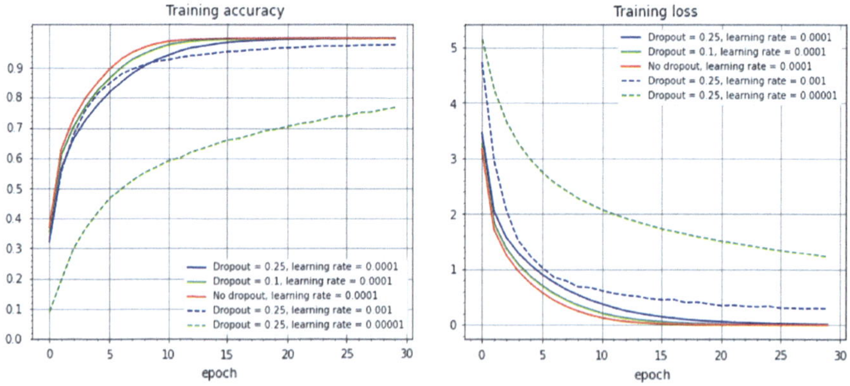

Fig. 1.8 VGG16 training metrics. The plot on the left shows the training accuracy, and the cross-entropy loss is shown on the right

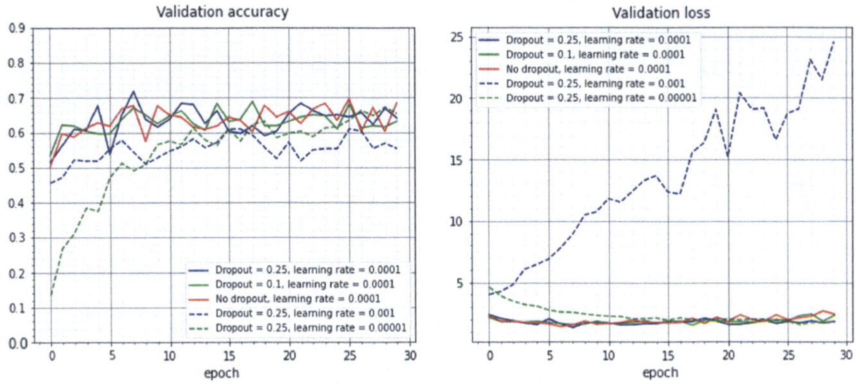

Fig. 1.9 VGG16 validation metrics. The plot on the left shows the accuracy, and the cross-entropy loss is shown on the right

dashed green curve) exceed 90% in their predictive accuracy over the training set indicating to some extend the effectiveness of the transfer learning.

Next we look at the metrics (both accuracy and cross-entropy loss function) over the validation set. Figure 1.9 shows the validation metrics of VGG16 architectures with all variants of the hyper-parameter settings. From this figure we can observe the following: (1) the learning rate of $lr = 0.0001$ is a kind of gold standard for this particular task, (2) reducing this learning rate by an order of magnitude, $lr = 0.00001$ (the green dashed curve) performed relatively well from the accuracy and loss perspectives, and even better when comparing its own performance in the training case (you may think of very low learning rate as being conservative and hence, a kind of regularization), (3) on the other hand, this small learning rate of $lr = 0.00001$ has relatively slower convergence than the other cases, (4) increasing the learning rate by an order of magnitude to $lr = 0.001$ has a detrimental effect giving much higher, and even worse, divergent loss (probably jumping over the minima of the error surface), (5) the predictive accuracy of all models oscillates between 60%–70%, which is lower than the training accuracy and that is of course expected, as the validation phase tests the generalization capability of the model on unseen data, and (6) it does not seem that the dropout has much effect on the predictive performance over the validation data, so the learning rate is the defining factor here.

1.5.2 VGG19

Figure 1.10 shows the training metrics of VGG19 architectures with all variants of the hyper-parameter settings.

The training accuracy and loss curves are very similar to those of the VGG16 model. The convergence rate is a little bit slower. This can be expected as the number

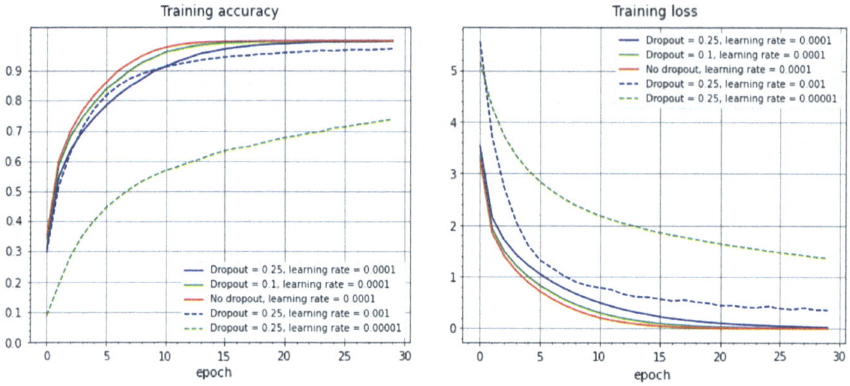

Fig. 1.10 VGG19 training metrics. The plot on the left shows the training accuracy, and the cross-entropy loss is shown on the right

of parameters in VGG19 is a lit bit larger than that of the VGG16 as shown in Table 1.2. For example, taking the red curve (no dropout with learning rate $lr = 0.0001$), we see that after 5 epochs, the training accuracy reaches 90% in the VGG16 case, however, it is still a bit less than 90% in the VGG19 case. Similarly, for the training loss, the VGG19 is a little bit lower than the VGG16 after 5 epochs. Other than training speed, the VGG19 training behavior is quite similar to that of VGG16.

Next we look at the metrics (both accuracy and cross-entropy loss function) over the validation set, Fig. 1.11. As expected the performance over the validation data is lower than that of the training data. Predictive accuracy ranges around 60%–65% for almost all models except the one with the largest learning rate of $lr = 0.001$ (dashed blue curve). This latter one also exhibits the worst evolution of the loss curve, the loss of this model is actually divergent. The large jumps of the learning rate causes swaying away into bad regions of the error surface. An interesting observation, which also applies in the VGG16 architecture, is that even though the loss of this latter model (dashed blue curve) diverges very quickly over the epochs, its predictive accuracy, remains stable over time. The loss function is chosen as cross-entropy, so this phenomenon can be attributed to the fact that over time the correct classifications remain as such, however, the confidence in these classifications keeps decreasing approaching the maximum entropy of 0.5. So we expect that increasing the number of epochs more will cause a sudden sharp decrease in accuracy as this confidence approaches the worst case of 0.5 and turning the correct classifications suddenly into misclassifications.

1.5.3 Xception

Figure 1.12 shows the training metrics of Xception architectures with all variants of the hyper-parameter settings. The training accuracy for Xception is much more stable

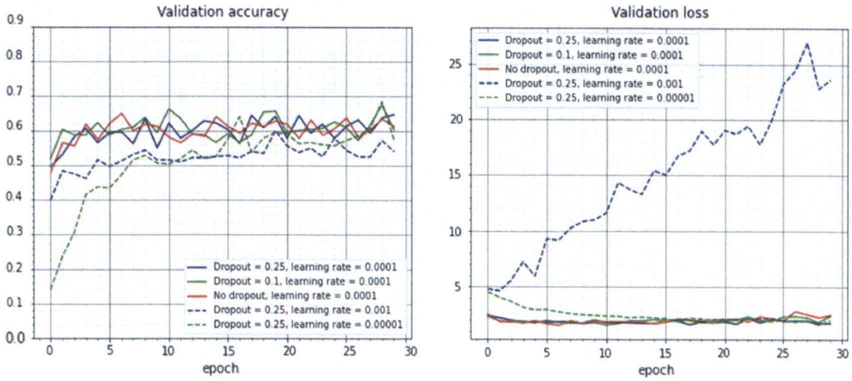

Fig. 1.11 VGG19 validation metrics. The plot on the left shows the accuracy, and the cross-entropy loss is shown on the right

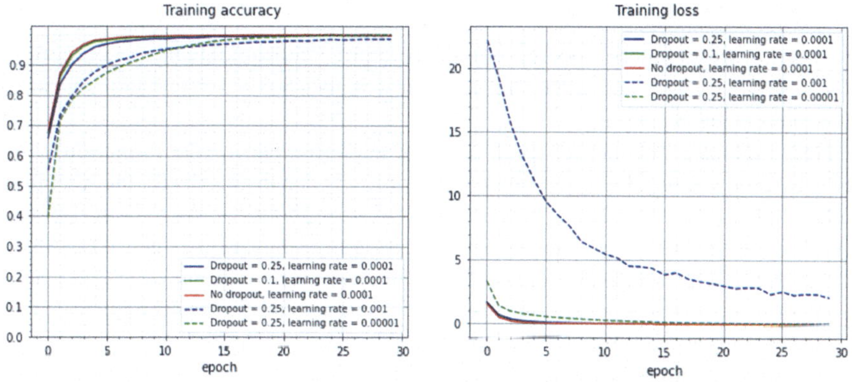

Fig. 1.12 Xception training metrics. The plot on the left shows the training accuracy, and the cross-entropy loss is shown on the right

than VGG16 and VGG19 across all variations of the dropout ratio and learning rate. All such variations are eventually equally performant, though some of them, with learning rates different from 0.0001 (the dashed curves), are initially slow. The start-off of the models are quite high from the accuracy perspective, at least 40% in the case of $lr = 0.00001$ (the dashed green curve) and 60% in all other cases, which is quite natural for the former as the too low learning rate is rather conservative and causes very slow jumps on the error surface until reaching convergence. An interesting observation about the same configuration (the dashed green curve) is that the behavior of its loss function during training is quite different from VGG16 and VGG19. In the latter architectures, the convergence of this configuration is the slowest. However, in Xception it is quite similar to the other best configurations, though it is the slowest with regarding to accuracy. This seemingly counterintuitive phenomenon, which is the converse to a similar phenomenon described above, can as well be explained by a converse argument. Though the number of misclassifications decreases slowly, the

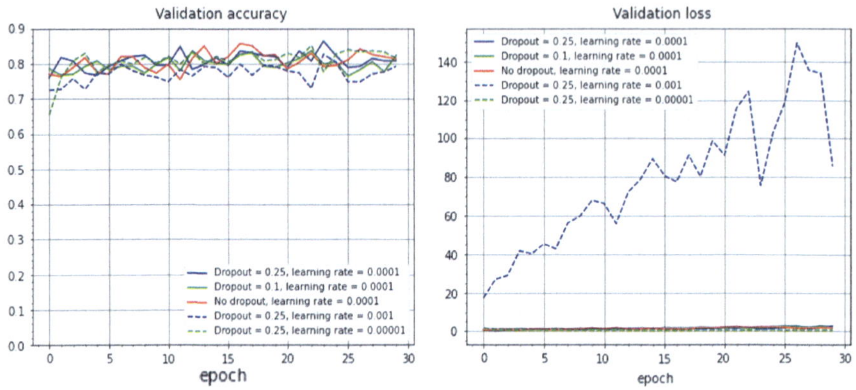

Fig. 1.13 Xception validation metrics. The plot on the left shows the accuracy, and the cross-entropy loss is shown on the right

cross-entropy loss function decreases quickly approaching quickly the misclassification error from a high ratio to 0.5 until suddenly it goes below 0.5 indicating turning into correct classifications.

Next we look at the metrics (both accuracy and cross-entropy loss function) over the validation set, Fig. 1.13. The first thing to notice is that the results for the Xception architecture is way much better than that of VGG16 and VGG19. We have the following observations: (1) accuracies of best Xception models vary between 80%–85%, better than VGG16 (60%–70%) and VGG19 (60%–65%), (2) Xception is very stable in the sense that it is more robust, than VGG16 and VGG19, regarding changes in the hyper-parameters (dropout ratio and learning rate), (3) in other words, all Xception variants are almost equally perfroamnt; Xception is less sensitive to changes in the hyper-parameters than VGG16 and VGG19, (4) the start-off performance for all Xception models are very high both from the perspectives of accuracy as well as the loss function, (5) the loss function of the model having the largest learning rate $lr = 0.001$ (the dashed blue) actually diverges, even though the accuracy boundedly oscillates close 80%; as before this can be attributed to the model sustaining correct classifications whilst reducing confidence in these correct classifications; with the cross-entropy loss function, the model seems to jump too much between rather bad minima, and (6) Xception architecture very clearly shows the contrast between shallow and deep learning/models; as shown from Table 1.2, Xception is an order of magnitude deeper than both VGG16 and VGG19, and at the same time it has an order of magnitude less parameters than these latter architectures; accordingly its predictive performance is higher and stable, and much more efficient regarding its computational demands.

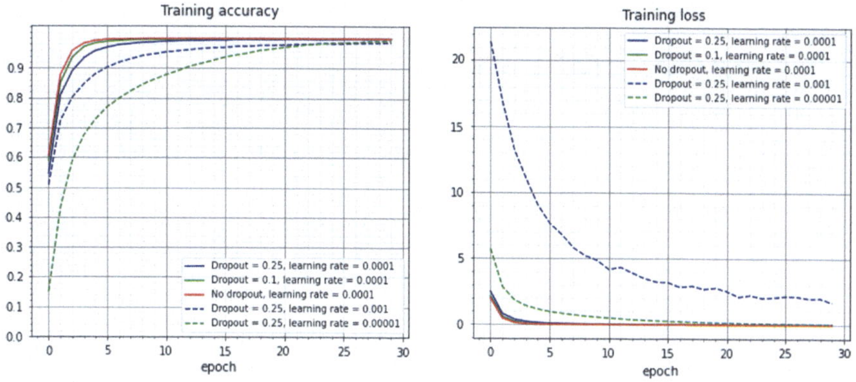

Fig. 1.14 DenseNet121 training metrics. The plot on the left shows the accuracy, and the cross-entropy loss is shown on the right

1.5.4 DenseNet121

Figure 1.14 shows the training metrics of DenseNet121 architectures with all variants of the hyper-parameter settings. We notice that the accuracy and loss metrics behave very closely to the Xception network (Fig. 1.12), especially in configurations with low learning and dropout rates. After few epochs the performance, of almost all configurations, jumps dramatically to almost its convergent values, for example, after just 5 epochs the training accuracy jumps above 90%. We can also notice, as well as observed from the Xception network, that higher dropout rates should be accompanied with higher learning rate as well. The dashed green curve (highest dropout rate and lowest learning rate) moves converges quite slowly, however, it eventually catches up with the other configurations after 25 epochs. Other configurations reached that performance quite early even before 5 epochs. And as noted in several cases above we find some discrepancy between the accuracy and loss metrics manifested by the fact that the loss function of the blue curve (dropout rate of 0.25 and learning rate of 0.001) behaves very badly and converges slowly, though the corresponding accuracy behaves well and converges rather fast to a high accuracy. As explained above, this can be attributed to the model sustaining correct classifications whilst reducing confidence in these correct classifications; with the cross-entropy loss function, the model seems to jump too much between rather bad minima. A general observation about DenseNet121 training performance is that the best configurations are those with no or low dropout rates. This can be attributed to the nature and philosophy behind such architectures where connections are made across non-consecutive layers and hence, co-learning of adjacent nodes in general is not that severe and higher dropout rates can cause a reverse worse results.

Now we look at the metrics (both accuracy and cross-entropy loss function) over the validation set, Fig. 1.15. As we say in the training metrics the validation metrics behave as well very close to that of the Xception metric. The only difference is in

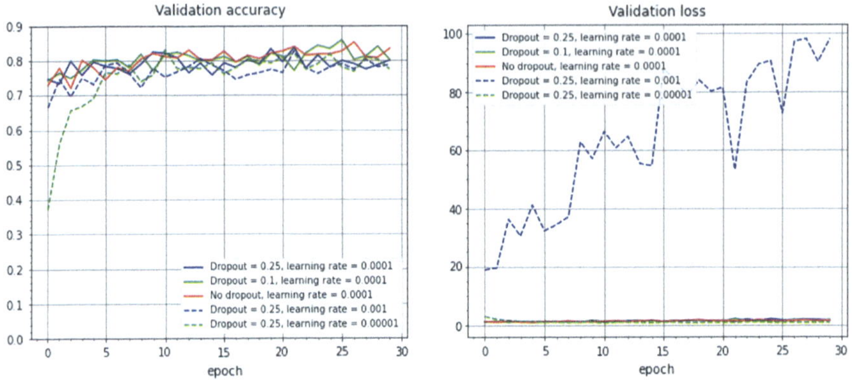

Fig. 1.15 DenseNet121 validation metrics. The plot on the left shows the accuracy, and the cross-entropy loss is shown on the right

the wildest configurations. The dashed green accuracy curve (dropout rate of 0.25 and learning rate of 0.00001) starts off badly in the case of DenseNet121 than that of Xception, whilst the dashed blue loss curve (dropout rate of 0.25 and learning rate of 0.001) behaves less wildly in the DenseNet121 case than that of the Xception network. Almost all configurations eventually converge in accuracy to around 80%.

What is most interesting about the DenseNet121 is its effectiveness when looking into its overall properties given in Table 1.2. It is one of the deepest networks investigated in this paper, however, it is the smallest one with respect to the number of parameters and memory requirements. These are so modest comparable to other architectures. The next one, namely Xception, is almost triple the memory requirements and number of parameters making DenseNet121 the most effective, especially when looking into its predictive performance both in training and validation.

1.5.5 ResNet50

Figure 1.16 shows the training metrics of ResNet50 architectures with all variants of the hyper-parameter settings. It is clear that these metrics are the worst among all the studied architectures above. Looking at the accuracy curve it is apparent that the configurations corresponding to the solid curves (low dropout and learning rates) seem to improve linearly over time and it might reach the performance of the above architectures if given more epochs to train (>30). However, even with this possibility its convergence rate is very slow, compared to the above architectures, especially DenseNet121 and Xception where the training accuracy exceeds 90% after just 5 epochs. The same remarks can also be said about the loss curve where it seems all configurations behave in general similar to the above architectures, however, at a very slow pace.

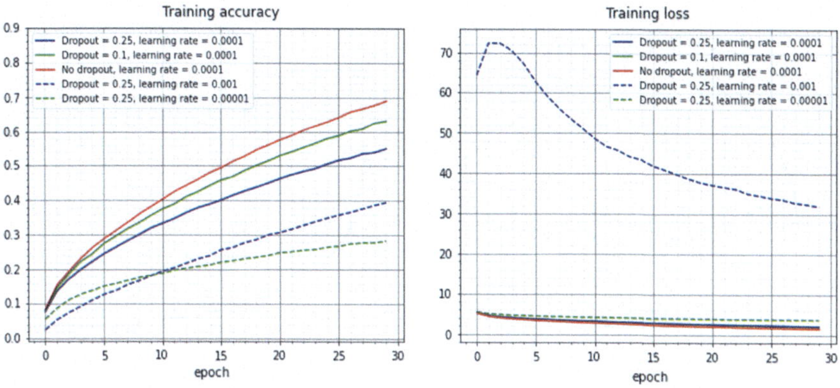

Fig. 1.16 ResNet50 training metrics. The plot on the left shows the accuracy, and the cross-entropy loss is shown on the right

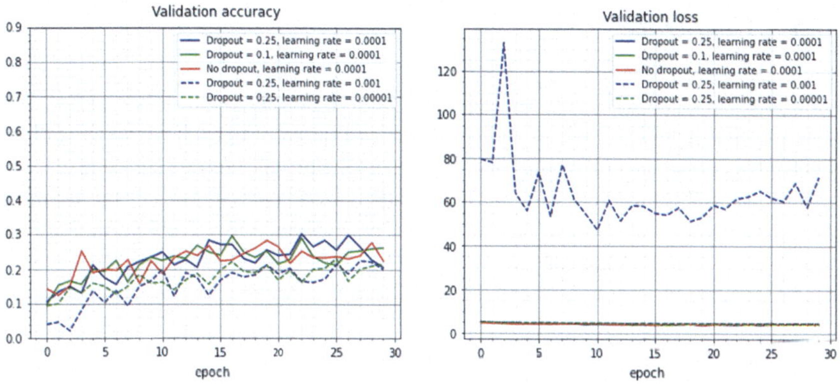

Fig. 1.17 ResNet50 validation metrics. The plot on the left shows the accuracy, and the cross-entropy loss is shown on the right

Figure 1.17 shows that validation metrics. As can be expected from the previous analysis on training the validation metrics are very bad. All configurations converge to an accuracy around 25%. It is worth noting from the training and validation that even though the loss curves seem to behave rather good the corresponding accuracies do not catch up. This can be explained by the fact that the loss value (cross-entropy loss) is very close to the value of correct classification, however, the corresponding classification is wrong. This shows the contrast between a continuous loss function and a corresponding accuracy function that is quite discrete. So minor changes in the loss function correspond to huge changes in the accuracy function. So it seems the loss curves jumps frequently between neighboring minima that are very close in value, however, the corresponding accuracy values make large leaps.

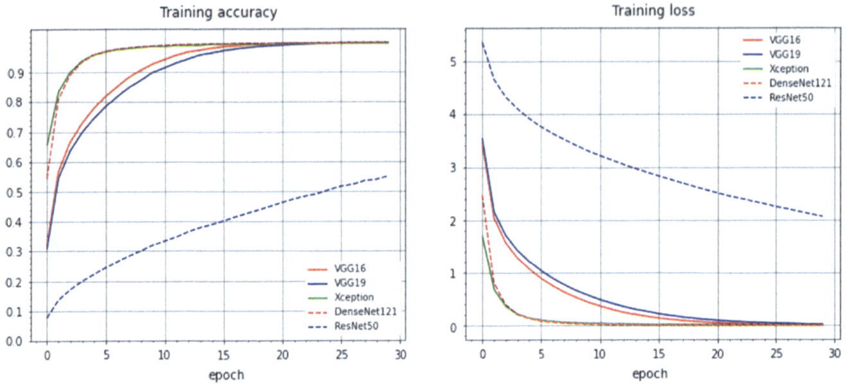

Fig. 1.18 Compare the training metrics across all architectures

1.5.6 *All at Once*

In the next set of experiments we compare the five architectures in one graph. We use both training and validation metrics in our comparison. In each we take the classification accuracy as well as the loss function as the proper metrics for comparison. In all the architectures we fix the same top supervised layers as well as the training hyper-parameters that showed the rather best performance in the above-mentioned experiments, namely, learning rate $lr = 0.0001$ and dropout ratio of 0.25. We use the cross-entropy loss function and the RMSProp optimizer across all architectures.

1.5.6.1 Training Metrics

Figure 1.18 shows the training accuracy and loss across the five deep architectures, VGG16, VGG19, Xception, DenseNet121, and ResNet50, over the Caltech-256 image dataset. It is clear that ResNet50 has the worst performance by far from the other architectures, though it seems it can catch up with them, however, very slowly much beyond 30 epochs. It is clear that the VGGs are pretty close to each other in performance, actually the VGG16 with lesser computational demands (Table 1.2), outperforms its cousin VGG19. The breakthrough actually occurs with the DenseNet121 network. It starts off with an accuracy of about 55%, it then converges very fast to closer than 100% of accuracy. This indicates that its transfer learning capability is high. Its loss function is also very much compatible with this behavior, indicating a smooth loss curve. The Xception network is very close in performance to DenseNet121, and it is even slightly better in initial epochs, however, the computational resources required for DenseNet121 (Table 1.2) is by far the lowest among the whole set of architectures verifying the significance and effectiveness of deep architectures (the role of depth in effectively representing mapping functions) and the effectiveness of the special connectivity of DenseNet121.

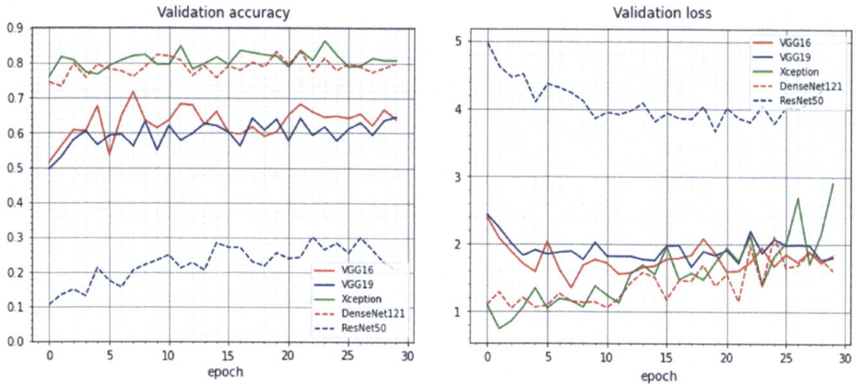

Fig. 1.19 Compare the validation metrics across all architectures

1.5.6.2 Validation Metrics

Figure 1.19 verify the superiority of both the DenseNet121 and the Xception architectures with their deeper construct and much less parameters. The computational requirements for DenseNet121 is almost one third of that of Xception, see Table 1.2. Looking at the accuracy plot we see three clusters of architectures. The first one, with the best performance, consists of both the DenseNet121 and Xception. The second best cluster consists of the VGGs, and the last and worst performing cluster is the ResNet50. The loss curves indicate that after few epochs DenseNet121 and Xception suffer from overfitting and hence declined loss even though the accuracy remains stable. This might be explained as mentioned above that the classifications remain correct, however, the cross-entropy increases with flattening of the probability distribution over the classes.

1.6 Conclusion and Future Work

In this paper we have studied the effectiveness of some of the state-of-the-art deep learning architectures in transfer learning in visual tasks. We choose five representative networks, two of them from the VGG family, namely, VGG16 and VGG19, and the others are the Xception, DenseNet121, and ResNet50 networks. These networks are already pretrained on the ImageNet dataset and tested for a new task using the Caltech-256 image set. The networks are just lightly fine tuned for the new task and tested for training and validation performance. The general conclusion is that two of these networks, namely, DenseNet121 and Xception, show very promising prospects in transfer learning over visual tasks, especially when considering the computational demands. Such networks has comparatively few parameters and low storage requirements, however, they show high stable predictive performance over training as well as validation phases. This apparent performance of the DenseNet121 and Xception networks show as well the conjectures that deep compositional networks are able to perform at a level that otherwise would need an exponentially more nodes in shallow networks.

In the future we plan to expand our comparative study to include more deep architectures in the literature such as the ResNet family, the DenseNet family, NasNet, etc. We will also use multiple other vision datasets for more comprehensive study.

Acknowledgements This work is Funded by the Science and Technology Development Fund STDF (Egypt); Project id: 42519 - "Automatic Video Surveillance System for Crowd Scenes".

References

1. Abdu-Aguye, M.G., Gomaa, W.: Novel approaches to activity recognition based on vector autoregression and wavelet transforms. In: 2018 17th IEEE International Conference on Machine Learning and Applications (ICMLA), pp. 951–954. IEEE (2018)
2. Abdu-Aguye, M.G., Gomaa, W.: Competitive feature extraction for activity recognition based on wavelet transforms and adaptive pooling. In: 2019 International Joint Conference on Neural Networks (IJCNN), pp. 1–8 (2019). https://doi.org/10.1109/IJCNN.2019.8852299
3. Abdu-Aguye, M.G., Gomaa, W.: Robust human activity recognition based on deep metric learning. In: Proceedings of the 16th International Conference on Informatics in Control, Automation and Robotics - Volume 1: ICINCO, pp. 656–663. INSTICC, SciTePress (2019). https://doi.org/10.5220/0007916806560663
4. Abdu-Aguye, M.G., Gomaa, W.: Versatl: versatile transfer learning for IMU-based activity recognition using convolutional neural networks. In: Proceedings of the 16th International Conference on Informatics in Control, Automation and Robotics - Volume 1: ICINCO, pp. 507–516. INSTICC, SciTePress (2019). https://doi.org/10.5220/0007916705070516
5. Abdu-Aguye, M.G., Gomaa, W., Makihara, Y., Yagi, Y.: On the feasibility of on-body roaming models in human activity recognition. In: Proceedings of the 16th International Conference on Informatics in Control, Automation and Robotics - Volume 1: ICINCO, pp. 680–690. INSTICC, SciTePress (2019). https://doi.org/10.5220/0007921606800690
6. Adel, O., Nafea, Y., Hesham, A., Goma, W.: Gait-based person identification using multiple inertial sensors. In: Proceedings of the 17th International Conference on Informatics in Control, Automation and Robotics - Volume 1: ICINCO, pp. 621–628. INSTICC, SciTePress (2020). https://doi.org/10.5220/0009791506210628
7. Agarwal, S., Roth, D.: Learning a sparse representation for object detection. In: Heyden, A., Sparr, G., Nielsen, M., Johansen, P. (eds.) Computer Vision - ECCV 2002, pp. 113–127. Springer, Heidelberg (2002)
8. Ashry, S., Elbasiony, R., Gomaa, W.: An LSTM-based descriptor for human activities recognition using IMU sensors. In: Proceedings of the 15th International Conference on Informatics in Control, Automation and Robotics - Volume 1: ICINCO, pp. 494–501. INSTICC, SciTePress (2018). https://doi.org/10.5220/0006902404940501
9. Ashry, S., Gomaa, W.: Descriptors for human activity recognition. In: 2019 7th International Japan-Africa Conference on Electronics, Communications, and Computations, (JAC-ECC), pp. 116–119 (2019)
10. Ashry, S., Gomaa, W., Abdu-Aguye, M.G., El-borae, N.: Improved IMU-based human activity recognition using hierarchical hmm dissimilarity. In: Proceedings of the 17th International Conference on Informatics in Control, Automation and Robotics - Volume 1: ICINCO, pp. 702–709. INSTICC, SciTePress (2020). https://doi.org/10.5220/0009886607020709

11. Ashry, S., Ogawa, T., Gomaa, W.: Charm-deep: continuous human activity recognition model based on deep neural network using IMU sensors of smartwatch. IEEE Sensors J. **20**(15), 8757–8770 (2020)
12. Baños, O., Damas, M., Pomares, H., Rojas, I., Tóth, M.A., Amft, O.: A benchmark dataset to evaluate sensor displacement in activity recognition. In: Proceedings of the 2012 ACM Conference on Ubiquitous Computing, pp. 1026–1035. ACM (2012)
13. Baykal, E., Dogan, H., Ercin, M.E., Ersoz, S., Ekinci, M.: Transfer learning with pre-trained deep convolutional neural networks for serous cell classification. Multimed. Tools Appl. **79**(21), 15593–15611 (2020). https://doi.org/10.1007/s11042-019-07821-9
14. Bengio, Y.: Deep learning of representations for unsupervised and transfer learning. In: Proceedings of ICML Workshop on Unsupervised and Transfer Learning, pp. 17–36 (2012)
15. Chikhaoui, B., Gouineau, F., Sotir, M.: A CNN based transfer learning model for automatic activity recognition from accelerometer sensors. In: International Conference on Machine Learning and Data Mining in Pattern Recognition, pp. 302–315. Springer (2018)
16. Chollet, F.: Xception: deep learning with depthwise separable convolutions. In: 2017 IEEE Conference on Computer Vision and Pattern Recognition (CVPR), pp. 1800–1807 (2017)
17. Deng, J., Dong, W., Socher, R., Li, L.J., Li, K., Fei-Fei, L.: ImageNet: a large-scale hierarchical image database. In: CVPR09 (2009)
18. Elkholy, A., Hussein, M., Gomaa, W., Damen, D., Saba, E.: Efficient and robust skeleton-based quality assessment and abnormality detection in human action performance. IEEE J. Biomed. Health Inform. **24**(1), 280–291 (2019)
19. Elkholy, A., Hussein, M.E., Gomaa, W., Damen, D., Saba, E.: A general descriptor for detecting abnormal action performance from skeletal data. In: Proceedings of the 39th Annual International Conference of the IEEE Engineering in Medicine & Biology Society (EMBC 17). JeJu Island, S. Korea (2017)
20. Elkholy, A., Makihara, Y., Gomaa, W., Ahad, M.A.R., Yagi, Y.: Unsupervised gei-based gait disorders detection from different views. In: 2019 41st Annual International Conference of the IEEE Engineering in Medicine and Biology Society (EMBC), pp. 5423–5426 (2019). https://doi.org/10.1109/EMBC.2019.8856294
21. Olivas, E.S., Guerrero, J.D.M., Martinez-Sober, M., Magdalena-Benedito, J.R., Serrano, L.: Handbook of Research on Machine Learning Applications and Trends - Algorithms, Methods, and Techniques. Information Science Reference, Hershey, PA (2009)
22. Fergus, R., Perona, P., Zisserman, A.: Object class recognition by unsupervised scale-invariant learning. In: 2003 IEEE Computer Society Conference on Computer Vision and Pattern Recognition, 2003. Proceedings, vol. 2, pp.II (2003)
23. Griffin, G., Holub, A., Perona, P.: Caltech-256 object category dataset. Technical Report 7694, Caltech (2007)
24. Gomaa, W.: Probabilistic approach to human activity recognition from accelerometer data. In: 2019 7th International Japan-Africa Conference on Electronics, Communications, and Computations, (JAC-ECC), pp. 63–66 (2019)
25. Gomaa, W.: Statistical and time series analysis of accelerometer signals for human activity recognition. In: 2019 14th International Conference on Computer Engineering and Systems (ICCES), pp. 351–356 (2019)
26. Gomaa, W.: Statistical metric-theoretic approach to activity recognition based on accelerometer data. In: Hassanien, A.E., Shaalan, K., Tolba, M.F. (eds.) Proceedings of the International Conference on Advanced Intelligent Systems and Informatics 2019, pp. 537–546. Springer International Publishing, Cham (2020)
27. Gomaa, W., Elbasiony, R., Ashry, S.: ADL classification based on autocorrelation function of inertial signals. In: 2017 16th IEEE International Conference on Machine Learning and Applications (ICMLA), pp. 833–837 (2017). https://doi.org/10.1109/ICMLA.2017.00-53
28. Gopalakrishnan, K., Khaitan, S.K., Choudhary, A., Agrawal, A.: Deep convolutional neural networks with transfer learning for computer vision-based data-driven pavement distress detection. Construct. Build. Mater. **157**, 322–330 (2017). https://doi.org/10.1016/j.conbuildmat.2017.09.110, http://www.sciencedirect.com/science/article/pii/S0950061817319335

29. Harley, A.W.: An interactive node-link visualization of convolutional neural networks. In: International Symposium on Visual Computing, pp. 867–877. Springer (2015)
30. Hasan, M.K., Aleef, T.A.: Automatic mass detection in breast using deep convolutional neural network and SVM classifier. CoRR **abs/1907.04424** (2019). http://arxiv.org/abs/1907.04424
31. Hassanein, A., Hussein, M., Gomaa, W.: Semantic analysis of crowded scenes based on non-parametric tracklet clustering. In: Proceedings of the 25th International Joint Conference on Artificial Intelligence IJCAI-16. New York City, USA (2016)
32. Hassanein, A.S., Hussein, M.E., Gomaa, W., Makihara, Y., Yagi, Y.: Identifying motion pathways in highly crowded scenes: Aa non-parametric tracklet clustering approach. Comput. Vision Image Understand. **191** (2020). https://doi.org/10.1016/j.cviu.2018.08.004, http://www.sciencedirect.com/science/article/pii/S1077314218301887
33. He, K., Zhang, X., Ren, S., Sun, J.: Deep residual learning for image recognition. In: 2016 IEEE Conference on Computer Vision and Pattern Recognition (CVPR), pp. 770–778 (2016)
34. Hinton, G.E., Salakhutdinov, R.R.: Reducing the dimensionality of data with neural networks. Science **313**(5786), 504–507 (2006)
35. Hu, D.H., Zheng, V.W., Yang, Q.: Cross-domain activity recognition via transfer learning. Pervasive Mobile Comput. **7**(3), 344–358 (2011). https://doi.org/10.1016/j.pmcj.2010.11.005, http://www.sciencedirect.com/science/article/pii/S1574119210001227. Knowledge-Driven Activity Recognition in Intelligent Environments
36. Huang, G., Liu, Z., Van Der Maaten, L., Weinberger, K.Q.: Densely connected convolutional networks. In: 2017 IEEE Conference on Computer Vision and Pattern Recognition (CVPR), pp. 2261–2269 (2017)
37. Goodfellow, I., Bengio, Y., Courville, A., Bengio, Y.: Deep Learning. Illustrated edn. Adaptive Computation and Machine Learning series. The MIT Press (2016)
38. ImageNet: http://www.image-net.org
39. Jordan, M.I., Mitchell, T.M.: Machine learning: trends, perspectives, and prospects. Science **349**(6245), 255–260 (2015)
40. Karpathy, A., Toderici, G., Shetty, S., Leung, T., Sukthankar, R., Fei-Fei, L.: Large-scale video classification with convolutional neural networks. In: Proceedings of the IEEE Conference on Computer Vision and Pattern Recognition, pp. 1725–1732 (2014)
41. Khan, M.A.A.H., Roy, N.: Transact: transfer learning enabled activity recognition. In: 2017 IEEE International Conference on Pervasive Computing and Communications Workshops (PerCom Workshops), pp. 545–550 (2017). https://doi.org/10.1109/PERCOMW.2017.7917621
42. Khan, M.A.A.H., Roy, N.: Untran: recognizing unseen activities with unlabeled data using transfer learning. In: 2018 IEEE/ACM Third International Conference on Internet-of-Things Design and Implementation (IoTDI), pp. 37–47 (2018). https://doi.org/10.1109/IoTDI.2018.00014
43. Li Fei-Fei, Fergus, R., Perona, P.: Learning generative visual models from few training examples: an incremental Bayesian approach tested on 101 object categories. In: 2004 Conference on Computer Vision and Pattern Recognition Workshop, pp. 178–178 (2004)
44. Lu, G., Hao, Q., Kong, K., Yan, J., Li, H., Li, X.: Deep convolutional neural networks with transfer learning for neonatal pain expression recognition. In: 2018 14th International Conference on Natural Computation, Fuzzy Systems and Knowledge Discovery (ICNC-FSKD), pp. 251–256 (2018). https://doi.org/10.1109/FSKD.2018.8687129
45. Masrour, T., El Hassani, I., Bouchama, M.S.: Deep convolutional neural networks with transfer learning for old buildings pathologies automatic detection. In: Ezziyyani, M. (ed.) Advanced Intelligent Systems for Sustainable Development (AI2SD'2019), pp. 204–216. Springer, Cham (2020)
46. Mostafa., A., Barghash., T.O., Assaf., A.A., Gomaa., W.: Multi-sensor gait analysis for gender recognition. In: Proceedings of the 17th International Conference on Informatics in Control, Automation and Robotics - Volume 1: ICINCO, pp. 629–636. INSTICC, SciTePress (2020). https://doi.org/10.5220/0009792006290636

47. Moustafa, A., Hussein, M., Gomaa, W.: Gate and common pathway detection in crowd scenes using motion units and meta-tracking. In: Proceedings of the International Conference on Digital Image Computing: Techniques and Applications (DICTA 2017), Sydney, Australia (2017)
48. Moustafa, A.N., Gomaa, W.: Gate and common pathway detection in crowd scenes and anomaly detection using motion units and LSTM predictive models. Multimed. Tools Appl. (2020). https://doi.org/10.1007/s11042-020-08840-7
49. Oquab, M., Bottou, L., Laptev, I., Sivic, J.: Learning and transferring mid-level image representations using convolutional neural networks. In: Proceedings of the IEEE Conference on Computer Vision and Pattern Recognition, pp. 1717–1724 (2014)
50. Pan, S.J., Yang, Q., et al.: A survey on transfer learning. IEEE Trans. Knowl. Data Eng. **22**(10), 1345–1359 (2010). https://doi.org/10.1109/TKDE.2009.191
51. Ponce, J., Berg, T.L., Everingham, M., Forsyth, D.A., Hebert, M., Lazebnik, S., Marszalek, M., Schmid, C., Russell, B.C., Torralba, A., Williams, C.K.I., Zhang, J., Zisserman, A.: Dataset Issues in Object Recognition, pp. 29–48. Springer, Heidelberg (2006)
52. Simonyan, K., Zisserman, A.: Very deep convolutional networks for large-scale image recognition. In: Bengio, Y., LeCun, Y. (eds.) 3rd International Conference on Learning Representations, ICLR 2015, Conference Track Proceedings, San Diego, CA, USA, 7–9 May 2015 (2015). URL http://arxiv.org/abs/1409.1556
53. Su, Y., Chiu, T., Yeh, C., Huang, H., Hsu, W.H.: Transfer learning for video recognition with scarce training data. CoRR **abs/1409.4127** (2014). http://arxiv.org/abs/1409.4127
54. Thrun, S., Pratt, L.: Learning to Learn: Introduction and Overview, pp. 3–17. Springer, Boston (1998)
55. Yosinski, J., Clune, J., Bengio, Y., Lipson, H.: How transferable are features in deep neural networks? In: Ghahramani, Z., Welling,M., Cortes, C., Lawrence, N., Weinberger, K.Q. (eds.) Advances in Neural Information Processing Systems, vol. 27, pp. 3320–3328. Curran Associates, Inc. (2014). https://proceedings.neurips.cc/paper/2014/file/375c71349b295fbe2dcdca9206f20a06-Paper.pdf
56. Zeiler, M.D., Fergus, R.: Visualizing and understanding convolutional networks. In: European Conference on Computer Vision, pp. 818–833. Springer (2014)

Chapter 2
Deep Reinforcement Learning: A New Frontier in Computer Vision Research

Sejuti Rahman, Sujan Sarker, A. K. M. Nadimul Haque, and Monisha Mushtary Uttsha

Abstract Computer vision has advanced so far that machines now can think and see as we humans do. Especially deep learning has raised the bar of excellence in computer vision. However, the recent emergence of deep reinforcement learning is threatening to soar even greater heights as it combines deep neural networks with reinforcement learning along with numerous added advantages over both. This, being a relatively recent technique, has not yet seen many works, and so its true potential is yet to be unveiled. Thus, this chapter focuses on shedding light on the fundamentals of deep reinforcement learning, starting with the preliminaries followed by the theory and basic algorithms and some of its variations, namely, attention aware deep reinforcement learning, deep progressive reinforcement learning, and multi-agent deep reinforcement learning. This chapter also discusses some existing deep reinforcement learning works regarding computer vision such as image processing and understanding, video captioning and summarization, visual search and tracking, action detection, recognition and prediction, and robotics. This work further aims to elucidate the existing challenges and research prospects of deep reinforcement learning in computer vision. This chapter might be considered a starting point for aspiring researchers looking to apply deep reinforcement learning in computer vision to reach the pinnacle of performance in the field by tapping into the immense potential that deep reinforcement learning is showing.

2.1 Introduction

Computer vision is an interdisciplinary field that attempts to enable computers to recognize and achieve a high-level understanding of an image or video [21]. From an engineering perspective, computer vision aims to automate the human vision system. For a vision system to work efficiently and seamlessly, multiple tasks need to be done simultaneously and in quick succession. They are mainly image acquisition, processing the acquired image, analyzing the image to understand the meaning,

S. Rahman (✉) · S. Sarker · A. K. M. N. Haque · M. M. Uttsha
Department of Robotics and Mechatronics Engineering, University of Dhaka, Dhaka, Bangladesh
e-mail: sejuti.rahman@du.ac.bd

M. A. R. Ahad and A. Inoue (eds.), *Vision, Sensing and Analytics: Integrative Approaches*, Intelligent Systems Reference Library 207, https://doi.org/10.1007/978-3-030-75490-7_2

and making rational and optimal decisions from that [34]. This is even harder than it sounds as replicating the complex resources embodied in the human brain in a machine is near impossible. The human brain can process multiple things at once at extraordinary speed. We do it daily without even thinking twice about it. Machines, on the other hand, even with its ever-expanding resources and technology, can hardly achieve a fraction of that level of performance. Thus even though computer vision shows great promise in the future, real-world applications are still limited.

Current algorithms have been optimized to recognize and localize objects in real-time. However, contextual understanding of an image is still troublesome. Human brains can do such tasks with ease because it is equipped with rich resources like complex processing capability and other sensory nodes. Most importantly, the brain can use its extensive past experiences as it has almost unlimited storage capacity. Traditional vision-based algorithms often have to compensate either for accuracy or speed. Faster computer vision algorithms, even with expensive resources, often lack in accuracy and precision. On the other hand, accurate algorithms take too long to render the use of computer vision in real-world scenarios highly improbable.

In recent times, however, breakthroughs in neural networks, especially deep learning [25] have seen computer vision systems thrive and be used in several complex real world applications such as real-time recognition [56], captioning [41], summarization [7] and many more. However, deep networks mostly help learn something and then predict the same thing in a new scenario. However, in an automated system, a machine is often required to learn from scratch in a dynamic environment and perform some actions. Also, deep networks are mostly supervised; that is, the training data must be labeled. Such labeled training data is hard to find. If every data were to be manually labeled, it would take a huge workforce and time to achieve a decent dataset. Most importantly, the ultimate goal of computer vision is to emulate the performance of human vision systems. Compared to the information of thousands of years of ancestral information stored in our DNAs, and the years of training we humans get from real-world images, no manually labeled dataset can match that amount. Thus, deep networks alone cannot achieve human-level performance in computer vision.

Since we cannot obtain nearly enough labeled data, something else had to be thought of. We need an algorithm that does not need constant supervision but can learn from the real world positive or negative feedback. That is where reinforcement learning (RL) comes in, which allows the machine or learning agent to adjust actions based on the rewards received in order to achieve its goal in the best possible way [67]. The agent receives positive or negative feedback from its interactions with the environment. RL not only allows the agent to learn without supervision but also enables the agent to generate some optimal actions from it. Using RL in computer vision is a relatively new concept. One of the main challenges is the huge amount of exploration required for RL algorithms to converge [14]. Thus, for image-based problems, a huge amount of data is required for representing the state space properly. Nevertheless, RL in some computer vision problems have seen success including parameter selection in edge detection algorithms [63], image classification problems [28], visual servoing [38] and many others.

Although RL can sometimes achieve superhuman performance levels, as we have stated, in most real-world scenarios with multiple agents and infinite state spaces, RL often cannot perform adequately. However, the current trend in artificial intelligence (AI), the amalgamation of RL, and deep neural networks (DNN), which is called deep reinforcement learning (DRL), has seen much success in various applications. Especially in computer vision, DRL has seen widespread success in various vision-based tasks in the real world where the vision system is coupled with different actions. Despite its success, DRL has not yet been used conclusively in computer vision, with only some successful works in the field.

This chapter aims to present the DRL methods and its applications and the challenges involved with DRL and its prospects, from a computer vision perspective. Thus the chapter is organized as follows. In Sect. 2.2, the readers are introduced to the core concepts of RL and envision how the amalgamation of different learning strategies of RL and computer vision techniques can be utilized to boost the performance of various computer vision tasks. In Sect. 2.3, we give an overview of DRL followed by Sect. 2.4, where we discuss some of the latest DRL techniques (e.g., attention-aware learning, deep progressive reinforcement learning, and multi-agent learning). Section 2.5 presents several applications of DRL in various computer vision-related tasks. In Sect. 2.6, we discuss some challenges in the field of DRL in vision-related tasks. Finally, we conclude by giving future research directions to develop more advanced algorithms in Sect. 2.7.

2.2 Reinforcement Learning

RL [67] is one of the hottest topics in machine learning. It mimics the learning procedure of the human brain as it modifies its actions according to the rewards received. However, learning about RL demands a few basic concepts about Markov decision processes (MDP) and the methods in solving them.

2.2.1 Basics

We humans learn by interacting with the environment we live in. From our birth, we interact with our environment and receive feedback. If we receive positive feedback from an action, we know it is a good thing to take that action. If the feedback is negative, we learn that it should not be repeated in that situation. Over time, we learn what to do, what not to do, i.e., a policy for our lives. In AI, we often look to build similar agents that learn from the environment through interactions and adjust its policies from the feedback it receives either as a positive or negative reward. Since the agent is reinforcing its policies according to the feedback it is receiving, it is rightfully named reinforcement learning.

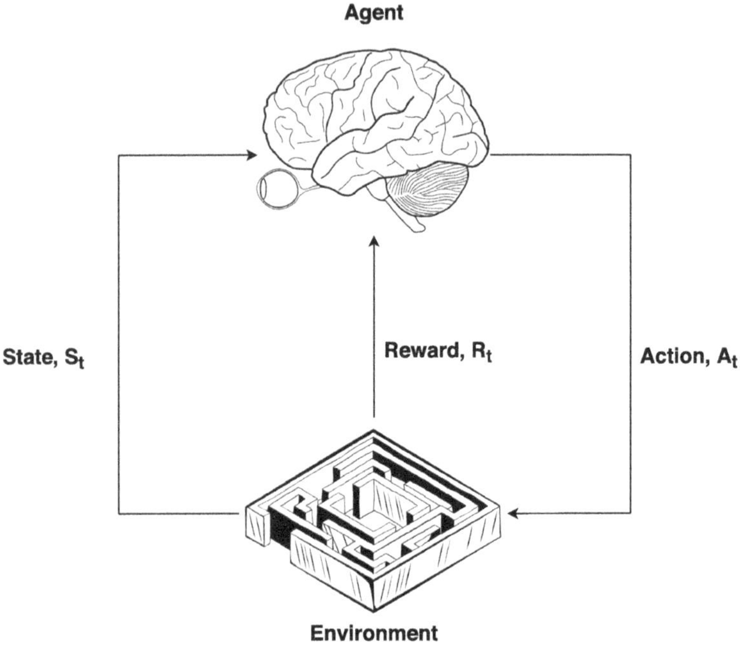

Fig. 2.1 Schematic diagram of reinforcement learning

The overall process is shown in Fig. 2.1 where an agent or learner takes in the state information of a particular time step, S_t interacts with the environment by taking some actions, A_t available in that time step according to what it has learned by then, receives the rewards or penalties, R_t, due to that action and modifies its policy accordingly. Note that the policy recommends the actions that provide the maximum utility, that is, the agent maximizes not only the immediate reward it receives but also the sum of the long term rewards that action can lead to. Thus, the goal of the agent is to maximize its expected rewards.

In many cases, the actions the agent takes might not have the outcome it desires. For example, if an agent decides to go right, there remains a chance that it might end up going right, or somewhere else entirely. This is because the environment with which the agent interacts may cause the agent's actions to be non-deterministic, i.e., multiple possible successor states can result from an action taken in a state. Such non-deterministic problems where the world adds a degree of uncertainty can be formulated using the MDP, which defines agent-environment interaction in terms of states, actions, and rewards. RL is the solution to the MDP formulation, a method to find the agent's optimal policy. Thus, if a problem is formulated well as an MDP, RL is a great framework to solve that.

An MDP is defined by the following properties:

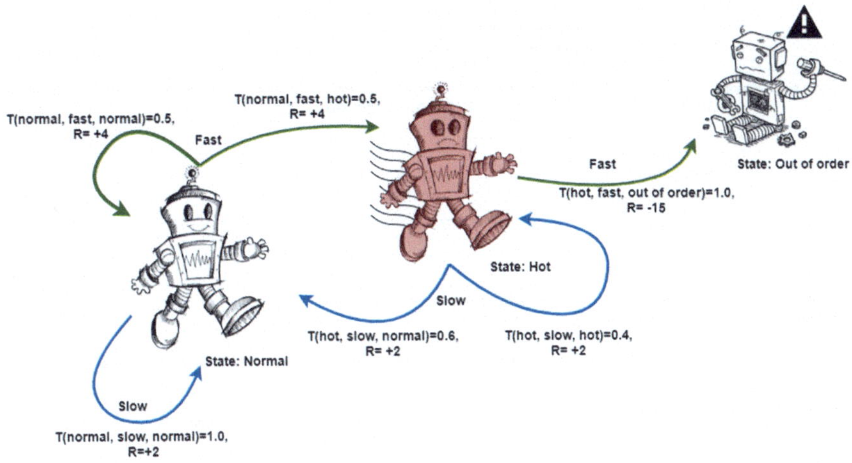

Fig. 2.2 Diagram for the robot runner problem where a robot runner must find the optimal policy for winning the marathon with the given states, actions, transitions and rewards

- A set of states, S. Here, state denotes an agent's configuration within its environment.
- A set of actions, A.
- A transition function $T(s, a, s')$, which is a probability function that gives the probability of the agent reaching state s' by taking an action $a \in A$ from state $s \in S$.
- A reward function $R(s, a, s')$, which represents the reward obtained by the agent after reaching state s' from state $s \in S$, taking an action $a \in A$.
- A start state, where the agent starts from.
- Possibly one or more terminal states, where the lifespan of the agent ends.

Let us consider the non-deterministic problem in Fig. 2.2, which can be mathematically formalized using the framework of an MDP. Suppose we have a robot runner taking part in a marathon as the learning agent, placed in a stochastic environment. The runner can be in one of the three possible states, $S = \{normal, hot, outoforder\}$ where it performs any of the two possible actions, $A = \{slow, fast\}$. It receives different rewards for every action, such as, for running slow from the hot state, it receives a reward of +2, for going fast, it receives a negative reward of -15. As the agent is in a stochastic environment, it is not always sure where it might go after taking action. Such as, while being in the normal state, if it takes fast action, there is half a chance that it will remain in the normal state and half a chance that it will heat up excessively and move to the hot state. So the transition functions for going in both normal and hot states are 0.5. However, when it is in a hot state and takes fast action, there is a 100% chance that it will be out of order from the heat and it will not be able to run anymore, that is, it reaches the terminal state. The goal of the robot runner is simple, to win the marathon, outpacing the other runners. So the

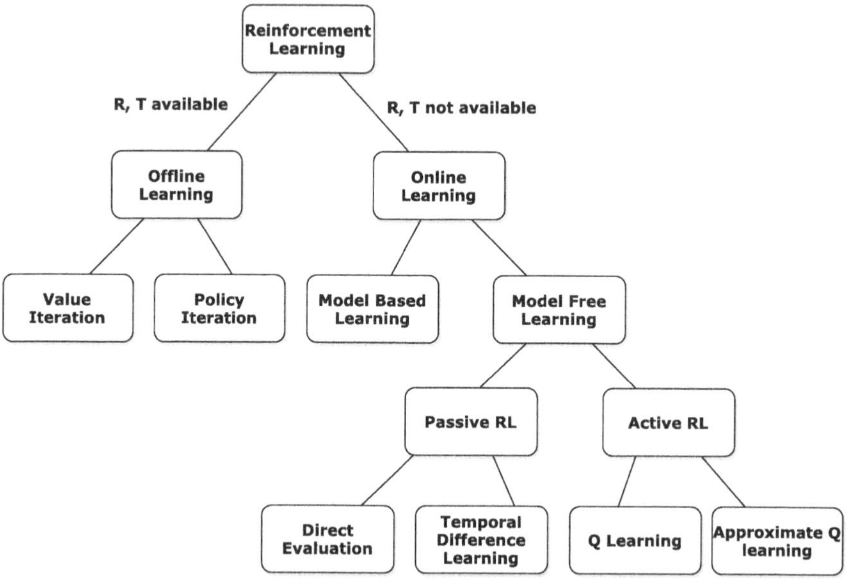

Fig. 2.3 Reinforcement learning algorithms used as solutions to MDPs

agent will want to take the fast action, but it will have to be wary lest it should be out of order. Thus, it will look for an optimal policy that will maximize the sum of expected rewards and win him the race.

2.2.2 Solving MDPs Using Reinforcement Learning Algorithms

In our robot-runner example, the agent is assumed to have complete knowledge of the environment. Therefore, it can learn the optimal behavior without having actual interaction with the environment. This is called offline learning where dynamic programming based approaches, e.g., value iteration and policy iteration, are used to learn the policy. On the other hand, if the agent does not know about the dynamics of the environment (i.e. T and R are unknown), the agent has to learn from actual experience. It explores the environment first, building up its knowledge about the world's dynamics, and then fix the policy accordingly. This approach is called online learning. Figure 2.3 presents the main approaches of RL algorithms that we will describe in the following sections.

2.2.2.1 Key Concepts

Before we go into details about the RL algorithms, we formally define some recurring terms.

- **Policy,** π - A mapping from the states of the environment to the actions to be taken while on those states.
- **Utility** - The utility of taking an action is the immediate rewards received for that action plus the sum of expected rewards after that, assuming that the agent abides by the optimal policy.
- **Optimal value of a state,** $V*(s)$- The optimal value $V*(s)$ of a state $s \in S$ is the expected value of utility the agent acquires if that agent acts optimally from thereon over the rest of its lifespan.
- **Optimal Q-value of a state,** $Q*(s, a)$- The optimal Q-value $Q*(s, a)$ of a state $s \in S$ is the expected value of utility the agent acquires by taking an action $a \in A$ from state $s \in S$, and acting optimally there on.
- **Optimal Policy,** $\pi*$ - An optimal policy is one that maximizes expected utility if followed.
- **Discount factor,** γ- It is a multiplication factor between 0 to 1, that is multiplied with the expected rewards. This is generally used to give more importance to the immediate rewards and also, for the convergence of the algorithm.
- **Bellman Equation** The equation is defined as

$$V*(s) = \max_a \sum_{s'} T(s, a, s')[R(s, a, s') + \gamma V*(s')] \tag{2.1}$$

We can define optimal Q value as follows -

$$Q*(s) = \sum_{s'} T(s, a, s')[R(s, a, s') + \gamma V*(s')] \tag{2.2}$$

So, the Bellman equation can be simplified to,

$$V*(s) = \max_a Q*(s, a) \tag{2.3}$$

It is clear that the optimal value $V*$ is the maximum expected utility of that state over all possible actions from that state.

2.2.2.2 Offline Learning

Value Iteration: Since the Bellman equation allows us to find the optimal values of a state over all possible actions, we can iteratively compute the values of all the states until convergence, i.e., find the optimal values for all the states. This is the main concept of value iteration. And the algorithm is as follows:

- Initialize $V_0(s)=0$ for all $s \in S$.
- For all $s \in S$, repeat until convergence

$$V_{k+1}(s) = \max_a \sum_{s'} T(s, a, s')[R(s, a, s') + \gamma V_k(s')] \tag{2.4}$$

This value update equation for value iteration is similar to the Bellman equation. However, whereas the Bellman equation represents the condition of optimality, the update equation simply represents the iterative method to update values till convergence, or optimality. In essence, when convergence is reached, for every state $s \in S$, $V_k(s) = V_{k+1}(s) = V^*(s)$.

For a search problem, finding optimal values is not enough. Agents require policies that act on the optimal values. That is where policy extraction comes from. Even though value iteration does not give the action that lead to the optimal value, we can look at the argmax of the optimal Q values for each state $s \in S$ and take the action that gives the optimal value as the optimal policy.

$$\pi^*(s) = \underset{a}{\text{argmax}}\, Q^*(s, a) = \underset{a}{\text{argmax}} \sum_{s'} T(s, a, s')[R(s, a, s') + \gamma V^*(s)] \tag{2.5}$$

Policy Iteration

Often it is seen that even though values take a long time to converge, the optimal policy that lead to those values converge long before that. So, it makes more sense not to wait for values to converge if we can extract the optimal policy much before the values converge. This is the main intuition behind policy iteration. Whereas in S number of states with A number of actions available, value iteration would require $O(S^2A)$ time for each update, policy iteration would only require $O(SA)$. The algorithm for policy iteration is as follows:

1. Define an initial policy.
2. Repeat until convergence
   - Evaluate the current policy using the following equation

$$V^\pi(s) = \sum_{s'} T(s, \pi(s), s')[R(s, \pi(s), s') + \gamma V^\pi(s')]$$

   - Improve the current policy to generate a better policy using the following equation

$$\pi_{i+1}(s) = \underset{a}{\text{argmax}} \sum_{s'} T(s, a, s')[R(s, a, s') + \gamma V^{\pi_i}(s)]$$

   When $\pi_{i+1} = \pi_i$, the policy will have converged and the problem will be solved.

2.2.2.3 Online Learning

Online learning is mainly of two types, model based learning and model free learning. Before we delve into the elaborate details of online learning, here we define some terminologies.

- **Sample**- During the online planning, at each iteration t, an agent starts from a state s, takes an action a that results in a successor state s' and some reward r. Each (s, a, s', r) tuple is known as a sample.
- **Episode** - An agent takes actions and collect samples until it reaches a terminal state. This collection of samples is defined as an episode. While exploring an environment agents generally go through many episodes where the episodes are independent of each other.
- **Exploration** - When the agent is taking possibly sub optimal actions to learn about unknown states, it is called exploration. When most of the states are unknown and the policy has not yet converged, the agent should explore more to learn about the state values.
- **Exploitation** - When the agent takes optimal actions according to the learnt policy, it is known as exploitation. When most of the state values are learnt and the agent has a converged, optimal policy, it should exploit more and take the optimal actions.

Model Based Learning
Model based learning is as its name suggests. The agent explores to build a model transition and/or reward function, whichever is unknown, based on a given policy and then define the problem as an MDP with the modeled transition and/or reward functions before solving it using value or policy iteration. This method has a lot of drawbacks though, as building a proper model would require extensive amount of explorations for accurate transition functions and rewards. A large state-space would result in a huge model that would require a lot of storage capacity of the agent.

Model Free Learning
There are ways to learn without building a model, which is called model free learning. Model free learning can be of two types, passive reinforcement learning and active reinforcement learning.

Passive Reinforcement Learning
Passive RL is where the agent learns the state values based on a given policy, rather than modeling the transition and reward functions on its own policy. There are two methods to do that. One is called direct evaluation, where the agent acts on the given policy and average the rewards received to estimate the value of each state. Here, the agent observes different episodes during training given the policy and averages the received rewards to get the average state value. Although such averages sometimes produce the correct state values, it can miss its mark as it ignores the state connections when computing the average value.

The other method is temporal difference learning where the agent learns from every sample exploration. This means that rather than averaging the rewards after some episodes to learn the state values, we keep a running average and tune the state values after each sample. Let us say that, a sample value from a state can be the sum of the reward obtained from that state, and the discounted utility obtainable from thereon under the fixed policy. That is,

$$sample = R(s, \pi(s), s') + \gamma V^{\pi}(s') \tag{2.6}$$

We update the state value from each sample with the equation

$$V^{\pi}(s) = (1 - \alpha)V^{\pi}(s) + (\alpha)sample \tag{2.7}$$

Where, α is the learning rate. Note that large value of the learning rate gives more importance to the newly found sample rather than the existing value, and smaller learning rate gives less emphasis to the samples and is prone to keeping the existing value of that state. Often at the beginning, learning rate is kept high because the existing value is probably not correct, and later, learning rate is gradually lowered to give more importance to the learned value, which can give converging averages. Even though temporal difference learning eventually learns the true values of all states under the policy it follows, it cannot generate the optimal policy. Computation of an optimal policy requires knowledge of the Q-values of states which cannot be computed in temporal difference learning as it does not know the values of T and R. Had we learned the Q values, this problem could have been solved. This is the intuition behind active reinforcement learning or Q learning.

Active Reinforcement Learning

Active RL, often called Q learning allows the agent to learn not just the state values, but the optimal values and policies, without needing to know the transition or reward functions. Q learning learns the Q values rather than the values, and thus, is able to generate optimal policy that gives the highest Q values. And since we do not know the transition or reward functions, we learn from samples on the go. The agent chooses it's own actions from each state, and learns the Q value for that state and action. It keeps a running average, similar to temporal difference learning. Only this time, rather than learning state values, the agent learns the Q values.

$$sample = R(s, a, s') + \gamma \max_{a} Q(s', a') \tag{2.8}$$

And update the Q value with,

$$Q(s, a) = (1 - \alpha)Q(s, a) + (\alpha)sample \tag{2.9}$$

Similar concept applies here. The learning rate α is kept high at the beginning to emphasize exploration, and learn more. α is then lowered gradually to give more emphasis to exploiting the learned policy. This is an algorithm that allows the agent to find the optimal policy by interacting with the world and exploring the rewards available. Let us look back to the example of Fig. 2.2, but this time we do not know

Table 2.1 Converged policy of the robot runner of Fig. 2.2

State	Optimal policy
Normal	Fast
Hot	Slow

the transition function or the reward. So the robot runner has to explore and find the state values himself. Assuming there is a discount factor of 0.9, and a learning rate of 0.5, by following the previous equations, we would get the Q values for each actions from each of the states. If the runner picks the max over the Q values for each state, he would end up with the optimal policy that will allow him to take the actions in each state that maximises the rewards and win the marathon. The optimal policy of Table 2.1 would be found if the Q values of each actions are calculated iteratively until they, or the policy itself, converges. In this case, the optimal policy would be to go fast when in normal state and slow in hot state, as these actions in the respective states would have the highest Q values.

The problem with Q learning is that the agent needs to explore the amount required to learn the optimal policy. He needs to learn how much to explore and when to start exploiting the learned policy. Also in the real world, there can be infinite amount of state spaces. Exploring and learning about all of them is not feasible. Even if that was possible, keeping Q values for all those states would be very expensive.

However, often the state information can be generalized. Learning an approximate and generalized Q value would then make much more sense than keeping infinite amount of Q values. This is called approximate Q learning. This can be done by describing the state with a feature vector and their subsequent weights rather than the state values or q values. Then the agent simply learns each state Q-value from the weighted sum of the features in that state.

$$Q(s, a) = w_1 f_1 + w_2 f_2 + ... + w_n f_n \tag{2.10}$$

The Q value is updated with difference between the sample and the existing Q-value.

$$difference = [R(s, a, s') + \gamma Q(s', a')] - Q(s, a) \tag{2.11}$$

$$Q(s, a) = Q(s, a) + \alpha[difference] \tag{2.12}$$

Here, the agent would need to tune the weights to the features as well. The updates to the weights would be using the following equation

$$w_i = w_i + \alpha[difference] f_i(s, a) \tag{2.13}$$

Even though using approximate Q learning allows the agent to work in infinite state spaces, the agent needs to depend on the approximate values. Thus the accuracy would be a question here. Also, choosing the right features will play a big part in the agent's success.

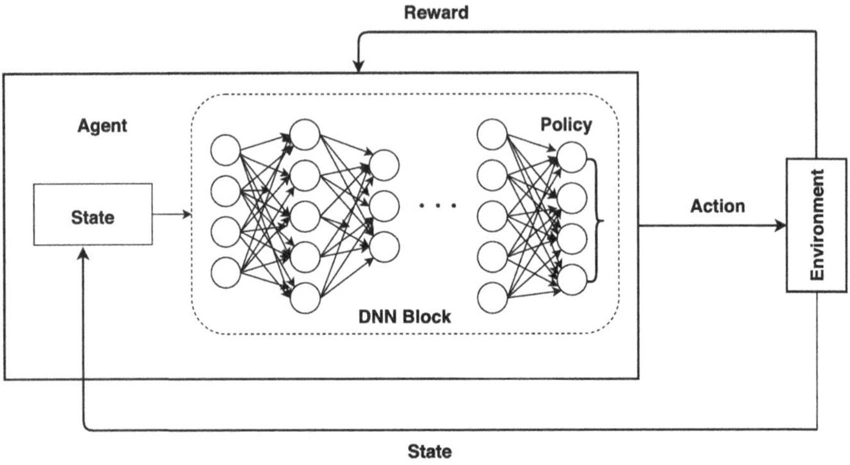

Fig. 2.4 Schematic diagram of deep reinforcement learning

2.3 Deep Reinforcement Learning

One of the hottest topics in artificial intelligence is DRL. It is the fusion of RL and neural networks. DRL enables the agent to learn proper values even in high dimensional state spaces, and it has even been able to beat humans in various tasks. DRL takes the deep neural network's ability to learn in high dimensional spaces and integrates the RL's capability of maximizing rewards in a dynamic environment. With its unlimited potential and promise, DRL has easily become one of the most potent research fields in artificial intelligence.

What makes deep DRL such an exciting fields is that it takes the best of two strong learning algorithms, neural networks and RL, and covers both of their weaknesses. Neural nets require a lot of labeled data, which is hard to get in real world scenarios. RL requires proper learning in order to perform well. In DRL though, one can use reinforcement learning combined with neural nets to allow the agent to experience the data available and make rational decisions from the experienced rewards in order to learn better features in higher dimensional state spaces and perform effectively as shown in Fig. 2.4.

One of the first examples of DRL is Deep Q-Network (DQN), first introduced in 2013 [45]. They applied this to the Atari games and obtained unprecedented results. At each time step, they took 4 raw 84x84 pixel images that are fed as input to a convolutional neural network as shown in Fig. 2.5. The output is the estimated value functions or Q values for each actions. The Q network then takes the optimal action accordingly. They used experience replay mechanism [42] that randomly samples the past experiences in order to have a smooth training distribution. They implemented stochastic gradient descent to update the weights. When the trained network was

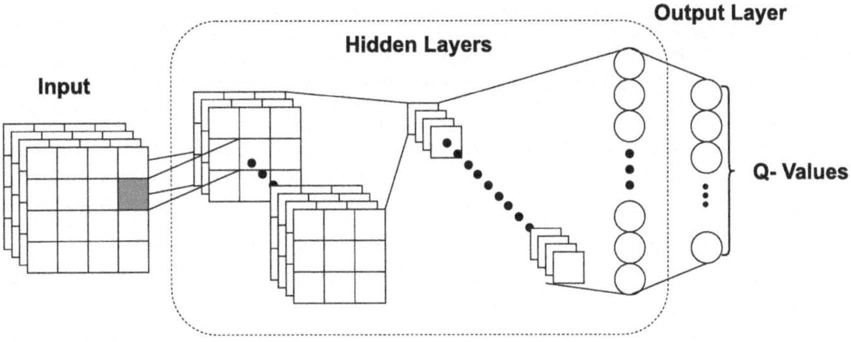

Fig. 2.5 Deep Q network architecture of [45]

implemented to seven of the Atari 2600 games, it beat all other algorithms to that date in six of them, and even outperformed human experts on three.

At the time, such a result was completely unprecedented. Machines beating humans was a thing of the sci-fi movies. Deep Q-network weaved that into reality. This was not a perfect algorithm though. In normal reinforcement learning algorithms, there exists a problem of overestimation. Since the algorithm takes the maximum of the Q values of all possible actions, it often overestimates the value function and thus introduces a maximization bias. Since Q learning learns estimate values from more estimate values, overestimation becomes a substantial problem. Such overestimation often leads to suboptimal policies. Van Hasselt et al. [69] illustrate this over Atari games and shows how a different implementation, Double Deep Q-network (DDQN), an updated version of Double Q network [29], estimates Q values closer to the true Q values than normal DQN. DDQN keeps two Q value estimates rather than one, and uses one estimate to update the other. Let us say that the two Q value estimates are Q_1 and Q_2. Then for updating Q_1, the value of Q_2 is used. And similarly for updating Q_2, the value of Q_1 is used. Thus, we land upon unbiased estimated Q values.

Although DQN works well in finite state spaces where the number of state action pairs is limited and the Q values can be stored without memory issues, it is not the most ideal choice for infinite state space problems. Even in large state spaces, DQN faces quite a challenge, explaining why it could not beat many of the Atari games. This problem is persistent on many of the value based algorithms where the network looks to learn the optimal value function which of each state action pairs and selects the pair leading to the highest value. So, for optimal decision making, it would need to store sufficient number of such values of the state action pairs. Thus, it clearly bows out of favor for real world applications where there could be infinite number of state action pairs. The most intuitive idea here would be a policy based learning algorithm such as Monte Carlo learning [58], where the policy is learned without a definite value function. It is also effective in continuous or stochastic space. However, it can only be used for episodic problems. Another issue is that a whole episode has to pass in order to iterate the policy given the returns.

It would be better to combine the value based method and the policy based method to make the best of both algorithms. This is the main idea behind the actor critic method. As the name suggests, there are two parts to this network, an actor network and a critic network. The actor network is a policy based network that controls how the agent operates. The critic is the value based network that evaluates the policies. One special thing about the actor critic method is it updates the value at each step much like temporal difference learning. The critic network approximates the value function at each time step rather than using the total reward after each episode like in Monte Carlo learning. In traditional actor critic algorithm, the actor takes the state as input and gives an action as the output. The critic takes the state and action pair and estimates the Q values for them.

However, we do not need not to know the exact Q value, rather we need to know which action corresponding to the highest Q value gives us the best result. Thus, rather than learning the Q values, the better option would be to learn the advantage value, which represents how good an action in that state is over all other possible actions. This is called advantage actor critic network or A2C. The successor of A2C is A3C, or asynchronous advantage actor critic model. This was developed by deepmind in 2016 [44] and blew everyone away with extraordinary results. A3C consists of multiple agents equipped with their own weights. They operate with different copies of the environment in parallel. They update a global network having shared weights periodically. After each update, the agents update their own weights with those of the global network. This allows A3C to cover more of the environment much quicker than other algorithms. In fact, its overwhelming performance at the time in RL problems made DQN and other similar algorithms obsolete. Even to this day, many DRL models implement actor critic network.

2.4 Deep Reinforcement Learning Methods

We have covered the basics of DRL. However, there are some special branches of DRL that has been gaining a lot of ground in the field of AI. This section covers a few of such methods.

2.4.1 Attention Aware Deep Reinforcement Learning

Attention means to focus. When our brain gives more attention to something, it can grasp more details of that object and perhaps get a better understanding. However, focusing on everything is not the way to go as unless the brain focuses on the important and correct things, it will not be able to learn or understand what it wants to. Rather it will waste its time and resources. But when we give our attention to the important parts, we can often get a better understanding or situation.

The concept of attention can be applied in machine learning as well. Attention in machine learning is simply focusing on the most important parts or features and learning them. For example, since the center of the image generally contains the most important information, focusing more on the center of the image seems the right way to go. This concept was first introduced in RNNs or recurrent neural networks, namely encoder decoder models [5]. To get a better understanding about this, let us briefly take a look at RNNs.

RNNs are required in order to handle sequential inputs or give sequential outputs. It can be thought of as multiple neural networks overlaid one after another, each handling one part of the input sequence. One might think that using normal neural networks repeatedly on the multiple inputs would yield the same result. However, this is not true because in that case, the relation between the inputs is overlooked. Recurrent neural networks takes the relations between the inputs using lateral connections between the networks. This means the decision of the network depends both on the present input as well as what it has learned in the past. This gives a better result in sequential inputs where some input is correlated with another. Normally in neural networks, the governing equation is a linear equation consisting of the weights and biases of that layer. However in RNN, there is an intermediary hidden state which acts as the memory of the network. The previous hidden unit is fed as an additional input to the next hidden unit as shown in the Fig. 2.6. Thus what has been already learned also gets taken into consideration in producing the output of the current unit. Unlike other neural networks, RNNs share weights in each time steps among all the units. There are three types of weights, U, V and W that are broadcast to all the units. The weights are updated in each time step through back propagation. There can be different types of RNNs. One to many RNN takes a single input and generates multiple outputs, many to one RNN takes multiple inputs but only gives a single output and many to many RNN that takes multiple inputs and gives multiple outputs.

One of the problems of RNN is that in long sequential inputs, the loss gradients for backward propagation may vanish or explode which is a common occurrence in any neural networks. This is a severe problem in the case of RNN due to the fact that the same weights are broadcast to all the branches. Also, in long input sequences, information from the earlier inputs are often lost. To solve these problems, gated recurrent neural networks were introduced. These networks have parameters called gates that allow the network to choose which values to remember and which values to forget. LSTM or Long Short Term Memory is a prime example of such gated RNNs. The original LSTM had 2 gates, input gate (i), output gate (o) and a cell state. Later on forget gate (f) was introduced [24]. As the name suggests, the input gate controls whether the cells are updated. The input and the previous hidden state is fed into a sigmoid function that gives values between 0 to 1. Here, 0 means the current input is not important, 1 means it is important.

$$i_t = \sigma(w_i[h_{t-1}, x_t] + b_i) \tag{2.14}$$

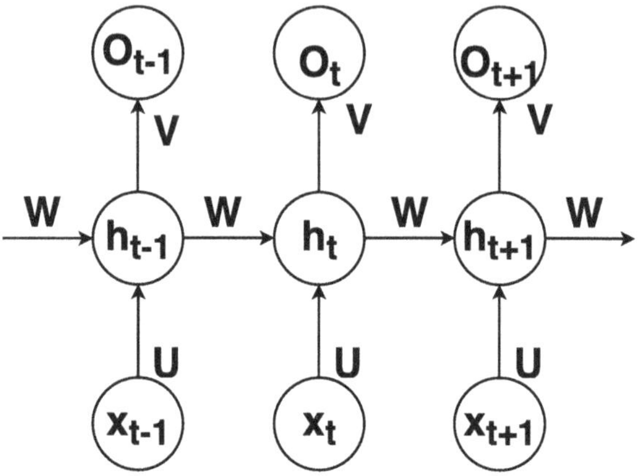

Fig. 2.6 Reccurent neural network structure

Forget gate controls whether the information should be forgotten or turned to 0, or kept. This is also done passing the current input and the previous hidden state to a sigmoid.

$$f_t = \sigma(w_f[h_{t-1}, x_t] + b_f) \tag{2.15}$$

Finally the output gate that gives the sigmoid activation at that time step, that helps to determine what the next hidden state will be.

$$f_t = \sigma(w_f[h_{t-1}, x_t] + b_f) \tag{2.16}$$

There is one more thing in LSTM that is called the cell state. Cell state is an interim value calculated at each time step in order to help with generating the memory or the next hidden state. The candidate cell state, $\tilde{c}$, is defined as,

$$\tilde{c} = tanh(w_c[h_{t-1}, x_t] + b_c) \tag{2.17}$$

Cell state is then calculated using the equation,

$$c_t = f_t * c_{t-1} + i_t * \tilde{c} \tag{2.18}$$

The Eq. 2.18 shows how the input and forget gates operate. The retention of the preceding cell state is regulated by the forget gate, and the input gate determines how much importance is given to the current input. The output gate helps to generate the next hidden state using the following equation,

$$h_t = o_t * tanh(c_t) \tag{2.19}$$

A special type of RNN structure is RNN encoder decoder network. An encoder decoder model can be thought of a many to one RNN combined with a one to many RNN. The inputs are encoded into a single vector using a many to one RNN and feed the output vector to the one to many RNN to be decoded. Like other RNNs, this has hidden states at each time steps, but only produces an output vector at the end.

$$c = f(h_1, h_2, ..., h_T) \tag{2.20}$$

Here, c is the output vector, $h_1, h_2, ..., h_T$ are hidden states at each time step and f is some non linear function.

The output vector c is then fed into a one to many RNN in order to generate output. Often the decoder is trained to predict what the next word should be, given the summary vector c and the previous outputs. What this means is that it takes the previous outputs and the summary vector c, and generates a probability distribution of the words to be given as output.

$$p(y) = \prod_{t=1}^{T} p(y_t | y_1, y_2, ..., y_{t-1}, c) \tag{2.21}$$

here, $y = y_1, y_2, ..., y_T$ are the cell predictions. The conditional probability is modeled as,

$$p(y_t | y_1, y_2, ..., y_{t-1}, c) = g(y_{t-1}, h_t, c) \tag{2.22}$$

where, g is some non-linear activation function. This novel approach was first used for machine translation [13] where it proved its effectiveness. It has since, become one of the most popular architectures in machine translations and language generation. This method is not without a flaw, however. Since the whole of the input sequence is encoded into a single vector, in long input sequences, some information might become diluted and lost. This is where the concept of attention comes into play.

In a mathematical sense, attention is provided using a context vector [5]. Here, the probability distribution is given as follows,

$$p(y_i | y_1, y_2, ..., y_{i-1}, x) = g(y_{i-1}, s_i, c_i) \tag{2.23}$$

where, s_i is the hidden state of the network at time i computed by,

$$s_i = f(s_{i-1}, y_{i-1}, c_i) \tag{2.24}$$

The context vector is computed using a sequence of annotations $h_1, h_2, ..., h_T$. Each of these annotations h_i contains information about the whole input sequence but focusing strongly around the i-th word. The context vector is a weighted sum of these annotations.

$$c_i = \sum_{j=1}^{T_x} \alpha_{ij} h_j \tag{2.25}$$

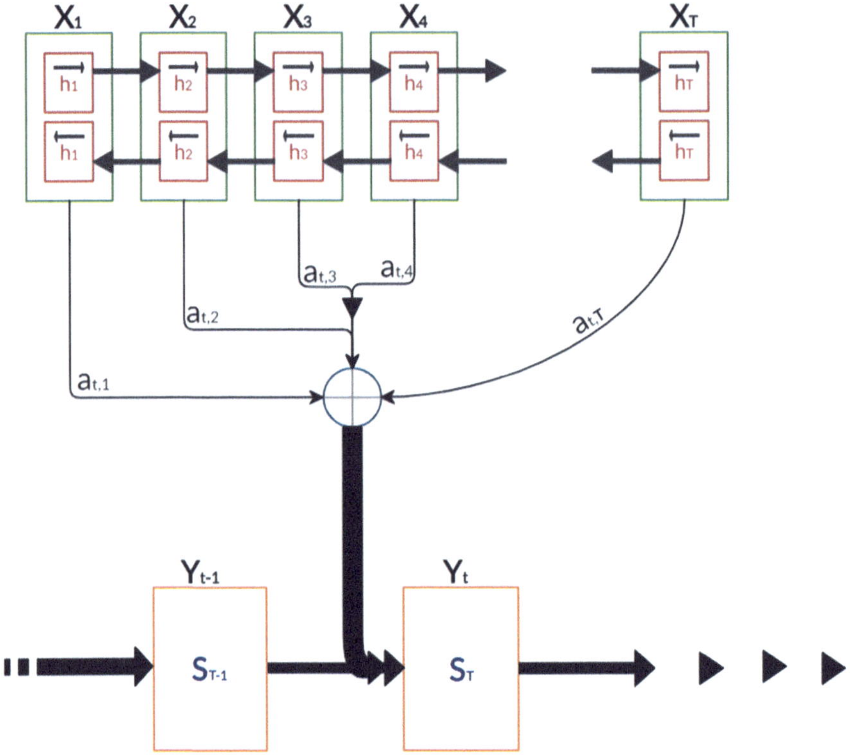

Fig. 2.7 Graphical illustration of attention model [5]

The weight α_{ij} of each annotation h_j is computed using

$$\alpha_{ij} = \frac{exp(e_{ij})}{\sum_{k=1}^{T_x} exp(e_{ik})} \tag{2.26}$$

where

$$e_{ij} = a(s_{i-1}, h_j) \tag{2.27}$$

is an alignment model. This shows how much inputs and outputs at i and j positions align (Fig. 2.7). The sum of the weights is equal to one, and ideally, the weights are going to be larger at the times corresponding to the inputs that are most related to the outputs. Thus, the information relating to the current output is given the most attention or importance. Thus attention aware RNNs perform better than regular RNNs.

Nowadays, this goes hand in hand with RNNs and its various applications of sequential inputs like video processing, natural language processing or generation etc. Attention mechanisms help to train the network faster and make the predictions more accurate. The model that has an attention mechanism built is often called an attention aware system. Attention aware deep reinforcement learning or ADRL is thus an RNN with attention mechanisms combined with RL in order to take sequential inputs and generate optimal policies from them. DRL networks that combine RL with RNNs are often recurrent policy networks. Recurrent policy networks generate policies for the agent to operate on sequential inputs such as video sequences or texts or speeches. When the RNN in recurrent policy network has an attention mechanism with it, it tends to train faster and make better predictions. Thus the policies generated converge faster and the agent performs better.

Many works have been done in this field. Most of them consist of sequential input systems like text understanding, text generation, image and video processing etc. as ADRL networks incorporate recurrent networks. For example, [54] used ADRL for video face recognition. They attempted to discard misleading images and find where the focus of attention lies. They formulated it into an MDP and divided the task into two sets, feature learning, where they input the whole video to learn various features using CNN and recurrent networks, and attention learning, where they evaluate the selected focus of attention using RL (Fig. 2.8). Cao et al. [9] applied ADRL for image enhancement, mainly face hallucination, which is a domain specific super-resolution problem. The goal is to create high-resolution image from a low-resolution one. Using ADRL with a CNN and an LSTM, they find the actions to select the patch of image to enhance. They call this recurrent policy network. Then they run a local enhancement network consisting of cascaded convolutional networks. This way, patch by patch, the whole of the image is enhanced. The effectiveness of such works show how potent ADRL can be (Fig. 2.9).

2.4.2 Deep Progressive Reinforcement Learning

Humans are adept at learning from previous experiences and applying them to the new ones. However, that has long been a problem for machines. Often it is seen that even though with deep learning, machines are able to learn how to get better at a task, when it faces a new task that is somewhat similar to the previous, it fails to perform adequately. The concept of learning continuously from the streams of data, and be able to positively transfer the learned model in order to perform well in other similar tasks is called continual learning. However, in most cases, this is improbable, as most networks often forget previously learned information when learning new information. This is called catastrophic forgetting and is one of the most significant issues in AI. In the case of RL, rather than forgetting the weights, the policy is updated, and thus overwritten for each task, which is called policy interference. Even though previously, it was thought that this is inevitable in the case of learning algorithms, recently, breakthroughs have emerged in this area. Catastrophic forgetting can be overcome by slowing down the learning for certain weights that are important for the previous tasks [37]. This is known as elastic weight consolidation. Another method of overcoming this problem is progressive neural networks.

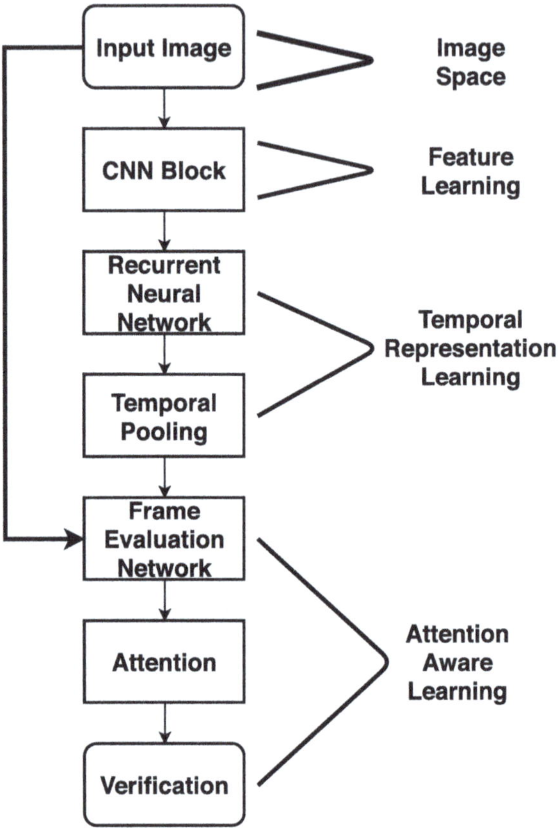

Fig. 2.8 ADRL architecture for video face recognition [54]

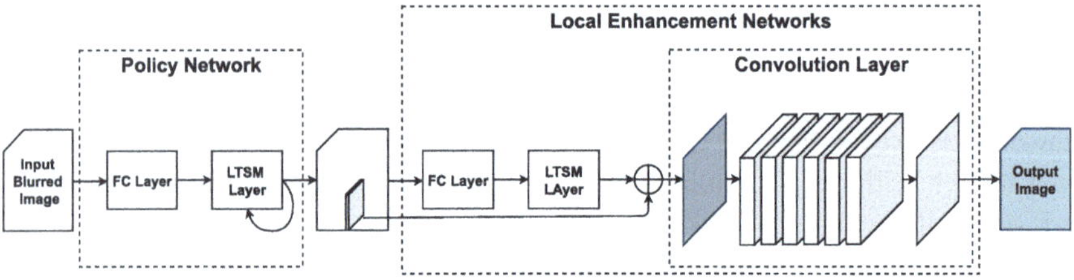

Fig. 2.9 ADRL architecture for attention aware face hallucination [9]

Progressive networks work for both neural networks and for RL. The key idea is to feed the network with lateral connections of the information about the previous tasks [60]. This way, the learned information about the previous task is preserved, and the newly learned policy or weights do not overwrite or replace the old ones. Note that this method allows the agent not to work with the policy or weights already learned, rather adds new capacity to the learned policies or weights. This does not necessarily help with transfer learning, but rather allows the agent to be inherently adept in multiple tasks by increasing its capacity. This is similar to the way humans learn different tasks. Rather than forgetting the previously learned task in order to learn a new one, we simply increase our capacity and learn the new task whilst remembering the key details of the previous tasks.

In terms of robotics, it is often necessary that the learned information be converted to some actions. Thus, progressive networks naturally transform into deep progressive reinforcement learning (DPRL) networks. Here the agent extracts the policies the same way as other DRL methods. However, rather than learning only one policy, lateral connections of the previously learned policies allow multiple policies to be learned, and that too, fairly quickly. DPRL has already been applied in skeletal action recognition [68] achieving great results. This holds major promise in the future of robotics paired with policy distillation methods [59] in creating an agent expert in multiple tasks.

2.4.3 Multi-agent Deep Reinforcement Learning

Multi agent systems have always been one of the most researched fields in artificial intelligence, mainly because it most resembles the real world environment, where multiple agents operate in the same environment. This has always been a concern for learning algorithms. Simply because with the addition of each agent, the problem space increases exponentially. However, this is an inevitable problem to be dealt with in deploying an agent in the real world scenarios. In an AI-driven world, there will always be multiple agents in the state space. The agents could be either cooperative or competitive or both. Either way, multi agent systems always have posed a significant level of difficulty, especially when the other agents are also intelligent agents and capable of learning. This is because the agent cannot properly predict what the other agents will do, simply because the other agents are also learning. We have seen the exponential growth of state space in adversarial games. However, in real world scenarios, there will be many more agents with an infinite number of actions. It is easy to grasp the infinite state space of such problems. Thus, RL algorithms come up short in such scenarios. There have been algorithms that deal with multi agent systems, given numerous constrictions in place. For example, cooperative agents in a defined space can have shared reward and transition functions, allowing them to limit the state space. Some algorithms are not scalable and are defined only for a specific number of agents. And selfishness of the agents has always been an issue.

Since DRL has been successfully applied to games, it makes sense to apply it to multi agent systems as well. Inherently, DRL does seem like the ideal solution for multi agent systems due to the ability of learning features of higher dimensions at lower complexities, and easily convert them to producing optimal policies. It has already proved its efficacy in adversarial games like Go [64]. Thus it should be able to somewhat replicate the results in real world scenarios as well, of course, given proper time to train. Obviously, there are problem areas that need to be addressed properly.

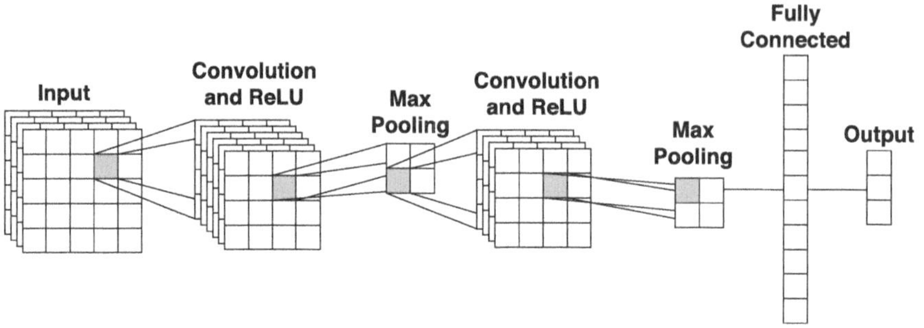

Fig. 2.10 DLCQN structure: four most recent frames are passed through two max pool and convolutional layers and then flattened. The flattened fully connected layer then goes into the Q network which gives three actions [10]

One of the main issues is the non-stationarity problem. As we have already said, in a multi agent system, the agent has to consider not only the environment it is in, but also the agents in it. Unlike in single agent systems, the environment is changing in each time step. The actions the agent takes has to consider in the other agents' actions as well. In a cooperative environment, the coordination of the actions of different agents has to be considered, whereas in competitive environments, that is not an issue. The rewards the agent is receiving are not only dependent on the agents action anymore, but partially dependent on the actions of the other agents in the environment as well. For example, in a cooperative environment, in a certain situation, the agent might need to coordinate its actions with other agents in order to achieve the best possible result. In other situations, such coordination might not be necessary for the best outcome possible. Thus, many works have been done in coordinated learning with a view to exploiting the degrees of independence by defining different levels of coordination. For example, Yu et al. [77] proposed an algorithm where the agent learns the degree of independence and thus can decide whether to coordinate actions or not. Castaneda [10] applied the concept using DQN and proposed the DLCQN or deep loosely coupled Q-network where the agent learns the independence degree and chooses whether to coordinate actions or act independently in order to deal with the non-stationarity problem (Fig. 2.10). Some other notable works that also deal with the non-stationarity problem in multi agent systems are lenient deep Q-network (LDQN) by [49], weighted double deep Q-network (WDDQN) by [82] etc.

Another big issue for any intelligent agent deployed in the real world environment is partial observability. In real world scenarios, often times the environment is not fully observable to the agent. Such environments can still be modeled using a partially observable markov decision process or POMDP. Even in DRL, DQN that deal with partial observability has been proposed. Deep recurrent Q-network or DRQN [30] is one such network that replaces the first post convolutional fully connected layer with an LSTM. Even though the agent could process a single frame, it could successfully integrate learned information through time and perform well in Atari games and other POMDP settings (Fig. 2.11). In the context of multi agent systems, the most

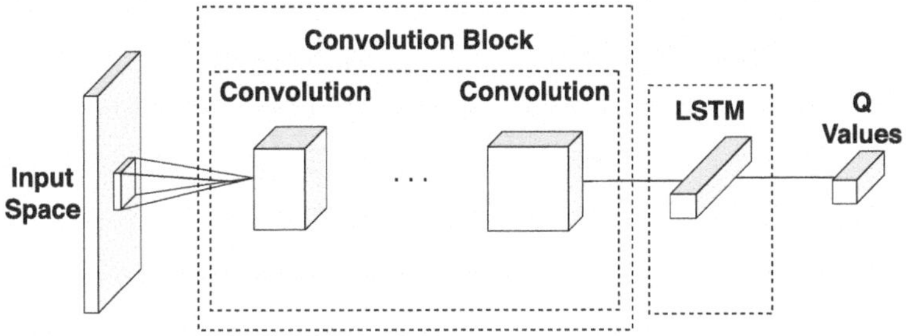

Fig. 2.11 Deep recurrent Q-network architecture [30]

naive approach would be to integrate the DRQN with independent Q learning that shares the rewards obtained with each agent in the environment. Partial observability problem can also be dealt with simply through distributive or coordinated learning. [20] improved upon the naive method and proposed a deep distributed recurrent Q-network or DDRQN. It offers the last action as the input of each agent, proposes an inter-agent weight sharing and thus, only one network is learned by the agents, and finally, turning off experience replay, simply because of the volatile nature of the environment in a multi agent system that renders past experiences obsolete and often misleading. Although this system performs well in various well known riddles, it assumes that all agents have the same set of actions. This, however, may not always be true. For example, one agent can operate on the ground, and another in the air in the same environment. Another system by [32] proposed deep policy inference Q-network or DPIQN for multi agent systems and deep recurrent policy inference Q-network or DRPIQN for multi agent systems with partial observability issues. They introduced policy features learned from raw observations of the agents by inferring the policies. The policy feature vector is incorporated in the Q network of both these architectures. Both of the networks are trained adaptively, which adjusts the attention to learn at different phases of training (Fig. 2.12).

Training is also a big concern in multi agent systems. The agents have to learn in an environment that is constantly changing, with its reward depending on the actions of other agents as well as its own. There are ways to deal with this however. One of the most basic one is independent Q learning that considers other agents as part of the environment and shares the rewards obtained among agents. However, this has scalability issues as with the increase in the number of agents, the complexity also increases. [17] used images for state representations and learning their features using a convolutional network. The learned features are then fed into a DQN that give out the policies. Agents can be trained one at a time, whilst keeping policies of the other agents fixed. One popular approach is centralized learning of decentralized policies, where several agents are trained in groups in a centralized method [46]. But they have decentralized policies so each agent can act by its own policy. Centralized policies can also be deployed, with the goal being that the agents then take joint actions from

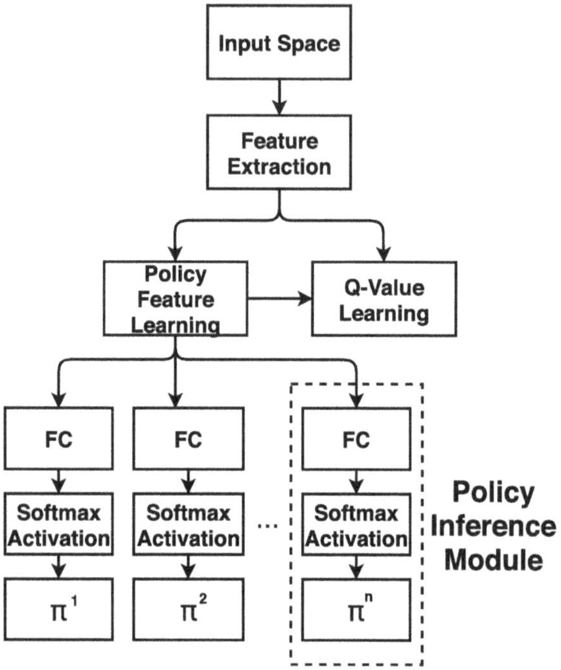

Fig. 2.12 Simplified architecture of DPIQN and DRPIQN [32]

joint observations. [27] show that a parameter sharing algorithm in such conditions can allow simultaneous training of agents.

Foerster et al. [19] introduced two methods for such centralized learning communication method called reinforced inter-agent learning or RIAL and differentiable inter-agent learning or DIAL. RIAL has a recurrent nature and looks to combine DRQN with an independent Q network. It treats communicating with other agents as a separate action and assigns reward for it accordingly. DIAL allows the flow of gradients through a channel from agent to agent and allows end to end backpropagation among agents.

2.5 Deep Reinforcement Learning Applications in Computer Vision

We have, thus far, discussed about DRL and various methods of implementing a DRL architecture. In this section, we discuss the various applications of DRL in the field of computer vision.

2.5.1 Visual Search and Tracking

Visual search, as the name suggests, is to use vision to search for a particular object in the environment. Humans use their sight to often seek out single or multiple objects in a much larger environment through the movement of their eyes. This has long been a field of interest in computer vision in order to mimic and improve upon the cognitive skills of humans. Often, such visual search is done on an environment with distractors such as noisy backgrounds in order to test the robustness of the cognitive abilities. Deep learning has seen a lot of success in vision search applications. [61] and other's works have been successful in such problems. So much so that it is being used commercially. However, such networks require a huge amount of labeled data, something we can avoid using DRL techniques. [3] implement DQN for visual search. They formed the problem of visual search as a POMDP and used DQN to solve the POMDP. The experiment was done on 6×6 images. There were three available actions, fixate on a pixel, positive response if object found or negative response otherwise. Positive rewards were given for correct response and negative rewards for the wrong ones. The speed vs. accuracy trade-off was dealt with by giving a small negative reward for each pixel fixation. They also added Gaussian white noise and spatial smearing for modeling the distractions. The DRL architecture consisted of 3 consecutive fully connected layers giving one output, an action. Their results show that their implementation yielded 96% accuracy in comparison to human's 96%. This proves that DRL can be a viable candidate for visual search problems. [39] used DRL in order to learn the path of humans eye movement when locating something in an image, i.e., visual search. Such eye movements are called saccades. They used decoupled DRL with an actor and a learner. They simulated visual search using a maze of digits from the MNIST database with the digit 9 being the target. They used 10×10 and 20×20 images of such a maze of digits. 5×5 observation windows are fed into the neural network, which generates the action (Go north, east, south or west). They validated the model in 3 types of mazes, random, circular, mazes of digits forming a path. All of them showed that the model could find and in the last case, follow the desired path.

Visual tracking takes visual search a step further. Here, not only do we have to locate an object, but track the object in multiple video frames, regardless of the change in the position of the object, viewpoint, camera angle etc. This also has seen some success with neural networks, especially CNNs and RNNs. However, DRL has also proved to be very effective. [79] implemented DRL using RNNs, or more precisely, LSTM for visual tracking. The task is formulated as a sequential decision making process. The video frames are given as input, and the output is a bounding box denoting the target location each time. The agent is given a scalar reward at each time frame. The architecture was evaluated with the Object Tracking Benchmark (OTB) and achieved state-of-the-art performance with an average bounding box overlap ratio of 0.562 along with 45 fps (FPS). [78] used three convolutional layers followed by three fully connected layers before the output action is fed into the RL network, as shown in Fig. 2.13. The network is pretrained using supervised learning as well

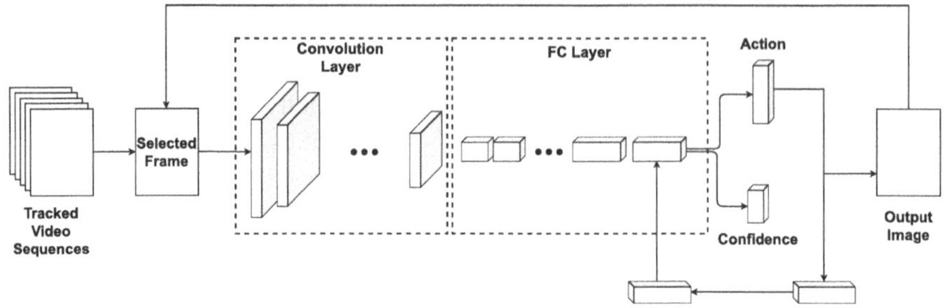

Fig. 2.13 Architecture of the Action-Decision Network [78]

as RL. This way, even partially labeled data can be fed into the network without any problems.

Like before, the output is a bounding box that follows the target. The agent decides on the actions that move the bounding box, preferably onto the target. The action space consists of 11 possible actions including translation, scaling and stopping. The reward is given at each time frame of either 1 or −1 depending on the Intersection over Union (IoU). The model was evaluated with OTB datasets. Its fast version achieved 15 FPS along with state of the art accuracy (IoU of 0.635). [15] use A3C model along with expert demonstrator for visual tracking. They used architecture similar to ResNet-18 along with fully connected layers to produce actions and reward the agent according to the IoU of the bounding box. In training, they merge both A3C and expert demonstrator. One half learns according to the A3C algorithm, the other half learns from the demonstrator in a supervised fashion, resulting in two algorithms, A3CT and A3CTD accordingly. Results show that A3CT and A3CTD can run in 90 and 50 FPS respectively, allowing real time applications.

2.5.2 *Video Captioning and Summarization*

Video captioning is another area of application for DRL. Traditionally, as with other computer vision applications, neural networks have seen a lot of success, especially sequential models. However, as we have seen in other fields, what neural networks can do, DRL can do better. Some of the DRL based video captioning has seen state of the art performance. [71] implemented hierarchical RL along with CNN and an encoder-decoder model for video captioning. It consists of a higher level sequence model that sets numerous sub goals, and the worker which selects actions to fulfill those goals and a critic layer that evaluates whether the goals are accomplished or not. The reward is set using the CIDEr score. However, instead of using the final CIDEr score for the whole caption, they used delta CIDEr scores for immediate rewards. For training the worker, they disabled the manager and only updated the policy of the worker. Similarly, for training the manager, they generated captions using greedy decoding and only update the manager's policy. The model is evaluated on the MSR VTT dataset and also, on their own Charades Captions dataset. The model obtained state-of-the-art results on both the datasets and managed to output semantically meaningful phrases. [6] also used actor-critic based attention model for video captioning. The actor predicts the captions given temporal segments of the video and the critic evaluates the quality of the generated captions. The model incorporates two worker actor-critic modules, and a global actor-critic module. The worker actor is fed feature representations and the critic is fed video features. The critic estimates Q values according to the similarity of the caption to the temporal segment. The Q values are then fed into an attention module to weight events for the final representation of the temporal segments, which the global module uses to generate captions. The rewards are computed from the similarity of the events representation and the ground truth captions. The actor network is trained by assuming conditional independence of the words generated. The critic network is trained using the difference of the reward and the value estimated by the critic network for a generated word at any random state. The model achieved state-of-the-art performances on both ActivityNet Captions and TACoS-Multilevel dataset. In the former, this implementation achieved an improvement on METEOR and CIDEr metrics as much as 18.1% and 10.2% respectively. On the latter, it improved by 10.4% and 5.9% respectively.

Video summarization is another field where DRL has the potential to outshine it's biggest competitor, DNNs, and it has shown that. Since DRL can employ unsupervised data, it can extract a lot of great benefits like end to end learning, easily available big datasets etc. [83] took these advantages and implemented a DRL based video summarization network called DR-DSN that outperformed all the state-of-the-art algorithms. Not only did it feature an end to end RL framework, but also designed two novel rewards, diversity reward and representative reward. In short, they proposed an RL based deep summarization network (DSN) that make the decision on the frames to appear in the video summary. To do that, they employed a CNN with a bidirectional RNN for state representations, as shown in Fig. 2.14. The reward is the sum of two rewards, diversity reward and representative reward. The diversity reward evaluates the degree of diversity by measuring the dissimilarity of selected frames. The representative reward measures the extent to which the summary represents the original video. They evaluated the model with two datasets, TVSum and SumMe. Their unsupervised algorithm, DR-DSN beat the state-of-the-art GAN $_{dpp}$ by 5.9% and 11.4% on SumMe and TVSum respectively.

The supervised algorithm, DR-DSN$_{sup}$ beat other algorithms by 3.2%~7.2% and 1%~12% on TVSum and SumMe respectively. The extent to which even the unsupervised algorithm beat the then state-of-the-art algorithms truly shows how potent DRL can be.

Taking video summarization a step further, it may be required that the summary be personalized according to the query of the users, i.e. query conditioned video summarization. [81] developed an algorithm to work out a mapping network (MapNet) that shows how related a video frame is to the query, and a summarization network (SummNet), that gives a personalized summary according to the query. In MapNet, the video shots are encoded using a ResNet and C3D architecture, and the queries are

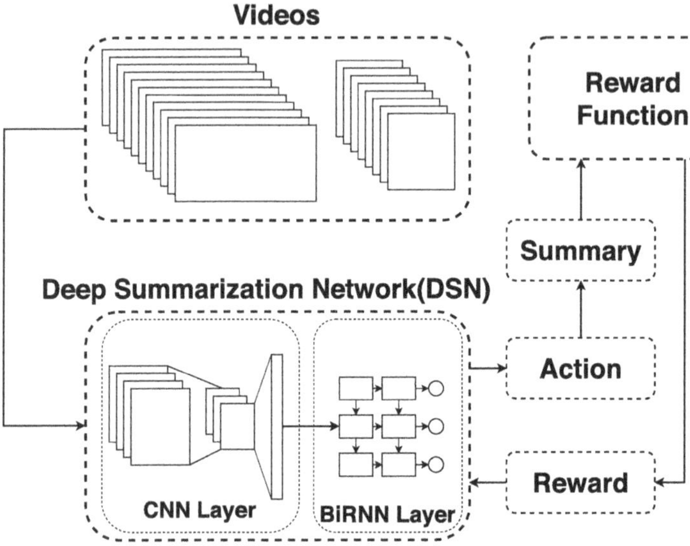

Fig. 2.14 Deep summarization network with diversity and representative rewards [83]

encoded using a Skip-gram model. In the SummNet, both the video shots and queries are encoded jointly into an embedded space, with an importance score for each video shot describing the likelihood of one of them appearing in the summary. Based on the importance score, a loss is allotted in order to guide the agent into taking better actions for a better summary. The policy is learned using three rewards, relatedness reward that measures the distance between the ground truth embedding and the predicted one, diversity reward that reduces the redundancy and representative reward that provides the most representative video shots in the summary. Their experiments show that the model outperforms all previous state-of-the-art algorithms by a large margin on all videos.

2.5.3 Image Processing and Understanding

A few years back, neural networks revolutionized image processing and understanding. However, DRL, like in other areas, is taking over as the go-to algorithms for this as well. Some of the DRL algorithms can do image enhancements and other processing actions at least as well as the state-of-the-art algorithms. We have already discussed about an attention aware face hallucination algorithm via DRL [9]. Like that, there are other image enhancement algorithms. [50] used DRL for color enhancement. They formulated color enhancement as an MDP problem. The state is defined as a combination of contextual features and color features. They adjust the color of the image by iteratively applying a global color adjusting action such as brightness

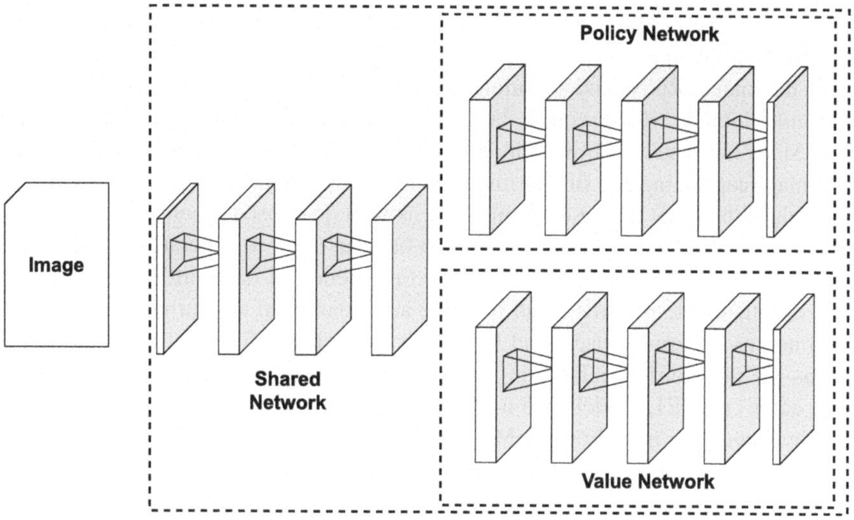

Fig. 2.15 Architecture of PixelRL [22]

change, contrast adjustment etc. They also introduce a "Distort-and-Recover" training scheme which only requires high-quality reference images rather than expensive paired databases. Their experiments showed that it is on par with competitive software on color enhancement. And because of their novel training scheme, the issue of collecting training data is less of a headache.

One very exciting work in image processing with DRL is PixelRL [22]. They implemented a novel approach of equipping each pixel with its own agent that can modify its pixel value. What is exciting about this is that it can be used for virtually any image processing application. Also, other algorithms often work globally. However, since PixelRL has an agent for each pixel, the processing techniques can be applied to each pixel. It is obvious that this is a multi agent RL problem. However, it is computationally improbable to consider for so many agents, as a 1000×1000 image can have a million agents. So, traditional multi agent RL algorithms will be rendered useless. So, rather than going for something like that, they used fully convolutional network or FCN and modified an A3C architecture to FCN form (Fig. 2.15). This way, all the agents have shared parameters, and so, the number of agents becomes less of an issue. Another novel feature is reward map convolution which allows the agent to take the future states of all the neighbouring pixels as well as it's own. This has proved to be an effecting learning method for PixelRL.

They tested this algorithm in four image processing applications, image denoising, restoration, local color enhancement and saliency driven image editing. In all the applications, it performed similar to or better than the state-of-the-art algorithms like CNN. However, as we have discussed, since this method works at the pixel level, and because unlike CNN or neural nets, it is easily interpretable, it's potential is intriguing, especially where supervised algorithms are difficult to implement.

Image understanding has been a difficult challenge because it not only requires the objects in the image to be recognized, but also provide a contextual understanding of the image. RNNs and attention aware models were often used to train image captioning models which require the feat of image understanding to generate captions. Many more neural network models have been developed that successfully deal with image captioning and similar image understanding applications. However, DRL based algorithms are taking over this field as well. [57] used a policy network and a value network that collaborates with each other in order to generate captions. The policy network is the confidence of the next prediction of the word. The value network calculates the rewards of all possible additions from the current state, which is defined as the image itself and the words predicted before that time step. The actions are the words available in their dictionary. Both the networks are trained using actor-critic RL model with a reward defined by visual semantic embedding. They evaluated the model on the MS COCO dataset. The model outperformed the state-of-the-art models in most evaluation metrics.

2.5.4 Action Detection, Recognition and Prediction

Action detection, recognition and prediction requires the feat of visual understanding through an extensive search over a video in a number of temporal scales and thus can be a complicated task. Like in other areas, neural networks have been successfully applied here as well. However, most of such algorithms have huge computational cost because it requires dense video analysis and deviates from how we humans observe them. DRL based algorithms are a possible solution to this. In action detection, [33] proposed a Self-Adaptive Proposal (SAP) that self-adaptively adjusts the temporal bounds. The model consists of a DQN where the state space is defined as the presentation of the current window and the history of actions taken. The state representation is done by using a C3D or convolutional 3D CNN model. The action space consists of seven actions over the temporal window with a reward calculated using IoU between the current temporal window and the ground truth. The agent tries to learn the optimal policy to precisely locate the action in the video. In order to refine the action detection even more, they deployed a regression network as well. They evaluated their model on THUMOS'14 dataset and found state-of-the-art performance with much less computational cost. [70] also used DRL with an actor-critic framework to train the model. But rather than going through the whole video sequence, they look to sparsely browse the video to spot only one frame per action rather than going through the whole video. This method greatly reduces the computational cost. With the actor-critic DRL framework along with GRU and CNN backbone encoding, this model outperformed the state of the art systems in both THUMOS'14 and ActivityNet while skipping almost one-fourth of the video.

Action recognition requires not only the action to be located, but also to be recognized. Neural networks with sliding window methods have been implemented before. However, since DNNs require extensive backpropagation is required to train the net-

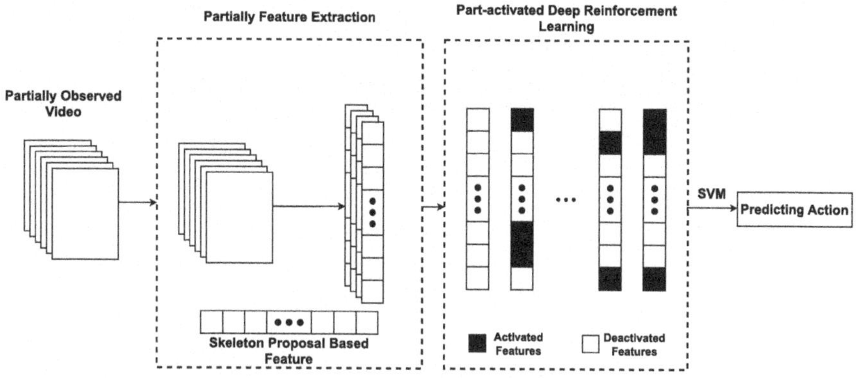

Fig. 2.16 Part activated DRL for action prediction [11]

work, which is computationally expensive in deep networks, especially in action recognition. However, DRL can be used to solve that issue. Also, DNNs are complicated and often not easily interpretable and does not reciprocate human vision. [72] looked to solve these issues by implementing an ADRL mechanism. It takes the sequence of video frames as input and extracts the features with a CNN model. The feature representation is then fed into an RL model that determines the attention, i.e. replaces back propagation and generates weights of the attention mechanism. The relevant features are then fed into LSTM networks to predict the recognized action. They evaluated the model on two datasets, HMDB-51 and ActivityNet, and found almost 20% improvement over the baseline attention model, and on per with the state-of-the-art with much less computational cost.

Action prediction is a much more complex problem than detection and recognition. Most algorithms take all of the frames to predict the action. But this method cannot avoid the noise of the current action. Also, there is the issue of mimicking humans. [11] proposed a part activated DRL or PA-DRL system that exploits the human body structure. From extracting the skeleton proposals, features in the local patches determined by the joint information are taken which contain the action related information. They devised an algorithm to only use those features for prediction. These are called part activated features, and thus, the name, PA-DRL. PA-DRL aims to learn the policy for activating only the action related parts and deactivate all other noises. They extracted 28 parts per frame and trained an actor network for each part. In every frame they selected region proposals of 20×20 pixels, and extracted the spatial features using a VGG network and fed into the actor network. The critic network was composed of linear SVM. They evaluated the model on UT Interaction #1 and UT Interaction #2 datasets. PA-DRL outperformed other state-of-the-arts in both the datasets (Fig. 2.16).

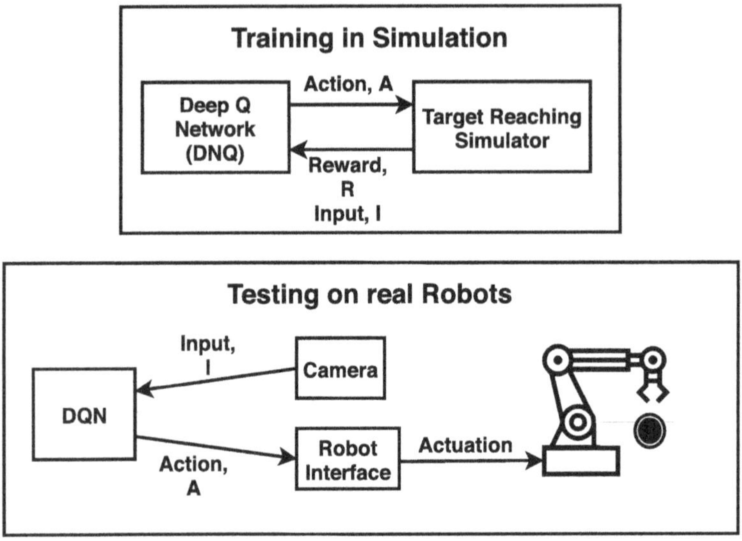

Fig. 2.17 System architecture of target reaching with DRL with Baxter's arm [80]

2.5.5 *Robotics*

The field of robotics is a challenging one in artificial intelligence. DNNs can achieve great results given enough data and training time. However, these are often not possible in real world environments. DRL thus, can be a viable option in robotics. Soon after the DQN was invented, scientist were looking to implement DRL in robotics. For example, [80] are one of the first to use image data with deep reinforcement learning for motion control. They used a 3 joint Baxter's manipulator equipped with external visual observations. The task is to reach a certain target. Visual images are fed into a DQN network in order to produce actions of navigating the manipulator. Rewards are given at each time step according to the distance of the end effector and the target (Fig. 2.17). For reaching the target, a final reward is also given. The experiment was simulated with some random noises added. They considered the task to be successful upon reaching within 16 cm radius of the target, or 15 pixels in the simulator screen. The results varied upon different settings. But it was obvious that DQN had the capability to cope with different kinds of noises. However, at the time, in real world scenario with a real arm and a camera, the success rate was 0. There were number of other works that performed in the similar manner. They showed promising result in simulated experiments, but failed to deliver on the promise in real world settings, especially in complex tasks. It seemed that DRLs success would only remain theoretical. However, [26] successfully implemented DQN in robotics in order to alleviate the limited and complicated training procedures. They used joint angles and end effector positions as state representations in order to manipulate a 7 DOF arm and JACO arm for complex door opening task. This proved that DRL is not just theoretical when it comes to robotics. But it seemed that robotic vision still had a long way to go with deep reinforcement learning to be successful.

However, recent works show very promising results in vision based robotic manipulation and control. [36] proposed a closed loop vision based robot navigation system called QT-Opt. It uses a monocular RGB camera mounted over the shoulder of the robot. It takes the RGB images as well as the gripper position as the state input to a DQN framework. The robot can take end effector Cartesian motion actions as well as opening and closing of the gripper. The robot received a reward for lifting up an object successfully. This setting seems similar to [80] until now. But what set them apart was that their goal was not just to maximize the reward, but also to generalize among different objects. This would require a large and diverse dataset. To train on such a dataset, they propose an off policy training method that is based on a generalization of Q learning they called QT-Opt. They also got rid of an explicit actor. Rather they used stochastic optimization for the critic to give action and target values. The manipulator is trained in an end to end manner, enabling self supervised labelling. The results showed staggering improvements over the then state-of-the-arts of success rates of 96%. [16] proposed a similar setting of a manipulator with a camera. They also look to deploy a generalized and self supervised model. But in their case, they go for sensory prediction from the raw sensor data like camera images. They collect unsupervised data from random actions and then train a video prediction model on those data. They deployed this model based approach in grasping and pushing tasks with 75% success rate. This approach holds a lot of promise regarding to the task of generalization. [76] deployed an autonomous robot navigation system equipped with a monocular camera rather than a LiDAR, which dramatically reduces the cost. They used a DDQN framework trained on a simulator. The trained model is then fed depth estimations as input from the monocular camera. All of these works show that DRL can play an effective role in robot vision. The works we are seeing today are already producing extraordinary results. But comparing to the potential DRL holds, they are possibly only a tip of the iceberg.

2.6 Research Challenges

Deep reinforcement learning is a relatively new and expansive field. Thus a lot of research is still going on with numerous obstacles to deal with. Most of these have existed since the dawn of artificial intelligence. Others although somewhat solved in other fields, remain a hurdle in DRL still. In this section, we briefly discuss some of the research challenges of DRL relating to computer vision.

2.6.1 Exploration Vs. Exploitation

Exploration vs exploitation has been one of the most fundamental obstacles since the dawn of artificial intelligence. How long should the agent keep taking suboptimal actions in order to explore the environment, when to start exploiting the policy and take optimal actions to lead to the best rewards, these are some of the ever existing dilemmas in artificial intelligence. And even to this date, this has not totally been solved. Each work in this area deals with this problem using some of the existing algorithms modified to deal with a specific environment. For example, epsilon greedy algorithm takes random actions at the beginning, and over time minimizes the number of such random actions [67]. This allows more exploration at the beginning and more optimal actions according to the policy at later periods. But due to the total randomness of the action choices at the beginning, it is prone to making the same mistake over and over again. So, it is not a very efficient algorithm. Nonetheless, it is used often in DQN.

Another algorithm to deal with this dilemma is exploration function. The function keeps a visit count for each state and adds more value to the state values if it is unexplored or less explored in order to increase the tendency to explore more. The more a state is explored, the lesser the added value becomes, thus removing the tendency to explore the already explored bad regions. There are numerous other algorithms that deal with this. In recent times, more notable ones are mainly ensemble approaches that allow deeper exploration by combining a number of base models into a more potent model. For example, the bootstrapped DQN [48] is a modified version of a DQN that builds an ensemble of K Q value functions. Every episode, the bootstrapped DQN randomly selects a single Q value function to follow. Upper confidence bound or UCB algorithm can also be used on the same ensemble as bootstrapped DQN for explorations [12]. However, rather than randomly selecting the Q value to follow, UCB takes the action of the highest confidence bound of all the bootstrapped Q values. There are many more algorithms to choose from. There are still no generalized algorithms that works best in every scenarios. So, this is still a hurdle to overcome in any DRL problems, simply due to the fact that learning about the environment plays such a crucial role in the agents performance.

2.6.2 Multi Agent Systems

This is one field that has seen loads of work done in the recent past. However, this still is a challenge in DRL. There are many issues regarding multi agent systems such as- infinite state spaces, non stationarity of the environment, partial observability of the agents, scalability of the system, difficulty in determining the performance measures, difficulty in training in multi agent environment etc. Each pose significant problems on its own. Some of these have already been discussed in this chapter before. As we have seen already, numerous works do deal with these problems, but none are without a catch or two. Even though there are plenty of works being done in this field like of [35] and others, the problem still exists when number of agents increase.

2.6.3 Real World Conditions

One of the primary goals of AI is to produce agents capable of operating in the real world. AI is progressing day by day in order to achieving that goal. Agents are often equipped with cameras to feed it visual inputs in hopes of extracting state information. Various real world conditions especially real time processing, noise, partial observability and non stationarity of the environment, however make this a daunting task.

Real time processing capability is required for extracting state information from the input images and making decisions on time. Although modern technologies have increased the processing capability by a considerable margin, real time processing of the images still proves to be a formidable task especially when it has to deal with other complexities like noise, motion etc. For example, if a race occurs between two robot runners and one robot has to run faster than the other, then with a faster speed the state spaces it would observe through its visual sensors, would be blurred and tilted. So, the further processing would have to be much advanced and critical to get a more clear observation of the state spaces which sometimes become impossible to do in time, due to greater amount of computational cost and time complexity. Even after getting the proper state space image, the features would still need to be extracted and then converted to actions. Processing all of these tasks in real time, is still a big challenge when it comes to deploying such DRL agents.

Again, the real world has virtually unlimited state spaces, it is easy to see the intimidating task of representing all the required states. There is also the issue of partial observability and non stationarity of the environment. Combine that with many real world noises and possibly in the midst of hundreds of humans, computer vision, especially robot vision has a long way to go. Let us go through another example. Nowadays AI equipped robots are being used for rescuing purposes from different places, such as in collapsed buildings. In these cases, the observability of the environment for the robot is very minimal most of the time. But because of the limitations in the robot's visual sensors, it becomes a more difficult challenge to get the whole picture of the incident and act accordingly to rescue properly without hurting both the people trapped over there and the robot itself. For example, removal of any collapsed part of the environment or moving anything in the environment can cause more hazard that the robot would not be even able to foresee. So the partial observability comes here at a greater cost. Again, considering the stationarity of the environment, in this case, a slight movement can change the whole environment. Thus, exploiting much easier stationary environment policy is rendered impossible in such cases, which is in fact, true for most real world scenarios.

2.6.4 *Transfer Learning*

Transfer learning is a concept of neural networks where the learned features are transferred from one problem to another. The idea is that the agent should be able to transfer the knowledge of one task onto a similar but different task. Also, especially for robotics applications of DRL, data is hard to get. One solution is running the agent through a simulation. But what is learned in the simulated environment has to be transferred to the agent operating in the real world. This is why transfer learning is all the more necessary in DRL. However, even though this is a common practice in neural nets, DRL has yet to see much success. One other aspect of this is continual learning where the agent builds on its knowledge continuously from one task to another. However, as we have discussed previously, catastrophic forgetting is a huge issue in this. The most notable works in these fields would be Deepmind's deep progressive net [60]. This has been covered in one of the previous sections. Some other notable works are policy distillation [59], which work on consolidating multiple policies into a single one. It also shows how to extract a policy from an RL agent and use it on a different, smaller network to extract expert level of performance. Another interesting approach is multitask deep reinforcement learning which allow multiple task learning. It is an approach of inductive transfer that improves generalization. Although this has been a subfield of machine learning for a long time, only a handful of works have been done in DRL such as multi task DRLs like [75] and [31], models with attentive DRL [8] etc. With potential algorithms like PixelRL [22], transfer learning can aid a lot in the advancement of DRL in computer vision.

2.7 Conclusion and Future Research Direction

We have discussed DRL and various methods of implementing a DRL network. We have also given a brief overview of the challenges and works that are being done to overcome them. However, the current works are only scratching the surface of the true potential of DRL. Only that has given some amazing results in the forms of Alpha Go, Alpha Zero, and many more. In the future, these will be perfected as time goes by, and the engines keep learning from experiences. For future research directions, the whole of DRL is a viable candidate because so little work has been done. Some of the most promising sub-fields of DRL are briefly discussed below -

2.7.1 *Deep Inverse Reinforcement Learning*

Inverse reinforcement learning, or IRL, is, as it sounds, the opposite of RL. In RL, we have a transition and reward function (or estimations of it) and try to find the optimal policy. In IRL however, we see an expert operate on the environment with an optimal policy. In this case, the agent tries to learn the reward function given the optimal policy or by observing an expert operating on the environment with an optimal policy [4]. This has been a sought after topic in AI for more than a decade. IRL has instigated many exciting applications in robotics and computer vision such as helicopter aerobatics [1], quadruped locomotion [55], autonomous driving [62], parking lot navigation [2] etc. However, the recent trend is to build deeper networks, deep inverse reinforcement learning or DIRL is an obvious yet exciting step to take because both deeper networks and IRL have seen so much success. DIRL is capable of learning much more complex nonlinear reward functions in large state spaces [73], which is ideal for real-world applications. Many more frameworks are proposed, like maximum entropy DIRL capable of life long learning scenarios with efficient training [74], and active task inference guided DIRL for applying in temporarily extended tasks [43]. The maximum entropy DIRL can be applied in numerous computer vision applications, such as pedestrian navigation learning. [18], autonomous navigation [66] etc.

2.7.2 *Robotic Vision: Visual Grasping and Navigation*

Vision-based grasping and navigation tasks are not particularly new in the field of robotics. However, most of the algorithms are quite complex and have some room for improvement. DRL algorithms have shown their potential in image processing, and the network is ideal for associating appropriate actions from visual inputs. So, visual-based grasping and navigation is, as it should be, a very promising application of DRL. Many works have been done relating to this field. [53] gave a simulated comparison of different algorithms on visual grasping. Their results show that comparatively simpler algorithms like Deep Q learning perform relatively better than other complex algorithms like DDQN. In visual navigation, works like [40, 65], and many more make DRL an up-and-coming candidate in these kinds of applications. However, new and intriguing prospects and approaches like progressive learning or an attention driven approach might give better results in both cases. All in all, it is still largely uncharted territory for DRL.

2.7.3 *Few Shot and Zero Shot Learning*

Few shot learning (FSL) and zero shot learning (ZSL) are among the most exciting fields in computer vision. It has been widely studied that bigger datasets are usually the way to go in deep learning for good performance. However, getting big enough datasets and labeling them is often impossible (e.g., cases of rare diseases). Thus, it would be great if we could somehow extract the required level of performance without needing to feed in a huge dataset. That is where the concept of FSL and ZSL comes from. FSL learns to predict classes never seen before, having seen only a few training examples. In the extreme, there might only be a single example of each class (one-shot learning) or no example at all (zero-shot learning). FSL and ZSL hold much potential in AI, especially when it comes to computer vision problems. DRL is already working on alleviating the need for labeled data. However, there are cases where getting any data can be difficult. Such cases could benefit greatly from FSL and ZSL algorithms. There have been numerous works merging FSL, ZSL with DRL. For example, [51] and [52] implement DRL with one-shot learning for change point detection and artificially intelligent classification system, respectively. [47] discussed zero-shot generalization of multi-task DRL. [23] proposed a deep attention convolutional neural network architecture with zero-shot reinforcement learning as well. With all of these works going on, FSL and ZSL hold much promise in DRL.

References

1. Abbeel, P., Coates, A., Ng, A.Y.: Autonomous helicopter aerobatics through apprenticeship learning. Int. J. Robot. Res. **29**(13), 1608–1639 (2010)
2. Abbeel, P., Dolgov, D., Ng, A.Y., Thrun, S.: Apprenticeship learning for motion planning with application to parking lot navigation. In: 2008 IEEE/RSJ International Conference on Intelligent Robots and Systems, pp. 1083–1090 (2008)
3. Acharya, A., Chen, X., Myers, C.W., Lewis, R.L., Howes, A.: Human visual search as a deep reinforcement learning solution to a POMDP. In: CogSci (2017)
4. Arora, S., Doshi, P.: A survey of inverse reinforcement learning: challenges, methods and progress (2018). *arXiv preprint* arXiv:1806.06877
5. Bahdanau, D., Cho, K., Bengio, Y.: Neural machine translation by jointly learning to align and translate (2014). *arXiv preprint* arXiv:1409.0473
6. Barati, E., Chen, X.: Critic-based attention network for event-based video captioning. In: Proceedings of the 27th ACM International Conference on Multimedia, pp. 811–817 (2019)
7. Basavarajaiah, M., Sharma, P.: Survey of compressed domain video summarization techniques. ACM Comput. Surv. (CSUR) **52**(6), 1–29 (2019)
8. Bram, T., Brunner, G., Richter, O., Wattenhofer, R. Attentive multi-task deep reinforcement learning (2019). *arXiv preprint* arXiv:1907.02874
9. Cao, Q., Lin, L., Shi, Y., Liang, X., Li, G.: Attention-aware face hallucination via deep reinforcement learning. In: Proceedings of the IEEE Conference on Computer Vision and Pattern Recognition, pp. 690–698 (2017)
10. Castaneda, A.O.: Deep reinforcement learning variants of multi-agent learning algorithms. Master's thesis, School of Informatics, University of Edinburgh (2016)
11. Chen, L., Lu, J., Song, Z., Zhou, J.: Part-activated deep reinforcement learning for action prediction. In: Proceedings of the European Conference on Computer Vision (ECCV), pp. 421–436 (2018)
12. Chen, R.Y., Sidor, S., Abbeel, P., Schulman, J.: UCB exploration via q-ensembles (2017). *arXiv preprint* arXiv:1706.01502
13. Cho, K., Van Merriënboer, B., Gulcehre, C., Bahdanau, D., Bougares, F., Schwenk, H., Bengio, Y.: Learning phrase representations using RNN encoder-decoder for statistical machine translation (2014). *arXiv preprint* arXiv:1406.1078
14. Dulac-Arnold, G., Mankowitz, D., Hester, T.: Challenges of real-world reinforcement learning (2019). *arXiv preprint* arXiv:1904.12901
15. Dunnhofer, M., Martinel, N., Luca Foresti, G., Micheloni, C.: Visual tracking by means of deep reinforcement learning and an expert demonstrator. In: Proceedings of the IEEE International Conference on Computer Vision Workshops (2019)

16. Ebert, F., Finn, C., Dasari, S., Xie, A., Lee, A., Levine, S.: Visual foresight: model-based deep reinforcement learning for vision-based robotic control (2018). *arXiv preprint* arXiv:1812.00568
17. Egorov, M.: Multi-agent deep reinforcement learning. In: CS231n: convolutional Neural Networks for Visual Recognition (2016)
18. Fahad, M., Chen, Z., Guo, Y.: Learning how pedestrians navigate: a deep inverse reinforcement learning approach. In: 2018 IEEE/RSJ International Conference on Intelligent Robots and Systems (IROS), pp. 819–826. IEEE (2018)
19. Foerster, J., Assael, I.A., De Freitas, N., Whiteson, S.: Learning to communicate with deep multi-agent reinforcement learning. In: Advances in Neural Information Processing Systems, pp. 2137–2145 (2016)
20. Foerster, J.N., Assael, Y.M., de Freitas, N., Whiteson, S.: Learning to communicate to solve riddles with deep distributed recurrent q-networks (2016). *arXiv preprint* arXiv:1602.02672
21. Forsyth, D.A., Ponce, J.: Computer Vision: A Modern Approach. Prentice Hall Professional Technical Reference (2002)
22. Furuta, R., Inoue, N., Yamasaki, T.: Pixelrl: fully convolutional network with reinforcement learning for image processing. IEEE Trans. Multimed. **22**(7), 1704–1719 (2019)
23. Genc, S., Mallya, S., Bodapati, S., Sun, T., Tao, Y.: Zero-shot reinforcement learning with deep attention convolutional neural networks (2020). *arXiv preprint* arXiv:2001.00605
24. Gers, F.A., Schmidhuber, J., Cummins, F.: Learning to forget: continual prediction with LSTM (1999)
25. Goodfellow, I., Bengio, Y., Courville, A.: Deep Learning. MIT press (2016)
26. Gu, S., Holly, E., Lillicrap, T., Levine, S.: Deep reinforcement learning for robotic manipulation with asynchronous off-policy updates. In: 2017 IEEE International Conference on Robotics and Automation (ICRA), pp. 3389–3396. IEEE (2017)
27. Gupta, J.K., Egorov, M., Kochenderfer, M.: Cooperative multi-agent control using deep reinforcement learning. In: International Conference on Autonomous Agents and Multiagent Systems, pp. 66–83. Springer (2017)
28. Harandi, M.T., Ahmadabadi, M.N., Araabi, B.N.: Face recognition using reinforcement learning. In: 2004 International Conference on Image Processing, 2004. ICIP 2004, vol. 4, pp. 2709–2712. IEEE (2004)
29. Hasselt, H.V.: Double q-learning. In: Advances in Neural Information Processing Systems, pp. 2613–2621 (2010)
30. Hausknecht, M., Stone, P.: Deep recurrent q-learning for partially observable MDPS. In: 2015 AAAI Fall Symposium Series (2015)
31. Hessel, M., Soyer, H., Espeholt, L., Czarnecki, W., Schmitt, S., van Hasselt, H.: Multi-task deep reinforcement learning with popart. In: Proceedings of the AAAI Conference on Artificial Intelligence, vol. 33, pp. 3796–3803 (2019)
32. Hong, Z.-W., Su, S.-Y., Shann, T.-Y., Chang, Y.-H., Lee, C.-Y.: A deep policy inference q-network for multi-agent systems. In: Proceedings of the 17th International Conference on Autonomous Agents and MultiAgent Systems, pp. 1388–1396. International Foundation for Autonomous Agents and Multiagent Systems (2018)
33. Huang, J., Li, N., Zhang, T., Li, G., Huang, T., Gao, W.: Sap: self-adaptive proposal model for temporal action detection based on reinforcement learning. In: Thirty-Second AAAI Conference on Artificial Intelligence (2018)
34. Jahne, B.: Computer Vision and Applications: A Guide for Students and Practitioners. Elsevier (2000)
35. Jiang, M., Hai, T., Pan, Z., Wang, H., Jia, Y., Deng, C.: Multi-agent deep reinforcement learning for multi-object tracker. IEEE Access **7**, 32400–32407 (2019)
36. Kalashnikov, D., Irpan, A., Pastor, P., Ibarz, J., Herzog, A., Jang, E., Quillen, D., Holly, E., Kalakrishnan, M., Vanhoucke, V., et al.: Qt-opt: scalable deep reinforcement learning for vision-based robotic manipulation (2018). *arXiv preprint* arXiv:1806.10293
37. Kirkpatrick, J., Pascanu, R., Rabinowitz, N., Veness, J., Desjardins, G., Rusu, A.A., Milan, K., Quan, J., Ramalho, T., Grabska-Barwinska, A., et al.: Overcoming catastrophic forgetting in neural networks. Proc. Natl. Acad. Sci. **114**(13), 3521–3526 (2017)

38. Kober, J., Oztop, E., Peters, J.: Reinforcement learning to adjust robot movements to new situations. In: Twenty-Second International Joint Conference on Artificial Intelligence (2011)
39. Kornuta, T., Rocki, K.: Utilization of deep reinforcement learning for saccadic-based object visual search (2016). *arXiv preprint* arXiv:1610.06492
40. Kulhánek, J., Derner, E., de Bruin, T., Babuška, R.: Vision-based navigation using deep reinforcement learning. In: 2019 European Conference on Mobile Robots (ECMR), pp. 1–8. IEEE (2019)
41. Li, S., Tao, Z., Li, K., Fu, Y.: Visual to text: survey of image and video captioning. IEEE Trans. Emerg. Top. Comput. Intell. **3**(4), 297–312 (2019)
42. Lin, L.-J.: Reinforcement learning for robots using neural networks (1992)
43. Memarian, F., Xu, Z., Wu, B., Wen, M., Topcu, U.: Active task-inference-guided deep inverse reinforcement learning (2020). *arXiv preprint* arXiv:2001.09227
44. Mnih, V., Badia, A.P., Mirza, M., Graves, A., Lillicrap, T., Harley, T., Silver, D., Kavukcuoglu, K.: Asynchronous methods for deep reinforcement learning. In: International Conference on Machine Learning, pp. 1928–1937 (2016)
45. Mnih, V., Kavukcuoglu, K., Silver, D., Graves, A., Antonoglou, I., Wierstra, D., Riedmiller, M.: Playing atari with deep reinforcement learning (2013). *arXiv preprint* arXiv:1312.5602
46. Nguyen, T.T., Nguyen, N.D., Nahavandi, S.: Deep reinforcement learning for multiagent systems: a review of challenges, solutions, and applications. IEEE Trans. Cybern. **50**(9), 3826–3839 (2020)
47. Oh, J., Singh, S., Lee, H., Kohli, P.: Zero-shot task generalization with multi-task deep reinforcement learning. In: Proceedings of the 34th International Conference on Machine Learning-Volume 70, pp. 2661–2670. JMLR. org (2017)
48. Osband, I., Blundell, C., Pritzel, A., Van Roy, B.: Deep exploration via bootstrapped DQN. In: Advances in Neural Information Processing Systems, pp. 4026–4034 (2016)
49. Palmer, G., Tuyls, K., Bloembergen, D., Savani, R.: Lenient multi-agent deep reinforcement learning. In: Proceedings of the 17th International Conference on Autonomous Agents and MultiAgent Systems, pp. 443–451. International Foundation for Autonomous Agents and Multiagent Systems (2018)
50. Park, J., Lee, J.-Y., Yoo, D., So Kweon, I.: Distort-and-recover: color enhancement using deep reinforcement learning. In: Proceedings of the IEEE Conference on Computer Vision and Pattern Recognition, pp. 5928–5936 (2018)
51. Puzanov, A., Cohen, K.: 2018 56th Annual Allerton Conference on Communication, Control, and Computing (Allerton), pp. 1047–1051 (2018)
52. Puzanov, A., Cohen, K.: Deep reinforcement one-shot learning for artificially intelligent classification systems (2018). *arXiv preprint* arXiv:1808.01527
53. Quillen, D., Jang, E., Nachum, O., Finn, C., Ibarz, J., Levine, S.: Deep reinforcement learning for vision-based robotic grasping: a simulated comparative evaluation of off-policy methods. In: 2018 IEEE International Conference on Robotics and Automation (ICRA), pp. 6284–6291. IEEE (2018)
54. Rao, Y., Lu, J., Zhou, J.: Attention-aware deep reinforcement learning for video face recognition. In: Proceedings of the IEEE International Conference on Computer Vision, pp. 3931–3940 (2017)
55. Ratliff, N., Bagnell, J.A., Srinivasa, S.S.: Imitation learning for locomotion and manipulation. In: 2007 7th IEEE-RAS International Conference on Humanoid Robots, pp. 392–397. IEEE (2007)
56. Redmon, J., Divvala, S., Girshick, R., Farhadi, A.: You only look once: unified, real-time object detection. In: Proceedings of the IEEE Conference on Computer Vision and Pattern Recognition, pp. 779–788 (2016)
57. Ren, Z., Wang, X., Zhang, N., Lv, X., Li, L.-J.: Deep reinforcement learning-based image captioning with embedding reward. In: Proceedings of the IEEE Conference on Computer Vision and Pattern Recognition, pp. 290–298 (2017)
58. Roy, N., McCallum, A.: Toward Optimal Active Learning Through Monte Carlo Estimation of Error Reduction, pp. 441–448. ICML, Williamstown (2001)

59. Rusu, A.A., Colmenarejo, S.G., Gulcehre, C., Desjardins, G., Kirkpatrick, J., Pascanu, R., Mnih, V., Kavukcuoglu, K., Hadsell, R.: Policy distillation (2015). *arXiv preprint* arXiv:1511.06295
60. Rusu, A.A., Rabinowitz, N.C., Desjardins, G., Soyer, H., Kirkpatrick, J., Kavukcuoglu, K., Pascanu, R.,Hadsell, R.: Progressive neural networks (2016). *arXiv preprint* arXiv:1606.04671
61. Shankar, D., Narumanchi, S., Ananya, H., Kompalli, P., Chaudhury, K.: Deep learning based large scale visual recommendation and search for e-commerce (2017). *arXiv preprint* arXiv:1703.02344
62. Sharifzadeh, S., Chiotellis, I., Triebel, R., Cremers, D.: Learning to drive using inverse reinforcement learning and deep q-networks (2016). *arXiv preprint* arXiv:1612.03653
63. Siebel, N.T., Grunewald, S., Sommer, G.: Creating edge detectors by evolutionary reinforcement learning. In: 2008 IEEE Congress on Evolutionary Computation (IEEE World Congress on Computational Intelligence), pp. 3553–3560. IEEE (2008)
64. Silver, D., Huang, A., Maddison, C.J., Guez, A., Sifre, L., Van Den Driessche, G., Schrittwieser, J., Antonoglou, I., Panneershelvam, V., Lanctot, M., et al.: Mastering the game of go with deep neural networks and tree search. Nature, **529**(7587), 484 (2016)
65. Skov, S.: Indoor visual navigation using deep reinforcement learning (2017)
66. Song, Y.: Inverse Reinforcement Learning for Autonomous Ground Navigation Using Aerial and Satellite Observation Data. Ph.D. thesis, Master's thesis, The Robotics Institute, Carnegie Mellon University (2019)
67. Sutton, R.S., Barto, A.G., et al.: Introduction to Reinforcement Learning, vol. 135. MIT press Cambridge (1998)
68. Tang, Y., Tian, Y., Lu, J., Li, P., Zhou, J.: Deep progressive reinforcement learning for skeleton-based action recognition. In: Proceedings of the IEEE Conference on Computer Vision and Pattern Recognition, pp. 5323–5332 (2018)
69. Van Hasselt, H., Guez, A., Silver, D.: Deep reinforcement learning with double q-learning. In: Thirtieth AAAI Conference on Artificial Intelligence (2016)
70. Vaudaux-Ruth, G., Chan-Hon-Tong, A., Achard, C.: Actionspotter: deep reinforcement learning framework for temporal action spotting in videos (2020). *arXiv preprint* arXiv:2004.06971
71. Wang, X., Chen, W., Wu, J., Wang, Y.-F., Yang Wang, W.: Video captioning via hierarchical reinforcement learning. In: Proceedings of the IEEE Conference on Computer Vision and Pattern Recognition, pp. 4213–4222 (2018)
72. Wang, Y., Wu, F.: Multi-agent deep reinforcement learning with adaptive policies (2019). *arXiv preprint* arXiv:1912.00949
73. Wulfmeier, M., Ondruska, P., Posner, I.: Deep inverse reinforcement learning. *ArXiv*, abs/1507.04888 (2015)
74. Wulfmeier, M., Ondruska, P., Posner, I.: Maximum entropy deep inverse reinforcement learning (2015). *arXiv preprint* arXiv:1507.04888
75. Yang, Z., Merrick, K.E., Abbass, H.A., Jin, L.:Multi-task deep reinforcement learning for continuous action control. In: IJCAI, pp. 3301–3307 (2017)
76. Yokoyama, K., Morioka, K.: Autonomous mobile robot with simple navigation system based on deep reinforcement learning and a monocular camera. In: 2020 IEEE/SICE International Symposium on System Integration (SII), pp. 525–530. IEEE (2020)
77. Yu, C., Zhang, M., Ren, F., Tan, G.: Multiagent learning of coordination in loosely coupled multiagent systems. IEEE Trans. Cybern. **45**(12), 2853–2867 (2015)
78. Yun, S., Choi, J., Yoo, Y., Yun, K., Young Choi, J.: Action-decision networks for visual tracking with deep reinforcement learning. In: Proceedings of the IEEE Conference on Computer Vision and Pattern Recognition, pp. 2711–2720 (2017)
79. Zhang, D., Maei, H., Wang, X., Wang, Y.-F.: Deep reinforcement learning for visual object tracking in videos (2017). *arXiv preprint* arXiv:1701.08936
80. Zhang, F., Leitner, J., Milford, M., Upcroft, B., Corke, P.: Towards vision-based deep reinforcement learning for robotic motion control (2015). *arXiv preprint* arXiv:1511.03791
81. Zhang, Y., Kampffmeyer, M., Zhao, X., Tan, M.: Deep reinforcement learning for query-conditioned video summarization. Appl. Sci. **9**(4), 750 (2019)

82. Zheng, Y., Meng, Z., Hao, J., Zhang, Z.: Weighted double deep multiagent reinforcement learning in stochastic cooperative environments. In: Pacific Rim International Conference on Artificial Intelligence, pp. 421–429. Springer (2018)
83. Zhou, K., Qiao, Y., Xiang, T.: Deep reinforcement learning for unsupervised video summarization with diversity-representativeness reward. In: Thirty-Second AAAI Conference on Artificial Intelligence(2018)

Chapter 3
Deep Learning for Data-Driven Predictive Maintenance

Muhammad Sohaib, Shiza Mushtaq, and Jia Uddin

Abstract With the advancement of technology, it is viable to obtain data from sensors in real-time from industrial equipment, engines, heavy machines, and manmade structures. The collected data in real-time can be further utilized to perform maintenance of assets before these entities get entirely worn-out which in technical terms known as predictive maintenance. The downtime of dilapidated instruments can lead to loss of revenues and can be a threat to the workers on the facilities. In time and precise prediction of such failures using the data acquired through sensors can alleviate the downtime of the equipment, hence, helps in cutting off the revenue losses and ensure the safety of the workers. If enough historical data associated with equipment is available, then artificial intelligence techniques like machine learning as well as deep learning may be utilized for identification of equipment failures in parts or as a whole. Deep learning algorithms have shown profound progress in the problem areas where practitioners and researchers had been eluded for several decades. This chapter provides an insight into the deep learning algorithms used for predictive maintenance. It also provides an overview of industrial sensors and future research aspects of sensors using techniques of deep learning for predictive maintenance.

3.1 Introduction

A sensor is a device that converts physical signals which are measured from the environment into electrical signals [1]. It is possible to repeatedly measure and store the physical quantities from the environment. The stored measurements can be utilized to study the behavior of these physical quantities. Moreover, a realistic system can

M. Sohaib · S. Mushtaq
Department Computer Science, Lashore Garrison University, Lahore, Pakistan

J. Uddin (✉)
Technology Studies Department, Endicott College, Woosong University, Daejeon, South Korea
e-mail: jia.uddin@wsu.ac.kr

M. A. R. Ahad and A. Inoue (eds.), *Vision, Sensing and Analytics: Integrative Approaches*, Intelligent Systems Reference Library 207, https://doi.org/10.1007/978-3-030-75490-7_3

also be devised to analyze the measurements taken from sensors in real-time to gain valuable insights and make important decisions.

Certain internal and external physical quantities are affected and their reading change during the operation of an asset due to the change in the working environment. These physical quantities include but not limited to internal oil temperature, and pressure, external temperature, and humidity, etc. Continuous monitoring of these variables to identify the drift of equipment from its normal condition and taking measures to evade them is referred as predictive maintenance [2].

Predictive maintenance provides reduced downtime of an asset, enhanced quality, limiting the revenue losses, reliability, and better safety of workers. In addition, the provision of early warnings to avoid catastrophic outcomes by recognizing uncommon behavior of equipment is the most important goal of predictive maintenance.

In the past few decades, data-driven predictive maintenance has been efficacious due to the improvised data acquisition techniques, application of different types of sensors, machine learning and deep learning, and availability of big data.

In summary, taking measurements through various types of sensors, storage of these measurements, data pre-processing, and analysis of the pre-processed data defines the complete predictive maintenance process [3].

3.2 Maintenance

The failure of industrial equipment, heavy machines, engines, and structures leads to unwanted downtime, enough economic damages, and put the worker's safety at risk. These issues can be mitigated if the maintenance of such assets is performed on a timely basis. In general, maintenance of an object can be defined as, servicing, functional checks, repairing or replacing of required components, machinery, artificial infrastructures, and supporting utilities in industrial, residential, business installations. In other words, the maintenance process can be defined as maintaining the equipment by troubleshooting problems either manually or through computerized diagnostic tools [4].

In the context of manufacturing and processing industries, the process of maintenance can be broadly categorized as reactive maintenance, preventive maintenance, predictive maintenance alternatively called condition-based maintenance, and proactive maintenance [5, 6].

In reactive maintenance, the ineffective or damaged components of an asset are repaired or replaced so that it may continue its function smoothly. Therefore, no need to take preventive measures and problem is tackled when it is reported. On the other hand, preventive maintenance is a schedule for periodic inspection of equipment. The main goal of preventive maintenance is to identify small problems as the equipment undergoes deterioration and addressing these issues before its complete collapse. Its main advantage is that the equipment under observation does not breaks down as the defected part is replaced with a new one on time. So, the equipment makes from one planed service to another without any failure due to fatigue, neglect, or wear. A few of the preventive maintenance actives include oil change, partial or complete work halts at specified intervals, use of lubricants, minor adjustments, worn parts replacement, etc. The objective of preventive maintenance is to take equipment from one planned operation to another without any failures. In this, the maintenance activities include partial or complete work halts after specified intervals, oil change, lubrication use, minor adjustment, and so on. Further, the workers can make a log of equipment wear and tear, so it is easy to replace the defected part with a new one easily without failure of the overall system.

Strategies of Predictive maintenance are adopted to identify the condition of equipment in service with the aim to estimate when the maintenance of equipment will be necessary. It helps in the reduction of maintenance costs as the maintenance is just carried out when it is required. Thus, it is also termed as condition-based maintenance because it is done when condition of the item is degraded. The hallmark point of predictive maintenance is that it allows scheduling of correcting activities and hence prevents sudden machinery failures. Key features of predictive maintenance are to provide correct information regarding lifetime of the equipment, enhanced plant safety, few accidents with increased environment protection, and good spare parts handling.

3.2.1 Predictive Maintenance

In predictive maintenance condition and performance monitoring of equipment during normal operation is performed to reduce the likelihood of failures. It is also known as condition-based maintenance because the equipment is continuously monitored to timely detect and identify its anomalous behavior. Predictive maintenance is an old field, but its history is not properly documented. However, it is speculated that it is operational in the industrial world since the 1990s.

Predictive maintenance aim is to predict a failure of equipment well before its actual failure, which is then followed by preventive measures such as regularly scheduled and corrective maintenance. The process of predictive maintenance is given in Fig. 3.1.

One of the key aspects of predictive maintenance is the continuous monitoring of equipment so that its optimal working can be ensured. Therefore, it is appropriate to say that without condition monitoring predictive maintenance is not feasible. Condition monitoring as three categories: periodic, online, and remote. In periodic condition monitoring behavior of equipment under observation is examined after fixed intervals. The periodic analysis of the data accumulated from the equipment can depict a certain trend that can define its health state. The benefit of periodic Condition Monitoring is the regular inspection of the devices. If something is wrong, it is detected and identified in time. Online condition monitoring is the utilization of

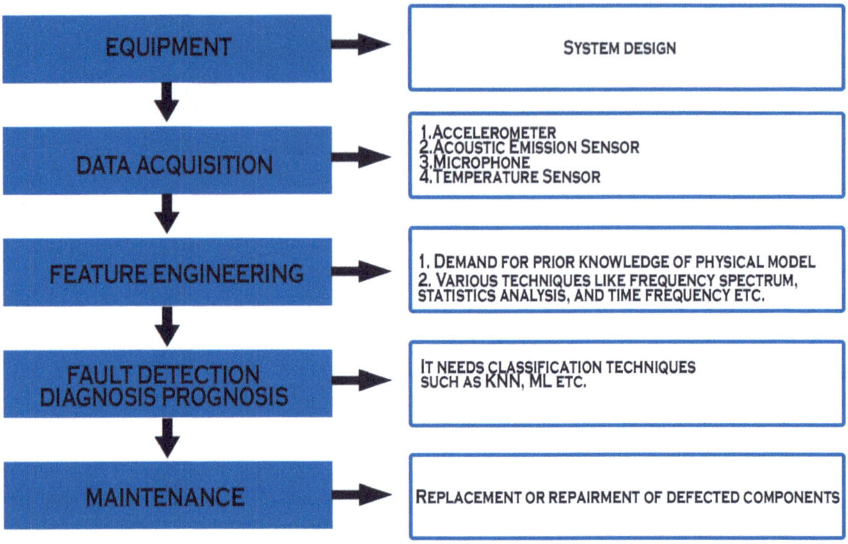

Fig. 3.1 Predictive maintenance process

measurement hardware and software to continuously monitor machines or production processes. Lastly, as the name suggests, remote condition monitoring is the monitoring of equipment from a remote location, with data transmitted for analysis.

Real life scenarios of predictive maintenance are:

- Recognizing that when fan get unbalanced.
- Recognizing when bearings demand lubrication.
- Alarming moment indication when oil replacement or when oil is contaminated.
- Finding misalignment between two rotating pieces of equipment.

Further, predictive maintenance can be of two types, i.e., Data-driven maintenance, and based on Model maintenance. Here the topic of discussion would be data-driven maintenance of industrial installations. The process of data driven maintenance consists of four crucial stages, i.e., data collection, fault detection and diagnostic, and prognosis. If properly implemented and executed, predictive maintenance is the foundation of an efficacious maintenance program.

3.2.1.1 Data Collection

To perform data-driven predictive maintenance, data associated with health states of equipment under observation is vital. It can be obtained through the data collection process in which information regarding target variables is gathered and measured in an established system. The gathered information can then be analyzed to answer questions related to the health states of the equipment and evaluate outcomes. Data collection is an important constituent of research in various disciplines including but not limited to humanities, physical and social sciences, engineering, computer science, and business. Further, data collection techniques may change depending upon the discipline. Nevertheless, the data collection process may vary but the significance of correct and honest data collection remains the same. In terms of data-driven predictive maintenance, data is collected via different types of sensors, i.e., temperature, accelerometer, acoustic, oil, etc. the measurements recorded through the sensors are further analyzed through signal processing, machine learning and deep learning techniques to infer momentous statistics about the conditions of the equipment.

3.2.1.2 Fault Detection and Diagnostic

Fault detection is the identification of anomalous patterns in the recorded data associated with the equipment under study. Fault detection does not just involve breathing the threshold level but understanding about the dynamics of the environment in which the equipment is operational and contextualizing the problem. It not only alerts the operators that there is a spike in the readings of given measurements but also indicates that a component is out of order.

Once the anomalies are being identified, the data can be further analyzed to pinpoint the type and location of fault which is known as fault diagnosis. Hence, the overall process of Fault Detection and Diagnostics (FDD) isolates anomalies and identifies the types of problems in the performance of essential equipment required in the manufacturing, and processing industries such as motors, boilers, chillers, motors, elevators, pumps, exhaust fans, etc. The application of FDD is beyond the manufacturing and processing industries. For instance, it can be applied to detect and identify anomalies in the man mad structures including buildings and bridges or to detect problems in the water, oil, and gas distribution networks, etc. Moreover, recent developments have empowered FDD systems to render anomalies into real-world faults and generate alerts to operators listing details about the failure of a component and how to resolve the problem.

3.2.1.3 Prognostic

Prognostics is the prediction of remaining useful time of equipment. Its purpose is to predict the time after that the complete system or one of its components will not give its desired performance [1]. The loss of smooth operation of a system is mostly assumed to the failure of the whole system or its components. Therefore, such a system can no longer be used to achieve the desired performance. The predicted time is known as the remaining useful life (RUL), which plays an important role for deciding alleviating failure. Prognostics can also be defined as predicting the future performance of equipment by evaluating the degree of deviance or deterioration of a system from its baseline operating conditions [2]. The concept of prognostics depends upon the exploration time to start prognosis (TSP), an examination of failure modes, detection of wear and aging of the components in the incipient stage, and fault conditions. Prognostic of a system can be helpful and effective if ample knowledge about the root cause of failure in the system is known. Therefore, it is essential to gather all the vital information on the possible failures in a product that includes information about the site, mode, cause, and mechanism of failure. This information is helpful to monitor and optimize the targeted parameters of a system. Potential uses for prognostics is in predictive maintenance. The field of studies that relates failure mechanisms to the lifecycle of a system is denoted as prognostics and health management (PHM). Prognostics techniques can be divided into data-driven, model-based, and hybrid approaches.

3.3 Sensors Commonly Used for Predicative Maintenance

There are different types of sensors that can be used for predictive maintenance of industrial equipment. This section provides insight into different kinds of sensors used for predictive maintenance, for instance, temperate, oil, accelerometers, acoustics emission sensors, and microphone. The selection of sensors is application-specific in which some sensors can be effective as compared to others. For instance, some sensors can detect bearing faults at the inception, much earlier as compared to others. This section covers different types of sensors commonly used in the predictive maintenance of assets. It is imperative to first confirm the type of potential fault in a system that may be encountered as predictive maintenance systems often utilizes a limited number of sensors. For instance, types of sensors that can detect fault signatures from a rotary machine are given in Fig. 3.2. Some of the commonly used sensors

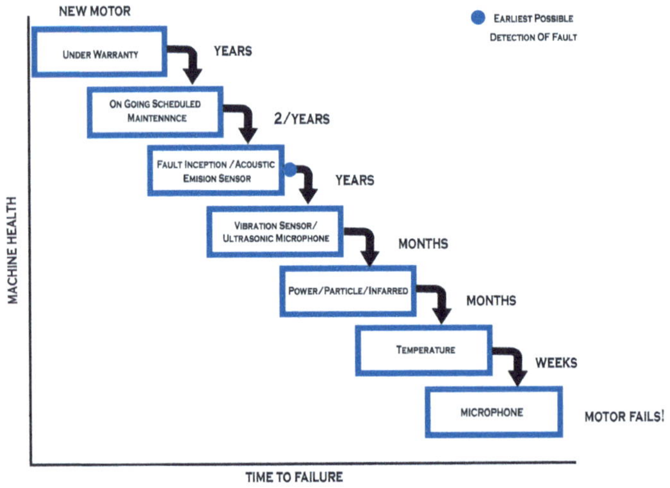

Fig. 3.2 A machine life cycle vs types of sensor that can detect fault signals [7, 8]

Table 3.1 Types and specification of sensors used in predictive maintenance [7, 8]

Measurement type	Sensor	Frequency	Application
Acoustic emissions	Piezo acoustic emission	100–400 kHz	Integrity testing of metallic structures, composite materials. Crack detection in hot reheat piping system. Plants and woods drying process monitoring, partial discharge detection
Acoustic emissions	Piezo acoustic emission	20–100 kHz	Corrosion detection in storage tanks (flat bottom), oil and water pipelines leakage detection, concrete crack detection, and partial discharge detection in low noise scenario
Vibration	Piezo accelerometer	Up to 30 kHz	Rotating machines
Vibration	MEMS accelerometer	Up to 20 kHz	Rotating machines
Sound pressure	Ultrasonic microphone	Up to 20 kHz	Pressure leaks, rotating machines
Sound pressure	Microphone	Up to 100 kHz	Pressure leaks, rotating machines and gearbox fault diagnosis, pump cavitation
Magnetic field	Magnetometer, hall, search coil		Rotor bar and end ring problems
Temperature	Infrared thermography	–	Heat source identification, change in load conditions, irregular turn off, power supply issues
Temperature	Thermocouple, RTD	–	Heat source identification, change in load conditions, irregular turn off, power supply issues
Oil quality	Particle monitor	-	Debris detection

in predictive maintenance are listed in Table 3.1 along with their specifications and types of faults they can detect.

3.4 Sensor Data Analysis and Machine Learning

Machine learning is a subfield of artificial intelligence where algorithms learn from input data without being explicitly programmed. These algorithms can automatically learn input data which is then used for predicting the out-of-sample data.

This property makes a machine learning algorithm useful for data-driven predictive maintenance of industrial equipment. This section provides an overview of the popular machine learning algorithms that have successfully been used in the predictive maintenance of industrial equipment. Moreover, it also describes the pros-cons of each machine learning algorithm in the context of predictive maintenance. Noteworthy, machine learning algorithms that have been used for data-driven predictive maintenance are listed as follow.

3.4.1 Logistic Regression

Logistic regression is a type of supervised classification algorithm. It was invented by David Cox in 1958. It is used for developing a regression model, the model performs prediction analysis which describes the data and its relationship between 1 dependent variable and 1 or many nominal, ordinal, interval/ratio level independent variables. There are three types of logistic regressions: 1) binomial, 2) multinomial, and 3) ordinal.

3.4.1.1 Binary or Binominal Logistic Regression

It is a binary response variable. The targeted variables can only have two possible types: 0 or 1, which can represent yes or no, scenarios.

3.4.1.2 Multinomial Logistic Regression

Multinomial logistic regression differs from binomial logistic regression based on the output variable values. The output in multinomial logistic regression may have greater than 2 possible discrete outputs that have no quantitative significance.

3.4.1.3 Ordinal Logistic Regression

The third type of ordinal deals with ordered categories. Regression coefficients are not easy to interpret in logistic regression as compare to linear regression. For example, a result card can be categorized in this form as excellent, very good, good, or bad. Logistic Regression assumes that the *logit* of the target variable is a linear function of the parameters. It outputs a variable between 0 and 1, and it is commonly used for binary classification. It is a discriminative model. It handles the linear solution.

3.4.2 k-Nearest Neighbors

k-NN that is abbreviation of k-nearest neighbors is one of the simplest machine learning techniques used for classification purposes. In k-NN, each data sample in the training set is represented by N-dimensional space represented by its "N" number of features [53]. Similarly, instances of the test set are also represented in a similar feature space. The goal of k-NN is to classify test set samples into appropriate classes based on neighborhood criterion calculated by using some similarity metrics. The classes of k nearest neighbors are determined and test samples is assigned with the class having majority votes. The matrices used to calculate the similarity among test samples and k nearest neighbors (training samples) include Euclidean distance, Mahalanobis distance, Manhattan distance. Furthermore, customized similarity metrics can also be designed which can effectively determine k nearest neighbors in the presence of outliers. For instance, Euclidian distance "D_E" between n^{th} test sample T_k and m^{th} train sample Tr_m can be calculated as:

$$D_E = \left[\sum_{n=1}^{N} (T_{n-Tr_m})^2 \right]^{\frac{1}{2}}, \quad \text{n} = 1, 2, 3, \ldots, \text{N}; \quad \text{m} = 1, 2, 3, \ldots, \text{M} \tag{3.1}$$

where N is the number of features and M represents the total number of training samples.

A schematic illustration of k-NN algorithms is presented in Fig. 3.3, which explains the whole process graphically. Due to its simple nature and easiness to implement the k-NN algorithm has been widely used in devising condition monitoring of equipment. To name a few it has been utilized in the condition monitoring

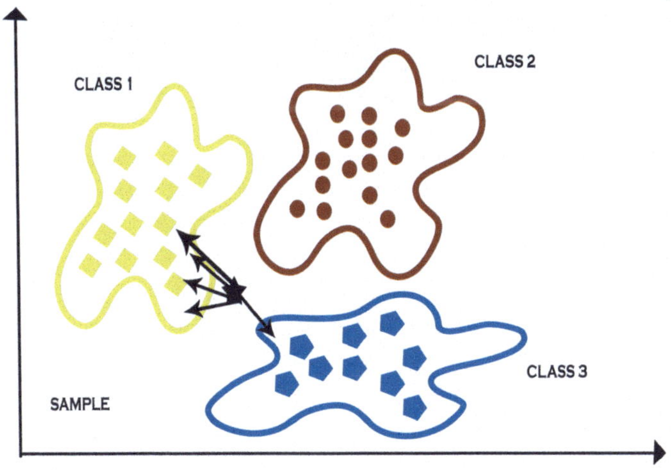

Fig. 3.3 A k-NN classifier [14]

of storage tanks [9], bearing and gearbox fault diagnosis [10, 11], shaft misalignment detection [12], structural health monitoring [13].

3.4.3 Artificial Neural Network

Artificial Neural Network (ANN) is adopted by the biological neural network that makes the human brain. The term "neural" in ANN is used to represent neuron and 'network' denotes a structure like a graph. ANNs are also referred to as neural nets, artificial neural systems. These networks learn information from data and perform tasks without being explicitly programmed according to task-related rules.

A shallow ANN is composed of three layers as can be seen in Fig. 3.4. The very first layer is the input layer that receives the original data as input into the system which is then further processed by following layers of artificial neurons. The following layers are the hidden layer, where artificial neurons get a set of defined inputs and generate an output via an activation function. The output layer, which is the last layer of ANN, gives the outputs for the program. Some applications of ANN in data-driven predictive maintenance can be seen in the form of condition monitoring [15], fault diagnosis [16], and prognosis health management of equipment [17].

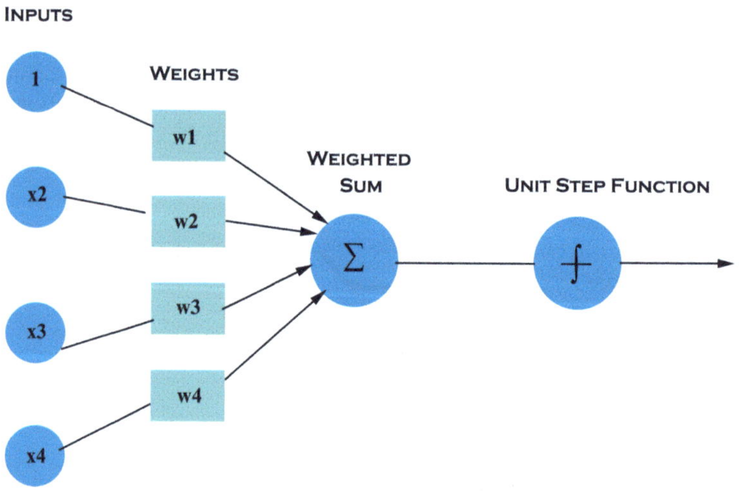

Fig. 3.4 The basic architecture of a feed-forward artificial neural network [18]

3.4.4 Support Vector Machine

Support-vector machine (SVM) or support-vector network, proposed by Vapnik [19] is a supervised learning model that is used for regression and classification. Statistic learning theory (SLT) is the basic building block of SVM. The goal of SVM is to find hyperplane based on support vectors that can effectively divide instances into different classes as given in Fig. 3.5. A hyperplane can be defined as a boundary that differentiates samples of different categories. This division can be performed in SVM either by finding hyperplane with maximum margins from support vectors or an optimal one that perform the categorization effectively.

SVM can be used in both binary and multiclass classification problems. In practice, multiclass problems are encountered frequently, therefore, there are multiple variants of SVM available to solve this sort of problem. A few of the variants of SVM includes one against one (OAO) approach, one against all (OAA) approach, decision directed acyclic graph (DDAG), and hierarchical structure for multiclass SVM [14]. SVMs have been considered as the gold standard for data-driven predictive maintenance in industrial applications. The concept of hyperplane makes SVMs different than other machine learning classifiers and enhances its classification performance as compared to other shallow networks. Moreover, the flexibility to use different types of kernel functions with SVM enhances its performance even on nonlinear and complex data. These kernel functions can also solve problems that are encountered while working with high dimensional data with features that are not easily separable [20]. It has been frequently used for the condition monitoring and fault diagnosis of rotary machines and gearbox [21–24], centrifugal pumps [25, 26], distillation columns [27, 28], ship engine fault [29], air-conditioning system [30].

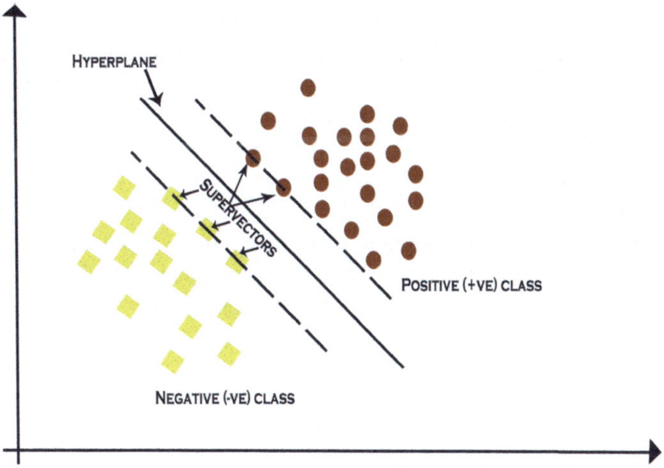

Fig. 3.5 A hyperplane separating two classes in support vector machine [31]

3.5 Limitations of Machine Learning Algorithms Used for Predictive Maintenance

Although machine learning algorithms have been extensively used in the development of a predictive maintenance mechanism there are some limitations associated with it. The reason behind the rendering of such algorithms for predictive maintenance is to automatically detect and diagnose any defect in equipment under observation. The detection of fault is also crucial to implement an effective prognostic strategy for equipment. A few of the limitations of machine learning in the context of predictive maintenance are listed as follow [32–35]:

1. Generalizability
   The implementation mechanism of machine learning is domain specific. It means each sort of application needs separate training and fine-tuning of the algorithm.
2. Domain Related Knowledge
   When using machine learning algorithms in predictive maintenance tasks expert knowledge about the problem domain is required. A feature engineering step is mandatory in the machine learning based fault detection, diagnostic, prognostic process. Feature engineering is a difficult process and requires a lot of expertise to generate hand-crafted features that can structure the dataset. And it can detect a growth in fault.
3. Learning ability, reliability and performance
   As the network architecture is simple for machine learning algorithms, therefore, such networks have limited learning capability. In general, these networks are referred to as shallow networks. In practice, in data-driven predictive maintenance the data used during the process is noisy, nonlinear, and complex. Machine learning algorithms are unable to handle irregularities, non-stationarity, nonlinearity of the data, which is often the case if data is from industrial equipment. Therefore, shallow networks have limited ability to provide data abstraction in the form of features that are used to predict faults. So, the overall performance of machine learning algorithms deteriorates when used with real time datasets for predictive maintenance.
4. Cross-domain Analysis
   Poor performance in cross-domain applications. If the nature of the application gets complex, satisfactory performance is not guaranteed. Maintenance actions are performed according to the failure prediction results.

3.6 Deep Learning Methods

Deep Learning is a subclass of machine learning algorithms which utilizes ANNs having stacked layers each containing several processing units [36]. This section provides an overview of the popular deep learning algorithms that have been used in predictive maintenance. Moreover, it highlights the advantage of deep learning over machine learning. Few of the deep learning algorithms worthy to mention in terms of predicative maintenance are described as follow.

3.6.1 Deep Artificial Neural Network

As the name suggests, a deep artificial neural network (DANN) is an architecture in which several layers each continuing multiple neurons. The simplest form of DANN is feed-forward DANN (FF-DANN) in which several fully connected layers of neurons are stacked together as presented in Fig. 3.6. These linkages are not all equal: every link can have a different quality or importance. The loads on these networks feeds the information on a system. Regularly the neural system units are also termed as hubs.

The first layer receives original inputs that travel through the whole network layer by layer in the forward direction. In this way, FF-DANN automatically learns approximation of the original input data which can be later used as features for regression as well as classification tasks.

A feedforward neural system is a naturally roused characterization calculation. It consists of several straightforward neuron-like individual units, sorted out in layers. Each unit in a layer relates to all the units in the preceding layer.

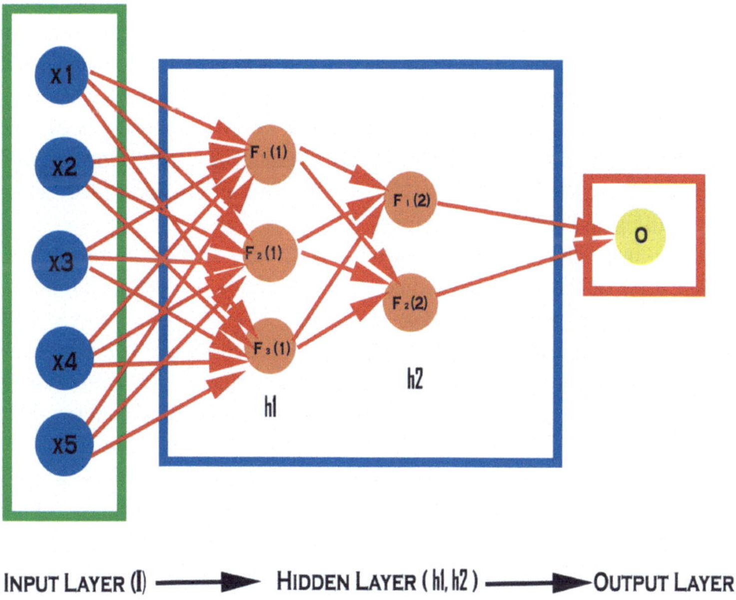

Fig. 3.6 The basic architecture of feed-forward deep artificial neural network (FF-DANN) [18]

DANN has been used for the fault diagnosis [37], leakage detection [38], and prognosis of industrial equipment [37]. Moreover, its application can also be seen in structural health monitoring [39].

3.6.2 Deep Convolutional Neural Network

Deep Convolutional Neural Networks (DCNN) are stacked ANN that used convolutional operation to extract information from the inputs. A backpropagation (BP) algorithm is used in DCNN to reduce the value of the cost role by adjusting weights and biases parameters of the network.

DCNN is a unique type of Neural Network that demonstrated exceptional performance in several computer vision and image processing competitions [40]. Many of the electrifying fields of operation of CNN involve image recognition and segmentation, object detection, video processing. Mainly thanks to multiple phases of removal, which can automatically get representative data, the learner-centered skill of DCNN. The accessibility of a great deal of device information has increased data analysis in DCNNs. There have been some exciting ideas for developments on CNNs, including the use of various triggering and loss functions, optimization of parameters, regularization, and architectural creativity. Nonetheless, technological advances achieve a major increase in the representative capability of the deep CNN. Significant attentiveness has been rewarded to the ideas of using space and the channel details, architectural depth and width, and multifaceted cognitive processes.

Likewise, there is also a concept of using a surface block as a supporting structure. This research thus focuses on the underlying taxonomy of the recently identified profound CNN architectures and thus divides recent developments into seven separate groups of CNN architectures. The usage of these seven categories is centered on the space, depth, multistage, distance, charts, channel boosting, and attention. In contrast, general knowledge of the CNN components is also provided, as well as current difficulties and CNN applications.

Due to the profound success in numerous fields, DCNN architectures have widely been used in the field of predictive maintenance to detect, diagnose faults in equipment, and perform prognostics.

Deep CNNs are traditional neural feedforward networks that use BP algorithms for the adjustment of the network's parameters to optimize the cost function. Fortunately, in four new aspects it is quite different from regular BP networks: local receptive fields, shared weights, pooling, and different layers combination. A basic deep neural network is given in Fig. 3.7. Moreover, the selection of network architecture is domain and application-specific and can be modified accordingly. In the field of predictive maintenance, researchers have utilized both one dimensional as well as two-dimensional DCNNN architectures. It has been under study for predictive maintenance in a number of domains including but not limited to bearing [35, 41], high-velocity oxy-fuel machine [42], and buildings [43].

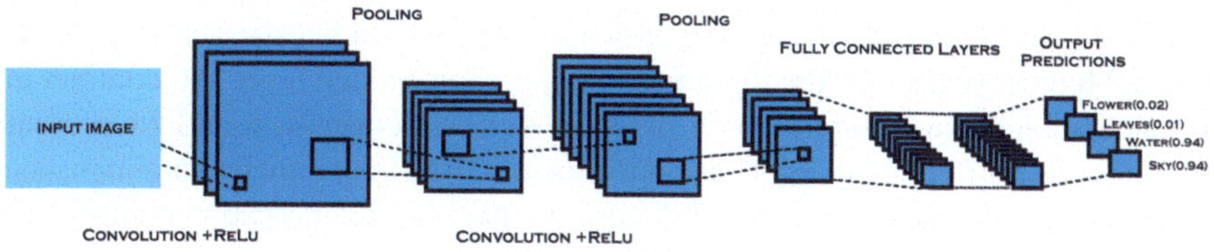

Fig. 3.7 Convolutional neural network architecture [36]

3.6.3 Deep Recurrent Neural Network

Deep recurrent neural networks (DRNN) have been successfully used for various tasks like for learning word embeddings, for language modeling, for online handwritten character recognition [45], and speech recognition [46]. Like a traditional DNN, it is made up of multiple hidden layers. A DRNN has recurrent (feedback) connections among hidden layers and works sequentially. Every concealed unit is linked to itself as well as to rest of the nodes in that hidden layer. Thus, DRNN means multiple recurrent hidden layers stacked to form a sequential network hierarchy.

DNN and DCNN are not much of a use for the processing of sequential information., e.g., if the occurrence of two inputs depends on each other and constitute to the next input in time and so on, in such a scenario it is better to adopt the recurrent neural network (RNN) architecture. The RNN network can be considered as the replication of the same recurrent unit throughout the network. The main goal behind using RNN is to explore the sequential nature of inputs and extract information for further usage. It predicts the next occurrence based upon the analyzed sequences in any application where data comes in sequential order. A 3-layered DRNN architecture is shown in Fig. 3.8 which takes input "s" at the time, t − 1, t, and t + 1 and predicts the respective output *P* at each time instance. Similarly, ct is the hidden state at time t, ct − 1

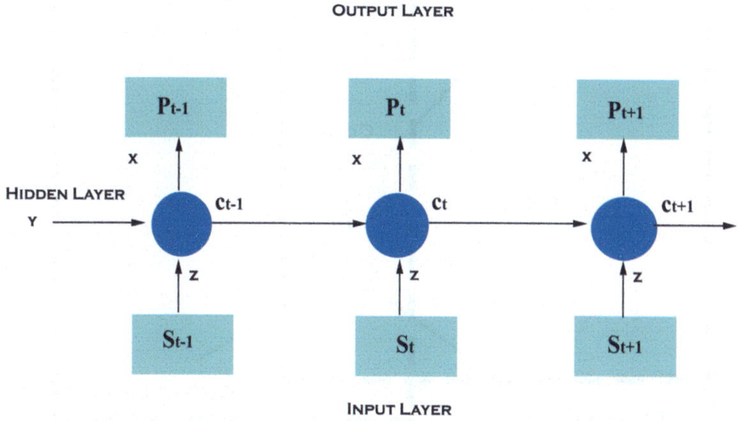

Fig. 3.8 Deep recurrent neural network architecture [50]

is the hidden state at time t − 1, and ct + 1 is the next hidden state at time t + 1. Here, all hidden states take the input from the previous hidden layer and contract in the calculation of final output Pt + 1 at time t + 1. Moreover, x, y, and z represent the weights for different layers. The application of DRNN in predictive maintenance can be found for induction motors [47], aero-engine [48], reciprocating compressor [49], etc.

3.6.4 Deep Auto-encoders

An autoencoder is a neural network used to acquire informative codes from the input data in an unaided way. There are multiple advantages associated with the usage of autoencoders which are listed as follows.

1. It can learn informative codes from raw data which can be used as features in regression and classification tasks.
2. It can be used for data dimensionality reduction.
3. There are variants of autoencoders which can help in the exploration of incomplete inputs (noisy data with missing values).
4. These can be helpful in fine-tuning of DNNs.

The least complex type of an autoencoder is a feedforward, a non-intermittent neural system like a multilayer perceptron (MLP) – having an input layer, an intermediate or code (hidden) layer, and output. In this way, autoencoder comprises of an encoder, and a decoder layer as shown in Fig. 3.9.

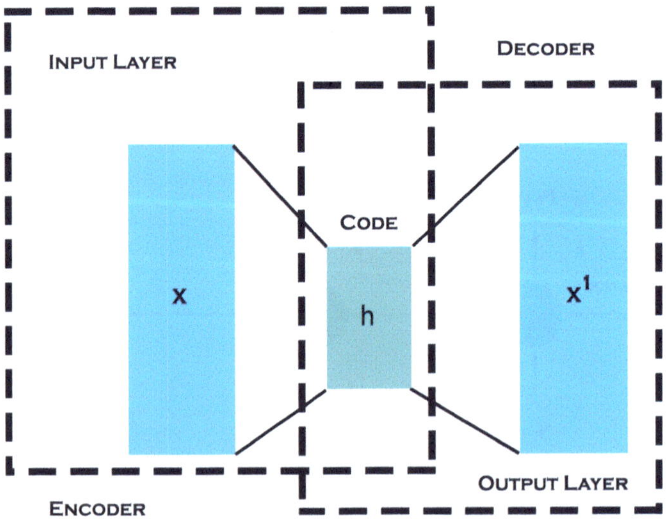

Fig. 3.9 Deep autoencoder architecture [53]

In each layer, there are multiple processing units (neurons). The number of processing units may be the same in all layers or may vary depending upon the nature of implementation. In practical scenarios, where dimensionality reduction of the input data is also intended, the number of processing units in the code layer is kept minor than the input layer. The hidden layer extracts prominent information from the input data which can be used as features for the categorization task. If multiple autoencoders are stacked, it creates a stacked autoencoder. The learning process of the autoencoder is unsupervised, whereas, if used in the categorization task, an additional layer is appended to the stacked autoencoders which is fine-tuned separately in a supervised manner, to determine the classes for instance under observation. Most often this supervised layer is of SoftMax classifier. Deep auto-encoders have been extensively used for the condition monitoring as well as fault diagnosis of rotary machine bearings [51], leakage detection of storage tank [31], fault detection of elevator system [52], etc.

3.6.5 Deep Belief Network

Deep Belief Network (DBN) is the graphical representation model for the inputs provided to the networks and is generative, i.e., it can generate all the possible value for a given scenario. It is constituted on the principles of machine learning, ANN, probability, and statistics. As in deep architecture, DBN consists of multiple layers with values, having relation between the layers but not among the values. Its main objective is to categorize the data.

DBN is made up of unsupervised neural networks such as Restricted Boltzmann Machines (RBMs). Multiple RBMs are stacked to form a DBNN. An illustration of a hidden layer constituted by RBM in a DBN is given in Fig. 3.10. In DBN each hidden layer of an RBM is an observable layer to the next one. The hidden layers of DBN are conditionally independent. The invisible layer of every sub-network is the visible layer of the next. The hidden or invisible layers are not interconnected and are conditionally independent. The predictive maintenance researchers and experts have utilized DBN for the fault detection and diagnosis of bearing [54], gearbox [55], complex chemical processes [56], etc.

3.7 Advantages of Deep Learning in Predictive Maintenance

The application of deep neural networks in various fields has achieved a remarkable feat. It has made the tasks possible which previously perceived to be impossible, such as the ability to handle big data as well as to characterize it in a meaningful manner. Its role does not differ in the field of predictive maintenance. In recent years,

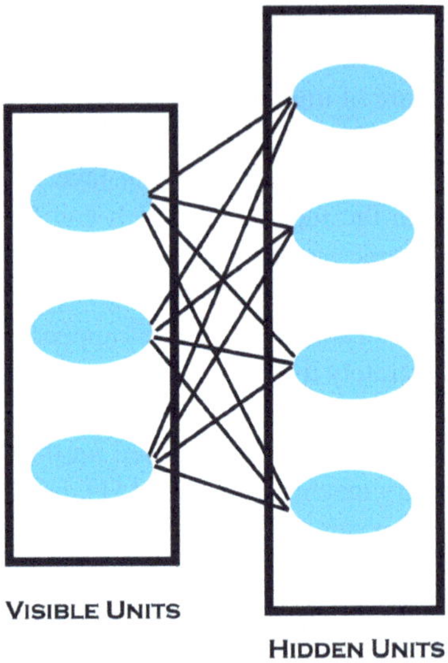

Fig. 3.10 Restricted Boltzmann machine [36]

a tendency has been evolved to apply deep learning techniques for fault detection [57], fault diagnosis [58], and its prognosis [59]. A few of the benefits that deep learning brings to the field of predictive maintenance are listed below:

1. The main advantage of incorporating a deep learning algorithm is the automated learning of structures from the fresh data. The nonlinear transformations in hierarchical order make it easy to extrapolate information from coarse data without the need for feature extraction and selection steps.
2. As the overhead of feature engineering and selection step is not mandatory, relatively it is easy to develop condition monitoring, fault detection and diagnosis, and prognosis strategy for predictive maintenance.
3. It is the age of data and data is the new gold, therefore, in the field of data-driven predictive maintenance, data plays an important role. Without it, nothing is possible in data-driven predictive maintenance. Through sensors, it is likely to recorded data regarding the health states of equipment on regular intervals for longer duration constituting datasets with huge records. Deep learning algorithms are suitable to tackle the challenge of big data analytics as compared to machine learning.
4. Deep learning algorithms are more suitable for transfer learning. It makes it feasible for the development of cross-domain data-driven predictive maintenance solutions.

5. An additional benefit that comes along with transfer learning is the evasion of the training process that saves lots of time and computation power.
6. The generalization power of deep learning based predictive maintenance strategies is greater as compared to machine learning ones.
7. Multi-task learning is possible in deep learning based predictive maintenance solutions. It helps to create multiple threads for different tasks instead of training a separate model for each one.
8. The larger number of layers and neurons in deep learning network, it allows the notion of complex problems which provides additional performance boost.
9. The most attractive part of applying deep learning in the domain of predictive maintenance is the fact that these networks can automatically extract the correct feature from the data while eliminating the need for manual feature engineering.
10. Use deep learning to predict failures when up to date, so that it covers any new event or behaviour.

3.8 Deep Learning for Predictive Maintenance

The goal of data-driven predictive maintenance is to automate the fault identification and prediction procedure with the help of artificial intelligence techniques most with focus on machine learning algorithms. However, in practice application of these techniques is not an easy task due to multiple reasons. First and foremost is the lack of availability of an efficient algorithm. Second, the data associated with an object under examination for the predictive maintenance is the composition of desired information coupled with noise. The addition of this extra information makes the exploration of the data challenging due to its non-linear, non-stationary, and complex nature. As stated in [50], the ability of traditional machine learning algorithms is usually limited to process complex data in raw form. For this reason, to develop an effective predictive maintenance system using conventional machine learning techniques a data engineering step is included in the predictive maintenance pipeline requires which considerable domain expertise. Through this additional step raw data can be transferred into carefully curated features that can be used as data descriptors in abstract form for the maintenance process. For the past few years there is a huge tendency towards using deep learning methods in data-driven predictive maintenance. One of the main reasons behind this is the ability of deep learning algorithms to learn from the raw data by their own. It means that if a deep learning technique has been incorporated in the predictive maintenance strategy, there is no need for in depth knowledge of the problem domain for feature engineering. Furthermore, a few additional advantages of deep learning algorithms as compare to traditional machine learning techniques in terms of predictive maintenance have been listed in Table 3.2 [41, 60–62].

Table 3.2 The advantages of different deep learning algorithms for predictive maintenance

Deep learning algorithm	Superiority for predictive maintenance
DANN	Degradation mapping, and failures identification, when enough history data could be obtained, and the complexity of target issue is relatively high
DCNN	It can be useful to use CNN as a diagnostic tool for data driven predictive maintenance when dealing with two-dimensional input data. The convolutional filters can extract useful local pattern from complex raw data in a robust manner. Further the implication of stacked convolutional layer enhances the fault diagnostic performance
DRNN	Through different variations of RNN deep sequential structures can be formed which provide the concept of memory unites and is useful to exploit temporal data while performing predictive maintenance
DSAE	DSAE are semi-supervised neural networks which are used to automatically mine fault signatures from the raw data. These networks are highly effective in discriminant information analysis about the health states of an object. Moreover, DSAE can be used for the dimensionality reduction
DBN	DBN is an energy-based architecture that can learn hidden information from the complex inputs. It is beneficial in the data-driven predictive maintenance when the decision is not solely based upon historical data. Moreover, it is also useful when the input data dimensions are large

3.9 Conclusion and Future Perspectives of Deep Learning Based Data-Driven Predictive Maintenance

3.9.1 Conclusion

With the advent of sensors, it became feasible to acquire data from industrial equipment. The acquired data is further analyzed for insights. If the analysis system is up to the mark that provides right perceptions without any postponement about the industrial or motorized equipment, it can be of an extra benefit to the expertise of the operators or engineers for taking a precautionary action. Due to the current advancements in ML and DL algorithms, accessibility of proper sensors and ubiquitous computing automated predictive maintenance is attainable.

It can be expected that the availability of data from industrial equipment will further improvise by the incorporation of IoT using different sensors. With the availability of humongous data, sophisticated big data analysis techniques are required to devise reliable data-driven predictive maintenance.

3.9.2 Future Perspectives of Deep Learning Based Data-Driven Predictive Maintenance

There is a still scope available for the improvement of deep learning based predictive maintenance. Some of the limitations that are faced by deep learning algorithms in terms of predictive maintenance are listed presented in the following subsections.

3.9.2.1 Enhanced Generalization

Although advanced deep learning techniques such as fine-tune transfer learning [63] and multitask learning [64] have brought a sense of generalization into the data-driven predictive maintenance strategies, but still these concepts must be explored in depth. Such concepts can be delved into for the implementation of domain-independent data-driven predictive maintenance.

3.9.2.2 Explainability

There is no doubt that the data handling and exploration strength of deep learning is far greater than that of machine learning. Its implication in the field of predictive maintenance has reduced a lot of overhead and complications that were faced by classical machine learning techniques. To name a few it can cope up with big data easily and can learn salient information from the inputs automatically without the requirement of domain-specific feature engineering step. Nevertheless, instead of enhanced capability, deep leering algorithms are like a black box. At present, there is no proper explanation that how deep learning algorithms accurately approximate the complex, nonlinear, and nonstationary data in an abstract way. Further, how the approximated codes which are also termed as features yield a better predictive maintenance performance than that of its predecessors. There is a need for explainable deep learning based predictive maintenance strategies.

3.9.2.3 Multimodal and Multi-sensors Data Fusion

The data fusion from multiple sensors and modalities is an interesting and possible extension of the deep learning-based data-driven predictive maintenance. Data fusion can provide in-depth details about the bearing defects that can help in the enhancement of bearing fault diagnosis models. Data fusion from multiple sensors is also a practical consideration as in practice for better performance multiple sensors are deployed on the concerned component to collect the data.

3.10 Glossary

Condition Monitoring: It is the constant monitoring of the parameters from a machine which are associate with its health indication. Any significant fluctuation in the monitored values of the parameters indicates the deviation of the machine from its normal state.
Data-Driven: It is an adjective which means that a given activity is based upon data.
Multimodal Data: Data taken through several modes of recordings within one application.
Multi-sensor: Data recorded through multiple sensors. Types of sensors may be the same or may differ.
Sensor: It is a device that takes a physical reading from the environment and converts it into electrical data or signals.
Support Vectors: The closest data point to the hyperplane are called support vectors. These entities define the location and orientation of the hyperplane.

References

1. Electronics Hub: What is a sensor? https://www.electronicshub.org/different-types-sensors. Accessed 15 Aug 2020
2. Butler, J., Smalley, C.: An introduction to predictive maintenance. Pharm. Eng. (2017). https://doi.org/10.1016/b978-0-7506-7531-4.x5000-3
3. Zhang, W., Yang, D., Wang, H.: Data-driven methods for predictive maintenance of industrial equipment: a survey. IEEE Syst. J. (2019). https://doi.org/10.1109/JSYST.2019.2905565
4. Mobley, R.K.: An Introduction to Predictive Maintenance, 2nd edn (2002)
5. Hemmerdinger, R.: Predictive maintenance strategy for building operations: a better approach. Schneider Electr. (2014)
6. Cheng, J.C.P., Chen, W., Chen, K., Wang, Q.: Data-driven predictive maintenance planning framework for MEP components based on BIM and IoT using machine learning algorithms. Autom. Constr. **112**, 103087 (2020)
7. E. by AspenCore: Choosing the most suitable predictive maintenance sensor (2020). https://www.embedded.com/choosing-the-most-suitable-predictive-maintenance-sensor/. Accessed 15 Aug 2020
8. Vallen Systeme: Acoustic emission sensors (2019). https://www.vallen.de/wp-content/uploads/2019/03/sov.pdf. Accessed 15 Aug 2020
9. Hasan, M.J., Kim, J.-M.: Fault detection of a spherical tank using a genetic algorithm-based hybrid feature pool and k-nearest neighbor algorithm. Energies **12**(6), 991 (2019)
10. Li, Z., Yan, X., Yuan, C., Li, L.: Gear multi-faults diagnosis of a rotating machinery based on independent component analysis and fuzzy k-nearest neighbor (2010). https://doi.org/10.4028/www.scientific.net/AMR.108-111.1033
11. Sharma, A., Jigyasu, R., Mathew, L., Chatterji, S.: Bearing fault diagnosis using weighted k-nearest neighbor (2018). https://doi.org/10.1109/ICOEI.2018.8553800
12. Gohari, M., Eydi, A.M.: Modelling of shaft unbalance: modelling a multi discs rotor using K-Nearest Neighbor and Decision Tree Algorithms. Meas. J. Int. Meas. Confed. (2020). https://doi.org/10.1016/j.measurement.2019.107253

13. Vitola, J., Pozo, F., Tibaduiza, D.A., Anaya, M.: A sensor data fusion system based on k-nearest neighbor pattern classification for structural health monitoring applications. Sensors (Switz.) (2017). https://doi.org/10.3390/s17020417
14. Wei, Y., Li, Y., Xu, M., Huang, W.: A review of early fault diagnosis approaches and their applications in rotating machinery. Entropy. (2019). https://doi.org/10.3390/e21040409
15. Sarma, D.V.S.S.S., Kalyani, G.N.S.: ANN approach for condition monitoring of power transformers using DGA (2004). https://doi.org/10.1109/tencon.2004.1414803
16. Zhang, Z.Y., Wang, K.S.: Wind turbine fault detection based on SCADA data analysis using ANN. Adv. Manuf. (2014). https://doi.org/10.1007/s40436-014-0061-6
17. Zhang, Z., Wang, Y., Wang, K.: Fault diagnosis and prognosis using wavelet packet decomposition, Fourier transform and artificial neural network. J. Intell. Manuf. (2013). https://doi.org/10.1007/s10845-012-0657-2
18. Yegnanarayana, B.: Artificial Neural Networks. PHI Learning Pvt. Ltd. (2009)
19. Cortes, C., Vapnik, V.: Support-vector networks. Mach. Learn. **20**(3), 273 (1995)
20. Orrù, P.F., Zoccheddu, A., Sassu, L., Mattia, C., Cozza, R., Arena, S.: Machine learning approach using MLP and SVM algorithms for the fault prediction of a centrifugal pump in the oil and gas industry. Sustainability (2020). https://doi.org/10.3390/su12114776
21. Li, X., Yang, Y., Pan, H., Cheng, J., Cheng, J.: A novel deep stacking least squares support vector machine for rolling bearing fault diagnosis. Comput. Ind. (2019). https://doi.org/10.1016/j.compind.2019.05.005
22. Li, Y., Zhang, W., Xiong, Q., Luo, D., Mei, G., Zhang, T.: A rolling bearing fault diagnosis strategy based on improved multiscale permutation entropy and least squares SVM. J. Mech. Sci. Technol. (2017). https://doi.org/10.1007/s12206-017-0514-5
23. Chen, F., Tang, B., Chen, R.: A novel fault diagnosis model for gearbox based on wavelet support vector machine with immune genetic algorithm. Meas. J. Int. Meas. Confed. (2013). https://doi.org/10.1016/j.measurement.2012.06.009
24. Sohaib, M., Kim, J.-M.: Hierarchical radial basis function based multiclass support vector machines and a hybrid feature pool for bearings fault diagnosis (2018). https://doi.org/10.1109/ICEE.2018.8566908
25. Panda, A.K., Rapur, J.S., Tiwari, R.: Prediction of flow blockages and impending cavitation in centrifugal pumps using Support Vector Machine (SVM) algorithms based on vibration measurements. Meas. J. Int. Meas. Confed. (2018). https://doi.org/10.1016/j.measurement.2018.07.092
26. Bordoloi, D.J., Tiwari, R.: Identification of suction flow blockages and casing cavitations in centrifugal pumps by optimal support vector machine techniques. J. Braz. Soc. Mech. Sci. Eng. (2017). https://doi.org/10.1007/s40430-017-0714-z
27. Taqvi, S.A., Tufa, L.D., Zabiri, H., Maulud, A.S., Uddin, F.: Multiple fault diagnosis in distillation column using multikernel support vector machine. Ind. Eng. Chem. Res. (2018). https://doi.org/10.1021/acs.iecr.8b03360
28. Liu, L., Liu, A.L.: Fault diagnosis of distillation column based on improved genetic algorithm optimization-based support vector machine. J. East China Univ. Sci. Technol. **37**, 228–233 (2011)
29. Cai, C., Zong, H., Zhang, B.: Ship diesel engine fault diagnosis based on the SVM and association rule mining (2016). https://doi.org/10.1109/CSCWD.2016.7566022
30. Sun, K., Li, G., Chen, H., Liu, J., Li, J., Hu, W.: A novel efficient SVM-based fault diagnosis method for multi-split air conditioning system's refrigerant charge fault amount. Appl. Therm. Eng. (2016). https://doi.org/10.1016/j.applthermaleng.2016.07.109
31. Sohaib, M., Islam, M., Kim, J., Jeon, D.-C., Kim, J.-M.: Leakage detection of a spherical water storage tank in a chemical industry using acoustic emissions. Appl. Sci. (2019). https://doi.org/10.3390/app9010196
32. Çınar, Z.M., Abdussalam Nuhu, A., Zeeshan, Q., Korhan, O., Asmael, M., Safaei, B.: Machine learning in predictive maintenance towards sustainable smart manufacturing in industry 4.0. Sustainability **12**(19), 8211 (2020)

33. Lv, F., Wen, C., Bao, Z., Liu, M.: Fault diagnosis based on deep learning. In: 2016 American Control Conference (ACC), pp. 6851–6856 (2016). https://doi.org/10.1109/ACC.2016.7526751
34. Lo, N.G., Flaus, J.-M., Adrot, O.: Review of machine learning approaches in fault diagnosis applied to IoT systems. In: 2019 International Conference on Control, Automation and Diagnosis (ICCAD), pp. 1–6 (2019)
35. Hasan, M.J., Sohaib, M., Kim, J.M.: 1D CNN-based transfer learning model for bearing fault diagnosis under variable working conditions (2019). https://doi.org/10.1007/978-3-030-03302-6_2s
36. Bengio, Y., Goodfellow, I., Courville, A.: Deep Learning, vol. 1. Citeseer (2017)
37. Pandarakone, S.E., Masuko, M., Mizuno, Y., Nakamura, H.: Deep neural network based bearing fault diagnosis of induction motor using fast Fourier transform analysis. In: 2018 IEEE Energy Conversion Congress and Exposition (ECCE), pp. 3214–3221 (2018)
38. Sohaib, M., Kim, J.-M.: Data driven leakage detection and classification of a boiler tube. Appl. Sci. (2019). https://doi.org/10.3390/app9122450
39. Khurjekar, I.D., Harley, J.B.: Uncertainty aware deep neural network for multistatic localization with application to ultrasonic structural health monitoring. arXiv Preprint. arXiv:2007.06814 (2020)
40. Khan, A., Sohail, A., Zahoora, U., Qureshi, A.S.: A survey of the recent architectures of deep convolutional neural networks. Artif. Intell. Rev. (2020). https://doi.org/10.1007/s10462-020-09825-6
41. Sohaib, M., Kim, J.-M.: Fault diagnosis of rotary machine bearings under inconsistent working conditions. IEEE Trans. Instrum. Meas. **69**(6), 3334–3347 (2019)
42. Ibrahim, K., Masrom, M.: Predictive maintenance of high-velocity oxy-fuel machine using convolution neural network. SSRN 3660305 (2020)
43. Özgenel, Ç.F., Sorguç, A.G.: Performance comparison of pretrained convolutional neural networks on crack detection in buildings. In: Proceedings of the International Symposium on Automation and Robotics in Construction, ISARC, vol. 35, pp. 1–8 (2018)
44. Missinglink.ai: The Complete Guide to Artificial Neural Networks: Concepts and Models
45. Ren, H., Wang, W., Liu, C.: Recognizing online handwritten Chinese characters using RNNs with new computing architectures. Pattern Recognit. (2019). https://doi.org/10.1016/j.patcog.2019.04.015
46. Lam, M.W.Y., Chen, X., Hu, S., Yu, J., Liu, X., Meng, H.: Gaussian process LSTM recurrent neural network language models for speech recognition (2019). https://doi.org/10.1109/ICASSP.2019.8683660
47. Xiao, D., Huang, Y., Qin, C., Shi, H., Li, Y.: Fault diagnosis of induction motors using recurrence quantification analysis and LSTM with weighted BN. Shock Vib. **2019** (2019)
48. Yuan, M., Wu, Y., Lin, L.: Fault diagnosis and remaining useful life estimation of aero engine using LSTM neural network. In: 2016 IEEE International Conference on Aircraft Utility Systems (AUS), pp. 135–140 (2016)
49. Cabrera, D., et al.: Bayesian approach and time series dimensionality reduction to LSTM-based model-building for fault diagnosis of a reciprocating compressor. Neurocomputing **380**, 51–66 (2020)
50. LeCun, Y., Bengio, Y., Hinton, G.: Deep learning. Nature **521**(7553), 436–444 (2015)
51. Sohaib, M., Kim, C.-H., Kim, J.-M.: A hybrid feature model and deep-learning-based bearing fault diagnosis. Sensors (Switz.) (2017). https://doi.org/10.3390/s17122876
52. Mishra, K.M., Krogerus, T.R., Huhtala, K.J.: Fault detection of elevator systems using deep autoencoder feature extraction. In: 2019 13th International Conference on Research Challenges in Information Science (RCIS), pp. 1–6 (2019)
53. Sohaib, M., Kim, J.-M.: Reliable fault diagnosis of rotary machine bearings using a stacked sparse autoencoder-based deep neural network. Shock Vib. (2018). https://doi.org/10.1155/2018/2919637
54. Shao, H., Jiang, H., Zhang, H., Liang, T.: Electric locomotive bearing fault diagnosis using a novel convolutional deep belief network. IEEE Trans. Ind. Electron. **65**(3), 2727–2736 (2017)

55. Chen, Z., Li, C., Sánchez, R.-V.: Multi-layer neural network with deep belief network for gearbox fault diagnosis. J. Vibroeng. **17**(5), 2379–2392 (2015)
56. Zhang, Z., Zhao, J.: A deep belief network based fault diagnosis model for complex chemical processes. Comput. Chem. Eng. **107**, 395–407 (2017)
57. Lee, K.P., Wu, B.H., Peng, S.L.: Deep-learning-based fault detection and diagnosis of air-handling units. Build. Environ. (2019). https://doi.org/10.1016/j.buildenv.2019.04.029
58. Yu, Y., Woradechjumroen, D., Yu, D.: A review of fault detection and diagnosis methodologies on air-handling units. Energy Build. (2014). https://doi.org/10.1016/j.enbuild.2014.06.042
59. Su, Y., Tao, F., Jin, J., Wang, T., Wang, Q., Wang, L.: Failure prognosis of complex equipment with multistream deep recurrent neural network. J. Comput. Inf. Sci. Eng. (2020). https://doi.org/10.1115/1.4045445
60. Chen, Z., Li, W.: Multisensor feature fusion for bearing fault diagnosis using sparse autoencoder and deep belief network. IEEE Trans. Instrum. Meas. (2017). https://doi.org/10.1109/TIM.2017.2669947
61. Rahhal, J.S., Abualnadi, D.: IOT based predictive maintenance using LSTM RNN estimator. In: 2020 International Conference on Electrical, Communication, and Computer Engineering (ICECCE), pp. 1–5 (2020)
62. Butte, S., Prashanth, A.R., Patil, S.: Machine learning based predictive maintenance strategy: a super learning approach with deep neural networks. In: 2018 IEEE Workshop on Microelectronics and Electron Devices (WMED), pp. 1–5 (2018)
63. Hasan, M.J., Islam, M.M.M., Kim, J.-M.: Acoustic spectral imaging and transfer learning for reliable bearing fault diagnosis under variable speed conditions. Measurement **138**, 620–631 (2019)
64. Cao, X., Chen, B., Zeng, N.: A deep domain adaption model with multi-task networks for planetary gearbox fault diagnosis. Neurocomputing **409**, 173–190 (2020)

Chapter 4
Multi-criteria Fuzzy Goal Programming Under Multi Uncertainty

Junzo Watada, Nureize Binti Arbaiy, and Qiuhong Chen

Abstract We still face situations where there are different goals and constraints that need to be optimized in mathematical programming problems, but their value cannot be easily determined. However, when uncertainty is included in the model, the translation of real-world problems into mathematical models becomes more difficult. Given the nature of the multi-objective conditions under hybrid uncertainty, it is important to adapt to the model. In some research studies, only purpose functions or constraints are taken into account vaguely. But in this research study, objectives, constraints, and coefficients were calculated based on fuzzy random data. Goal and constraint functions should be constructed by analyzing past data, then determining constraint functions based on volatile hybrid environments. In the building constraint function, "about equal to" should be converted into fuzzy mathematical symbols and figures to express. The final step is provided to find the best answer using scalable methods and maximum operators to complete multi-criteria goal programming. The algorithm uses the concept of satisfaction to multi-objective optimization. The problem model was also developed by means of a fuzzy random regression approach. From that, we emphasize that the proposed method has significant advantages in solving multi-objective problems in which fuzzy random information coexisting.

J. Watada (✉)
Waseda University, Tokyo, Japan
e-mail: watada@waseda.jp

N. B. Arbaiy
Universiti Tun Hussein Onn Malaysia, Parit Raja, Malaysia
e-mail: nureize@uthm.edu.my

Q. Chen
International Society of Management Engineers, Kitakyushu, Japan
e-mail: chenqiuhong86@gmail.com

Spiber, Inc., Tsuruoka, Japan

M. A. R. Ahad and A. Inoue (eds.), *Vision, Sensing and Analytics: Integrative Approaches*, Intelligent Systems Reference Library 207, https://doi.org/10.1007/978-3-030-75490-7_4

Keywords Fuzzy random data · fuzzy multi-criteria LP · Fuzzy random regression model · Max-min method · Scalable index method · Multi uncertainty

4.1 Introduction

In recent years, fuzzy mathematics theories have developed widely and been used in various fields. In many practical cases, a satisfaction approach should be a much better solution strategy, given the inherent problems with optimization. The goal programming (GP) model (Charnes and Cooper [12]), built on the basis of a satisfaction theory, is one of the solutions to solve the mathematical programming problem with multiple goals. The GP model helps the decision-maker to simultaneously analyse multiple objectives and choose the most rewarding action from among a range of feasible actions. In GP, the satisfying principle is to follow the assessment actions of decision-makers to achieve a set of defined goals and to achieve these goals as adequately as possible. The satisfying idea communicates the idea that if their aims are properly accomplished in that specific decision situation, the decision-makers feel appropriately contented [3, 7, 23].

The objects of critique were different aspects of GP, such as the complexity of deciding the exact goal values and the close absence of the decision maker in the modelling process. To fix the inaccuracy present in the GP model, Fuzzy goal programming (FGP) uses fuzzy set theory (Zadeh [39]). Therefore, FGP offers better tools to represent topics that include fuzzy objectives. In such issues, the inaccuracy typically relates to the target goals, but may also be extracted from other aspects of the model, such as its system constraints. FGP will however, treat only fuzzy values which are given in its model. Randomness occurring in the environment is beyond the limits of what can be handled by the current FGP model. Since fuzziness and randomness often occur in real-life scenarios, it is difficult to use current systematic methods to deal with such situations, although it is crucial to consider such multi-hold uncertainties.

Given that the data, goals or conditions are not specified, uncertainty must be transformed into certainly by several methods in order to obtain an optimal solution. The primary factor to consider carefully is the type of procedures used and how they are modified. The fuzzy model should be first adapted compared to the conventional GP and the needs of decision makers must be addressed. Most analysis is currently being used to address fuzzy objectives., constraints or coefficients, but the solution process does not take place simultaneously. This analysis therefore takes all the sections into account simultaneously and seeks a fair and suitable solution.

The remaining consists of the following sections: The preliminary brief concept of GP and fuzzy programming in Sect. 4.2. In Sect. 4.3 the basic methodology is explained as the FGP. And Sect. 4.4 explains the proposed methodology and in Sect. 4.5 its brief application is given. Finally the conclusions are drawn in Sect. 4.6.

4.2 Preliminary Preparation

A fuzzy set is a set whose elements have degrees of membership. Fuzzy sets were introduced simultaneously by Lotfi A. Zadeh and Dieter Klaua in 1965 [39], [22]as an extension of the classical notion of set. In classical set theory, the membership of elements in a set is assessed in binary terms—an element either belongs or does not belong to the set. By contrast, fuzzy set theory permits the gradual assessment of the membership of elements in a set; this is described with the aid of a membership function valued in real unit interval [0, 1]. Classical sets are special cases of fuzzy sets, if memership function take only values 0 or 1. Fuzzy set theory used to denote classical sets as crisp sets. The fuzzy set theory is used to deal with incomplete and imprecise information such as bioinformatics.

Fuzzy sets can be applied, for example, to the field of genealogical research. When an individual is studied in vital records such as birth records for possible ancestors, a number of issues should be encapsulated in a membership function. A fuzzy set is denoted by a pair (U, m) with $m : U \to [0, 1]$ where U indicates a set. For each $x \in U$ the value $m(x)$ means the grade of membership of x in (U, m). For a finite set $U = \{x_1, x_2, \cdots, x_n\}$ the fuzzy set (U, m) is often denoted by $\{m(x_1)/x_1, m(x_2)/x_2, \cdots, m(x_n)/x_n\}$. Let $x \in U$. Then x is called not included in the fuzzy set (U, m) if $m(x) = 0$. x is called fully included if $m(x) = 1$ [21]. The set $\{x \in U|m(x) > 0\}$ is called the support of (U, m) and the set $\{x \in U|m(x) = 1\}$ is called its kernel. The function m is called the membership function of the fuzzy set (U, m).

4.2.1 Fuzzy Number

A fuzzy number is fuzzy set $A \subset R$ which is convex, normalized, at least segmentally continuous, and has membership function of its functional value $\mu_A(x) = 1$ at precisely one element.

4.2.2 Membership Function

A membership function on X for any set X is a map from X to the real unit interval [0, 1]. A membership functions on X define a fuzzy subset of X. The membership function which represents a fuzzy set A is usually written as μ_A. For an element x of X, the value $\mu_A(x)$ is called the membership degree of x to the fuzzy set A. The membership degree $\mu_{\tilde{A}}(x)$ means the grade of membership of the element x to the fuzzy set A. The value 0 denotes that x is not a member of the fuzzy set; the value 1 expresses that x is fully a member of the fuzzy set $\tilde{A}$. The values between 0 and 1 express fuzzy members, which belong to the fuzzy set only partially. Decision theory defines "a capacity" as a function ν from S, the set of subsets of some set, into [0, 1], as ν is a set-wise monotone function which is normalized

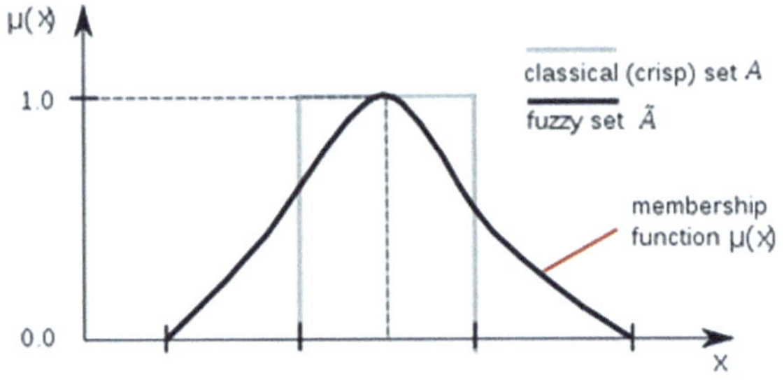

Fig. 4.1 Membership function [40]

(i.e. $\nu(\emptyset) = 0, \nu(\Omega) = 1$. This definition generalizes a probability measure. The definition weakens the probability axiom of count ability. The capacity is applied as a subjective measure of the likelihood of an event, and the "expected value" of an outcome. This certain capacity can be taken the Choquet integral over the capacity (Fig. 4.1).

4.2.3 GP

The GP model (Charnes and Cooper [13]; Jones and Tamiz, 2010 [18], 2016 [19]) is organised as a linear target system with positive and negative deviations between each objective and target or aspiration level, respectively. In the decision problem, this decides the most satisfactory point for the set of goals. The best compromises that decision makers can make are the solutions achieved through the GP model (Schniederjans [34]; Jones and Romero [20]). For this reason, GP is based on the "satisfaction" principle. The GP model can be analyzed in two parts, set limits, and objective functions. For each associated objective, goal constraints are established by considering the target variables involved. The GP model using all deterministic values is written as follows.

$$\max \quad V(\mu) = \sum_{i=1}^{m} \mu_i$$

$$\text{subject to} \quad \mu_i = \frac{G_i(\mathbf{X}) - L_i}{g_i - L_i}$$

$$\mathbf{AX}_i \leq \mathbf{b} \qquad \mathbf{A} = \begin{bmatrix} a_{11} & \cdots & a_{1m} \\ a_{21} & \cdots & a_{2m} \\ \vdots & & \vdots \\ a_{n1} & \cdots & a_{nm} \end{bmatrix}, \mathbf{X}_i = \begin{bmatrix} X_{i1} \\ X_{i2} \\ \vdots \\ x_{im} \end{bmatrix}, \mathbf{b} = \begin{bmatrix} b_1 \\ b_2 \\ \vdots \\ b_n \end{bmatrix}, \tag{2.1}$$

$$\mu_i \leq 1$$

$$\mathbf{X}_i \geq \mathbf{0}, \mu_i \geq 0.$$

$$i = 1, 2, \cdots, m,$$

where $V(u)$ is fuzzy achievement function. Note that $\mathbf{AX} \leq \mathbf{b}$ is the crisp system constraints in vector.

4.2.4 Fuzzy GP

Classical GP is designed with objective features, constraints, and target values, all of which are deterministic values. It is difficult to achieve the precise value for creating a model if the expertise of experts is imprecise or inaccessible. Fuzzy values in the GP description are used in certain fuzzy and imprecise circumstances. In the GP model, the inaccurate values reflect the decision maker's fuzziness or tolerance and also the imprecision of expert expertise.

Fuzzy GP is a GP with fuzzy factors, coefficients, data, parameters, limitations or conditions. A business, for example, wants to construct a new factory, but faces several requirements and environments that are unpredictable. This involves circumstances where there is no reference information, economies are unpredictable, costs are difficult to mange, etc. In this situation, in order to construct a new and productive plant, the decision maker needs to develop an appropriate plan. But how do decision makers define these needs with the aim of achieving a goal (building a new plant) and making a profit? Situations such as this require the use of a solution model for multi-objective optimization such as GP.

4.2.5 Notation

Let us note that we do not emphasize a fuzzy set or a fuzzy number by using $\sim$ such as $\tilde{A}$ for a fuzzy set because the meaning is clear in the context.

4.3 Fuzzy Multi-criteria Linear Programming

4.3.1 Purposes of this Research

What kind of this research can solve?

First, when we face a problem and do not know how to get the right target, we can use historical data to get a proper linear regression model to formulate a target. Second, even when the historical data is unclear or uncertain, the target (goal) can also be built. Third, solve some unspecified problems, such as constraints, unspecified or changing circumstances.

4.3.2 Definitions of Fuzzy Sets

Definition 3.1 Given some universe Γ, let Pos be possibility measure defined on the power set $P(\Gamma)$ of Γ, let $\Re$ be the set of real numbers, a function $Y : \Gamma \to \Re$ is said to be a fuzzy variable defined on Γ, the possibility distribution μ_Y of Y is defined by $\mu_Y(t) = Pos(Y = t), t \in \Re$, which is the possibility of event $(Y = t)$, for fuzzy variable Y with possibility distribution $\mu_Y(t)$, the possibility, necessity, credibility of event $\{Y \leq t\}$ are given, as follows

$$\begin{aligned} Pos\{Y \leq t\} &= \sup r \overset{\mu_Y}{\leq} \tau\ (t) \\ Nec\{Y \leq t\} &= 1 - \sup r \overset{\mu_Y}{\geq} \tau\ (t) \\ Cr\{Y \leq t\} &= \frac{1}{2}\big(1 + \sup r \overset{\mu_Y}{\leq} \tau\ (t) - \sup r \overset{\mu_Y}{\geq} \tau\ (t)\big) \end{aligned} \tag{3.1}$$

Credibility is defined as average of the possibility and the necessity measure, $cr(\cdot) = (pos(\cdot) + nec(\cdot))/2$, and it is a self dual function. The credibility measure denotes a certain measure which aggregates the two extreme cases such as the possibility (express a level of overlap and being highly optimistic in this sense) and necessity(articulating a degree of inclusion and being pessimistic in its nature) [4]. Based on the credibility measure, the expected value of a fuzzy variable is presented as follow.

Definition 3.2 Let Y be a Fuzzy variable. The expected value of Y is defined as

$$E[Y] = \int_0^{\infty} Cr(Y \geq t)dr - \int_{-\infty}^{0} Cr(Y \leq t)dr \tag{3.2}$$

Definition 3.3 Assume $Y = (a^l, a, a^r)$ is a triangular fuzzy variable whose possibility distribution is

$$\mu_Y(x) = \begin{cases} \dfrac{x - a^l}{a^r - a^l}; & ; a^l \leq x \leq a^r \\ \dfrac{a^r - x}{a^r - a^l}; & a^l \leq x \leq a^r \\ 0; & \text{otherwise} \end{cases} \tag{3.3}$$

From (3.3), We determine the expected value of Y.

$$E[Y] = \frac{a^l + 2a + a^r}{4} \tag{3.4}$$

For more theoretical results on fuzzy random variables, we may refer to Baoding Liu [30], Jaime Gil-Aluja [16], Wang [37] (Fig. 4.2),

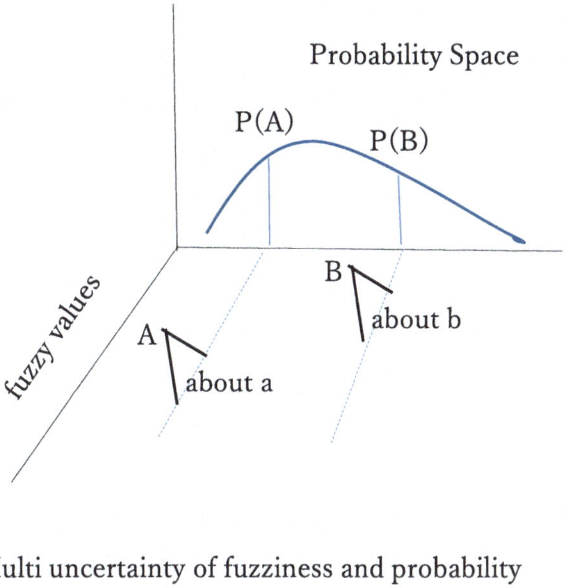

Fig. 4.2 Fuzzy random data. Note that fuzzy numbers occur with probability [37]

Definition 3.4 Suppose that (Ω, Σ, pr) is a probability space, F_υ is a collection of fuzzy variable defined on possibility space $(\Gamma, P(\Gamma), Pos)$, a fuzzy random variable is a mapping $X : \Omega \rightarrow F_\upsilon$ such that for any Borel subset B of $\Re$, $Pos\{X(\omega) \in B\}$ is a measurable function of ω. e.g. Let V be a random variable defined on probability space (Ω, Σ, pr), define that for every $\omega \in \Omega$, $X(\omega) = (V(\omega) + 2, V(\omega) - 2, V(\omega) + 6)_\tau$ which is a triangular fuzzy variable defined on some possibility space $(\Gamma, P(\Gamma), Pos)$, X is a triangular fuzzy random variable [36].

Definition 3.5 Let X be a fuzzy random variable defined on a probability space (Ω, Σ, pr). The expected value of X is defined as

$$E[\xi] = \int [\int_0^\infty [\inf_0^\infty Cr\{Y \geq t\} - \inf_{-\infty}^0 Cr\{Y \leq t\} dr p(\omega) d\omega \tag{3.5}$$

Suppose V is a discrete fuzzy random variable, which takes $V_1 = 3$ with probability 0.2, $V_2 = 6$ with probability 0.8, we calculate the expected value of X. From definition 3.3, we can know the following:

$$X(V_1) = (5, 1, 9)T; \quad pro = 0.2,$$
$$X(V_2) = (8.4.12)T, \quad pro = 0.8,$$

Then from the equation, we can calculate the following:

Table 4.1 Fuzzy random input-output data

Sample	Output	Input					
1	Y_1	X_{11}	X_{12}	$\cdots$	X_{1j}	$\cdots$	X_{1k}
2	Y_2	X_{21}	X_{22}	$\cdots$	X_{2j}	$\cdots$	X_{2k}
$\vdots$	$\vdots$	$\vdots$	$\vdots$		$\vdots$		$\vdots$
i	Y_i	X_{i1}	X_{i2}	$\cdots$	X_{ij}	$\cdots$	X_{ik}
$\vdots$	$\vdots$	$\vdots$	$\vdots$		$\vdots$		$\vdots$
n	Y_n	X_{n1}	X_{n2}	$\cdots$	X_{nj}	$\cdots$	X_{nk}

$$E(X(V_1)) = 5,$$
$$E(X(V_2)) = 8,$$
$$E(X) = 0.2 * E(X(V_1)) + 0.8 * E(X(V_2)) = 7.4$$

Definition 3.6 Let X be a fuzzy random variable defined a probability space with expected value e, the variance of x is defined as $Var[X] = E[(X - e)^2]$, If the variables are symmetrical triangular fuzzy number, we can know the following:

$$Var[X] = \frac{(a^r - a^l)^2}{24} \tag{3.6}$$

4.3.3 Building the Goal Model by Fuzzy Random Regression

$$Y_i = \sum_{j=1}^{k} \tilde{A}_j X_{ij} \tag{3.7}$$

means Y and X are fuzzy numbers with probability, which is so-called fuzzy random number. We denote fuzzy linear regression model as same in 3.8 as follows,

$$Y_i = \sum_{j=1}^{k} \tilde{A}_j X_{ij} \tag{3.8}$$

Let us denote y_i the output data, Y_i the estimated value, X input data, we can need to from the relationship among data to get the estimated value Y_i

$$Y_i = \sum_{j=1}^{k} \tilde{A}_j X_{ij} \underset{FR}{\supset} y_i \tag{3.9}$$

$\underset{FR}{\supset}$ is a fuzzy random inclusion relation, so we can change the equation to

Table 4.2 Input-output data with confidence interval

Sample	Output	Input		
	$I[e_Y, \sigma_Y]$	$I[e_{X_1}.\sigma_{X_1}]$	$\cdots$	$I[e_{X_k}\sigma_{X_k}]$
1	$I[e_{Y_1}, \sigma_{Y_1}]$	$I[e_{X_{11}}\sigma_{X_{11}}]$	$\cdots$	$I[e_{X_{1k}}\sigma_{X_{1k}}]$
$\vdots$	$\vdots$	$\vdots$		$I[e_{X_{1k}}\sigma_{X_{1k}}]$
i	$I[e_{Y_i}, \sigma_{Y_i}]$	$I[e_{X_{i1}}\sigma_{X_{i1}}]$	$\cdots$	$I[e_{X_{ik}}\sigma_{X_{ik}}]$
$\vdots$	$\vdots$	$\vdots$		$I[e_{X_{1k}}\sigma_{X_{1k}}]$
n	$I[e_{Y_n}, \sigma_{Y_n}]$	$I[e_{X_{n1}}\sigma_{X_{n1}}]$	$\cdots$	$I[e_{X_{nk}}\sigma_{X_{nk}}]$

$$\min_{\tilde{A}_k; k=1\cdots k} \tag{3.10}$$

This relationship is critical, so we should change that to watada and wang' s method which is expected value regression model [5],

$$\begin{gathered}\min \\ \min \tilde{A}_j;\ j = 1 \cdots k\end{gathered} \tag{3.11}$$

means the fuzzy inclusion relation realized at level h, then we employ confidence-interval based inclusion, which combine the expectation and variance of fuzzy random variable and fuzzy inclusion relation satisfied at level h, so the Table 4.1 can be changed into Table 4.2.

We define the confidence interval which is induce by expectation and variance of a fuzzy random variable. When we consider the one sigma confidence interval of each fuzzy random variable, we can express it as one-sigma interval [2],

$$I[e_x, \sigma_x] \simeq [E(x) - \sqrt{Var}, E(x) + \sqrt{Var}] \tag{3.12}$$

So, we can change the fuzzy random regression model to,

$$\begin{aligned}
\min_{\tilde{A}_j; j=1\cdots k} \quad & J(\tilde{A}) = \sum_{j=1}^{k}(\tilde{A}_j^r - \tilde{A}_j^l) \\
\text{subject to} \quad & \tilde{A}_j^r \geq \tilde{A}_l^; \\
& Y_i = \sum_{j=1}^{k} \tilde{A}_j I[e_x, \sigma_x] \underset{h}{\supset} I[e_y, \sigma_y] \\
& ; i = 1, 2, \cdots, n \\
& ; j = 1, 2, \cdots, k
\end{aligned} \tag{3.13}$$

In this model, we can get $\tilde{A}_k$ by LINGO, and build the goal equations.

4.4 Formulate Constraints Equations

4.4.1 Definition of Constraints

4.4.1.1 Linear- Fuzzy Constrain Equations

In a program, some constraints are not determined and can be changed in a range or it will be one result with a probability and another with other probability. In this case, we should consider the uncertainties and changes when we build the model. some fuzzy language should be changed into mathematic symbol and number, such as "about", "about equal to", we use fuzzy symbol to express, such as the constraints' coefficients are fuzzy random variable, we change that to expected value. The constraints model as,

$$\text{subject to} \quad S(\mathbf{X})_1 \sum_{j=1}^{k} \tilde{A}_j x_{1j} \prec g_l; \quad S(\mathbf{X})_2 \sum_{j=1}^{k} \tilde{A}_j x_{2j} \approx g_2; \tag{4.1}$$

4.4.1.2 Solving Fuzzy Constrain Equations

– 1 Weighting Method
Weight is a concept relative to a specific goal, indicator, or standard. Weights are used to assess the level of importance by comparing them with others. Weight is, in the process of evaluation, giving different factors with different weights to indicate different importance. For example, a teacher evaluates the final score from three aspects, attendance, intermediate exam score, final exam score. But the in linear-linear criterion programming, different purposes should be connected using weights. If the importance is the same, the weights are the same, if not, give them different weights.

Give the three linear equations weights to connect them into an equation.

$$\max Y_i = 30\% * Y_1 + 30\% * Y_2 + 30\% * Y_3 \tag{4.2}$$

– 2 Change the fuzzy coefficients to determined number a_k is the fuzzy random data $\begin{pmatrix} \tilde{a}_k^l & \tilde{a}_k^r \\ p^l & p^r \end{pmatrix}$, we calculate the expectation use the Definitions 3.3 and 3.4. Example, a fuzzy random number like $\tilde{a}_1 = (0.3/2, 0.7/7)$, from the distribution, the fuzzy random variable takes $(0.3/(0.3/(4, 0, 8)), 0.7/(0.5/(9, 5, 13)))$, then we can calculate the expected value $\tilde{a}_1$ is as follows:

$$E(\tilde{a}_1) = 0.3 * \frac{0 + 2 * 4 + 8}{4} + 0.7 * \frac{5 + 2 * 9 + 13}{4} = 7.5$$

- 3 Alter the constraints goals $\tilde{g}_i$
  To obtain random triangular data, manager experience is required. For example, we give the price around 800. This 800 always changes in the range according to the total supply-demand. Based on the data and reference experience of 10 years, we can find out that 3 years is 700 to 800, 7 years is 800 to 900. So, we show that about 800 is 700 with a probability of 0.3 and 900 with a probability of 0.7. After knowing how to obtain blurred random data, we can calculate the expected value to obtain a coordination solution to determine the upper and lower limits.
- 4. Transform of Fuzzy Symbols
  Scalable index method is setting a scalable number l_i to enlarge the range of uncertain number to an acceptable range. e.g. approximately equal to 10, set the scalable index is 1, the range was changed to 9 or 11, any number in the range [29, 38] is acceptable. The scalable index is bigger, the range is wider, the accuracy is lower. Translate to $\prec, \succ \approx$ to $<, >=$. $\leq, \geq$

$$\begin{aligned} \max \tilde{Y}_i \quad & \tilde{Y}_i = \sum_{k=1}^{k} \tilde{A}_k X_{ik} \\ \text{subject to} \quad & S(X)_1 : \sum_{j=1}^{k} \tilde{a}_j X_{1j} \leq E(g_1) + l_1; \\ & S(X)_2 : \sum_{j-1}^{k} \tilde{a}_j X_{2j} = E(\tilde{g}_2) + l_2; \end{aligned} \tag{4.3}$$

Calculate Eqns (4.2) and (4.3), the boundary can be obtained.

4.4.2 Solution

4.4.2.1 Max-Min Method

Here, give two kinds of linear GP models $f(x)$ are the goals, $s(x)$ are the constraint equations, b_i and d_i are interval values.

- 1. Maximize goals and constraints equation smaller than a fuzzy number.

$$
\begin{aligned}
\max \quad & f_i(\mathbf{x}) = \sum_{i=1}^{n}\sum_{j=1}^{k} \tilde{A}_j X_{ij} \\
\text{subject to} \quad & S(\mathbf{X}) = \begin{cases} \tilde{a}_{ij} x_{ij} \prec [b_i, d_i] \\ \mathbf{X} \geq \mathbf{0} \end{cases}
\end{aligned}
\tag{4.4}
$$

$M_G(X)$ is goal equations $f(x)$'s membership function,

$$
M_G(\mathbf{x}) = \frac{\sum_{i=1}^{n}\sum_{j=1}^{k} \tilde{A}_{ij} x_{ij} - G_0}{d_0} \tag{4.5}
$$

$$
d_0 = G_1 - G_2
$$

G_1 is the maximum value and G_0 is the minimum value. $M_G(X)$ is constraints equations $s(x)$'s membership function,

$$
M_G(\mathbf{x}) = \begin{cases} 1 & ; \sum_{i=1}^{n}\sum_{j=1}^{k} \tilde{a}_j x_{ij} \leq b_i \\ 1 - \dfrac{\sum_{i=1}^{n}\sum_{j=1}^{k} \tilde{a}_i x_{ij} - b_i}{d_i} & ; b_i < \sum_{i=1}^{n}\sum_{j=1}^{k} \tilde{a}_j x_{i,j} \leq b_i + d_i \\ 0 & ; \sum_{i=1}^{n}\sum_{j=1}^{k} \tilde{a}_j x_{i,j} \geq b_i + d_i \end{cases} \tag{4.6}
$$

Introduce the auxiliary variable λ to connect $M_G(X)$ and $M_C(X)$.

$$
\begin{aligned}
\max \quad & \lambda \\
\text{subject to} \quad & \begin{cases} 1 - \dfrac{\sum_{i=1}^{n}\sum_{j=1}^{k} \tilde{a}_j x_{ij} - b_i}{d_i} \geq \lambda \\ \dfrac{\sum_{i=1}^{n}\sum_{j=1}^{k} \tilde{A}_j X_{ij} - G_0}{d_0} \geq \lambda \end{cases}
\end{aligned}
\tag{4.7}
$$

– 2. Minimize goals and constraints equation is bigger than a fuzzy number.

$$\begin{aligned} &\min \quad f_i(\mathbf{x}) = \sum_{i=1}^{n}\sum_{j=1}^{k}\tilde{A}_j X_{ij} \\ &\text{subject to} \ . \ s(\mathbf{x}) = \begin{cases} \tilde{a}_j x_{ij} \succ [b_i, d_i] \\ \mathbf{x} \geq \mathbf{0} \end{cases} \end{aligned} \tag{4.8}$$

$$M_G(\mathbf{x}) = \frac{G_0 - \sum_{i=1}^{n}\sum_{j=1}^{k}\tilde{A}_j X_{ij}}{d_0}, \qquad d_0 = G_1 - G_0 \tag{4.9}$$

G_1 is the maximum value and G_0 is the minimum value. $M_C(X)$ is constraints equations $S(x)$'s membership function,

$$M_G(\mathbf{x}) = \begin{cases} 1 & ; \sum_{i=1}^{n}\sum_{j=1}^{k}\tilde{a}_j x_{ij} \geq b_i \\ 1 + \dfrac{\sum_{i=1}^{n}\sum_{j=1}^{k}\tilde{a}_j x_{ij} - b_i}{d_i} & ; d_i - d_i \geq \sum_{i=1}^{n}\sum_{j=1}^{k}\tilde{a}_j x_{ij} x_{ij} \\ 0 & ; \sum_{i=1}^{n}\sum_{j=1}^{k}\tilde{a}_j x_{ij} x_{ij} < d_i - d_i \end{cases} \tag{4.10}$$

Introduce the auxiliary variable λ to connect $M_G(X)$ and $M_C(X)$.

$$\begin{aligned} &\max \quad \lambda \\ &\text{subject to} \ . \begin{cases} 1 + \sum_{i=1}^{n}\sum_{j=1}^{k} a_i x_{ij} \geq \lambda \\ \dfrac{G_0 - \sum_{i=1}^{n}\sum_{j=1}^{k} A_i X_{ij}}{d_1} \geq \lambda \end{cases} \end{aligned} \tag{4.11}$$

– 3. Maximize and minimize goals and constraint equations have $\prec$, $\succ$ at the same time.

$$
\text{Evaluation} \begin{cases} \max f_i(\mathbf{x}) = \sum_{i=1}^{n}\sum_{j=1}^{k} \tilde{A}_j X_{ij} \\ \min f_i(\mathbf{x}) = \sum_{i=1}^{n}\sum_{j=1}^{k} \tilde{A}_j X_{ij} \end{cases}
$$
$$
\text{subject to} \;.\; S(\mathbf{x}) = \begin{cases} \tilde{a}_j x_{ij} \succ [b_i, d_i] \\ \tilde{a}_j x_{ij} \prec [b_i, d_i] \\ \mathbf{x} \geq \mathbf{0} \end{cases} \tag{4.12}
$$

Introduce the auxiliary variable λ to connect $M_G(X)$ and $M_C(X)$.

$$
\begin{aligned} &\max \quad \lambda \\ &\text{subject to} \begin{cases} 1 - \dfrac{\sum_{i=1}^{n}\sum_{j=1}^{k} \tilde{a}_j x_{ij} - b_i}{d_i} \geq \lambda \\ 1 + \dfrac{\sum_{i=1}^{n}\sum_{j=1}^{k} \tilde{a}_j x_{ij} - b_i}{d_i} \geq \lambda \\ \dfrac{\sum_{i=1}^{n}\sum_{j=1}^{k} \tilde{A}_j X_{ij} - G_0}{d_0} \geq \lambda \\ G_0 - \dfrac{\sum_{i=1}^{n}\sum_{j=1}^{k} \tilde{A}_j X_{ij}}{d_0} \geq \lambda \end{cases} \end{aligned} \tag{4.13}
$$

4.4.2.2 Constructing Membership Function

- 1. Goals' membership function
  From Eqs. (4.12) and (4.13), we get the boundary of solution, so, the difference is $d_0 = f_1 - f_0$,

$$
\begin{aligned} &\max \quad \lambda \\ &\text{subject to} \quad \mu(Y_i) = \begin{cases} 1 & : Y_i \leq f_0 + d_0 \\ \dfrac{1}{d_0}[Y_i - f_0] \; ; & f_0 + d_0 < Y_i \leq f_0 : d_0 \\ 0 & ; Y_i > f_0 \end{cases} \end{aligned} \tag{4.14}
$$

– 2. Constraints' membership function

$$\mu(Y_i) = \begin{cases} 1 & : S(X) \leq E(g_i) \\ 1 - \frac{1}{l_i}[S(X) - E(g_i)] & ; E(g_i) < S(X) \leq E(g_i) + l_i \\ 0 & ; S(X) > E(g_i) + l_i \end{cases} \tag{4.15}$$

is scalable index, $s(x)$ are constraints'equations, $E(g_i)$ is expected value of g_i.
– 3. Link goals and constraints'membership function
we introduce an auxiliary variable λ to get a new equivalent model by referring the max-min method

$$\begin{array}{ll} \max & \lambda \\ \text{subject to} & \begin{cases} 1 - \frac{1}{l_i}[S(X) - E(g_i)] \geq \lambda \\ \frac{1}{d_0}[Y_i - f_0] \geq \lambda \end{cases} \end{array} \tag{4.16}$$

4.5 Application

4.5.1 Case Introduction

A company wants to build a new factory, but it has no its own clear historical data and just has some similar data comes from the other factories as reference. In this case, how to build reasonable working time plan, production plan and costs plan under some manager's needs? The other factories' data about product amounts, working time, price, and cost from the 1st year to the 11th year are in Table 4.3, production capacity and production capacity constraints in Table 4.4. By regression model, we need to find some law to formulate the working time goal, price purpose and cost target of the 12th year. Then building the constraint's equations and transform the fuzzy coefficients, fuzzy symbols and constraints' fuzzy goals to determined numbers. Next, get the membership function by max-min method and scalable index method. Finally, solve the equations by LINGO.

Given reference data in Tables 4.3 and 4.4, manager want to get some targets:

– 1 maximize profits as much as possible over 1800 m
– 2 produce x_1 at least 150
– 3 produce x_2 at least 100
– 4 make working time of 1.2 close to max limit time.

Table 4.3 Original data of product amounts, working time, price, and cost

Year	Product amount		Working time Process 1 and 2	Price A and B Per dozen	Cost A and B Per dozen
	A	B			
	$\tilde{X}_1$	$\tilde{X}_2$	$\tilde{Y}_1$	$\tilde{Y}_2$	$\tilde{Y}_3$
1st year	0.5 0.5 (60, 70)	0.5 0.5 (50, 55)	0.1 0.9 (1500, 1550)	0.1 0.9 (80000, 85000)	0.5 0.5 (20000, 25000)
2nd year	0.4 0.6 (70, 75)	0.5 0.5 (60, 75)	0.1 0.9 (1600, 1750)	0.1 0.9 (85000, 90000)	0.5 0.5 (20000, 28000)
3rd year	0.3 0.7 (80, 95)	0.6 0.4 (85, 95)	0.6 0.4 (1700, 1750)	0.2 0.8 (95000, 98000)	0.6 0.4 (25000, 28000)
4th year	0.3 0.7 (60, 70)	0.4 0.6 (85, 95)	0.6 0.4 (1800, 1850)	0.2 08 (95000, 110000)	0.6 0.4 (25000, 30000)
5th year	0.5 0.5 (90, 115)	0.2 0.8 (75, 90)	0.3 0.7 (1800, 1950)	0.3 07 (110000, 115000)	0.8 0.2 (25500, 30000)
6th year	0.7 0.3 (130, 145)	0.2 0.8 (95, 110)	0.3 0.7 (1900, 1950)	0.1 09 (1100000, 120000)	0.1 0.9 (25500, 30000)
7th year	0.7 0.3 (150, 165)	0.1 0.9 (120, 130)	0.8 0.2 (2000, 2150)	0.1 09 (120000, 125000)	0.2 0.8 (30500, 31000)
8th year	0.6 0.4 (135, 160)	0.1 0.9 (140, 160)	0.8 0.2 (2000, 2250)	0.1 09 (1200, 130000)	0.5 0.5 (355000, 38000)
9th year	0.5 0.5 (160, 180)	0.3 0.7 (150, 170)	0.5 0.5 (2100, 2250)	0.5 05 (125000, 135000)	0.5 0.5 (36500, 40000)
10the year	0.5 0.5 (185, 215)	0.5 0.5 (160, 180)	0.1 0.9 (22000, 22500)	0.7 03 (125000, 145000)	0.4 0.6 (405000, 430000)
11th year	0.5 0.5 (215, 235)	0.5 0.5 (180, 190)	0.1 0.9 (2300, 2350)	0.2 08 (135000, 145000)	0.2 0.8 (42500, 45000)

Table 4.4 Production constraints

	Product	
Product amount	X_1(about)	X_2(about)
Price (per dozen)	90,000	60,000
Cost (per dozen)	30,000	20,000
Process 1 time (per dozen)	2	3
Process 2 time (per dozen)	6	3

4.5.2 Solution

4.5.2.1 Solving Steps

(1) using the regression model to get the coefficients by considering series fuzzy random data to build the goal equations.
(2) construct constraints equations including the fuzzy random data, fuzzy symbols, fuzzy goal values.
(3) transform the fuzzy random data by getting expectation, change the fuzzy symbols like $\prec$, $\succ\approx$ to $<, >, =$. by adding tolerance to goal values.
(4) *o* gets the boundary numbers, give goal equations same weights when the importance of goals are the same. If not, the weights are different.
(5) formulate the membership functions by maximum and minimum method (Fig. 4.3).

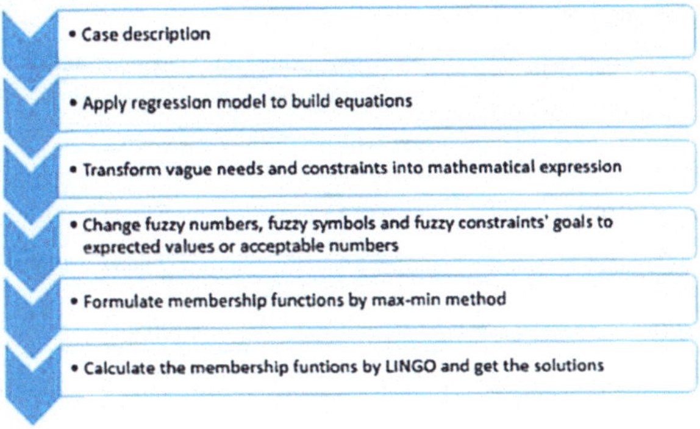

Fig. 4.3 Flowchart of the process

4.5.3 Construction of Goal Equations

We need find the law or relationship among the data to construct goals by using regression model. In a linear regression model, the relationship is coefficients, so, The processes of building goals are how to get the coefficients. Here, the linear regression models as follow,

$$\begin{aligned} \tilde{Y}_1 &= \tilde{A}_1 X_{11} + \tilde{A}_2 X_{12} \\ \tilde{Y}_2 &= \tilde{A}_1 X_{21} + \tilde{A}_2 X_{2,2} \\ \tilde{Y}_3 &= \tilde{A}_1 X_{31} + \tilde{A}_2 X_{32} \end{aligned} \tag{5.1}$$

From Definitions 3.3 and 3.4, we calculate the expectation of Tables 4.3 and 4.4. Example

A fuzzy random number like $\tilde{a}_1 = (0.3/2.0.7/7)$, from the distribution, the fuzzy random variable takes $(0.3/(4, 0, 8).0.7(9, 5, 13))$, then we can calculate the expected value $\tilde{a}_1$ is $E(\tilde{a}_1) = 0.3 * \dfrac{0 + 2 * 4 + 8}{4} + 0.7 * \dfrac{5 + 2 * 9 + 13}{4} = 7.5$

According to the functions (3.12) and (3.13), variance value of Table 4.5 can be calculated (Tables 4.6 and 4.7).

From function (3.12) and Table 4.6 can get the interval of expected values. Using Eqn (3.13) and LINGO, we can get the coefficients, calculation processes of Y_1 are as follow and Y_2 and Y_3 are the same (Table 4.8).

$$\begin{aligned} \min \quad & J(\mathbf{A}) = A_1^r - A_1^l + A_{i2}^r - A_{i2}^l \\ \text{subject to} \quad & A_1^l * I_{X_1}^l + A_2^l * I + X_2{}^l \le I^l_{\tilde{Y}_1} \\ & A_1^r * I_{X_1}^r + A_2^r * I + X_2{}^r \ge I^r_{\tilde{Y}_1} \end{aligned}$$

Table 4.5 Expected values

Year	Product amount		Working time Process 1 and 2	Price A and B Per dozen	Cost A and B Per dozen
	A	B			
	$E(X_1)10$ $E(X_2)10$		$E(\tilde{Y}_1)100$	$E(\tilde{Y}_2)$	$E(\tilde{Y}_3)$
1st year	75.0	62.5	1625	855	235
2nd year	83.0	77.5	1590	905	250
3rd year	100.5	79.0	1820	984	272
4th year	83.0	101	1920	1080	280
5th year	112.5	97.0	2005	1150	274
6th year	144.5	117	2035	1200	301
7th year	164.5	139	2130	1255	319
8the year	155.0	168	2150	1260	377.5
9th year	185.0	174	2275	1310	392.5
10th year	210.0	175	2345	1320	430
11th year	235.0	195	2445	1440	455

Table 4.6 Variance values Year

Year	Product amount				Working time Process 1 and 2	
	A		B			
	$V(X_1)$	$\sqrt{V}$	$V(X_2)$	$\sqrt{V}$	$V(\tilde{Y}_1)$	$\sqrt{V}$
1st year	4.17	2.04	1.04	1.02	104.17	10.21
2nd year	1.04	1.02	9.38	3.06	937.5	30.62
3rd year	9.38	3.06	4.17	2.04	104.17	10.21
4th year	4.17	2.04	4.17	2.04	104.17	10.21
5th year	26.04	5.10	9.38	3.06	937.5	30.62
6th year	9.38	3.06	9.38	3.06	104.17	10.21
7th year	9.38	3.06	4.17	2.04	937.5	30.62
8th year	9.38	3.06	16.67	4.08	2604.17	51.03
9th year	16.6	7 4.08	16.67	4.08	937.5	30.62
10th year	37.5	6.12	16.67	4.08	104.17	10.21
11th year	16.67	4.08	4.17	2.04	104.17	10.21

Table 4.7 Variance values Year

Year	Product amount				Price A and B Per dozen		Cost A and B Per dozen	
	A		B					
	$V(X_1)$	$\sqrt{V}$	$V(X_2)$	$\sqrt{V}$	$V(\tilde{Y}_2)$	$\sqrt{V}$	$V(Y_3)$	$\sqrt{V}$
1st year	4.17	2.04	1.04	1.02	1041666.7	1020.62	1041666.7	1020.62
2nd year	1.04	1.02	9.38	3.06	1041666.7	1020.62	2666666.7	1632.99
3rd year	9.38	3.06	4.17	2.04	375000	612.37	375000	612.37
4th year	4.17	2.04	4.17	2.04	9375000	3061.86	1041666.7	1020.62
5th year	26.04	5.10	9.38	3.06	1041666.7	1020.62	843750	918.56
6th year	9.38	3.06	9.38	3.06	4166666.7	2041.24	843750	918.56
7th year	9.38	3.06	4.17	2.04	1041666.7	1020.62	10416.7	102.06
8th year	9.38	3.06	16.67	4.08	4166666.7	2041.24	260416.7	510.31
9th year	16.67	4.08	16.67	4.08	4166666.7	2041.24	510416.7	714.43
10th year	37.5	6.12	16.67	4.08	16666666.7	4082.48	260416.7	510.31
11th year	16.67	4.08	4.17	2.04	4166666.7	2041.24	260416.7	510.31

Table 4.8 Interval of expectation year

Year	Product amount		Working time Process 1 and 2	Price A and B (Per dozen)	Cost A and B (Per dozen)
	A	B			
	$[A^L, A^R]$	$[B^L, B^R]$	$[Y_1^L, Y_1^R]$	$[Y_2^L, Y_2^R]$	$[Y_3^L, Y_3^R]$
1st year	72.96	61.48	1614.79	84479.38	22479.38
	77.04	63.52	1635.21	86520.62	24520.62
2nd year	81.98	74.44	1559.38	89479.38	23367.01
	84.02	80.56	1620.62	91520.62	26632.99
3rd year	97.44	76.96	1809.79	97787.63	26587.63
	103.56	81.04	1830.21	99012.37	27812.37
4th year	80.96	98.96	1909.79	104938	26979.38
	85.04	103.04	1930.21	111061.9	29020.62
5th year	107.4	93.94	1974.38	113979.4	26481.44
	117.6	100.06	2035.62	116020.6	28318.56
6th year	141.44	113.94	2024.79	117958.8	29181.44
	147.56	120.06	2045.21	122041.2	31018.56
7th year	161.44	136.96	2099.38	124479.4	31797.94
	167.56	141.04	2160.62	126520.6	32002.06
8th year	151.94	163.92	2098.97	123958.8	37239.69
	158.06	172.08	2201.03	128041.2	38260.31
9th year	180.92	169.92	2244.38	128958.8	38535.57
	189.08	178.08	2305.62	133041.2	39964.43
10th year	203.88	170.92	2334.79	127917.5	42489.69
	216.12	179.08	2355.21	136082.5	43510.31
11th year	230.92	192.96	2434.79	141958.8	44989.69
	239.08	197.04	2455.21	146041.2	46010.31

Table 4.9 Coefficients' values

A_{11}		Average value	A_{12}		Average value
A_{11}^L	A_{11}^R	$\frac{(A_{11}^L + A_{11}^R)}{2}$	A_{12}^L	A_{12}^R	$\frac{(A_{12}^L + A_{12}^R)}{2}$
0.1967247	18.09021	9.05	12.61811	3.802590	8.21
A_{21}			A_{22}		
A_{21}^L	A_{21}^R	$\frac{(A_{21}^L + A_{21}^R)}{2}$	A_{22}^L	A_{22}^R	$\frac{(A_{22}^L + A_{22}^R)}{2}$
0.1967247	733.4759	366.74	735.6903	472.5069	604.1
A_{31}		Average value	A_{32}		Average value
A_{31}^L	A_{31}^R	$\frac{(A_{31}^L + A_{31}^R)}{2}$	A_{32}^L	A_{32}^R	$\frac{(A_{32}^L + A_{32}^R)}{2}$
0.647363	269.3576	134.68	59.34057	226.7865	143.06

$$
\left\{
\begin{array}{l}
72.96 * \overset{\sim l}{A_1} + 61.49 * \overset{\sim l}{A_2} \leq 1614.79 \\
81.98 * \overset{\sim l}{A_1} + 74.44 * \overset{\sim l}{A_2} \leq 1559.38 \\
97..44 * \overset{\sim l}{A_1} + 76.96 * \overset{\sim l}{A_2} \leq 1809.79 \\
80.96 * \overset{\sim l}{A_1} + 98.96 * \overset{\sim l}{A_2} \leq 1909.79 \\
107.4 * \overset{\sim l}{A_1} + 93.94 * \overset{\sim l}{A_2} \leq 1974.38 \\
141.44 * \overset{\sim l}{A_1} + 113.94 * \overset{\sim l}{A_2} \leq 2024.79 \\
161.44 * A_{11}^{l} + 136.96 * A_{12}^{l} \leq 2099.38 \\
151.94 * \overset{\sim l}{A_1} + 163.92 * \overset{\sim l}{A_2} \leq 2098.97 \\
180.92 * \overset{\sim l}{A_1} + 169.92 * \overset{\sim l}{A_2} \leq 2244.38 \\
203.88 * \overset{\sim l}{A_1} + 170.92 * \overset{\sim l}{A_2} \leq 2334.79 \\
230.92 * \overset{\sim l}{A_1} + 192.96 * \overset{\sim l}{A_2} \leq 2434.79 \\
\\
77.04 * \overset{\sim r}{A_1} + 63.52 * \overset{\sim r}{A_2} \geq 1635.21 \\
84.02 * \overset{\sim r}{A_1} + 80.56 * \overset{\sim r}{A_2} \geq 162.62 \\
103.56 * \overset{\sim r}{A_1} + 81.04 * \overset{\sim r}{A_2} \geq 1830.21 \\
865.04 * \overset{\sim r}{A_1} + 103.04 * \overset{\sim r}{A_2} \geq 1930.21 \\
117.6 * \overset{\sim r}{A_1} + 100.06 * \overset{\sim r}{A_2} \geq 2035.62 \\
147.56 * \overset{\sim r}{A_1} + 120.06 * \overset{\sim r}{A_2} \geq 2045.21 \\
167.56 * \overset{\sim r}{A_1} + 141.04 * \overset{\sim r}{A_2} \geq 2160.62 \\
158.06 * \overset{\sim r}{A_1} + 172.08 * \overset{\sim r}{A_2} \geq 2201.0 \\
189.09 * \overset{\sim r}{A_1} + 178.08 * \overset{\sim r}{A_2} \geq 2305.62 \\
216.12 * \overset{\sim r}{A_1} + 179.08 * \overset{\sim r}{A_2} \geq 2335.21 \\
239.08 * \overset{\sim r}{A_1} + 197.04 * \overset{\sim r}{A_2} \geq 2455.21
\end{array}
\right.
$$

From above, the coefficients are obtained. Here, we choose the average number as the final value to construct the three goals.

$$
\begin{aligned}
Y_1 &= 9.05X_1 + 8.21X_2 \\
Y_2 &= 366.74X_1 + 604.1X_2 \\
Y_3 &= 134.68X_1 + 143.06X_2
\end{aligned}
\tag{5.2}
$$

Table 4.10 Production constraints

	Product	
	A	*B*
Product amount	X_1	X_2
Price (per dozen)	[85000, 95000; 0.3, 0.7]	[58000, 62000; 0.2, 0.8]
Cost (per dozen)	[26000, 36000; 0.5, 0.5]	[18000, 22000; 0.7, 0.3]
Process 1 time(per dozen)	[1.8, 2.2; 0.4, 0.6]	[2.7, 3.3; 0.6, 0.4]
Process 2 time(per dozen)	[5.5, 6.5; 0.3, 0.7]	[2.8, 3.2; 0.9, 0.1]

Table 4.11 Expectation of production constraints

	Product	
	A	*B*
Product amount	X_1	X_2
Price (per dozen)	92000	961200
Cost (per dozen)	31000	19200
Process 1 time(per dozen)	2.04	2.94
Process 2 time(per dozen)	6.2	2.84

4.5.4 Formulation of Constraints' Equations

Table 4.10 comes from Table 4.9 based on the references and experience, [85000, 95000; 0.3, 0.7] means that x_1's price is in the interval of 85000 with probability 0.3 to 95000 with probability 0.7. Here, we also can get the expected values by using the Definitions 3.3 and 3.4. Given the reference data in Tables 4.10 and 4.11, manager wants his requirements can be satisfied.

- 1 maximize profits as much as possible over 1800m.
- 2 product x_1 at least 150.
- 3 product x_2 at least 100.
- 4 make working time of products 1 and 2 close to max limit time.

Manger's needs are also fuzzy number and need to be transformed to fuzzy random data based on experience.

- 1 maximize profits are [17000000, 19000000; 0.5, 0.5]
- 2 product x_1 is [140, 160; 0.3, 0.7]
- 3 product x_2 is [80,120; 0.5, 0.5]
- 4 make working time of products 1 and .2 as [800,1000; 0.5,0.5] and [1600,2000; 0.4,0.6]

Under the production constraints and manager's requirements, five constraints equations are built, numbers and symbols are fuzzy expression.

$$S_i(\mathbf{x}) \begin{cases} g_1: \text{about } 60000 * X_1 + \text{about } 40000 * X_2 \succ \text{about } 18000000 \\ g_2: X_1 \succ \text{about } 150 \\ g_3: X_2 \succ \text{about } 100 \\ g_4: \text{about } 2 * X_1 + \text{about } 3 * X_2 \prec \text{about } 900 \\ g_5: \text{about } 6 * X_1 + \text{about } 3 * X_2 \prec \text{about } 1800 \end{cases} \tag{5.3}$$

Next, numbers are changed to expected values and these equations will be used to got the lower limits.

$$S_i(\mathbf{x}) \begin{cases} g_1: 61000 * X_1 + 42000 * X_2 \succ 18000000 \\ g_2: X_1 \succ 154 \\ g_3: X_2 \succ 100 \\ g_4: 2.04 * X_1 + 2.94 * X_2 \prec 900 \\ g_5: 6.2 * X_1 + 2.84 * X_2 \prec 1840 \end{cases} \tag{5.4}$$

Then, the scalable indexes are introduced to solve the fuzzy symbols (undetermined relation). Set a scalable number li to enlarge the range of uncertain number to an acceptable range. (e.g. approximately equal to 10, set the scalable index is 1, the range was changed to 9 or 11, any number in the range [29], [38] is acceptable). We confer scalable index 1000000 to g_1, 20 to g_2, 20 to g_3, 100 to g_4, 200 to g_5, so, the new constraints equations are,

$$S_i(\mathbf{x}) \begin{cases} g_1: 61000 * X_1 + 42000 * X_2 \geq 19000000 \\ g_2: X_1 \geq 174 \\ g_3: X_2 \geq 120 \\ g_4: 2.04 * X_1 + 2.94 * X_2 \leq 1000 \\ g_5: 6.2 * X_1 + 2.84 * X_2 \leq 2040 \end{cases} \tag{5.5}$$

Here, these equations are to be taken as upper limits.

4.5.5 Solution of the Multi-criteria Linear Goal Models

Three goal models are defined by the following:

$$\begin{aligned} Y_1 &= \min_{X_1,X2} 9.05 * X_1 + 8.21 * X_2; \\ Y_2 &= \max_{X_1,X2} 366.74 * X_1 + 604.1 * X_2; \\ Y_3 &= \min_{X_1,X2} 134.68 * X_1 + 143.06 * X_2; \end{aligned} \tag{5.6}$$

In order to satisfy the tree goals simultaneously, we confer everyone the same weights 0.333 and combine them into one equation,

$$\max_{X_1,X2} 0.333 * (9.05 * X_1 + 8.21 * X_2) + 0.333 * (366.74 * X_1 + 604.1 * X_2)$$
$$= -0.333 * (134.68 * X_1 + 143.06 * X_2); \tag{5.7}$$

then, we can get upper and lower limits after connecting the goal equations and constraint equations.

$$\begin{aligned} \max \quad & \{-0.333 * (9.05 * X_1 + 8.21 * X_2) + 0.333 * (366.74 * X_1 + 604.1 * X_2) \\ & -0.333 * (134.68 * X_1 + 143.06 * X_2)\}; \\ \text{Subject to} \quad & 61000 * x_1 + 42000 * x_{\geq} 18000000; \\ & x_1 \geq 154; \\ & x2 \geq 100; \\ & 2.04 * x1 + 2.94 * x2 <= 900; \\ & 6.2 * x1 + 2.84 * x2 <= 1840; \end{aligned} \tag{5.8}$$

$$Y_1 = \min_{X_1,X_2} 9.05 * X_1 + 8.21 * X_2;$$

$$Y_2 = \max_{X_1,X_2} 366.74 * X_1 + 604.1 * X_2;$$

$$Y_3 = \min_{X_1,X_2} 134.68 * X_1 + 143.06 * X_2;$$

$$\begin{aligned} \max_{X_1,X_2} \quad & \{-0.333 * (9.05 * X_1 + 8.21 * X_2) + 0.333 * (366.74 * X_1 + 604.1 * X_2) \\ & -0.333 * (134.68 * X_1 + 143.06 * X_2); \} \\ \text{Subject to} \quad & 61000 * x_1 + 42000 * x_2 \geq 19000000; \\ & x_1 \geq 174; \\ & x_2 \geq 120; \\ & 2.04 * x_1 + 2.94 * x_2 \leq 1000; \\ & 6.2 * x_1 + 2.84 * x_2 \leq 2040; \end{aligned} \tag{5.9}$$

By calculation, the upper and lower limits are

$$\begin{aligned} (X_1 &= 161.4350, X_2 = 194.1063) \\ (X_1 &= 174.0000, X_2 = 219.4014) \\ Y_1 &= [3054.68, 3375.97] \\ Y_2 &= [176468.36, 196352.3] \\ Y_3 &= [49512.12, 54821.68] \end{aligned}$$

The fuzziness are $d_1 = 321.29$, $d_2 = 19883.94$, $d_3 = 5309.56$. By using max-min method and Equations (4.4) and (4.5),

$$\mu_{G_i}(\mathbf{x}) = \begin{cases} \dfrac{1}{321.29}[3054.68 - f(Y_1)] \\ \\ \dfrac{1}{19883.94}[f(Y_2) - 176468.36] \\ \\ \dfrac{1}{5309.56}[49512.12 - f(Y_1)] \end{cases} \tag{5.10}$$

By combining the two membership function by introducing auxiliary variable λ to get a new equivalent model is defined as follows:.

$$\begin{aligned} \max \quad & \lambda \\ \text{subject to} \quad & \begin{cases} \dfrac{1}{321.29}[3054.68 - f(Y_1)] \geq \lambda \\ \\ \dfrac{1}{19883.94}[f(Y_2) - 176468.36] \geq \lambda \\ \\ \dfrac{1}{5309.56}[49512.12 - f(Y_1)] \geq \lambda \\ \\ 1 + \dfrac{1}{1000000}[s_1(x) - 18000000] \geq \lambda \\ \\ 1 + \dfrac{1}{20}[s_2(x) - 154] \geq \lambda \\ \\ 1 + \dfrac{1}{20}[s_3(x) - 100] \geq \lambda \\ \\ 1 + \dfrac{1}{100}[s_4(x) - 100] \geq \lambda \\ \\ 1 + \dfrac{1}{200}[s_5(x) - 100] \geq \lambda \end{cases} \end{aligned} \tag{5.11}$$

By calculation, $x_1 = 144.5988$, $x_2 = 221.7770$ and $\lambda = 0.5299402$ are obtained. Submit the values into the following equations:

$$\begin{aligned} &Y_1 = 9.05 * X_1 + 8.21 * X_2; \\ &Y_2 = 366.74 * X_1 + 604.1 * X_2; \\ &Y_3 = 134.68 * X_1 + 143.06 * X_2, \\ &Y_1 = 3129.44,\ Y_2 = 187007.9,\ Y_3 = 51202.6 \end{aligned}$$

The upper and lower limits are obtained as

$$Y1 = [3054.68, 3375.97],$$
$$Y2 = [176468.36, 196352.3],$$
$$Y3 = [49512.12, 54821.68].$$

These results are acceptable.

4.6 Conclusions

In our daily life, many tasks or programs exist and require us to solve and decide. We studied the FGP mainly and solved the following three types of problem from two sides. First, when we face a program and don't know how to get a suitable target, we could obtain reasonable linear regression model from the past data to formulate target problems. Second, even if the past data cannot clarify the present situation well which should be uncertain, we can solve and build the model considering such uncertainty.

Third, solve some undetermined problems, like the constrains, conditions are not determined or changing.

Two sides are goal and constraint, making a vague goal to be clear by building a linear model from old fuzzy data; letting all the constraints or conditions in a certain range. In this process, the expected value and scalable index will be used to get acceptable boundary. Utilizing max-min method to construct membership function to connect goals and constraints equations. The final purpose is to obtain a reasonable solution in the boundary. If we want to get an accurate solution, the scalable difference should be small. That is to say, the scalable index of errors is smaller, the accuracy is higher.

Many methods can solve this kind of problem, such as constraint method, multi-layer sequence method, efficacy coefficient method, evaluation function method, square weighted method, linear weighting method, all of these methods have the same problem of accuracy. Improving the accuracy and applying this into non-linear programming, finance marketing are future works.

References

1. Alvarez-Alvarez, A., Trivino, G., Cordon, O.: Human gait modeling using a genetic fuzzy finite state machine. IEEE Trans. Fuzzy Syst. **20**(2), 205-223
2. Arbaiy, N., Watada, J.: Approximation of goal constraint coefficients in fuzzy goal programming. In: 2010 Second International Conference on Computer Engineering and Applications (ICCEA), vol. 1, pp. 161–165 (2010)
3. Ct, A.R.N., Cooper, W.W.: Programming with Linear Fractional Functionals. Nay c Q. **9**, 181–186 (1962)
4. Nureize, A., Watada, J.: Building fuzzy random objective function for interval fuzzy goal programming. In: IEEE International Conference on Industrial Engineering and Engineering Management (IEEM), pp. 980–984 (2010)

5. Arbaiy, N., Watada, J.: Multi-level multi-objective decision problem through fuzzy random regression based objective function. In: IEEE International Conference on Fuzzy Systems (FUZZ), pp. 557–563 (2011)
6. Belarbi, K., Megri, F.A.: Stable model-based fuzzy predictive control based on fuzzy dynamic programming. IEEE Trans. Fuzzy Syst. **15**(4), 746–754 (2007)
7. Charnes, A., Cooper, W.W.: Programming with linear fractional functionals. Naval Res. Logistics Q. **9**, 181–186 (1962). https://doi.org/10.1002/nav.3800090303
8. Chang, C.-T.: Binary behavior of fuzzy programming with piecewise linear membership functions. IEEE Trans. Fuzzy Syst. **15**(4), 710–717 (2007)
9. Chang, C.-T.: An approximation approach for representing S-shaped membership functions. IEEE Trans. Fuzzy Syst. **18**(2), 412–424 (2010)
10. Charnes, A., Cooper, W.W., Ferguson, R.O.: Optimal estimation of executive compensation by linear programming. Manage. Sci. **1**(2), 103–194 (1955)
11. Charnes, A., Cooper, W.W.: Management Models and Industrial Applications of Linear Programming. Part I and II Wiley, New York (1961)
12. Charnes, A., Cooper, W.: Programming with linear fractional functions. Naval Res. Logistics Q. **9**(3–4), 181–186 (1962)
13. Charnes, A., Cooper, W.W.: Goal programming and multiple objective optimizations: part 1. Eur. J. Oper. Res. **1**(1), 39–54 (1977)
14. Chen, H.-H., Chen, L.-H.: A fuzzy goal programming approach for solving the decentralized bi-level optimization problem with imprecise cooperation relations. In: 2010 40th International Conference on Computers and Industrial Engineering (CIE), pp. 1–5 (2010)
15. Gao, Y., Zhang, G., Ma, J., Lu, J.: A $lambda$-cut and goal-programming-based algorithm for fuzzy-linear multiple-objective bilevel optimization. IEEE Trans. Fuzzy Syst. **18**(1), 1–13 (2010)
16. Gil-Aluja, J.: Handbook of Management Under Uncertainty. Springer, Heidelberg (2013). https://doi.org/10.1007/978-1-4613-0285-8
17. Ignizio, J.P., Cavalier, T.M.: Linear Programming. Prentice-Hall, Inc. (1994)
18. Jones, D., Tamiz, M.: Practical Goal Programming, vol. 141. Springer, New York (2010). https://doi.org/10.1007/978-1-4419-5771-9
19. Jones, D., Tamiz, M.: A review of goal programming. In: Multiple Criteria Decision Analysis, pp. 903–926. Springer, New York (2016)
20. Jones, D., Romero, C.: Advances and new orientations in goal programming. In: New Perspectives in Multiple Criteria Decision Making, pp. 231–246. Springer, Cham (2019)
21. Joseph, G.G.: The Crest of the Peacock/the non-European Roots of Mathematics. Princeton, Princeton University Press (2000)
22. Klaua, D.: Über einen Ansatz zur mehrwertigen Mengenlehre. Monatsb. Deutsch. Akad. Wiss. Berlin 7, 859–876 (1965). In: An Early Approach Toward Graded Identity and Graded Membership in Set Theory. Fuzzy Sets Syst. **161**(18), 2369–2379 (1965)
23. Jsh, K.: A survey of goal programming. Omega **1**(2), 193–205. https://doi.org/10.1016/0305-0483(73)90023-6
24. Lachhwani, K.: On solving multi-level multi objective linear programming problems through fuzzy goal programming approach. Opsearch **51**(4), 624–637 (2014)
25. Li, S., Hu, C.: An interactive satisfying method based on alternative tolerance for multiple objective optimization with fuzzy parameters. IEEE Trans. Fuzzy Syst. **16**(5), 1151–1160 (2008)
26. Li, S., Yang, Y., Teng, C.S.: Fuzzy goal programming with multiple priorities via generalized varying-domain optimization method. IEEE Trans. Fuzzy Syst. **12**(5), 596–605 (2004)
27. Liang, R., Gao, J.: Dependent-chance programming models for capital budgeting in fuzzy environments. Tsinghua Sci. Technol. **13**(1), 117–120 (2008)
28. Liang, T.-F., Cheng, P.-Y., Shen, K.-H.: Application of fuzzy sets to aggregate production planning with multiproducts and multitime periods. IEEE Trans. Fuzzy Syst. **19**(3), 465–477 (2011)

29. Liu, B.: Dependent-chance programming with fuzzy decisions. IEEE Trans. Fuzzy Syst. **7**(3), 354–360 (1999)
30. Liu, B.: Theory and Practice of Uncertain Programming. Springer, Heidelberg (2009)
31. Mirakhorli, A.: Multi-objective optimization of reverse logistics network with fuzzy demand and return-product using an interactive fuzzy goal programming approach. In: 2010 40th International Conference on Computers and Industrial Engineering (CIE), pp. 1-6 (2010)
32. Romero, C.: Handbook of Critical Issues in Goal Programming. Elsevier, Amsterdam (2014)
33. Sakawa, M., Yauchi, K.: An interactive fuzzy satisficing method for multiobjective nonconvex programming problems with fuzzy numbers through coevolutionary genetic algorithms. IEEE Trans. Syst. Man Cybern. B Cybern. **31**(3), 459–467 (2001)
34. Schniederjans, M.: Goal Programming: Methodology and Applications: Methodology and Applications. Springer, Heidelberg (2012)
35. Tiwari, R.N., Dharmar, S., Rao, J.R.: Fuzzy goal programming an additive model. Fuzzy Sets Syst. **24**(1), 27–34 (1987)
36. Wang, B., Wang, S., Watada, J.: Fuzzy-portfolio-selection models with value at-risk. IEEE Trans. Fuzzy Syst. **19**(8), 758–769 (2011)
37. Wang, S., Watada, J.: Fuzzy Stochastic Optimization. Springer, New York (2013). ISBN 976-4419-9559-9
38. Xu, Y., Huang, C., Da, Q., Liu, X.: Linear goal programming approach to obtaining the weights of intuitionistic fuzzy ordered weighted averaging operator. J. Syst. Eng. Electron. **21**(6), 990–994 (2010)
39. Zadeh, L.: A.Fuzzy sets. Inf. Control **8**(3), 338–353 (1965)
40. wikipedia, Membersip function,@, Wikipedia (Mathematics), Wikimedia Foundation, Inc. Accessed 31 Jan 2021

Chapter 5
Skeleton-Based Human Action Recognition on Large-Scale Datasets

Tonmoy Hossain, Sujan Sarker, Sejuti Rahman, and Md Atiqur Rahman Ahad

Abstract Skeleton-based Human Action Recognition (SHAR) is one of the most trending research topics in computer vision, which relies on the investigation of multi-modal data acquired from different sensory devices. Due to its faster execution speed, skeleton-based automated SHAR systems are widely adopted in real-time applications such as surveillance systems, behavior analysis, gesture recognition, and security systems. To build such an efficient system, the recognition model needs to be trained on a large-scale multi-modal dataset to accurately identify multi-class actions. However, the recognition of multi-class actions with higher accuracy requires the extraction of the spatio-temporal discriminative features, which is a challenging task. Moreover, the traditional handcrafted feature-based Machine Learning (ML) models when dealing with a large dataset, fail to preserve spatio-temporal correlations, resulting in poor recognition accuracy. While exploring the recent literature, a paradigm shift in SHAR research from traditional ML-based models to deep learning models is observed, which intrigues us to hypothesize that *deep learning is the future avenue for SHAR research on large datasets*. To establish it, we perform an exhaustive study on the recent works published in 2017 onwards that utilize deep learning models in SHAR on large skeletal datasets, notably alleviating the tedious task of feature engineering and extraction. The comparative performance analysis of the recent models shows that the successor models outperform its predecessors in

T. Hossain
Department of Computer Science and Engineering, Ahsanullah University of Science and Technology, Dhaka, Bangladesh
e-mail: tonmoyhossain.cse@ieee.org

S. Sarker · S. Rahman
Department of Robotics and Mechatronics Engineering, University of Dhaka, Dhaka, Bangladesh
e-mail: sujan@du.ac.bd

M. A. R. Ahad (✉)
Department of Electrical and Electronic Engineering, University of Dhaka, Dhaka, Bangladesh
e-mail: atiqahad@du.ac.bd

Department of Intelligent Media, Osaka University, Suita, Japan

M. A. R. Ahad and A. Inoue (eds.), *Vision, Sensing and Analytics: Integrative Approaches*, Intelligent Systems Reference Library 207, https://doi.org/10.1007/978-3-030-75490-7_5

terms of recognition accuracy, confirming the prospects of the deep learning-based SHAR research on large skeletal datasets.

5.1 Introduction

Skeleton-based Human Action Recognition (SHAR) relies on the analysis of human skeletal data, which can be captured through a variety of depth sensors or extracted from the direct RGB videos. Skeletal data can also be obtained by placing inexpensive markers in different places of the human body. Skeleton models the entire structure of a human body with joints, and connections between joints while excluding most details of the actual scene. Thus, a skeleton-based system preserves the subject's privacy and ensures faster execution speed, especially when dealing with large datasets. As a result, SHAR research draws significant attention in recent times and facilitates the development of real-time applications [1], including health-care, human-computer interaction, chaos detection, robotics, etc. However, practitioners in this arena face many challenges while developing an efficient action recognition model. One of the difficulties is the occurrence of unfavorable conditions such as—occlusions, cluttered background, and object redundancies in the video samples of a skeletal dataset. RGB-based action recognition requires extensive processing, thereby higher processing cost. Moreover, similar types of actions such as drinking-eating, jumping-skipping, punching-kicking, etc. need to be identified discretely to appropriately understand the actual activity of a particular person. By far, the most vital challenge to build a real-time SHAR application is to train the model on a large (multi-modal, multi-variation, and ample amount of diverse video samples) skeletal dataset. For that purpose, researchers are continuously exploring to establish any efficient real-time SHAR model.

A noteworthy development in the SHAR field has taken place in the past few years. Earlier, traditional machine learning (ML) or depth sequence-based models achieved distinguished results [2] but these models have trained on small-scale datasets (i.e., limited number of samples and action classes) that is inappropriate to engulf a highly-varied large domain of activities and any real-time HAR-based application. In a real-time scenario, the performance of these ML-based models is highly intermittent and unreliable. These models need to train on a large dataset where the sequences are from various perspectives. Moreover, hand-crafted features such as—coordinates of key points, velocity, and co-occurrence features are not sufficient to appropriately recognize the multi-modal actions. Deep learning features (such as, joint distance and orientation metrics, and temporal step matrix) outperform the performance of using hand-crafted features because deep learning features can distinctly identify multi-variation action more precisely [3]. Hence, we observe that the machine learning-based models are not optimal and appropriate to work on a large-scale multi-modal dataset. Then the research expanded to the deep learning-based SHAR model. In the past few years, researchers are trying to establish a robust deep learning-based SHAR model. Initially, global co-occurrence learning methods are developed to build the relationship between deep learning features. After that, different ensemble Long Short-Term Memory (LSTM) methods are adapted and achieved better results than the traditional ones. However, LSTM takes an extensive amount of time, and because of that, Multi-Task Learning Network (MTLN) with the attention mechanism is introduced. For performance analysis, though NTU-60 dataset by far the most employed large dataset to train these models, multi-class large datasets should be operated to evaluate these models' performance.

In this chapter, we aim to establish the hypothesis—*Deep learning-based models are the future avenue for skeleton-based SHAR research on large datasets*. In particular, we try to investigate some key questions here: (i) What characteristics of large skeletal datasets make it challenging to train the model? (ii) Why do deep learning-based models outperform traditional handcrafted feature-based machine learning models? (iii) How do the existing deep model architectures outperform its predecessors in terms of recognition accuracy? (iv) Why are the deep learning-based models sustainable future research in this field? To answer these questions, we provide an introductory and explanatory overview of the skeleton-based deep learning action recognition approaches for large skeletal datasets. Different from the existing works [4, 5], we focus our study on providing a compact insight into the necessity of deep learning models on large multi-modal skeletal datasets to build real-time applications.

The rest of the chapter is structured as follows: In Sect. 5.2, we briefly discuss the attributes and structures of large and mostly used skeletal datasets. We summarize the existing deep learning architectures with advantages and limitations in Sect. 5.3. In Sect. 5.4, a comparative performance analysis of the existing models performing on a large skeletal dataset is characterized. Finally, we conclude the chapter, shedding light on the insights developed in this study with future research directions in Sect. 5.5.

5.2 Large-Scale Skeletal Datasets

A large-scale dataset can be defined as the dataset, which have a considerable number of multi-class skeleton data in different perspectives. The skeletal data can be extracted from RGB-D videos by pose estimation methods, such as OpenPose [6]. These datasets are developed based on views, modalities, and classes. There are more than 80 SHAR dataset is available where most of them are belongs to the small-scale category such as—Kinect Activity Recognition Dataset (KARD) dataset (18 activities) [7], UTKinect dataset (10 activities) [8], Florence3D dataset (9 activities) [9], etc. But for real-time application it is necessary to train a model on large-scale multimodal dataset. For that purpose, this chapter specifically concentrates on the large-scale dataset. In this section, we briefly describe the most familiar and challenging large scale skeleton datasets for SHAR along with their core attributes.

5.2.1 NTU-60 Dataset

NTU RGB-D is a large-scale dataset, having two versions, namely—NTU-60 [10] and NTU-120 [11] classes. NTU-60 has 60 distinct activity classes. These activities are subdivided into three categories—Daily Actions (40), Medical Conditions (9). and Two Person Interactions or Mutual Actions (11). The NTU-60 dataset has 56880 video samples (40 subjects, ages between 10 and 35) in 80 different camera viewpoints. During its inception, it was the largest dataset on skeleton. This datasets contain daily-life RGB videos (1920 $\times$ 1080), 3D skeleton data (25 body joints), depth map sequences, and Infrared videos (512 $\times$ 424) for each sample. Three concurrent Kinect V2 sensors are employed to capture the videos. The size of this dataset is approximately 1.3 TB.

5.2.2 NTU-120 Dataset

NTU-120 dataset is the updated version of the NTU-60 dataset, increasing the activity classes to 120 different daily life actions. NTU-120 [11] is splitted into three categories: daily actions (82), mutual or two-person interactions (26), and medical conditions (12). This dataset has 56880 + 57660 (114540) videos from 106 subjects (ages between 10 and 57) in 155 distinct camera viewpoints. NTU-120 dataset consists of the same types of data modalities as NTU-60 such as—skeleton data, RGB videos, and Infrared videos. Besides, this dataset is built using three Kinect V2 sensors. It is one of the most gigantic and complex dataset in the action recognition field to work on. Furthermore, the size of this dataset is almost 2.3 TB.

5.2.3 Kinetics-400 Dataset

Kinetics-400 dataset is a large-scale RGB action recognition dataset. From this RGB dataset, skeletal data are extracted by OpenPose to work on SHAR system [12]. This dataset is built by collecting the YouTube video samples of daily life activities. Approximately 306425 short trimmed videos are integrated with 400 action types. Each class consists of at least 400 video clips. Using OpenPose toolbox [13], an estimated 2D pose of 18 joints is generated for every frame to work on the SHAR files ([6, 14–25]).

5.2.4 Other Challenging Datasets

Besides the NTU-60, NTU-120, and Kinetics-400 datasets, we have analyzed some of the large-scale datasets. In the following, a brief analysis of these datasets is described.

- *UESTC RGB-D Varying-view Action Database:* Based on aerobic exercises with multi-view, cross-view, and arbitrary-view, the UESTC dataset contains 40 classes, from 40 subjects, with 25600 RGB-D actions [26]. This dataset is built by two Kinect V2 cameras to capture from 9 different viewpoints. The captured actions from the cameras covered three data modalities—RGB video, depth map sequences, and skeleton sequences.

- *KIMORE Dataset:* Based on the rehabilitation conditions, the KIMORE dataset (KInematic MOvement REhabilitation) is developed by [27]. It consists of RGB, depth videos and skeletal data from 78 subjects (average age of 35 years, 44 healthy subjects and 34 subjects with motor dysfunctions problems). Five different physical exercises are covered in this dataset—lifting of the arms, lateral tilt of the trunk, arms are extended, rotation of the pelvis on the transverse plane, rotation of the trunk, and squatting. A Microsoft Kinect V2 is operated as an RGB-D sensor to capture the data. Moreover, Kinect V2 is used to capture the 3D points of 25 body joints.

- *UWA3D Multiview Activity II Dataset:* It is a multi-view dataset [28] consists of 30 activity classes of 10 subjects. Front, top, left, and right view images are captured continuously by one Kinect sensor to build this dataset. Then the classes are segmented.

- *PKU Multi-Modal Dataset (PKU-MMD):* PKU-MMD is a large-scale multi-modal action dataset [29], consisting of 28000 action instances captured by Kinect V2 sensors from three camera views. The dataset has two versions, where the first one is developed by 66 subjects and has 51 actions (1076 video sequences having four modalities (RGB, Depth videos, skeleton, and infrared radiation)). The second version has 2000 short video sequences of 49 action classes by 13 subjects, having RGB and skeleton data. Developed for small-margin action detection tasks, the second version is more challenging for the following reasons—action intervals are shorter than the first version, more occlusions, and having concurrent actions.

- *Multi-Modal Action (MMAct) Dataset:* Introduced by Kong et al. [30], this dataset considers seven data modalities such as—RGB, keypoints, acceleration, gyroscope, orientation, Wi-Fi, and pressure data. These are captured by a commercial surveillance camera, smart glass, smartphone, and smartwatch in five different views. There are 20 subjects acted for 37 actions (12 simple actions, 16 complex actions, and 9 desk-related actions) consisting of more than 36000 sequences

Having occlusions and irregular walking patterns, this dataset becomes very challenging than most of the existing skeleton-based datasets.

- *MSR Dataset:* MSR dataset can be broken down into two sections—MSR Action 3D dataset and MSR DailyActivity 3D Dataset. In Action 3D dataset, a depth camera is used to capture the depth sequences consists of 20 distinct classes with 567 sample videos to form this action dataset [31]. In the DailyActivity 3D dataset, a Kinect device is employed to build this dataset. A total of 16 classes is covered with 320 activity sequences [32].

- *UTD-MHAD Dataset:* Considering the practicality or non-intrusiveness factors, only one Kinect camera is used with an inertial sensor [33]. Consists of 27 different classes, this dataset is well known in skeleton-based action recognition research. A total of 861 data sequences is captured covering four data modalities—skeleton joint position, RGB videos, depth videos, and the signals of inertial sensors.

- *Northwestern-UCLA Dataset:* Three data modalities have been covered in this dataset, which are the human skeleton, depth, and RGB. Captured by three Kinect cameras, the authors covered ten classes with 1475 data sequences [34].

A summary of the described large-scale multimodal dataset is characterized in Table 5.1. Considering the attributes, which define a large-scale multi-modal dataset—the number of skeleton samples, classes, data modalities, occlusions, cluttered background, and other adverse factors we can infer that NTU-60, NTU-120, and Kinetics-400 are among the most challenging multi-modal datasets. All the real-time deep learning-based SHAR applications should evaluate the models on these datasets. In the following, we discuss the skeleton-based deep learning models in which these datasets are trained for performance evaluation.

5.3 Deep Learning-Based Methods for SHAR

Extraction of the skeletal spatial-temporal features by maintaining the relationship of latent dependencies, and employing those into a deep learning network is a challenging task in action recognition. Besides, training the model on a large dataset with a considerable amount of time is also a critical issue to build a real-time SHAR application. Deep learning models have proven effective to deal with the challenges above while, developing a robust SHAR application. In this section, we thoroughly discuss the state-of-the-art deep architectures for large skeletal datasets.

Table 5.1 A summary of some of the prominent large-scale datasets

Dataset	Action classes	Video samples	Subjects	Camera viewpoints	Body joints	Data modality
NTU-60 [10]	60	56880	40	80	25	4 (RGB videos, 3D skeleton data, depth map sequences, and Infrared videos)
NTU-120 [11]	120	114540	106	155	25	4 (RGB videos, 3D skeleton data, depth map sequences, and Infrared videos)
Kinectice-400 [6]	400	306425	—	—	—	—
UESTC [26]	40	25600	40	9	25	3 (RGB video, depth map sequences, and skeleton sequences)
KIMORE [27]	5	—	78	—	25	3 (RGB, depth videos and skeletal data)
UWA3D [28]	30	900	10	1	—	3 (RGB, depth videos and skeletal data)
PKU-MMD [29]	51	1076	66	3	25	4 (RGB, depth videos, skeleton, and infrared radiation)
	49	2000	13	3	25	2 (RGB and skeleton data)
MMAct [30]	37	36000	20	5	—	7 (RGB, keypoints, acceleration, gyroscope, orientation, Wi-Fi, and pressure data)
Northwestern UCLA [34]	10	1475	10	3		3 (Human Skeleton, Depth, and RGB)

5.3.1 Co-occurrence Feature Learning-Based Methods for SHAR

The texture features of the neighboring joints within a convolutional kernel is known as co-occurrence features. A convolutional co-occurrence feature learning framework is used for skeleton-based action recognition, where the features are trained hierarchically [35]. For multi-person activities such as—hand-shaking, hugging, etc., two fusion strategies (Early fusion: all joints of the persons are stacked for input, and Late fusion: inputs of multiple people pass through the same network by combining their convolutional feature maps) are integrated into the training phase of the network. Operating in a large dataset, the early fusion stacked all the co-occurrence features, which elevated the learning process, and the late fusion performed well in handling the stacked feature maps.

Focusing on the co-occurrence relation between the features, Li et al. [36] introduced an algorithm named Spatio-Temporal-Unit Feature Enhancement (STUFE) to learn the skeletal spatial-temporal co-occurrence feature globally. The authors adapted the Hierarchical Co-occurrence Network (HCN) algorithm combining with an Active Coordinate System Conversion (ACSC) method, which automatically learned the transformation of skeleton samples. There was a performance gap after implementing the HCN algorithm with the actual application because of operating the method on a different dataset. Furthermore, the hyperparameter is set using a trial and error process, which is not efficient if this model is employed on a multi variational large dataset.

Moreover, this model is optimized with Stochastic Gradient Descent (SGD) methods, while the SGD methods have an adverse effect as SGD uses an identical learning rate for all the hyperparameters. To solve this issue, different types of optimizing methods working along a different axis such as—Adam, RMSProp, and AdaGrad can be accustomed [37].

5.3.2 LSTM-Based Methods for SHAR

LSTM networks are useful to capture the spatio-temporal features of a skeletal sequence. These networks can overcome the long-term dependency problem while extracting the dynamic temporal dependencies between the features. Fundamentally, it is integrated with a graph-based network to extract the variation and stochasticity in the data [38]. Also, attention enhanced methods can be incorporated with the LSTM to learn the semantic representation of the features [39]. However, the attention model adds additional weight parameters to the model, which increases the computation time that is incompetent for the training of a multi-varying large dataset. To extract the spatial structure information and temporal dynamic features, Si et al. [40] proposed a Spatial Reasoning and Temporal Stack Learning (SR-TSL) model consists of a Spatial Reasoning Network (SRN) and Temporal Stack Learning Network (TSLN). The SRN architecture extracts the high-level spatial information, while the TSLN architecture can build detailed dynamic skeleton features. The TSLN structure assembles with multiple skip-clip LSTM networks by combining the velocity and position networks.

To extract the most informative features among all the discriminative features, the attention mechanism performs efficiently by introducing a global memory cell to handle the descriptive features. Liu et al. [41] integrated the LSTM network with a context-aware attention model to target the relevant features. It enhanced the performance of the model as the model did not consider any irrelevant features in a large-scale dataset. In another work, Huang et al. [42] proposed a Long-Short Graph Memory (LSGM) to develop the discriminative ability of the features and improve the competence to build the skeleton sequences. Though this model is developed to improve the feature representation, it is computationally expensive. It divides the skeleton sequence into a 3D coordinate system where the data of each axis is passed through a hybrid architecture of Fully-Connected (FC) Bi-LSGM (FC+forward and backward LSGM+FC+FC) and temporal attention. Moreover, a basic LSTM takes extensive time to train the model, also the overfitting and dropout issues are much higher than the MTLN or RNN architectures. In a nutshell, modifying the LSTM network to build the LSGM-based action recognition can be extremely limited and inefficient in terms of dataset size, time and space complexities.

Integrating a View Adaptive Recurrent Neural Network (VA-RNN) with an LSTM structure (VA-LSTM), Zhang et al. [43] overcame the significant view variation problem. It automatically operated on the observation viewpoints and maintained its continuity throughout the training process. This architecture can be partitioned into two networks: view adaption subnetwork and LSTM network. At each time slot, two LSTM layers worked parallelly with an FC layer to produce an appropriate observation viewpoint. In another work, Lee et al. [44] focused on the feature representation of skeleton joints and proposed an ensemble method based on LSTM networks. The authors transformed the raw skeleton into a definitive coordinate system for the robustness and long-term dependency between the features. Different types of augmenting operations are performed on the coordinate system to extract the salient features from the transformed input data. Then, Temporal Sliding LSTM (TS-LSTM) networks consisting of short-term, medium-term, and long-term LSTMs are employed, allowing robustness to the temporal dynamics.

5.3.3 Multi-Task Learning (MTLN)-Based Approaches for SHAR

MTLN is a network in which different learning tasks are carried out at the same time, while traversing the similarities and the discrepancies across the task. Ke et al. [45] proposed a SHAR method emphasizing on the 3D trajectories of skeleton joints and learning the clips generated from skeleton sequences.

Each skeleton sequence is transformed into three sub-clips where each of the clips subsisted with seven frames. This transformation facilitates the learning of spatial-temporal features using a deep network architecture. An MTLN architecture is then employed to process all the frames in parallel, incorporating the skeletal spatio-temporal correlation. Moreover, this method learns the features effectively by developing intra-frame relationships. An MTLN framework was proposed to simultaneously estimate 3D human poses and classify actions from video sequences [46]. With a throughput above 100 frames per second and operating on the NTU-60 dataset, this model used extra post data from different datasets (MPII [47] and H36M [48]) for training purposes. It is not appropriate as the results of the existing alternative models are trained on the NTU-60 dataset individually.

5.3.4 Graph Convolutional Network Based Approaches for SHAR

A Directed Graph-based Neural Network (DGNN) for action recognition is developed by Shi et al. [17]. In this chapter, the joint and bone information are integrated to take advantage of bone and joint association. The author exemplified the data as a Directed Acyclic Graph (DAG) focused on kinematic dependencies of the skeletal features. Furthermore, the motion features are exploited and embedded with the spatial information to augment the model's performance. This model further developed into a two-stream adaptive model focusing on the topology of the network [16]. The authors highlighted the manual settings of the topology, which is identical in all layers over the graph network architecture that is not optimal for the hierarchical GCN and inept in optimizing large datasets. To capture the long-term spatial-temporal relationship, Cho et al. [19] proposed three versions of the Self-Attention Network (SAN): SAN-V1 (baseline model encoding the input features), SAN-V2 (learn the movement of a particular person), and SAN-V3 (learn the different modalities) by leveraging the long-range correlations. Figure 5.1 depicts a basic architecture of a self attention network. A SAN is developed by position embedding layer, self attention layer(s), average layer, and fully connected layer. Firstly, all the data-input (x) is fed into the position embedding layer. A position embedding layer is employed to figure out the joint motion features. Secondly, a number of self attention layer(s) is introduced considering the needs of the model. The output or joint motion features ($z_1, z, ..., z_n$) are concatenated to find the global average. Finally, a fully connected layer (FC) is employed to categorize the specific action classes. The three variants of the SAN in Cho et al. [19] were exercised on the output of the position embedding and concatenated the variants' results.

An ensemble method can be defined as a hybrid method, integrating multiple learning algorithms to achieve better performance. A dynamic SHAR method is developed, modeling the regular patterns of human movements for anomaly detection [49]. Firstly, skeleton motion decomposition is employed to extract the dynamic patterns and relationships between the features. Secondly, these distinct patterns and relationships are modeled into the proposed Message-Passing Encoder-Decoder Recurrent-Network (MPED-RNN) to extract the decoupled spatio-temporal features. In terms of operating on the real-time dataset, this MPED-RNN architecture achieves favorable results. However, it worked on a limited number of skeletal samples.

To condense the most informative frames and dispense with the ambiguous or redundant frames, Tang et al. [50] embraced Deep Progressive Reinforcement Learning (DPRL) for action recognition. The authors focused on the postures of the human body as a topology of a graph-based architecture, employed a graph-based CNN, integrating Frame Distillation Network (FDNet) and Spatio-Temporal Graph-based Convolutional Neural Network (ST-GCN) to extract the dependencies between the joints. Here, the FDNet condensed a fixed number of keyframes with the help of DPRL from the input. After that, the organized output was fed into the GCN for the reorganization of the action label. In another hybrid method, an encoder-decoder

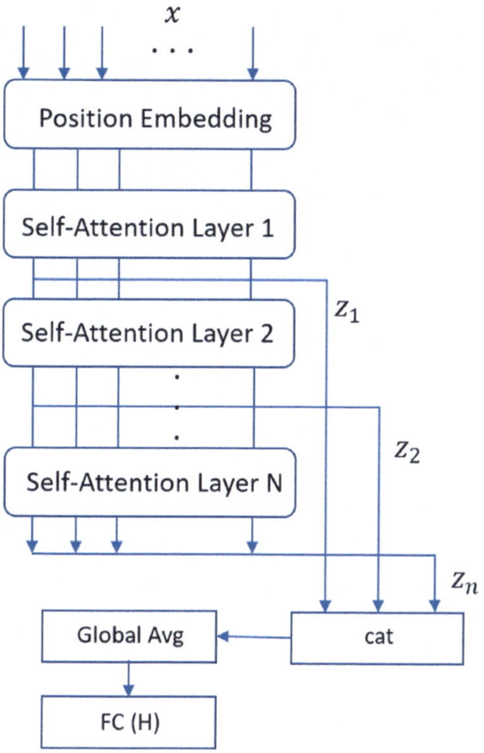

Fig. 5.1 A basic composition of self attention network

structure is introduced, termed as the A-link interference module for extracting the action-specific latent dependencies [15]. Modifying the existing GCN model [50] and the A-link interference module, the author proposed an Actional-Structural Graph Convolutional Network (AS-GCN) to capture the spatial-temporal features. It can suppress the latent dependencies among the features. Furthermore, a future pose prediction head is added to capture the most detailed patterns with self-supervision architecture.

5.3.5 Adversarial Approaches

Baek et al. [51] introduced a model to generate new entries to the database and complete the existing database by synthesizing data in the skeleton space. A Hand Pose Generator (HPG), is developed by the author, trained to synthesize the depth

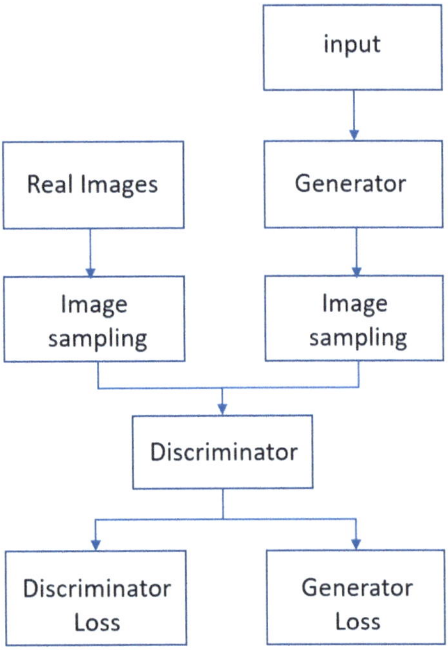

Fig. 5.2 A fundamental architecture of generative adversarial network

map because the skeleton data entries do not align with the corresponding depth map entries. The authors adopted Generative Adversarial Network (GAN) consisting of a Hand Pose Discriminator (HPD) that finds out the appropriate hand pose features. It is shown that the model outperformed the alternative skeleton data synthesizing models, but testing data should be increased as 90% data is used for training, which directly affects the dropout section and sometimes overfits the model. This model also followed an evaluation protocol that is not similar to the settings of the trained dataset [31] which might be inappropriate to evaluate the proposed GAN-based model. Figure 5.2 represents an elemental GAN architecture adapted by the authors working on SHAR, consists of two essential parts—Generator (develop new action classes) and Discriminator (try to classify the action classes).

Developed by Ian Goodfellow [52], Generative Adversarial Network (GAN) is an unsupervised learning method to generate data and learn the density distribution of a data. Training of a 3D hand pose estimator (HPE) depends on the availability of an extensive dataset covering all types of diverse camera shapes and perspectives. Baek et al. [51] introduced a model to generate new entries to the database to complete the existing database by synthesizing data in skeleton space. Because the skeleton data entries do not align with the corresponding depth map entries, a depth map entry is needed. Hence, a hand pose generator (HPG) is trained to synthesize the depth map. The authors adopted Generative Adversarial Network (GAN) consisting of a hand pose discriminator (HPD) that finds out the appropriate hand poses. Incorporating with the capability of HPD, refinement of the initial skeleton in testing has been comprehensively carried out by the model [53].

5.3.6 Hidden Markov Model (HMM)

HMM is an analytical Markov Model where the modeling system is done by using the Markov Process. In this model, learning can take place straight from the raw data. Aggregating the body parts extracts using Kinect V2 and the modified star skeleton algorithm, a Hidden Markov Model is developed for human action recognition [8, 54, 55]. At first, video frames consisting of human actions are aggregated and converted the gestures to Skeleton Joint Mapping (SJM). Then, these observation sequences from SJM are used to train the HMM where Baum-Welch and Forward-Backward algorithms are exercised to extract the optimal parameters for HMM.

5.3.7 Joint-Centered Local Information (JOLO)

Named after **Jo**int-aligned optical F**lo**w or **Jo**int-centered **lo**cal information (JOLO), the authors worked on the subtle motion differences to fully characterize human motion [25]. This model simultaneously employs the joint-centered light-weight information and human pose skeleton in a two-stream GCN (JOLO-GCN) to capture the sophisticated motion around each joint. The two-stream GCN is operated to measure skeleton sequence and Joint Aligned Optical Flow Patches sequence independently. After that, the predicted scores from these two sequences are accumulated into a final predicted score to classify each distinct action class using linear blending.

Studying these existing architectures, researchers adopted graph-based deep learning models to build SHAR systems and operated their models on large-scale datasets. For this purpose, most of the authors operated their models on NTU-60, NTU-120, and Kinect-400 datasets. Though the authors achieved excellent results on the NTU datasets, performances on Kinect-400 are still not up to the mark because of the extensive number of videos and multi-variational classes. At the beginning of adapting deep learning models, Co-occurrence Feature (CoF) learning methods are explored because of their spatial-temporal feature relationship factors. Then, LSTM-based methods have shown better results by managing the affinities of all connections throughout the network. Afterward, different versions of RNN methods are operated to overcome the extensive training time and complexities of LSTM-based methods. After that, the attention models integrating with RNN or LSTM grab the attention that can extract the relevant discriminative skeletal features from a large-scale dataset. Recently, a GCN model focusing on mining the joint-centered light-weight information is developed which worked efficiently on detecting subtle motion differences. In summary, there are enormous opportunities for the researchers to explore on this perspective in human action recognition research.

5.4 Performance Analysis and Discussion

In this section, a comprehensive performance analysis of the deep learning models (considering the number of videos, action classes, and variations), reported by the authors, is presented in which three mostly employed large datasets—NTU-60, NTU-120, and Kinetics-400 are exploited. The performances of the models are evaluated by the Cross-Subject (CS) and Cross-View (CV) accuracy, as mentioned in the NTU-60 and NTU-120 dataset by Shahroudy et al. [10] and Liu et al. [11]. CS is a evaluation metrics where the number of subjects is splitted for the purpose of the model and CV considers the camera viewpoint of the dataset for the evaluation. In calculating CS, among the 40 subjects of NTU-60 dataset, video samples of 20 subjects (40320 samples) are used to train the model, and the rest of them (16560 samples) are used for testing purposes. For enumerating CV, the training data is collected from cameras at view 2 and 3 while the testing data is aggregated from the camera at view 1.

In Table 5.2, we have summarized the performance of the deep learning model operating on the NTU-60 dataset. Reviewing the advanced deep learning architectures, graph-based methods are mostly adapted to build a SHAR model. Among all of the recent models, an ensemble method integrating the Graph Convolutional Network (GCN) with the joint-centered local information (JOLO-GCN) [25] outperformed all the existing deep learning models in terms of CS accuracy (93.8%) and CV accuracy (98.1%). The reason is that it can adapt the high-order spatial-temporal features, mitigating the redundant feature dependencies as well as distinctly distinguish subtle action differences by considering joint-aligned optical flow patches. At the beginning of developing deep learning models for SHAR, multi-task learning, RNN, LSTM based methods are mostly adapted to recognize distinct actions, taking an extensive amount of time to extract all the spatio-temporal features. After that, considering the issues in these models, the attention-based mechanisms integrating graph-based CNN and higher-order features are developed.

In Table 5.3, we have characterized the performances of the deep learning-based action recognition approaches based on the NTU-120 dataset. Several deep learning models such as Scale Selection Network (SSN), Geometric Pose Transition Dual-Stream CNN (GPTDS-CNN), Shift Graph Convolutional Network (SGCN), Graph Vertex Feature Encoder (GVFE), MT-CNN (Multi-Task CNN) with RotClips, Soft RNN, and Multi-scale Graph Convolutional Operation (MS-G3D) are trained on NTU-120 dataset. Among all the deep learning models, the JOLO-GCN model [25] accomplished the optimal result with 87.6% and 89.7% CS and CV accuracies, respectively. It is the joint-aligned optical flow information of the graph convolution between the features of all the 120 classes in the NTU-120 dataset that elevated the performance of the JOLO-GCN model.

The performances of the deep learning models trained on the Kinetics-400 dataset are represented in Table 5.4, where Top-1 (the accuracy of the actual class accurately matches the most desirable classes) and Top-5 (the accuracy of the actual action class matches the five most desirable action classes) accuracies are used for the performance evaluation by the authors. The JOLO-GCN model [25] outperformed

Table 5.2 Performance analysis (in %) of deep learning-based SHAR approaches on NTU-60 dataset. Handcrafted methods are not mentioned as the accuracies are very low and insignificant. CS: Cross-Subject; CV: Cross-View

Method	Result (%)	
	CS	CV
Temporal Sliding - LSTM (TS-LSTM) (2017) [44]	74.6	81.25
Convolutional Neural Network (CNN) (2017) [56]	75.2	83.1
View Adaptive - LSTM (VA-LSTM) (2017) [43]	79.4	87.6
CNN + Multi-Task Learning Network (2017) [57]	79.57	84.83
Spatial Temporal - Graph Convolutional Network (ST-GCN) (2018) [6]	81.5	88.3
Deep Progressive Reinforcement Learning (DPRL) + GCN (2018) [50]	83.5	89.8
Spatial Reasoning and Temporal Stack Learning (SR-TSL) (2018) [40]	84.8	92.4
View Adaptive - Recurrent Neural Network (VA-RNN) (2019) [58]	79.4	87.6
Bayesian Long Short-Term Memory (2019) [38]	81.8	89.0
Ensemble ResNet (2019) [59]	86.68	91.79
Actional-Structural - Graph Convolutional Network (AS-GCN) (2019) [15]	86.8	94.2
Attention enhanced graph convolutional - LSTM (2019) [39]	89.2	95.0
2 stream - Adaptive Graph Convolutional Networks (2s-AGCN) (2019) [16]	88.5	95.1
Feature-level auto-encoder (FW-AEC) (2020) [60]	50.7	76.1
Self Attention Network (SAN) (2020) [19]	87.2	92.7
Long-short graph memory network (LSGM) (2020) [42]	84.71	91.74
Spatio Temporal-Unit Feature Enhancement (STUFE) + ACSC (2020) [36]	86.9	92.5
STGCN + Co-occurrence Feature Learning (2020) [61]	84.7	90.2
4-stream Shift Graph Convolutional Network (4s-SGCN) (2020) [62]	90.7	96.5
Multi-scale Graph Convolutional Operation (2020) [20]	91.5	96.2
Modular Network (FUSION) + CNN (2020) [63]	91.6	94.5
Graph Convolutional Networks (2020) [64]	91.7	96.8
Motion Networks (2021) [65]	80.19	85.77
Shallow Graph Convolutional Network (2021) [21]	90.1	96.2
Cross-Channel Graph Convolutional Network (2021) [22]	88.87	95.33
Sparse Graph Convolutional Network (2021) [23]	91.3	96.9
Multi-scale Mixed Dense Graph Convolutional Network (2021) [24]	90.8	96.5
Joint-centered local information (JOLO-GCN) (2021) [25]	**93.8**	**98.1**

Table 5.3 Performance Analysis of deep learning-based SHAR approaches on NTU-120 dataset. CS: Cross-Subject; CV: Cross-View

Method	Result (%)	
	CS	CV
Temporal Sliding - LSTM + feature fusion (2017) [41]	58.2	60.9
Global Context-Aware Attention (GCAA) - LSTM (2017) [66]	58.3	59.2
MTLN (2017) [57]	58.4	57.9
Skeleton Visualization (2017) [67]	60.3	63.2
Soft RNN (2018) [68]	36.3	44.9
Scale Selection Network (SSN) (2019) [69]	59.9	62.4
2 stream GCAA LSTM (2017) [41]	61.2	63.3
MT-CNN with RotClips (2018) [45]	62.2	61.8
Graph Vertex Feature Encoder (GVFE) + AGCN (2019) [70]	78.3	79.8
Geometric Pose Transition Dual-Stream (GPTDS) - CNN (2020) [71]	76.3	78.9
4-stream Shift Graph Convolutional Network (2020) [62]	85.9	87.6
Graph Convolutional Network (2020) [64]	86.4	89.4
Multi-scale Graph Convolutional Operation Net (2020) [20]	86.9	88.4
Sparse Graph Convolutional Network (SGCN) (2021) [23]	67.5	68.8
Multi-scale Mixed Dense GCN (2021) [24]	86.8	88.0
Motion Networks (2021) [65]	84.16	85.53
Joint-centered local information - GCN (JOLO-GCN) (2021) [25]	**87.6**	**89.7**

all the existing models in terms of measuring the Top-1 (38.3%) and Top-5 (62.3%) accuracies. Analyzing Table 5.2, Table 5.3, and Table 5.4, we can infer that the JOLO-GCN model achieved the optimal results on NTU-60, NTU-120, and Kinetics-400 datasets.

To the best of our knowledge, there are barely any machine learning-based SHAR architectures that are employed in a large-scale dataset, such as NTU-60, NTU-120, Kinetics-400 or others. Machine learning model efficiently worked on small-scale datasets such as Weng et al. [2] achieved 94.8% performing on MSR-Action3D dataset (consists of 567 depth map sequences, covering 20 class acquired from 10 subjects) [72], or 98% accuracy operating on UTKinect dataset (consists of 10 action classes acquired from 10 subjects) [73]. This model achieved higher accuracy because of the small number of class, and video samples but to develop a real-time application, a model should operate on a large-scale multi-class dataset to extract and learn every kind of spatio-temporal features. From this, we can infer that machine learning models are not appropriate to develop a real-time SHAR application.

Studying the performance of the models, the performance of the deep learning architectures operating on the Kinetics-400 dataset is relatively lower than NTU-60 and NTU-120 datasets. It is worth noting that the performance of the deep

Table 5.4 Performance analysis of deep learning-based SHAR approaches on Kinetics-400 dataset

Method	Result (%)	
	Top-1	Top-5
ST-GCN (2018) [6]	30.7	52.8
Body-Part Level Hybrid Model (BPLHM) (2019) [74]	33.4	56.2
Spatio-Temporal Graph Routing-GCN (STGR-GCN) (2019) [14]	33.6	56.1
AS-GCN (2019) [15]	34.8	56.5
2s-AGCN (2019) [16]	36.1	58.7
Directed Graph Neural Network (DGNN) (2019) [17]	36.9	59.6
Advanced CA-GCN (2019) [18]	34.1	56.6
SAN (2020) [19]	35.1	55.7
Multi-scale Graph Convolutional Operation (2020) [20]	38.0	60.9
Shallow GCN (2021) [21]	37.1	60.0
Cross-Channel GCN (2021) [22]	36.7	58.9
Sparse GCN (2021) [23]	36.6	54.8
Multi-scale Mixed Dense GCN (2021) [24]	37.6	60.4
JOLO-GCN (2021) [25]	**38.3**	**62.3**

learning architectures on the Kinetics-400 dataset is comparatively lower than NTU-60 and NTU-120 datasets, since working with 400 classes by maintaining the spatio-temporal dependencies is an ambitious task rather than administering with 40/60/120 classes.

5.5 Conclusion and Future Direction

Due to its emerging application potential, researchers have been given significant efforts in developing the skeleton-based human activity recognition systems powered by a wide variety of learning models, ranging from traditional machine learning to advanced deep learning models. These models require to be trained on large scale multi-modal datasets, which is facilitated by making datasets such as NTU-60, NTU-120, and Kinetics-400 datasets available in the SHAR research community. However, the machine learning methods could not leverage the multi-modal features of such large datasets, which is essential to identify all the distinct actions. The extensive literature study corroborates that the deep learning models are more prominent than the traditional ML-based models and outperform those in terms of recognition accuracy and execution speed, especially for the sizeable skeletal dataset. The comparative performance analysis on three mostly explored large dataset reveals that the highest accuracies for cross-subject in case of NTU-60, NTU-120, and Kinetics-400 are 91.7%, 86.9%, and 38.0%, respectively whereas for cross-view these accuracies are

96.8%, 88.4%, and 60.9%, respectively, which are achieved using deep learning-based models.

Here, we conclude that deep learning-based models are future ventures for SHAR research on large datasets. Though these deep learning models achieve the highest performance so far, there is still scope of improvement in this field. Some of the important issues related to the large skeleton-based datasets for human activity recognition are mentioned below:

- Different deep networks and their variants can be embedded to extract more appropriate features. The existing models can also be improved by adjusting the network parameters to capture various spatio-temporal dynamics.
- Other directions include the data augmentation techniques in the existing models and incorporating more contextual information required to SHAR.
- The above analysis indicates that there is still much scope of improvement in deep learning-based human action recognition on extensive skeletal data. The existing models can also be significantly improved by tuning and embedding the network blocks' variants, thus superseding the current performance on large datasets, which is nearly impossible for traditional ML-based approaches.

References

1. Mokari, M., Mohammadzade, H., Ghojogh, B.: Recognizing involuntary actions from 3D skeleton data using body states. arXiv preprint arXiv:1708.06227 (2017)
2. Weng, J., Weng, C., Yuan, J.: Spatio-temporal Naive-Bayes nearest-neighbor (ST-NBNN) for skeleton-based action recognition. In: Proceedings of the IEEE Conference on Computer Vision and Pattern Recognition, pp. 4171–4180 (2017)
3. Asadi-Aghbolaghi, M., Bertiche, H., Roig, V., Kasaei, S., Escalera, S.: Action recognition from RGB-D data: comparison and fusion of spatio-temporal handcrafted features and deep strategies. In: Proceedings of the IEEE International Conference on Computer Vision Workshops, pp. 3179–3188 (2017)
4. Dang, L.M., Min, K., Wang, H., Piran, M.J., Lee, C.H., Moon, H.: Sensor-based and vision-based human activity recognition: a comprehensive survey. Pattern Recogn. **108**, 107561 (2020)
5. Jegham, I., Khalifa, A.B., Alouani, I., Mahjoub, M.A.: Vision-based human action recognition: an overview and real world challenges. Forensic Sci. Int.: Digit. Invest. **32**, 200901 (2020)
6. Yan, S., Xiong, Y., Lin, D.: Spatial temporal graph convolutional networks for skeleton-based action recognition. In: Thirty-Second AAAI Conference on Artificial Intelligence (2018)
7. Gaglio, S., Re, G.L., Morana, M.: Human activity recognition process using 3-D posture data. IEEE Trans. Hum.-Mach. Syst. **45**(5), 586–597 (2014)
8. Xia, L., Chen, C.C., Aggarwal, J.K.: View invariant human action recognition using histograms of 3D joints. In: 2012 IEEE Computer Society Conference on Computer Vision and Pattern Recognition Workshops (CVPRW), pp. 20–27. IEEE (2012)
9. Seidenari, L., Varano, V., Berretti, S., Bimbo, A., Pala, P.: Recognizing actions from depth cameras as weakly aligned multi-part bag-of-poses. In: Proceedings of the IEEE Conference on Computer Vision and Pattern Recognition Workshops, pp. 479–485 (2013)
10. Shahroudy, A., Liu, J., Ng, T.T., Wang, G.: NTU RGB+D: a large scale dataset for 3D human activity analysis. In: IEEE Conference on Computer Vision and Pattern Recognition, June 2016

11. Liu, J., Shahroudy, A., Perez, M., Wang, G., Duan, L.-Y., Kot, A.C.: NTU RGB+D 120: a large-scale benchmark for 3D human activity understanding. IEEE Trans. Pattern Anal. Mach. Intell. **42**, 2684–2701 (2019)
12. Carreira, J., Zisserman, A.: Quo Vadis, action recognition? A new model and the kinetics dataset. In: Proceedings of the IEEE Conference on Computer Vision and Pattern Recognition, pp. 6299–6308 (2017)
13. Cao, Z., Hidalgo, G., Simon, T., Wei, S.-E., Sheikh, Y.: OpenPose: realtime multi-person 2D pose estimation using part affinity fields. arXiv preprint arXiv:1812.08008 (2018)
14. Li, B., Li, X., Zhang, Z., Fei, W.: Spatio-temporal graph routing for skeleton-based action recognition. Proc. AAAI Conf. Artif. Intell. **33**, 8561–8568 (2019)
15. Li, M., Chen, S., Chen, X., Zhang, Y., Wang, Y., Tian, Q.: Actional-structural graph convolutional networks for skeleton-based action recognition. In: Proceedings of the IEEE Conference on Computer Vision and Pattern Recognition, pp. 3595–3603 (2019)
16. Shi, L., Zhang, Y., Cheng, J., Lu, H.: Two-stream adaptive graph convolutional networks for skeleton-based action recognition. In: Proceedings of the IEEE Conference on Computer Vision and Pattern Recognition, pp. 12026–12035 (2019)
17. Shi, L., Zhang, Y., Cheng, J., Lu, H.: Skeleton-based action recognition with directed graph neural networks. In: Proceedings of the IEEE Conference on Computer Vision and Pattern Recognition, pp. 7912–7921 (2019)
18. Zhang, X., Xu, C., Tao, D.: Context aware graph convolution for skeleton-based action recognition. In: Proceedings of the IEEE/CVF Conference on Computer Vision and Pattern Recognition, pp. 14333–14342 (2020)
19. Cho, S., Maqbool, M., Liu, F., Foroosh, H.: Self-attention network for skeleton-based human action recognition. In: The IEEE Winter Conference on Applications of Computer Vision, pp. 635–644 (2020)
20. Liu, Z., Zhang, H., Chen, Z., Wang, Z., Ouyang, W.: Disentangling and unifying graph convolutions for skeleton-based action recognition. In: Proceedings of the IEEE/CVF Conference on Computer Vision and Pattern Recognition, pp. 143–152 (2020)
21. Yang, W., Zhang, J., Cai, J., Zhiyong, X.: Shallow graph convolutional network for skeleton-based action recognition. Sensors **21**(2), 452 (2021)
22. Xie, J., et al.: Cross-channel graph convolutional networks for skeleton-based action recognition. IEEE Access **9**, 9055–9065 (2021)
23. Ahmad, T., Jin, L., Lin, L., Tang, G.Z.: Skeleton-based action recognition using sparse spatio-temporal GCN with edge effective resistance. Neurocomputing **423**, 389–398 (2021)
24. Xia, H., Gao, X.: Multi-scale mixed dense graph convolution network for skeleton-based action recognition. IEEE Access **9**, 36475–36484 (2021)
25. Cai, J., Jiang, N., Han, X., Jia, K., Lu, J.: JOLO-GCN: mining joint-centered light-weight information for skeleton-based action recognition. In: Proceedings of the IEEE/CVF Winter Conference on Applications of Computer Vision, pp. 2735–2744 (2021)
26. Ji, Y., Xu, F., Yang, Y., Shen, F., Shen, H.T., Zheng, W.-S.: A large-scale varying-view RGB-D action dataset for arbitrary-view human action recognition. arXiv preprint arXiv:1904.10681 (2019)
27. Capecci, M., et al.: The KIMORE dataset: kinematic assessment of movement and clinical scores for remote monitoring of physical rehabilitation. IEEE Trans. Neural Syst. Rehabil. Eng. **27**(7), 1436–1448 (2019)
28. Rahmani, H., Mahmood, A., Huynh, D., Mian, A.: Histogram of oriented principal components for cross-view action recognition. IEEE Trans. Pattern Anal. Mach. Intell. **38**(12), 2430–2443 (2016)
29. Liu, C., Hu, Y., Li, Y., Song, S., Liu, J.: PKU-MMD: a large scale benchmark for continuous multi-modal human action understanding. arXiv preprint arXiv:1703.07475 (2017)
30. Kong, Q., Wu, Z., Deng, Z., Klinkigt, M., Tong, B., Murakami, T.: MMAct: a large-scale dataset for cross modal human action understanding. In: Proceedings of the IEEE International Conference on Computer Vision, pp. 8658–8667 (2019)

31. Li, W., Zhang, Z., Liu, Z.: Action recognition based on a bag of 3D points. In: 2010 IEEE Computer Society Conference on Computer Vision and Pattern Recognition-Workshops, pp. 9–14. IEEE (2010)
32. Wang, J., Liu, Z., Wu, Y., Yuan, J.: Mining actionlet ensemble for action recognition with depth cameras. In: 2012 IEEE Conference on Computer Vision and Pattern Recognition, pp. 1290–1297. IEEE (2012)
33. Chen, C., Jafari, R., Kehtarnavaz, N.: UTD-MHAD: a multimodal dataset for human action recognition utilizing a depth camera and a wearable inertial sensor. In: 2015 IEEE International conference on image processing (ICIP), pp. 168–172. IEEE (2015)
34. Wang, J., Nie, X., Xia, Y., Wu, Y., Zhu, S.-C.: Cross-view action modeling, learning and recognition. In: Proceedings of the IEEE Conference on Computer Vision and Pattern Recognition, pp. 2649–2656 (2014)
35. Zhu, W., et al.: Co-occurrence feature learning for skeleton based action recognition using regularized deep LSTM networks. In: Thirtieth AAAI Conference on Artificial Intelligence (2016)
36. Li, S., Jiang, T., Huang, T., Tian, Y.: Global co-occurrence feature learning and active coordinate system conversion for skeleton-based action recognition. In: The IEEE Winter Conference on Applications of Computer Vision, pp. 586–594 (2020)
37. Dean, J., et al.: Large scale distributed deep networks. In: Advances in Neural Information Processing Systems, pp. 1223–1231 (2012)
38. Zhao, R., Wang, K., Su, H., Ji, Q.: Bayesian graph convolution LSTM for skeleton based action recognition. In: Proceedings of the IEEE International Conference on Computer Vision, pp. 6882–6892 (2019)
39. Si, C., Chen, W., Wang, W., Wang, L., Tan, T.: An attention enhanced graph convolutional LSTM network for skeleton-based action recognition. In: Proceedings of the IEEE Conference on Computer Vision and Pattern Recognition, pp. 1227–1236 (2019)
40. Si, C., Jing, Y., Wang, W., Wang, L., Tan, T.: Skeleton-based action recognition with spatial reasoning and temporal stack learning. In: Proceedings of the European Conference on Computer Vision (ECCV), pp. 103–118 (2018)
41. Liu, J., Shahroudy, A., Xu, D., Kot, A.C., Wang, G.: Skeleton-based action recognition using spatio-temporal LSTM network with trust gates. IEEE Trans. Pattern Anal. Mach. Intell. **40**(12), 3007–3021 (2017)
42. Huang, J., Xiang, X., Gong, X., Zhang, B., et al.: Long-short graph memory network for skeleton-based action recognition. In: The IEEE Winter Conference on Applications of Computer Vision, pp. 645–652 (2020)
43. Zhang, P., Lan, C., Xing, J., Zeng, W., Xue, J., Zheng, N.: View adaptive recurrent neural networks for high performance human action recognition from skeleton data. In: Proceedings of the IEEE International Conference on Computer Vision, pp. 2117–2126 (2017)
44. Lee, I., Kim, D., Kang, S., Lee, S.: Ensemble deep learning for skeleton-based action recognition using temporal sliding LSTM networks. In: Proceedings of the IEEE International Conference on Computer Vision, pp. 1012–1020 (2017)
45. Ke, Q., Bennamoun, M., An, S., Sohel, F., Boussaid, F.: Learning clip representations for skeleton-based 3D action recognition. IEEE Trans. Image Process. **27**(6), 2842–2855 (2018)
46. Luvizon, D., Picard, D., Tabia, H.: Multi-task deep learning for real-time 3D human pose estimation and action recognition. IEEE Trans. Pattern Anal. Mach. Intell. (2020)
47. Andriluka, M., Pishchulin, L., Gehler, P., Schiele, B.: 2D human pose estimation: new benchmark and state of the art analysis. In: Proceedings of the IEEE Conference on computer Vision and Pattern Recognition, pp. 3686–3693 (2014)
48. Ionescu, C., Papava, D., Olaru, V., Sminchisescu, C.: Human3.6M: large scale datasets and predictive methods for 3D human sensing in natural environments. IEEE Trans. Pattern Anal. Mach. Intell. **36**(7), 1325–1339 (2013)
49. Morais, R., Le, V., Tran, T., Saha, B., Mansour, M., Venkatesh, S.: Learning regularity in skeleton trajectories for anomaly detection in videos. In: Proceedings of the IEEE Conference on Computer Vision and Pattern Recognition, pp. 11996–12004 (2019)

50. Tang, Y., Tian, Y., Lu, J., Li, P., Zhou, J.: Deep progressive reinforcement learning for skeleton-based action recognition. In: Proceedings of the IEEE Conference on Computer Vision and Pattern Recognition, pp. 5323–5332 (2018)
51. Baek, S., Kim, K.I., Kim, T.-K.: Augmented skeleton space transfer for depth-based hand pose estimation. In: Proceedings of the IEEE Conference on Computer Vision and Pattern Recognition, pp. 8330–8339 (2018)
52. Goodfellow, I., et al.: Generative adversarial nets. In: Advances in Neural Information Processing Systems, pp. 2672–2680 (2014)
53. Gammulle, H., Denman, S., Sridharan, S., Fookes, C.: Fine-grained action segmentation using the semi-supervised action GAN. Pattern Recogn. **98**, 107039 (2020)
54. Lv, F., Nevatia, R.: Recognition and segmentation of 3-D human action using HMM and multi-class AdaBoost. In: European Conference on Computer Vision, pp. 359–372. Springer (2006)
55. Wu, D., Shao, L.: Leveraging hierarchical parametric networks for skeletal joints based action segmentation and recognition. In: Proceedings of the IEEE Conference on Computer Vision and Pattern Recognition, pp. 724–731 (2014)
56. Rahmani, H., Bennamoun, M.: Learning action recognition model from depth and skeleton videos. In: Proceedings of the IEEE International Conference on Computer Vision, pp. 5832–5841 (2017)
57. Ke, Q., Bennamoun, M., An, S., Sohel, F., Boussaid, F.: A new representation of skeleton sequences for 3D action recognition. In: Proceedings of the IEEE Conference on Computer Vision and Pattern Recognition, pp. 3288–3297 (2017)
58. Zhang, P., Lan, C., Xing, J., Zeng, W., Xue, J., Zheng, N.: View adaptive neural networks for high performance skeleton-based human action recognition. IEEE Trans. Pattern Anal. Mach. Intell. **41**(8), 1963–1978 (2019)
59. Nie, Q., Wang, J., Wang, X., Liu, Y.: View-invariant human action recognition based on a 3D bio-constrained skeleton model. IEEE Trans. Image Process. **28**(8), 3959–3972 (2019)
60. Su, K., Liu, X., Shlizerman, E.: Predict & cluster: unsupervised skeleton based action recognition. In: Proceedings of the IEEE/CVF Conference on Computer Vision and Pattern Recognition, pp. 9631–9640 (2020)
61. Tian, D., Lu, Z.-M., Chen, X., Ma, L.-H.: An attentional spatial temporal graph convolutional network with co-occurrence feature learning for action recognition. Multimed. Tools Appl. **2020**, 1–19 (2020)
62. Cheng, K., Zhang, Y., He, X., Chen, W., Cheng, J., Lu, H.: Skeleton-based action recognition with shift graph convolutional network. In: Proceedings of the IEEE/CVF Conference on Computer Vision and Pattern Recognition, pp. 183–192 (2020)
63. Main de Boissiere, A., Noumeir, R.: Infrared and 3D skeleton feature fusion for RGB-D action recognition. arXiv preprint arXiv:2002.12886 (2020)
64. Dong, J., et al.: Action recognition based on the fusion of graph convolutional networks with high order features. Appl. Sci. **10**(4), 1482 (2020)
65. Wang, H., Baosheng, Yu., Xia, K., Li, J., Zuo, X.: Skeleton edge motion networks for human action recognition. Neurocomputing **423**, 1–12 (2021)
66. Liu, J., Wang, G., Hu, P., Duan, L.-Y., Kot, A.C.: Global context-aware attention LSTM networks for 3D action recognition. In: Proceedings of the IEEE Conference on Computer Vision and Pattern Recognition, pp. 1647–1656 (2017)
67. Liu, M., Liu, H., Chen, C.: Enhanced skeleton visualization for view invariant human action recognition. Pattern Recogn. **68**, 346–362 (2017)
68. Jian-Fang, H., Zheng, W.-S., Ma, L., Wang, G., Lai, J., Zhang, J.: Early action prediction by soft regression. IEEE Trans. Pattern Anal. Mach. Intell. **41**(11), 2568–2583 (2018)
69. Liu, J., Shahroudy, A., Wang, G., Duan, L.-Y., Kot, A.C.: Skeleton-based online action prediction using scale selection network. IEEE Trans. Pattern Anal. Mach. Intell. **42**(6), 1453–1467 (2019)
70. Papadopoulos, K., Ghorbel, E., Aouada, D., Ottersten, B.: Vertex feature encoding and hierarchical temporal modeling in a spatial-temporal graph convolutional network for action recognition. arXiv preprint arXiv:1912.09745 (2019)

71. Huynh-The, T., Hua, C.-H., Tu, N.A., Kim, D.-S.: Learning geometric features with dual-stream CNN for 3D action recognition. In: ICASSP 2020-2020 IEEE International Conference on Acoustics, Speech and Signal Processing (ICASSP), pp. 2353–2357. IEEE (2020)
72. Kong, Y., Fu, Y.: Bilinear heterogeneous information machine for RGB-D action recognition. In: Proceedings of the IEEE Conference on Computer Vision and Pattern Recognition, pp. 1054–1062 (2015)
73. Li, X., Zhang, Y., Zhang, J.: Improved key poses model for skeleton-based action recognition. In: Pacific Rim Conference on Multimedia, pp. 358–367. Springer (2017)
74. Zhang, X., Xu, C., Tian, X., Tao, D.: Graph edge convolutional neural networks for skeleton-based action recognition. IEEE Trans. Neural Netw. Learn. Syst. **31**(8), 3047–3060 (2019)

Chapter 6
Sensor-Based Human Activity and Behavior Computing

Anindya Das Antar, Masud Ahmed, and Md Atiqur Rahman Ahad

Abstract A major goal of human activity and behavior recognition (HAR) is to recognize activities and behaviors from a series of action data for different subjects under different environmental conditions. The use of wearables, smart devices, and vision-based systems enables the collection of human action and behavior data for greater health benefits, rehabilitation, elderly care, and monitoring. This chapter overviews the state-of-the-art in activity and behavior recognition based on sensor data. First, we provided a general architecture of HAR describing primary data sources and pre-processing techniques. Then, we describe robust feature extraction and selection strategies with a comparative analysis of statistical and deep learning-based models. Additionally, we provide information about more than 100 benchmark datasets and repositories in this domain. The paper concludes with a discussion of major issues and challenges, highlighting open issues and scopes that need to be approached in prospective research.

6.1 Introduction

Human Activity Recognition (HAR) is a method to detect, recognize, and interpret the activities and behaviors of individuals based on the information obtained from camera, mobile devices, smart wearables, and their instrumented environments [1–6]. The purpose of HAR is to detect human activities in controlled and uncontrolled settings. HAR has a wide range of application opportunities, that include health care [7, 8], ambient assisted living (AAL) [9], rehabilitation treatment [10], elderly care

---

A. D. Antar · M. Ahmed · M. A. R. Ahad (✉)
Department of Electrical and Electronic Engineering, University of Dhaka, Dhaka, Bangladesh
e-mail: atiqahad@du.ac.bd

A. D. Antar
Department of CSE, University of Michigan, Ann Arbor, MI, USA

M. A. R. Ahad
Department of Intelligent Media, Osaka University, Suita, Japan

M. Ahmed
Department of Information Systems, University of Maryland Baltimore County, Baltimore, MD, USA

M. A. R. Ahad and A. Inoue (eds.), *Vision, Sensing and Analytics: Integrative Approaches*, Intelligent Systems Reference Library 207, https://doi.org/10.1007/978-3-030-75490-7_6

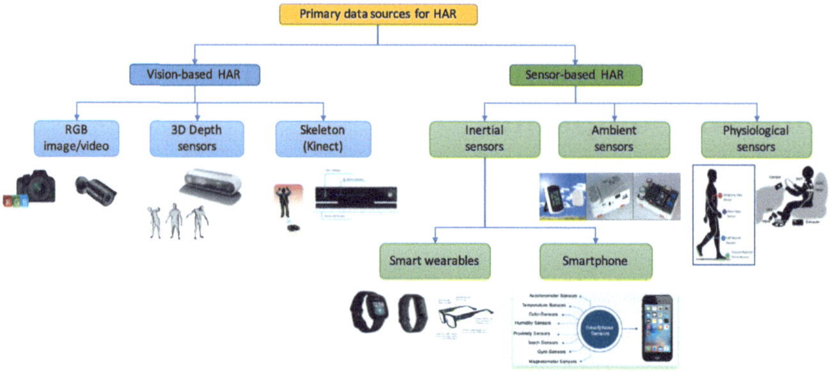

Fig. 6.1 Primary data sources for Human Activity Recognition (HAR)

[11], smart homes with assistive technology [12, 13], behavior analysis [14], security-based surveillance [15, 16], energy-efficient devices [17, 18], fitness tracking [19, 20], gait analysis [21–23], monitoring transportation activities [24–27], sensor-based COVID-19 analysis [28–30], sports or game controls [31], and so on.

There are two primary input data sources in HAR systems to detect and analyze complex and basic human activities and behaviors: 1) RGB, Depth, and Skeleton data from images or videos [2, 32, 33] and 2) sensor data from ambient/environmental sensors (temperature, pressure, CO_2, humidity, PIR), Inertial sensors (sensors embedded in smartphone or smart wearables including smart watch, wristband, belt, glass, clothes, bracelets, etc.), and physiological sensors (EEG, ECG, EMG, etc. for detecting physiological signals like heart beat, blood pressure, brain electrical activity, skin conductance, and so on) [3, 34–36] as shown in Fig. 6.1. Environmental factors also play a role in human activities, which can range from simple to complex and can be categorized into atomic actions with a sequence of various activities, person–object interactions, person-person interactions, and groups of activities [37].

There are several major challenges [37] in vision-based HAR including low-level challenges (occlusions, cluttered background, shadows) [38, 39], view changes [40], scale variance [41], privacy and computational requirements [42], and intra-class variability or inter-class similarity of actions [43]. In response to a major advancement in sensor-enabled smart devices (smartphones, smart wearables, and environmental sensors), researchers sought to design user interfaces that reason about and act upon metrics derived from people's behavior and activities to improve their well-being and quality of life. Because of the rapid development in microelectronics and sensor technology, state-of-the-art sensors (accelerometers, gyroscopes, magnetometers, pressure sensors, barometers, GPS, etc.) are usually small, easy to integrate with wearable devices, highly sensitive, and have a strong anti-interference ability, making them suitable for practical situations [1]. This indicates an increasing need for research and applications that utilize wearable and smartphone sensors to recognize human behavior and activities.

Despite these developments in the field of HAR, challenges still exist due to complex and varying daily activity patterns, intra-subject/inter-subject variability for

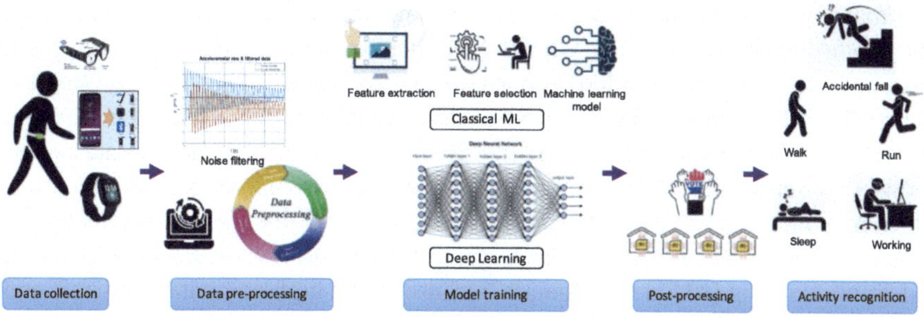

Fig. 6.2 A generic system architecture pipeline for any sensor-based human activity recognition approach

similar activities, performance and privacy tradeoff, energy consumption and computational efficiency in portable devices, and difficulties regarding data annotation. Considering these challenges, any sensor-based HAR system design follows the system architecture pipeline shown in Fig. 6.2 for recognizing daily human activities and behaviors. This framework consists of five parts: 1) data collection, 2) data pre-processing, 3) model training, 4) post-processing, and 5) activity recognition.

In this data-driven approach, the first step is to collect time-series activity and behavior data for the training, validation, and testing (or, evaluation) phase from sensors embedded into smartphones, smart wearable devices, and medical equipments [44, 45]. For analyzing the temporal pattern that exists in sensor-data, it is necessary to filter intrinsic and environmental noises in the pre-processing stage. Usage of median filter, Butterworth low-pass filter, discrete wavelet package shrinkage filter, and Kalman filter is common to remove noises from acceleration and gyroscope sensor data [46–49]. It is also important to consider Signal-to-noise ratio (SNR or S/N), correlation coefficient (R) between filtered and reference signal, waveform delay, filter size, window length of median filter, etc. factors while choosing noise filters. For the case of statistical classification methods, existing research studies also demonstrated segmentation techniques (splitting time-series data into time windows) for extracting hand-crafted features [24, 50, 51]. In the learning phase, an activity or behavior recognition model is trained leveraging the extracted features from the training dataset. For hand-crafted features in the case of statistical classification, it is common to apply different feature selection and dimensionality reduction methods including wrapper method, correlation method, Relief-F, game-theory, Factor Analysis, t-SNE, ISOMAP, ICA, PCA, etc. for reducing the number of input variables and computational cost of the predictive model [52–54].

HAR research also includes Deep Learning models (Deep Neural Network (DNN), Convolutional Neural Network (CNN), Recurrent Neural Network (RNN), Stacked Autoencoder (SAE), etc.) and hybrid models (CNN-RNN, Convolutional Long short-term memory (ConvLSTM), RNN + LSTM, CNN + SAE, etc.) showing comparable performance only in the case of large datasets [55, 56]. Though deep learning-based models can automatically learn the features reducing human effort in hand-crafted feature engineering, these models lack explainability and interpretability (black-box models). The increased computational cost for training these models also makes it challenging to deploy in real-time applications. On the other hand, linear, non-linear, and ensemble Classic Machine Learning (CML) models (Support Vector Machine (SVM), Random Forest (RF), Decision Tree (DT), Logistic Regression (LR), k-Nearest Neighbors (KNN), etc.) perform better in the case of small datasets with lower data dimensionality. It is also possible to incorporate expert knowledge during problem formulation and selecting human-engineered features, which may increase model explainability and interpretability. Researchers also used transfer learning [57], reinforcement learning [58], unsupervised learning [59, 60], and active learning [61] in this domain because of the high computational cost and lengthy training period related to the processing of massive real-time sensor data. Finally, after the classification tasks, researchers showed various post-processing techniques (majority voting, Mod technique, etc.) for improving model performance and explainability in sensor-based HAR domain [24, 62–64].

We present a comprehensive summary of the state of the art to recognize, compute, understand, and explain human activity and behaviors making use of ambient, wearable, and smartphone sensors. The rest of the paper is structured as follows: In Sect. 6.2 and 6.3, we briefly review feature extraction and feature selection techniques and give a thorough discussion of different factors during data collection. We summarize a list of more than 100 existing benchmark datasets and repositories in Sect. 6.4. Section 6.5 gives an overview of classification problems, their various performance metrics, and various performance evaluation criteria. In Sect. 6.6, we mention the importance of deep learning-based models in the field of sensor-based HAR. Finally, we sum up this chapter by providing future research directions with the insights gained to further improve the performance of human activity and behavior recognition utilizing sensors in Sect. 6.7.

6.2 Statistical Learning-Based Approaches

After preprocessing step, we will get noise-free clean data. These data represent the value of different axes of the sensor over time and can not be fed to statistical classifiers directly. Hence after segmentation, it is required to extract features from the processed data. In practical life, individuals usually perform identical activities in several distinct patterns. This reason may produce a wide variance in the extracted features. Therefore, it is important to extract robust, relevant features that will help the classifier to distinguish different activities from each other. If the set of features provides a slight contrast between the same behavior across different subjects, then we can consider that as a perfectly extracted feature set.

6.2.1 Feature Extraction

Several research works have been conducted regarding feature extraction. Based on the researches [3], we can divide these features into four categories, e.g., time domain features, frequency domain features, wavelet analysis-based features, heuristic features.

Time Domain Features: Time domain features are widely used features in the field of statistical classifier based human activity recognition. Different kinds of time domain features have been thoroughly investigated in various research [24, 51, 65–70]. Different types of features play different roles. Mean calculate the average signal value over the window, which helps the classifier to distinguish whether the signal is dynamic or static [71]. Standard deviation estimates the signal variation over the window, which represents the range of the potential variance of the signal [72]. Standard deviation is the root square value of variance. Therefore, if the standard deviation is taken under consideration, then we can ignore the variance feature. Another feature that measures the variance of the signal is the median absolute deviation (MAD) [73]. The MAD feature is commonly used to detect epileptic seizures automatically [65]. Besides, maximum and minimum features are used to define the boundary the signal.

These are other statistical features that also used in several research. The Kurtosis function represents the peak of a frequency distribution curve. By using this feature we can determine whether the distribution of the signal is elevated or plain [74]. Using skewness function, the degree of asymmetric information of the signal distribution can be determined [75]. The interquartile range reflects the signal's spatial dispersion [65].

Frequency Domain Features: Frequency domain features are extracted by implementing Fast Fourier Transformation (FFT). FFT features are transformed sensor data in the frequency domain. Different types of frequency domain features are explored in the field of human activity recognition such as median frequency [76], subsets of several FFT coefficients [77], spectral energy [78], entropy [79], etc. Entropy can distinguish between basic and complicated acceleration patterns in different activities. Frequency domain features are powerful features to distinguish activities based on spectral analysis. FFT analysis can be a good choice for identifying complex activities with several commanding frequency elements.

Wavelet Domain Features: Wavelet domain features usually represent both time domain and frequency domain characteristics. By using wavelets, we will get information of temporal and frequency that is more accurately localized. These features are extracted by employing discrete wavelet transformers (DWT). In the field of human activity recognition, wavelet analysis is utilized in three cases, classification of the transitional moment from one activity to another [80], signal augmentation [81], and extraction of time-frequency features that are then used for activity detection [80, 82].

6.2.2 Feature Selection Technique

The feature extraction process aims to generate useful information that will be fed to the classifier. It may happen that several correlated data were retrieved during the extraction process. This will increase the size of the model resulting in computational resource wastage. Also removing unimportant and abundant features can improve the performance of the classifiers. There are several methods utilized in the field of human activity recognition for feature selection, such as (i) Principal Component Analysis and Independent Component Analysis [83], (ii) Using a low variance approach to exclude a function [84], (iii) The approach of the missing value ratio [84], (iv) Univariate feature selection [85], (v) Select-k-Best appraoch [86], (vi) Select percentile apporach [87], (vii) Generic univariate selection [85], (viii) Recursive feature elimination [88], (ix) L1-based feature selection [89], (x) t-Distributed Stochastic Neighbor Embedding (t-SNE) [90], (xi) Uniform Manifold Approximation and Projection (UMAP) [91], (xii) Factor analysis [92], (xiii) Isometric Feature Mapping [93], (xiv) Elimination of highly correlated features [94], (xv) Tree-based feature selection [95].

Two commonly used approaches for feature selection are PCA and ICA. ICA is a statistical and analytical tool, while PCA concentrates on retrieving powerful patterns from the original data based on variance. Furthermore, in some works [96], the Discrete Cosine Transform (DCT) was used in conjunction with autoregressive model coefficients [97], resulting in better recognition accuracies.

6.2.3 Statistical Classifiers and Factors

After extracting features, the next step is feeding the data into the classifiers. It is important to choose the classifier carefully for getting a good accuracy result. Statistical classifiers are commonly applied approaches in the field on accelerometer and gyroscope sensor-based human activity tracking [98]. For classification following statistical classifiers are usually utilized by recent research works, Decision Tree, K-Nearest Neighbours, Naive Bayes, Support Vector Machine, Logistic regression, Random Forest. During classification, few factors need to be handled carefully. E.g.,

- **Feature Normalization:** The extracted features may vary in terms of magnitude one from another, which will create difficulties for few statistical classifiers. Therefore it is necessary to normalize all the features on the same scale. Four commonly used feature normalization techniques are recalling to min-max normalization, mean normalization, standardization, scaling to unit length.
- **Overfitting:** Overfitting occurs when the estimator models the training data too accurately. When the model is conscious of the input and noise in the training data, overfitting occurs. Overfitting has the effect of allowing the model to learn the noises in the training data as concepts. The issue is that new data does not have these noises. As a consequence, these new environment data may be classified inaccurately.
- **Underfitting:** Underfitting happens when the model is unable to adequately be trained on available data. A machine learning algorithm that underfits is not a good model. As a result, the output on training data is subpar.

Solutions to the Overfitting and Underfitting Data: Overfitting occurs where the model is evaluated on the same data used for training. As the model has already seen the data during the training phase, the accuracy will be high during the testing phase. But for unseen data, the model can not provide such high accuracy. To prevent this problem, we have to evaluate the model on separate data. There are two commonly used techniques for this purpose.

- **Resampling technique:** The most well-known and effective resampling technique is K-fold cross-validation. The whole dataset is divided into K subsets. The model is then trained for K times treating only one set as the testing set and rests as the training set. Finally, the average accuracy is computed of these K outcomes.
- **Train Test Split:** To estimate the accuracy, we can cut some data from the training set or use completely different data as the testing set. Then, the learning models are evaluated on the testing dataset to get an understanding of how the model would do on unseen results.

6.3 Data Collection for HAR

Data collection is one of the most important parts of the field of machine learning-based HAR. There are lots of challenges to deal with [39, 50, 99–102]. Factors that need to be considered during the data collection phase, are described below.

- **Activity pattern:** What type of activity data need to be collected for the research purpose. The activities can be partitioned into three basic patterns, static activities, dynamic activities, postural transition activities
- **Data collection environment:** In which environment the data need to be collected, inside laboratory environment, inside house environment, or outside environment.
- **Subject dependency:** Whether the collected data should be subject dependent or independent.
- **Sensor requirement:** What sort of sensors need to used to achieve the goal. There are some constraints to consider when choosing a sensor, output signal, computation cost, energy consumption, data processing time, optimum performance, frequency response, sensitivity, etc.

In the field of HAR, few primary sensors are usually used. Some of other sensors provide data by transforming data from these primary sensors. The primary sensors used in this sector are as following.

- **Accelerometer:** An accelerometer can detect acceleration in the human body. It has also been reported that the use of accelerometry for long-term tracking of human activity is increasingly growing [103]. The operating theory is following Newton's second law of motion.
- **Gyroscope:** A gyroscope can measure the angular movement of the smartphone. The gyroscope provides output as roll, yaw, and pitch. The device is useful for

detecting quick transitions or movement of the user. The use of both an accelerometer and a gyroscope has a positive effect on the recognition of human activity [104, 105].

- **Magnetometer:** The tri-axis magnetometer detects the influence of the earth's magnetic field around the three principal axes using the Hall effect on a miniature scale.
- **Pressure sensor:** The air pressure near the device can be detected using a pressure sensor. The pressure sensor is used to distinguish between transportation modes of differing altitude ranges, such as buses and subway trains.

Other types of sensors in smart devices such as light sensors, Wi-Fi, temperature sensors, barometers, GPS, Bluetooth, etc. are also used for human activity recognition. Based on these sensors several devices are embedded such as ESense [100, 106], MITes (MIT Environmental Sensors) [107], MICA2DOT [108], Intel Mote [109], Luna Nurse [110], etc.

6.4 Sensor-Based Benchmark Datasets in HAR

For training a model to recognize complex human activities and behavior, data is collected from various modalities including wearable, environmental, physiological, and smartphone sensors, and stored in the datasets. In the process of collecting data, researchers face numerous obstacles, including technical issues, questions of anonymity and corporate authority, etc. [50]. Although these challenges exist, a large number of databases for sensor-dependent activity and behavior recognition are currently accessible. It is important to select appropriate dataset(s) in the evaluation phase to find out efficient pre-processing, feature extraction, feature selection, classification, and post-processing techniques [111]. An ideal dataset, which is diverse (in terms of participant number, age, gender, race, etc.) and representative of real-world scenarios (varying sensor orientation, device position, etc.) can help to build a robust model with efficient performance in real-time or near real-time.

6.4.1 Database Repositories

Here, we provide descriptions of benchmark datasets from well-known databases and repositories. These datasets include data related to Activities of Daily Living (ADL), fall detection, healthcare and medical activities, sports and gaming activities, fitness activities, sleep monitoring, and so on. In these datasets, data has mainly been collected through mobile phones, smart wearable accessories (smartwatch, bracelet, clothes, etc.), and several environmental sensing-based smart home devices.

6.4.1.1 UCI Machine-Learning Repository

University of California Irvine (UCI) machine learning repository includes a large number of datasets, data generators, and domain theories. These datasets help the research community with the empirical analysis of different machine learning (ML) algorithms [112]. Publicly available datasets in this repository related to sensor-based HAR include Heterogeneity Human Activity Recognition (HHAR) [113] dataset, UCI Battery-less Wearable Sensor (UCIBWS) [114] dataset, Activity Recognition system based on Multisensor data fusion (AReM) [115] dataset, Human Activity Recognition (HAR) [116] dataset, Human Activities and Postural Transitions (HAPT) [117] dataset, Activity recognition from Single Chest-mounted accelerometer (Single Chest) [118] dataset, OPPORTUNITY activity recognition (OPPORTUNITY) [119] dataset, Activities of Daily Living (ADLs) recognition using binary sensors [120] dataset, REALDISP activity recognition (REALDISP) [121] dataset, User Identification From Walking Activity (UIFWA) [118] dataset, PAMAP2 Physical Activity Monitoring (PAMAP2) [122] dataset, Daily and Sports Activities (DSA) [123] dataset, Dataset for ADL Recognition with Wrist-worn accelerometer (Wrist ADL) [123] dataset, Indoor user movement prediction from RSS data (RSS) [124] dataset, MHEALTH dataset (MHEALTH) [125] dataset, WISDM smartphone and smartwatch activity biometrics (WISDM) [126] dataset, Wearable Stress and Affect Detection (WESAD) [127] dataset, and Simulated Falls and Daily Living Activities (SFDLA) [128] dataset. Some of these datasets, such as HHAR, OPPORTUNITY, PAMAP2, and SFDLA, contain missing values.

6.4.1.2 Pervasive System Research Group

This research group[1] deals with pervasive systems that represent networks of collaborative sensing, computation, and reasoning components that constantly interact with the environment in activities embedded in the environment, but are invisible to the surface. Publicly available sensor-based datasets from their repository are the Smoking Activity dataset [129], Complex Human Activities dataset [130], Physical Activity Recognition dataset [131], and Sensors Activity dataset [132].

6.4.1.3 Human Activity Sensing Consortium (HASC)

Researchers who belong to the Human Activity Sensing Consortium (HASC)[2] have a strong interest in the field of human activity and behavior recognition (HAR). Their goal is to collect and distribute large-scale human activity corpus. This repository contains the HASC-2010-corpus dataset [133], HASC-2011-corpus dataset [134], HASC-2012-corpus dataset [135], HASC-IPSC: Indoor Pedestrian Sensing Corpus

---

[1] http://ps.ewi.utwente.nl/Datasets.php.

[2] http://hasc.jp/.

dataset [136], HASC-PAC2016: Pedestrian Activity Corpus dataset [137], and ASC-BDD: Ballroom Dance dataset [138]. A number of these datasets are publicly available, including HASC2010corpus, HASC-IPSC, and HASC-BDD dataset.

6.4.1.4 UCR Archive for Time Series Classification

UCR[3] repository is hosted by the University of California, Riverside, and contains electrocardiogram (ECG) data, robots' surface detection data, etc. collected utilizing different devices, environmental sensors, and instruments [139].

6.4.1.5 PhysioBank Repository

Sensor data from smartphones and wearable devices are included in the datasets from this repository named PhysioBank.[4] These datasets also incorporate ECG data, EEG data, human gait data at various stages, and healthcare data with multivariate time series characteristics [140].

6.4.1.6 CRAWDAD Repository

Mobile devices and wireless networks make up the majority of the data collected in the Community Resource for Archiving Wireless Data At Dartmouth (CRAWDAD)[5] domain. In general, there is no restriction on the format of the data in this repository. Occasionally, this domain contributes location or mobility data for users of mobile devices leveraging embedded sensors such as accelerometers, GPS, gyroscopes, etc. [141].

6.4.2 Sensor-Based Datasets for Healthcare

Human activity recognition systems use many types of sensors to prevent, manage, and treat patients [7]. Wearable, environmental, and smartphone sensor-based HAR systems can continuously observe heart attack patients, pregnant women, and elderly people living alone at home for clinical events that may affect their quality of life [3]. Few sensor-based datasets under this category include the Daphnet Freezing of Gait (FoG) dataset [142], Nursing Activity dataset [143], Nursing Care Records dataset [144], and Predicting Parkinson's Disease dataset [145]. Publicly available datasets

---

[3] http://www.cs.ucr.edu/~eamonn/time_series_data/.

[4] https://www.physionet.org/physiobank.

[5] http://crawdad.org/.

from this group include Daphnet Freezing of Gait (Daphnet FoG) and Predicting Parkinson's Disease (PPD).

6.4.3 *Sensor-Based Datasets for Physical Fitness and Sports Activities*

The utilization of sensor-based human activity recognition (HAR) in both academic and practical fields has gained interest and can serve as a valuable resource in healthy lifestyle coaching, fitness tracking, and sports training areas for instance [146]. Few publicly available datasets in this area are the Body Attack Fitness dataset [147], Swimmaster dataset [148], Fitbit dataset [149], and Body Weight Training (BWT) dataset [150].

6.4.4 *Sensor-Based Datasets for Household Activities*

Sensors detecting human and environmental interactions are increasingly common with the advancement in the field of Internet of Things (IoT) and Human-Computer Interaction (HCI). Though challenging, these ambient sensors (radar, sound sensors, pressure sensors, temperature sensors, accelerometers or gyroscopes mounted on objects, etc.) are normally installed in smart environments, where users pass most of their time. Few datasets collected using these environmental sensors are MIT PlaceLab dataset [151], CMU Multi-Modal Activity Database (CMU-MMAC) [152], Ubicomp dataset [153], Activity Recognition with Ambient Sensing (ARAS) dataset [154], Smart Home dataset [155], Center for Advanced Studies in Adaptive Systems (CASAS) dataset [156], Cooking Activity dataset,[6] Home Sensors dataset [157], and Daily Routine dataset [158]. Most of these datasets can be found online, except for Activity Recognition with Ambient Sensing (ARAS) [154].

6.4.5 *Sensor-Based Datasets for Device Usage Activities*

Nowadays, many people use smart devices (smartphones, tablets, smartwatches, etc.) for various daily applications. Thus, research on the activities of smart device users may be a promising field of study. With the help of user-agreements, works in this area protect users' privacy, which keeps track of their device usage and helps them choose the best plan or app for them based on their history of device usage. Under this domain, available datasets are the Activity Classification dataset [159], Reality

[6]Cooking Activity Challenge with *International Conference on Activity and Behavior Computing* (ABC), 2020 https://abc-research.github.io/cook2020.

Mining dataset [160], Mobile Data Challenge (MDC) dataset [161], Device Analyzer dataset [162], and Insight for Wear dataset [163].

6.4.6 Sensor-Based Datasets for Transportation Activities

Sensor data collected through multimodal activities such as transportation and locomotion have been the subject of several recent studies. Data can be collected from wearable devices and smartphone sensors while users visit different transportation modes (car, bus, train, subway, bike, etc.) and application domains include smart parking, traffic monitoring, finding optimal transportation modes, and so on. Sussex Huawei Locomotion (SHL) dataset is a very popular dataset in this domain for the classification of locomotion and transportation modes [164]. There are other transportation datasets in this domain including dataset by Yang et al. [165], Geolife dataset [166], dataset by Wang et al. [167], dataset by Reddy et al. [168], dataset by Siirtola et al. [169], dataset by Hemminki et al. [170], dataset by Zhang et al. [171], dataset by Xia et al. [172], dataset by Widhalm et al. [173], dataset by Jahangiri et al. [174], dataset by Su et al. [175], HTC dataset [176], US Transportation dataset [177], Transportation Mode Detection (TMD) dataset [178] and Sussex-Huawei Locomotion-Transportation (SHL) dataset [164]. Among these datasets, TMD, SHL, Geolife, and US Transportation datasets are publicly available.

6.4.7 Sensor-Based Datasets for Fall Detection

Due to various applications including death, injury, and accident prevention, researchers are increasingly interested in accidental fall detection among older people, pregnant women, and patients [179]. As a consequence, access to fall detection databases that are publicly available can establish a comprehensive and well-organized evaluation of fall detection methods. Datasets in this domain include DLR dataset [180], MobiFall dataset [181], UR dataset [182], TST dataset [183], Cogent dataset [184], Gravity dataset [185], DMPSBFD dataset [186], UMA dataset [187], MobiAct dataset [188], UniMiB dataset [186], and SisFall dataset [189].

6.4.8 Smart Wearable Sensing-Based Datasets

Smart wearable devices (smartwatch, smartglass, smart clothes, smart jewelry, etc.) and sensors (accelerometer, gyroscope, magnetometer, etc.) must consider consumer preferences. For behavior tracking, low weight, fashionable, and convenient wearable devices with integrated sensors were used in several databases. Few available datasets in this category are the University of Southern California Human Activity Dataset

(USC-HAD) [190], UC Berkeley WARD (Wearable Action Recognition Database) [191], Skoda mini checkpoint dataset [192], Human-Computer Interaction gestures dataset (HCI) [147], PPS Grouping dataset [193], University of Dhaka – Mobility Dataset (DU-MD) [67, 194], Human Gait Database (HuGaDB) [195], OU-ISR Gait dataset [196, 197], and UTD Multimodal Human Action Dataset (UTD-MHAD) [198]. With 745 subjects, the OU-ISR inertial sensor-based gait database [197, 199] by the Institute of Scientific and Industrial Research (ISIR), Osaka University is the current largest inertial sensor-based gait database. Among them 388 subjects are males and 357 are females.

6.4.9 Smart Device-Based Datasets (Mobilephone/Tablets)

Many people spend most of their time in a day with smart devices (smartphones or tablets). It is easy to track user activities and behaviors using data from embedded sensors (accelerometer, gyroscope, pressure sensor, GPS, light sensor, etc.) of smartphones or other smart devices (e.g., smart watch, smart glass, smart wristband, smart clothings, etc.). Datasets in the domain of HAR using smartphone sensors can be found in UCI ML repository [112], Human Activity Sensing Consortium (HASC) repository [133], HASC Indoor Pedestrian Sensing Corpus (HASC-IPSC) [136] repository, and so on. Few popular datasets in this domain are Wireless Sensor Data Mining (WISDM) dataset [200], Smart Devices dataset [201], Transportation Mode Detection (TMD) dataset [178], SHL (Sussex-Huawei Locomotion-Transportation) dataset, etc.

6.5 Overview of Classification Problems and Performance Measures in HAR

In general, four common types of classification tasks exist in the domain of HAR, including binary, multi-class, multi-label, and hierarchical classification [3]. The input is generally classified into one of two non-overlapping groups in a binary classification task. In this case, we need to consider response time [202] and the selection of feature subsets [203] for optimal performance. Input data is classified into one of three or more classes for multi-class classification utilizing K-nearest neighbors (KNN), Decision trees (DT), Support Vector Machines (SVM), Deep Neural Networks (DNN), etc. algorithms. This method is famous for classifying complex human activities and behaviors. Multilabel classification differs from multi-class classification, as multilabel classification algorithm predicts multiple labels for each instance. A further feature of multiclass classification is that it offers mutually exclusive class predictions, as opposed to the multilabel classification which offers distinct classification tasks for each predicted label. The hierarchical classification algorithms classify the input into only one class. Then this algorithm further divides the output into subclasses or group into superclasses. The hierarchy must be predefined beforehand and cannot be altered during the classification process.

While evaluating model predictions it is necessary to choose optimal performance evaluation metrics to justify the performance, explainability, interpretability, and effectiveness of a trained model. In general, accuracy, precision, recall, F1 score, Balanced Classification Rate (BCR), Balanced Error Rate (BER), Matthews Correlation Coefficient, Discriminant Power Normalized Likelihood Index, Cohen's kappa Coefficient are common performance metric for model evaluation. Researchers also utilize visualization techniques, Receiver Operating Characteristics (ROC) curves with AUC score, confusion matrix, Cumulative Gains charts, Lift charts, and so on. It is also necessary to consider bias-variance tradeoff while justifying a model's ability to minimize bias and variance. High-bias models are oversimplified, ignore training data, and lead to a high error on both training and test data. Contrary to this, models with high variance tend to pay more attention to training data and don't generalize on unseen data. In the case of the training dataset, high variance models perform well but show high error rate on test data.

If our model has very few parameters and is too simple, then it may have low variance and high bias. On the contrary, a model with a large number of parameters will have high variance and low bias. To find the perfect balance, we need to consider parameters that don't overfit (low bias and high variance) or underfit (low variance and high bias) the data.

6.6 Deep Learning-Based Approaches

Classical handcrafted features along with machine learning-based methods have been explored more in the domain of human activity recognition based on sensors. Vision-based domain, action or activity recognition, temporal action segmentation, action understanding, tracking and related works are mainly done based on deep learning-based models in the last several years. These are extremely successful in image and vision domain in very large-scale datasets.

Skeleton-based activity recognition approaches are still dominated by feature-based and classical machine learning-based methods on smaller datasets. There are over 50 datasets based on skeleton data, which have 10 to 20 classes by less number of subjects. Therefore, the methods are dominated by classical learning schemes. On the other hand, there are only a few large datasets on skeleton domain, and deep learning-based models are very successful on these. For example, one the very widely-explored large dataset is the NTU-60 having 60 classes from many subjects. Another recent very large skeleton-based dataset is the NTU-120 covering 120 diverse classes by many subjects. These are created based on Kinect sensors. On the contrary, the Kinect-400 dataset is basically a video-based dataset having 400 classes and it is a very challenging dataset. By using OpenPose, skeleton data are created from the video data. On these large datasets, classical machine learning or handcrafted features are found to be much less successful. So, these datasets' success are dominated by various deep learning-based models. Some of the successful models are:

- Convolutional co-occurrence feature learning framework [204],
- Spatio-Temporal-Unit Feature Enhancement (STUFE) based on Hierarchical Co-occurrence Network (HCN) [205],
- LSTM-based networks [206–208],
- Spatial Reasoning and Temporal Stack Learning (SR-TSL) model [209],
- Long-Short Graph Memory (LSGM) [210],
- Multi-Task Learning (MTLN)-based approaches [211, 212],
- Message-Passing Encoder-Decoder Recurrent-Network (MPED-RNN) [213],
- View Adaptive Recurrent Neural Network (VA-RNN) with an LSTM structure (VA-LSTM) [214],
- Ensemble method based on LSTM networks [215],
- Deep Progressive Reinforcement Learning (DPRL) [216],
- Actional-Structural Graph Convolutional Network (AS-GCN) [216],
- Directed Graph-based Neural Network (DGNN) [217],
- Self-Attention Network (SAN) [218],
- Hand Pose Generator and Discriminator based on Generative Adversarial Network (GAN) [219],
- Joint-aligned optical Flow or Joint-centered Local information (JOLO) [220].

Similar to the skeleton domain, sensor-based activity recognition methods with larger datasets are also explored based on deep learning methods. However, there is no champion or clear-winner so far on the sensor domain. Many recent methods are published either on handcrafted methods, or deep learning-based methods. In the sensor domain, the performances of deep learning-based methods are found to be not as dominant as in the image or vision domain. Though deep learning-based approaches are explored in various sensor-based activity analysis, the dominance is still not absolute, similar to the vision and image domain. For example, in some recent competitions of challenges on sensor-based activity challenges globally, we find that the results are not fully obvious. For example,

- OU-ISIR Wearable Sensor-based Gait Challenge: Age and Gender (GAG) at the 12th IAPR International Conference on Biometrics (ICB), Greece, 2019: It is based on sensor data to understand age and gender from gait or walking data based on a large dataset [221]. The training set has 745 subjects (388 males and 357 females – age range: 2 to 78 years old), whereas the test dataset has 58 subjects (39 males and 19 females). In this competition, there were ten teams with 67 different solutions. Of them, deep learning-based methods dominated as top results. But feature-based methods with classical machine learning methods also performed well.
- Sussex-Huawei Locomotion (SHL) challenge at the ACM UbiComp in Singapore (in 2018), UK (in 2019), and Mexico (in 2020): The SHL dataset is a very large-scale dataset and each year, the complexity was added [222]. Even though the dataset is a very large one, still, a number of classical methods have demonstrated

good results along with deep learning model-based approaches. Apart from various classical approaches, a series of deep learning-based methods produced good results. Some of the approaches are:

- CNN-LSTM
- Bidirectional LSTM
- DNN Spectrogram
- CNN + FFT Spectrogram
- Gated Recurrent Network (GRU)
- Dual ConvLSTM
- ConvLSTM
- CNN + LSTM + fANOVA
- Machine learning + Deep learning + Hidden markov Model (HMM)
- Independently RNN

- Cooking Activity Challenge at the 2nd International Conference on Activity and Behavior Computing (ABC) with the 9th International Conference on Informatics, Electronics and Vision (ICIEV), Japan, 2020: It is a cooking activity-related dataset having 9 micro activities (e.g., add, mix, cut, open, etc.) and 3 macro activities (cereal, fruit salad, and sandwich) [223]. It has motion capture data (not explored for the challenge), accelerometer data and OpenPose data. In this challenge, the range of testing accuracy for deep learning-based methods is smaller, though both deep learning and machine learning methods have smaller median in the testing accuracy. It is more logical as the dataset is relatively smaller in size.
- Nurse Care Activity Recognition Challenge at ACM UbiComp in UK (2019), and Mexico (2020): In this sensor-based dataset, classical methods become the winners profoundly [224]. For example, the challenge in 2020, top 3 teams are based on handcrafted features and classical machine learning models.

In these sensor-based challenges, various methods were introduced based on handcrafted features, or on deep learning-based models. However, there is no clear winner between classical approaches vs. deep learning approaches. Therefore, we can say that classical approaches can still perform superior over deep learning-based methods on sensor-based datasets, especially, if the datasets are relatively smaller in size.

6.7 Conclusion and Future Direction: Challenges to Overcome

In this chapter, we summarize the core aspects of human activity and behavior recognition based on wearable, smartphone, and environmental sensors. Vision-based action or activity recognition is another major research direction – based on video data of actions. Some of the actions are atomic or short, whereas some of these actions consist of multiple people, multiple interactions (person–object interactions, person-person interactions, etc.), and groups of micro activities. The background, illumination variations, presence of shadow, presence of multiple persons, objects, etc. along with the very well-known issue of privacy – turn the video-based action analysis more challenging. For the sensor-based (mainly wearable sensors and embedded sensors in smartphone and smart devices) activity recognition and analysis have different perspectives and scopes than video-based domain. Individualism is a core issue and privacy is a much less concerning point for the sensor-based HAR domain, compared to the vision-based HAR domain. Though data from smartphones and wearable sensors also have privacy issues and security concerns, research is going on to ensure hardware and software security through data encoding. Therefore, sensor-based human activity recognition has evolved tremendously for over a decade, especially due to the excessive use of IoT sensors, WiFi, Internet connectivity, and low-cost and lightweight IoT devices.

There are several datasets in sensor-based human activity recognition for various perspectives (healthcare, fall detection, gaming activity monitoring, ambient assisted living (AAL), building smart homes, etc.). However, the depth, difficulties, diversity, number of subjects, number of classes, and other challenging issues within the datasets are relatively less than video-based activity recognition datasets. Till-to-date, most of the datasets contain fewer classes (e.g., 10 or 15 classes) by fewer subjects, and lack diversity in terms of age, gender, race, data collection environment, and other important factors. On the other hand, due to the availability of many YouTube videos, video-based datasets for activity analysis have even 600/1000 classes by a huge number of subjects in various constrained and unconstrained environments. However, there is a trade-off in terms of computational capability or cost to do research: sensor-based HAR can be achieved by normal computer or even on edge devices with less computational and memory capacity. However, for video-based HAR, we need GPU and more computational cost due to much larger datasets.

Sensor-based HAR has another challenge on temporal action segmentation. The research works on this arena have more focus on the video domain than the sensor domain. Therefore, we need to explore more on temporally segment an action more smartly. Handcrafted features can still dominate different datasets on sensors. However, on large datasets – deep learning-based methods are explored and have proven to be successful.

How to get human-object interactions or two-persons interactions based on sensors? These are difficult challenging areas for sensor-based HAR. Wearable sensors are attached to individuals and the data are not shared. Moreover, sensors are different and data collection processes can be varied from one to another. Unless we make a system for a constrained environment or case (for example, elderly persons in a smart-house, or sports activities), then data sharing and understanding interactions among more than one subject or object will be a very daunting task.

Multi-modality can bring a new dimension and can provide more information to explore some tasks. For example, let's consider a system to develop for patients with Parkinson's disease, or stroke, or alike. We need to collect data from video or skeleton of some actions to understand. Video can be difficult due to privacy issues. Convincing subjects/patients and doctors will be difficult too. Therefore, we can consider skeleton data that preserves much less personal information as video on privacy issues. However, skeleton data is found to be erroneous from Kinect V2 or recent depth/skeleton sensors. From erroneous data, we can not get minute differences between two similar kinds of actions (e.g., move left and right hands separately as two video/skeleton classes). If one hand has a stroke problem or partially paralyzed than the other – we can get better and differentiable information from both hands' data. However, if we would like to explore the improvement level of the paralyzed side from last week to this week – then can we get or collect any subtle difference between the actions of last week and this week? This is very difficult from noisy skeleton data. To overcome this challenge, if we can attach any wearable sensor on the wrist of the person's paralyzed hand as a band (similar to the patient's identification band) – then the wearable sensor can provide more information. Then a multimodal system can decipher more subtle differences. There are more scopes for multi-modality in this domain. Note that some patients may not like to wear anything wearable – as it might be painful or uncomfortable to wear on paralyzed or trouble parts of the body. Therefore, very light and comfortable sensors can be devised. Can we explore multiple sensors for a patient? Challenging indeed! However, [225] did an experiment on combinations of sensors on various locations. They concluded that a more suitable sensor location on-body can be the combination of right thigh, right lower leg, and back waist. But the work is done on some rehabilitation-related actions, performed by 16 healthy subjects, not any real patients. Moreover, this is not a proven issue in a very profound and robust manner. Therefore, more works to be done in these areas.

There are many methods for smartphone-based activity recognition. However, we do not keep phones always with us. Also, the location of smartphones and orientations of sensors are highly varied from person to person while performing different activities. A man usually carries a phone in his pants pocket or hand. A woman can keep the phone in her purse or vanity bag. When we work - the phones are usually not attached to the body but on the desk, etc. Therefore, we can not get seamless data. Other wearable sensors are used but the performances vary from one to another. Data modalities are different too. These are real-life challenges regarding sensor-based HAR. The orientation or location of sensors can be an issue to explore.

As we mentioned above that most of the sensor-based datasets are smaller in size in terms of the number of subjects and classes. To mitigate this shortcoming, we can merge multiple datasets into a large dataset. Then it can be more complex, challenging, and realistic. How much data can be considered sufficient in this domain? By varying the data sizes, we can try to explore more information. However, it is not easy to merge. The sensors are different, locations of the sensors are not always known, subjects are highly dominated by young students and mainly male, classes are mostly simple actions (e.g., walk, run, stand, sit, stair up, stair down, etc.). Accelerometer data, gyroscope data, magnetometer data, or other sensor data can be available in different datasets. According to [226], it is challenging to differentiate between accelerometer and magnetometer data. Multi-sensors data may improve performances. Can we explore audio sensors? Earable sensors? Pressure sensors? These should be explored based on various applications. Pressure sensors can be essential for transportation- and locomotion-based analysis (e.g., subway, train).

Though much progress on sensor-based human activity analysis and recognition has been made by industry and academia in the last decade, more challenges are in hand to explore. This chapter highlights some of these and we hope that we can progress more – so that we can build smarter systems, methods to help more people.

References

1. Wang, H., Zhao, J., Li, J., Tian, L., Tu, P., Cao, T., An, Y., Wang, K., Li, S.: Wearable sensor-based human activity recognition using hybrid deep learning techniques. Secur. Commun. Netw. **2020** (2020)
2. Dang, L.M., Min, K., Wang, H., Piran, Md.J., Lee, C.H., Moon, H.: Sensor-based and vision-based human activity recognition: a comprehensive survey. Pattern Recog. 107561 (2020)
3. Ahad, M.A.R., Antar, A.D., Ahmed, M.: IoT Sensor-Based Activity Recognition. Springer (2020)
4. Ahad, M.A.R., Antar, A.D., Shahid, O.: Vision-based action understanding for assistive healthcare: a short review. In: CVPR Workshops, pp. 1–11 (2019)
5. Abdel-Salam, R., Mostafa, R., Hadhood, M.: Human activity recognition using wearable sensors: review, challenges, evaluation benchmark. arXiv preprint arXiv:2101.01665 (2021)
6. Abdul Lateef Haroon, P.S., et al.: Human activity recognition using machine learning approach. J. Robot. Control (JRC) **2**(5), 395–399 (2021)
7. Zhou, X., Liang, W., Kevin, I., Wang, K., Wang, H., Yang, L.T., Jin, Q.: Deep-learning-enhanced human activity recognition for internet of healthcare things. IEEE Internet of Things J. **7**(7), 6429–6438 (2020)
8. Meng, L., Zhang, A., Chen, C., Wang, X., Jiang, X., Tao, L., Fan, J., Wu, X., Dai, C., Zhang, Y., et al.: Exploration of human activity recognition using a single sensor for stroke survivors and able-bodied people. Sensors **21**(3), 799 (2021)
9. Sankar, S., Srinivasan, P., Saravanakumar, R.: Internet of things based ambient assisted living for elderly people health monitoring. Res. J. Pharm. Technol. **11**(9), 3900–3904 (2018)
10. Zhang, W., Caixia, S., He, C.: Rehabilitation exercise recognition and evaluation based on smart sensors with deep learning framework. IEEE Access **8**, 77561–77571 (2020)
11. Schrader, L., Toro, A.V., Konietzny, S., Rüping, S., Schäpers, B., Steinböck, M., Krewer, C., Müller, F., Güttler, J., Bock, T.: Advanced sensing and human activity recognition in early intervention and rehabilitation of elderly people. J. Population Ageing 1–27 (2020)
12. Irvine, N., Nugent, C., Zhang, S., Wang, H., Ng, W.W.Y.: Neural network ensembles for sensor-based human activity recognition within smart environments. Sensors **20**(1), 216 (2020)
13. Fahad, L.G., Tahir, S.F.: Activity recognition and anomaly detection in smart homes. Neurocomputing **423**, 362–372 (2021)
14. Batchuluun, G., Kim, J.H., Hong, H.G., Kang, J.K., Park, K.R.: Fuzzy system based human behavior recognition by combining behavior prediction and recognition. Expert Syst. Appl. **81**, 108–133 (2017)
15. Pirbhulal, S., Wu, W., Muhammad, K., Mehmood, I., Li, G.: de Albuquerque VHC: mobility enabled security for optimizing IoT based intelligent applications. IEEE Netw. **34**(2), 72–77 (2020)
16. Abdulla, A.I., Abdulraheem, A.S., Salih, A.A., Sadeeq, M.A.M., Ahmed, A.J., Ferzor, B.M., Sardar, O.S., Mohammed, S.I.: Internet of things and smart home security. Technol. Rep. Kansai Univ. **62**(5), 2465–2476 (2020)
17. Shi, J., Zuo, D., Zhang, Z.: An energy-efficient human activity recognition system based on smartphones. In: 2020 7th International Conference on Soft Computing & Machine Intelligence (ISCMI), pp. 177–181. IEEE (2020)
18. Rawat, K.: Human activity recognition based on energy efficient schemes. Master's thesis, University of Twente (2020)

19. Tarafdar, P., Bose, I.: Recognition of human activities for wellness management using a smartphone and a smartwatch: a boosting approach. Decis. Support Syst. **140** (2021)
20. Ishii, S., Yokokubo, A., Luimula, M., Lopez, G.: Exersense: physical exercise recognition and counting algorithm from wearables robust to positioning. Sensors **21**(1), 91 (2021)
21. Martindale, C.F., Christlein, V., Klumpp, P., Eskofier, B.M.: Wearables-based multi-task gait and activity segmentation using recurrent neural networks. Neurocomputing **432**, 250–261 (2021)
22. Ngo, T.T., Ahad, M.A.R., Antar, A.D., Ahmed, M., Muramatsu, D., Makihara, Y., Yagi, Y., Inoue, S., Hossain, T., Hattori, Y.: OU-ISIR wearable sensor-based gait challenge: age and gender. In: 2019 International Conference on Biometrics (ICB), pp. 1–6. IEEE (2019)
23. Ahad, M.A.R., Ngo, T.T., Antar, A.D., Ahmed, M., Hossain, T., Muramatsu, D., Makihara, Y., Inoue, S., Yagi, Y.: Wearable sensor-based gait analysis for age and gender estimation. Sensors **20**(8), 2424 (2020)
24. Antar, A.D., Ahmed, M., Ishrak, M.S., Ahad, M.A.R.: A comparative approach to classification of locomotion and transportation modes using smartphone sensor data. In: Proceedings of the 2018 ACM International Joint Conference and 2018 International Symposium on Pervasive and Ubiquitous Computing and Wearable Computers, pp. 1497–1502 (2018)
25. Ahmed, M., Antar, A.D., Hossain, T., Inoue, S., Ahad, M.A.R.: POIDEN: position and orientation independent deep ensemble network for the classification of locomotion and transportation modes. In: Adjunct Proceedings of the 2019 ACM International Joint Conference on Pervasive and Ubiquitous Computing and Proceedings of the 2019 ACM International Symposium on Wearable Computers, pp. 674–679 (2019)
26. Friedrich, B., Lübbe, C., Hein, A.: Analyzing the importance of sensors for mode of transportation classification. Sensors **21**(1), 176 (2021)
27. Wang, L., Gjoreski, H., Ciliberto, M., Lago, P., Murao, K., Okita, T., Roggen, D.: Summary of the Sussex-Huawei locomotion-transportation recognition challenge 2020. In: Adjunct Proceedings of the 2020 ACM International Joint Conference on Pervasive and Ubiquitous Computing and Proceedings of the 2020 ACM International Symposium on Wearable Computers, pp. 351–358 (2020)
28. Liang, S.H.L., Saeedi, S., Ojagh, S., Honarparvar, S., Kiaei, S., Jahromi, M.M., Squires, J.: An interoperable architecture for the internet of Covid-19 things (IoCT) using open geospatial standards-case study: workplace reopening. Sensors **21**(1), 50 (2021)
29. Magesh, S., Niveditha, V.R., Rajakumar, P.S., Natrayan, L., et al.: Pervasive computing in the context of Covid-19 prediction with AI-based algorithms. Int. J. Pervasive Comput. Commun. (2020)
30. Zhang, H., Cai, Y., Zhang, H., Leung, C.: A hybrid framework for smart and safe working environments in the era of Covid-19. Int. J. Inf. Technol. **26**(1) (2020)
31. Smith, P.D., Bedford, A.: Automatic classification of locomotion in sport: a case study from elite netball. Int. J. Comput. Sci. Sport **19**(2), 1–20 (2020)
32. Franco, A., Magnani, A., Maio, D.: A multimodal approach for human activity recognition based on skeleton and RGB data. Pattern Recogn. Lett. **131**, 293–299 (2020)
33. Shaikh, M.B., Chai, D.: RGB-D data-based action recognition: a review (2021)
34. Tavakoli, M., Carriere, J., Torabi, A.: Robotics, smart wearable technologies, and autonomous intelligent systems for healthcare during the Covid-19 pandemic: an analysis of the state of the art and future vision. Adv. Intell. Syst. **2**(7), 2000071 (2020)
35. Formica, D., Schena, E.: Smart sensors for healthcare and medical applications (2021)
36. Demrozi, F., Pravadelli, G., Bihorac, A., Rashidi, P.: Human activity recognition using inertial, physiological and environmental sensors: a comprehensive survey. IEEE Access (2020)
37. Aggarwal, J.K., Xia, L.: Human activity recognition from 3D data: a review. Pattern Recogn. Lett. **48**, 70–80 (2014)
38. Weinland, D., Özuysal, M., Fua, P.: Making action recognition robust to occlusions and viewpoint changes. In: European Conference on Computer Vision, pp. 635–648. Springer (2010)

39. Ahad, M.A.R.: Computer Vision and Action Recognition: A Guide for Image Processing and Computer Vision Community for Action Understanding, vol. 5. Springer (2011)
40. Holte, M.B., Moeslund, T.B., Nikolaidis, N., Pitas, I.: 3D human action recognition for multi-view camera systems. In: 2011 International Conference on 3D Imaging, Modeling, Processing, Visualization and Transmission, pp. 342–349. IEEE (2011)
41. Chen, M., Hauptmann, A.: Mosift: recognizing human actions in surveillance videos (2009)
42. Donahue, J., Hendricks, L.A., Guadarrama, S., Rohrbach, M., Venugopalan, S., Saenko, K., Darrell, T.: Long-term recurrent convolutional networks for visual recognition and description. In: Proceedings of the IEEE Conference on Computer Vision and Pattern Recognition, pp. 2625–2634 (2015)
43. Poppe, R.: A survey on vision-based human action recognition. Image Vis. Comput. **28**(6), 976–990 (2010)
44. Liu, L., Peng, Y., Liu, M., Huang, Z.: Sensor-based human activity recognition system with a multilayered model using time series shapelets. Knowl.-Based Syst. **90**, 138–152 (2015)
45. Kanjo, E., Younis, E.M.G., Ang, C.S.: Deep learning analysis of mobile physiological, environmental and location sensor data for emotion detection. Inf. Fusion **49**, 46–56 (2019)
46. Mathie, M.: Monitoring and interpreting human movement patterns using a triaxial accelerometer. University of New South Wales Sydney (2003)
47. Huang, M., Zhao, G., Wang, L., Yang, F.: A pervasive simplified method for human movement pattern assessing. In: 2010 IEEE 16th International Conference on Parallel and Distributed Systems (ICPADS), pp. 625–628. IEEE (2010)
48. Liu, R., Zhou, J., Liu, M., Hou, X.: A wearable acceleration sensor system for gait recognition. In: 2nd IEEE Conference on Industrial Electronics and Applications, ICIEA 2007, pp. 2654–2659. IEEE (2007)
49. Wen, T., Wang, L., Gu, J., Huang, B.: An acceleration-based control framework for interactive gaming. In: Annual International Conference of the IEEE Engineering in Medicine and Biology Society, EMBC 2009, pp. 2388–2391. IEEE (2009)
50. Antar, A.D., Ahmed, M., Ahad, M.A.R.: Challenges in sensor-based human activity recognition and a comparative analysis of benchmark datasets: a review. In: 2019 Joint 8th International Conference on Informatics, Electronics & Vision (ICIEV) and 2019 3rd International Conference on Imaging, Vision & Pattern Recognition (icIVPR), pp. 134–139. IEEE (2019)
51. Ahmed, M., Antar, A.D., Ahad, M.A.R.: An approach to classify human activities in real-time from smartphone sensor data. In: 2019 Joint 8th International Conference on Informatics, Electronics & Vision (ICIEV) and 2019 3rd International Conference on Imaging, Vision & Pattern Recognition (icIVPR), pp. 140–145. IEEE (2019)
52. Zhang, M., Sawchuk, A.A.: A feature selection-based framework for human activity recognition using wearable multimodal sensors. In: BodyNets, pp. 92–98 (2011)
53. Wang, A., Chen, G., Xi, W., Liu, L., An, N., Chang, C.-Y.: Towards human activity recognition: a hierarchical feature selection framework. Sensors **18**(11), 3629 (2018)
54. Wang, Z., Wu, D., Chen, J., Ghoneim, A., Hossain, M.A.: A triaxial accelerometer-based human activity recognition via EEMD-based features and game-theory-based feature selection. IEEE Sens. J. **16**(9), 3198–3207 (2016)
55. Zhang, L., Wu, X., Luo, D.: Real-time activity recognition on smartphones using deep neural networks. In: 2015 IEEE 12th International Conference on Ubiquitous Intelligence and Computing and 2015 IEEE 12th International Conference on Autonomic and Trusted Computing and 2015 IEEE 15th International Conference on Scalable Computing and Communications and Its Associated Workshops (UIC-ATC-ScalCom), pp. 1236–1242. IEEE (2015)
56. Bengio, Y.: Deep learning of representations: looking forward. In: International Conference on Statistical Language and Speech Processing, pp. 1–37. Springer (2013)
57. Cook, D., Feuz, K.D., Krishnan, N.C.: Transfer learning for activity recognition: a survey. Knowl. Inf. Syst. **36**(3), 537–556 (2013)
58. Ijjina, E.P., Chalavadi, K.M.: Human action recognition in RGB-D videos using motion sequence information and deep learning. Pattern Recogn. **72**, 504–516 (2017)

59. Seyfioğlu, M.S., Özbayoğlu, A.M., Gürbüz, S.Z.: Deep convolutional autoencoder for radar-based classification of similar aided and unaided human activities. IEEE Trans. Aerosp. Electron. Syst. **54**(4), 1709–1723 (2018)
60. Nguyen, T.N., Lee, S., Nguyen-Xuan, H., Lee, J.: A novel analysis-prediction approach for geometrically nonlinear problems using group method of data handling. Comput. Methods Appl. Mech. Eng. **354**, 506–526 (2019)
61. Hossain, H.M.S., Al Hafiz Khan, Md.A., Roy, N.:. Active learning enabled activity recognition. Pervasive Mob. Comput. **38**, 312–330 (2017)
62. Grzeszick, R., Lenk, J.M., Rueda, F.M., Fink, G.A., Feldhorst, S., ten Hompel, M.: Deep neural network based human activity recognition for the order picking process. In: Proceedings of the 4th International Workshop on Sensor-Based Activity Recognition and Interaction, pp. 1–6 (2017)
63. Gil-Martin, M., San-Segundo, R., Fernandez-Martinez, F., Ferreiros-López, J.: Improving physical activity recognition using a new deep learning architecture and post-processing techniques. Eng. Appl. Artif. Intell. **92** (2020)
64. Kwon, H., Tong, C., Haresamudram, H., Gao, Y., Abowd, G.D., Lane, N.D., Ploetz, T.: IMU-Tube: automatic extraction of virtual on-body accelerometry from video for human activity recognition. Proc. ACM Interact. Mob. Wearable Ubiquit. Technol. **4**(3), 1–29 (2020)
65. Rafiuddin, N., Khan, Y.U., Farooq, O.: Feature extraction and classification of EEG for automatic seizure detection. In: 2011 International Conference on Multimedia, Signal Processing and Communication Technologies (IMPACT), pp. 184–187. IEEE (2011)
66. Hossain, T., Goto, H., Ahad, M.A.R., Inoue, S.: A study on sensor-based activity recognition having missing data. In: 2018 Joint 7th International Conference on Informatics, Electronics & Vision (ICIEV) and 2018 2nd International Conference on Imaging, Vision & Pattern Recognition (icIVPR), pp. 556–561. IEEE (2018)
67. Saha, S.S., Rahman, S., Rasna, M.J., Zahid, T.B., Mahfuzul Islam, A.K.M., Ahad, M.A.R.: Feature extraction, performance analysis and system design using the du mobility dataset. IEEE Access **6**, 44776–44786 (2018)
68. Saha, S.S., Rahman, S., Rasna, M.J., Hossain, T., Inoue, S., Ahad, M.A.R.: Supervised and neural classifiers for locomotion analysis. In: Proceedings of the 2018 ACM International Joint Conference and 2018 International Symposium on Pervasive and Ubiquitous Computing and Wearable Computers, pp. 1563–1570 (2018)
69. Hossain, T., Islam, Md.S., Ahad, M.A.R., Inoue, S.: Human activity recognition using earable device. In: Adjunct Proceedings of the 2019 ACM International Joint Conference on Pervasive and Ubiquitous Computing and Proceedings of the 2019 ACM International Symposium on Wearable Computers, pp. 81–84 (2019)
70. Hossain, T., Ahad, M.A.R., Tazin, T., Inoue, S.: Activity recognition by using lorawan sensor. In: Proceedings of the 2018 ACM International Joint Conference and 2018 International Symposium on Pervasive and Ubiquitous Computing and Wearable Computers, pp. 58–61 (2018)
71. Veltink, P.H., Bussmann, H.B.J., De Vries, W., Martens, W.L.J., Van Lummel, R.C.: Detection of static and dynamic activities using uniaxial accelerometers. IEEE Trans. Rehabil. Eng. **4**(4), 375–385 (1996)
72. Pirttikangas, S., Fujinami, K., Nakajima, T.: Feature selection and activity recognition from wearable sensors. In: International Symposium on Ubiquitious Computing Systems, pp. 516–527. Springer (2006)
73. Lavanya, B., Gayathri, G.S.: Exploration and deduction of sensor-based human activity recognition system of smart-phone data. In: 2017 IEEE International Conference on Computational Intelligence and Computing Research (ICCIC), pp. 1–5. IEEE (2017)
74. Bulling, A., Blanke, U., Schiele, B.: A tutorial on human activity recognition using body-worn inertial sensors. ACM Comput. Surv. (CSUR) **46**(3), 1–33 (2014)
75. Casale, P., Pujol, O., Radeva, P.: Human activity recognition from accelerometer data using a wearable device. In: Iberian Conference on Pattern Recognition and Image Analysis, pp. 289–296. Springer (2011)

76. Foerster, F., Smeja, M., Fahrenberg, J.: Detection of posture and motion by accelerometry: a validation study in ambulatory monitoring. Comput. Hum. Behav. **15**, 571–583 (1999)
77. Preece, S.J., Goulermas, J.Y., Kenney, L.P.J., Howard, D., Meijer, K., Crompton, R.: Physiological Measurement
78. Englehart, K., Hudgins, B., Parker, P., Stevenson, M.: Time-frequency representation for classification of the transient myoelectric signal. In: Proceedings of the 20th Annual International Conference of the IEEE Engineering in Medicine and Biology Society, vol. 20, Biomedical Engineering Towards the Year 2000 and Beyond (Cat. No.98CH36286), vol. 5, pp. 2627–2630 (1998)
79. Bao, L., Intille, S.S.: Activity recognition from user-annotated acceleration data. In: Pervasive Computing, pp. 158–175. Springer, Heidelberg (2004)
80. Nyan, M.N., Tay, F.E.H., Seah, K.H.W., Sitoh, Y.Y.: Classification of gait patterns in the time-frequency domain. J. Biomech. **39**(14), 2647–2656 (2006)
81. Najafi, B., Aminian, K., Paraschiv-Ionescu, A., Loew, F., Bula, C.J., Robert, P.: Ambulatory system for human motion analysis using a kinematic sensor: monitoring of daily physical activity in the elderly. IEEE Trans. Biomed. Eng. **50**(6), 711–723 (2003)
82. Sekine, M., Tamura, T., Togawa, T., Fukui, Y.: Classification of waist-acceleration signals in a continuous walking record. Med. Eng. Phys. **22**(4), 285–291 (2000)
83. Mantyjarvi, J., Himberg, J., Seppanen, T.: Recognizing human motion with multiple acceleration sensors. In: IEEE International Conference on Systems, Man, and Cybernetics, vol. 2, pp. 747–752. IEEE (2001)
84. Hira, Z.M., Gillies, D.F.: A review of feature selection and feature extraction methods applied on microarray data. Adv. Bioinform. **2015** 2015
85. Guyon, I., Elisseeff, A.: An introduction to variable and feature selection. J. Mach. Learn. Res. **3**(Mar), 1157–1182 (2003)
86. Miao, J., Niu, L.: A survey on feature selection. Procedia Comput. Sci. **91**, 919–926 (2016)
87. Forman, G.: An extensive empirical study of feature selection metrics for text classification. J. Mach. Learn. Res. **3**(Mar), 1289–1305 (2003)
88. Chen, X., Jeong, J.C.: Enhanced recursive feature elimination. In: Sixth International Conference on Machine Learning and Applications (ICMLA 2007), pp. 429–435. IEEE (2007)
89. Talenti, L., Luck, M., Yartseva, A., Argy, N., Houzé, S., Damon, C.: L1 logistic regression as a feature selection step for training stable classification trees for the prediction of severity criteria in imported malaria. arXiv preprint arXiv:1511.06663 (2015)
90. van der Maaten, L., Hinton, G.: Visualizing data using t-SNE. J. Mach. Learn. Res. **9**(Nov), 2579–2605 (2008)
91. McInnes, L., Healy, J., Melville, J.: UMAP: uniform manifold approximation and projection for dimension reduction. arXiv preprint arXiv:1802.03426 (2018)
92. Ali, M.U., Ahmed, S., Ferzund, J., Mehmood, A., Rehman, A.: Using PCA and factor analysis for dimensionality reduction of bio-informatics data. arXiv preprint arXiv:1707.07189 (2017)
93. Shi, H., Yin, B., Zhang, X., Kang, Y., Lei, Y.: A landmark selection method for l-isomap based on greedy algorithm and its application. In: 2015 54th IEEE Conference on Decision and Control (CDC), pp. 7371–7376. IEEE (2015)
94. Hall, M.A.: Correlation-based feature selection for machine learning (1999)
95. Baranauskas, J.A., Netto, O.P., Nozawa, S.R., Macedo, A.A.: A tree-based algorithm for attribute selection. Appl. Intell. **48**(4), 821–833 (2018)
96. He, Z., Jin, L.: Activity recognition from acceleration data based on discrete consine transform and SVM. In: IEEE International Conference on Systems, Man and Cybernetics, SMC 2009, pp. 5041–5044. IEEE (2009)
97. He, Z.-Y., Jin, L.-W.: Activity recognition from acceleration data using AR model representation and SVM. In: International Conference on Machine Learning and Cybernetics, vol. 4, pp. 2245–2250. IEEE (2008)
98. The aware home. http://awarehome.imtc.gatech.edu. Accessed 17 Mar 2021
99. Ahad, M.A.R.: Motion History Images for Action Recognition and Understanding. Springer (2012)

100. Hossain, T., Islam, Md.S., Ahad, M.A.R., Inoue, S.: Human activity recognition using earable device. In: Proceedings of the 2019 ACM International Joint Conference on Pervasive and Ubiquitous Computing and Proceedings of the 2019 ACM International Symposium on Wearable Computers, pp. 81–84. ACM (2019)
101. Hossain, T., Ahad, M.A.R., Tazin, T., Inoue, S.: Activity recognition by using lorawan sensor. In: 2018 ACM International Joint Conference on Pervasive and Ubiquitous Computing and the 2018 International Symposium on Wearable Computers (UbiComp/ISWC) (2018)
102. Ahmed, M., Antar, A.D., Ahad, M.A.R.: An approach to classify human activities in real-time from smartphone sensor data. In: 2019 Joint 8th International Conference on Informatics, Electronics Vision (ICIEV) and 2019 3rd International Conference on Imaging, Vision Pattern Recognition (icIVPR), pp. 140–145 (2019)
103. Zheng, Y., Wong, W.-K., Guan, X., Trost, S.: Physical activity recognition from accelerometer data using a multi-scale ensemble method. In: Twenty-Fifth Annual Conference on Innovative Applications of Artificial Intelligence. IAAI (2013)
104. Reyes-Ortiz, J.-L., Oneto, L., Ghio, A., Sama, A., Anguita, D., Parra, X.: Human activity recognition on smartphones with awareness of basic activities and postural transitions. In: International Conference on Artificial Neural Networks, pp. 177–184 (2014)
105. Bahle, G., Gruenerbl, A., Lukowicz, P., Bignotti, E., Zeni, M., Giunchiglia, F.: Recognizing hospital care activities with a coat pocket worn smartphone. In: 6th International Conference on Mobile Computing, Applications and Services (MobiCASE), pp. 175–181. IEEE (2014)
106. Kawsar, F., Min, C., Mathur, A., Montanari, A.: Earables for personal-scale behavior analytics. IEEE Pervasive Comput. **17**(3), 83–89 (2018)
107. Tapia, E.M., Marmasse, N., Intille, S.S., Larson, K.: Mites: wireless portable sensors for studying behavior. In: Proceedings of Extended Abstracts Ubicomp 2004: Ubiquitous Computing (2004)
108. Mica2dot wireless microsensor mote. https://www.willow.co.uk/mpr5x0-_mica2dot_series.php. Accessed 22 Mar 2019
109. Kling, R.M., et al.: Intel mote: an enhanced sensor network node. In: International Workshop on Advanced Sensors, Structural Health Monitoring, and Smart Structures, pp. 12–17 (2003)
110. Luna nurse. http://www.g-mark.org/award/describe/41326?locale=en. Accessed 17 Mar 2021
111. De-La-Hoz-Franco, E., Ariza-Colpas, P., Quero, J.M., Espinilla, M.: Sensor-based datasets for human activity recognition-a systematic review of literature. IEEE Access **6**, 59192–59210 (2018)
112. Lichman. UCI machine learning repository (2013). http://archive.ics.uci.edu/ml. Accessed 14 Feb 2021
113. Blunck, H., Bhattacharya, S., Stisen, A., Prentow, T.S., Kjærgaard, M.B., Dey, A., Jensen, M.M., Sonne, T.: Activity recognition on smart devices: dealing with diversity in the wild. GetMobile: Mob. Comput. Commun. **20**(1), 34–38 (2016)
114. Torres, R.L.S., Ranasinghe, D.C., Shi, Q., Sample, A.P.: Sensor enabled wearable RFID technology for mitigating the risk of falls near beds. In: 2013 IEEE International Conference on RFID (RFID), pp. 191–198. IEEE (2013)
115. Palumbo, F., Gallicchio, C., Pucci, R., Micheli, A.: Human activity recognition using multisensor data fusion based on reservoir computing. J. Ambient Intell. Smart Environ. **8**(2), 87–107 (2016)
116. Anguita, D., Ghio, A., Oneto, L., Parra, X., Reyes-Ortiz, J.L.: A public domain dataset for human activity recognition using smartphones. In: ESANN (2013)
117. Reyes-Ortiz, J.-L., Oneto, L., Samà, A., Parra, X., Anguita, D.: Transition-aware human activity recognition using smartphones. Neurocomputing **171**, 754–767 (2016)
118. Casale, P., Pujol, O., Radeva, P.: Personalization and user verification in wearable systems using biometric walking patterns. Pers. Ubiquit. Comput. **16**(5), 563–580 (2012)
119. Chavarriaga, R., Sagha, H., Calatroni, A., Digumarti, S.T., Tröster, G., Millán, J.D.R., Roggen, D.: The opportunity challenge: a benchmark database for on-body sensor-based activity recognition. Pattern Recogn. Lett. **34**(15), 2033–2042 (2013)

120. Ordóñez, F.J., de Toledo, P., Sanchis, A.: Activity recognition using hybrid generative/discriminative models on home environments using binary sensors. Sensors **13**(5), 5460–5477 (2013)
121. Baños, O., Damas, M., Pomares, H., Rojas, I., Tóth, M.A., Amft, O.: A benchmark dataset to evaluate sensor displacement in activity recognition. In: Proceedings of the 2012 ACM Conference on Ubiquitous Computing, pp. 1026–1035. ACM (2012)
122. Reiss, A., Stricker, D.: Introducing a new benchmarked dataset for activity monitoring. In: 2012 16th International Symposium on Wearable Computers (ISWC), pp. 108–109. IEEE (2012)
123. Altun, K., Barshan, B., Tunçel, O.: Comparative study on classifying human activities with miniature inertial and magnetic sensors. Pattern Recogn. **43**(10), 3605–3620 (2010)
124. Bacciu, D., Barsocchi, P., Chessa, S., Gallicchio, C., Micheli, A.: An experimental characterization of reservoir computing in ambient assisted living applications. Neural Comput. Appl. **24**(6), 1451–1464 (2014)
125. Banos, O., Garcia, R., Holgado-Terriza, J.A., Damas, M., Pomares, H., Rojas, I., Saez, A., Villalonga, C.: mhealthdroid: a novel framework for agile development of mobile health applications. In: International Workshop on Ambient Assisted Living, pp. 91–98. Springer (2014)
126. Weiss, G.M., Yoneda, K., Hayajneh, T.: Smartphone and smartwatch-based biometrics using activities of daily living. IEEE Access **7**, 133190–133202 (2019)
127. Schmidt, P., Reiss, A., Duerichen, R., Marberger, C., Van Laerhoven, K.: Introducing wesad, a multimodal dataset for wearable stress and affect detection. In: Proceedings of the 2018 on International Conference on Multimodal Interaction, pp. 400–408. ACM (2018)
128. Özdemir, A., Barshan, B.: Detecting falls with wearable sensors using machine learning techniques. Sensors **14**(6), 10691–10708 (2014)
129. Shoaib, M., Scholten, H., Havinga, P.J.M., Incel, O.D.: A hierarchical lazy smoking detection algorithm using smartwatch sensors. In: 2016 IEEE 18th International Conference on e-Health Networking, Applications and Services (Healthcom), pp. 1–6. IEEE (2016)
130. Shoaib, M., Bosch, S., Incel, O.D., Scholten, H., Havinga, P.J.M.: Complex human activity recognition using smartphone and wrist-worn motion sensors. Sensors **16**(4), 426 (2016)
131. Shoaib, M., Scholten, H., Havinga, P.J.M.: Towards physical activity recognition using smartphone sensors. In: 2013 IEEE 10th International Conference on and 10th International Conference on Autonomic and Trusted Computing (UIC/ATC), Ubiquitous Intelligence and Computing, pp. 80–87. IEEE (2013)
132. Shoaib, M., Bosch, S., Incel, O.D., Scholten, H., Havinga, P.J.M.: Fusion of smartphone motion sensors for physical activity recognition. Sensors **14**(6), 10146–10176 (2014)
133. Hasc2010 corpus. http://hasc.jp. Accessed 27 Mar 2019
134. Kawaguchi, N., Yang, Y., Yang, T., Ogawa, N., Iwasaki, Y., Kaji, K., Terada, T., Murao, K., Inoue, S., Kawahara, Y., et al.: Hasc2011corpus: towards the common ground of human activity recognition. In: Proceedings of the 13th International Conference on Ubiquitous Computing, pp. 571–572. ACM (2011)
135. Kawaguchi, N., Watanabe, H., Yang, T., Ogawa, N., Iwasaki, Y., Kaji, K., Terada, T., Murao, K., Hada, H., Inoue, S., et al.: Hasc2012corpus: large scale human activity corpus and its application. In: Proceedings of the Second International Workshop of Mobile Sensing: From Smartphones and Wearables to Big Data, pp. 10–14 (2012)
136. Kaji, K., Watanabe, H., Ban, R., Kawaguchi, N.: HASC-IPSC: indoor pedestrian sensing corpus with a balance of gender and age for indoor positioning and floor-plan generation researches. In: Proceedings of the 2013 ACM Conference on Pervasive and Ubiquitous Computing Adjunct Publication, pp. 605–610. ACM (2013)
137. Ichino, H., Kaji, K., Sakurada, K., Hiroi, K., Kawaguchi, N.: HASC-PAC2016: large scale human pedestrian activity corpus and its baseline recognition. In: Proceedings of the 2016 ACM International Joint Conference on Pervasive and Ubiquitous Computing: Adjunct, pp. 705–714. ACM (2016)

138. Matsuyama, H., Hiroi, K., Kaji, K., Yonezawa, T., Kawaguchi, N.: Ballroom dance step type recognition by random forest using video and wearable sensor. In: Proceedings of the 2019 ACM International Joint Conference on Pervasive and Ubiquitous Computing and Proceedings of the 2019 ACM International Symposium on Wearable Computers, pp. 774–780. ACM (2019)
139. Chen, Y., Keogh, E., Hu, B., Begum, N., Bagnall, A., Mueen, A., Batista, G.: The UCR time series classification archive (2015). www.cs.ucr.edu/~eamonn/time_series_data 2016
140. Goldberger, A.L.: Physiobank, physiotoolkkit, and physionet: components of a new research resource for complex physiologic signals. Circulation **101**(23), e215–e220 (2000)
141. Kotz, D., Henderson, T.: Crawdad: a community resource for archiving wireless data at dartmouth. IEEE Pervasive Comput. **4**(4), 12–14 (2005)
142. Bachlin, M., Plotnik, M., Roggen, D., Maidan, I., Hausdorff, J.M., Giladi, N., Troster, G.: Wearable assistant for Parkinson's disease patients with the freezing of gait symptom. IEEE Trans. Inf Technol. Biomed. **14**(2), 436–446 (2010)
143. Inoue, S., Ueda, N., Nohara, Y., Nakashima, N.: Recognizing and understanding nursing activities for a whole day with a big dataset. J. Inf. Process. **24**(6), 853–866 (2016)
144. Inoue, S., Lago, P., Hossain, T., Mairittha, T., Mairittha, N.: Integrating activity recognition and nursing care records: the system, deployment, and a verification study. In: Proceedings of the ACM on Interactive, Mobile, Wearable and Ubiquitous Technologies, vol. 3, no. 86 (2019)
145. Predicting Parkinson's disease progression with smartphone data. https://www.kaggle.com/c/3300/download/Participant%20Description.xls. Accessed 27 Mar 2019
146. Tian, Y., Zhang, J., Chen, L., Geng, Y., Wang, X.: Selective ensemble based on extreme learning machine for sensor-based human activity recognition. Sensors **19**(16), 3468 (2019)
147. Forster, K., Roggen, D., Troster, G.: Unsupervised classifier self-calibration through repeated context occurences: is there robustness against sensor displacement to gain? In: International Symposium on Wearable Computers, ISWC 2009, pp. 77–84. IEEE (2009)
148. Bächlin, M., Förster, K., Tröster, G.: Swimmaster: a wearable assistant for swimmer. In: Proceedings of the 11th International Conference on Ubiquitous Computing, pp. 215–224. ACM (2009)
149. Crowd-Sourced Fitbit Datasets (2016). Accessed 27 Mar 2019
150. Takata, M., Nakamura, Y., Fujimoto, M., Arakawa, Y., Yasumoto, K.: Investigating the effect of sensor position for training type recognition in a body weight training support system. In: Proceedings of the 2018 ACM International Joint Conference on Pervasive and Ubiquitous Computing and Proceedings of the 2018 ACM International Symposium on Wearable Computers, pp. 1–5. ACM (2018)
151. Tapia, E.M., Intille, S.S., Lopez, L., Larson, K.: The design of a portable kit of wireless sensors for naturalistic data collection. In: International Conference on Pervasive Computing, pp. 117–134. Springer (2006)
152. De la Torre, F., Hodgins, J., Bargteil, A., Martin, X., Macey, J., Collado, A., Beltran, P.: Guide to the Carnegie Mellon university multimodal activity (CMU-MMAC) database. Robotics Institute, p. 135 (2008)
153. Chen, L., Nugent, C.D., Biswas, J., Hoey, J.: Activity Recognition in Pervasive Intelligent Environments, vol. 4. Springer (2011)
154. Alemdar, H., Ertan, H., Incel, O.D., Ersoy, C.: Aras human activity datasets in multiple homes with multiple residents. In: Proceedings of the 7th International Conference on Pervasive Computing Technologies for Healthcare, pp. 232–235. ICST (Institute for Computer Sciences, Social-Informatics and Telecommunications Engineering) (2013)
155. Gani, Md.O., Saha, A.K., Ahsan, G.M.T., Ahamed, S.I., Smith, R.O.: A novel framework to recognize complex human activity. In: 2017 IEEE 41st Annual Computer Software and Applications Conference (COMPSAC), pp. 948–956. IEEE (2017)
156. Cook, D.J.: Learning setting-generalized activity models for smart spaces. IEEE Intell. Syst. **27**(1), 32–38 (2012)

157. Tapia, E.M., Intille, S.S., Larson, K.: Activity recognition in the home using simple and ubiquitous sensors. In: International Conference on Pervasive Computing, pp. 158–175. Springer (2004)
158. Huynh, T., Fritz, M., Schiele, B.: Discovery of activity patterns using topic models. In: Proceedings of the 10th International Conference on Ubiquitous Computing, pp. 10–19. ACM (2008)
159. Activity classification. https://www.kaggle.com. Accessed 28 Mar 2019
160. Eagle, N., Pentland, A.S.: Reality mining: sensing complex social systems. Pers. Ubiquit. Comput. **10**(4), 255–268 (2006)
161. Laurila, J.K., Gatica-Perez, D., Aad, I., Bornet, O., Do, T.-M.-T., Dousse, O., Eberle, J., Miettinen, M., et al.: The mobile data challenge: big data for mobile computing research. In: Pervasive Computing, number EPFL-CONF-192489 (2012)
162. Wagner, D.T., Rice, A., Beresford, A.R.: Device analyzer: large-scale mobile data collection. ACM SIGMETRICS Perform. Eval. Rev. **41**(4), 53–56 (2014)
163. Rawassizadeh, R., Tomitsch, M., Nourizadeh, M., Momeni, E., Peery, A., Ulanova, L., Pazzani, M.: Energy-efficient integration of continuous context sensing and prediction into smartwatches. Sensors **15**(9), 22616–22645 (2015)
164. Gjoreski, H., Ciliberto, M., Wang, L., Morales, F.J.O., Mekki, S., Valentin, S., Roggen, D.: The university of Sussex-Huawei locomotion and transportation dataset for multimodal analytics with mobile devices. IEEE Access (2018)
165. Yang, J.: Toward physical activity diary: motion recognition using simple acceleration features with mobile phones. In: Proceedings of the 1st International Workshop on Interactive Multimedia for Consumer Electronics, pp. 1–10. ACM (2009)
166. Zheng, Yu., Xie, X., Ma, W.-Y.: Geolife: a collaborative social networking service among user, location and trajectory. IEEE Data Eng. Bull. **33**(2), 32–39 (2010)
167. Wang, S., Chen, C., Ma, J.: Accelerometer based transportation mode recognition on mobile phones. In: 2010 Asia-Pacific Conference on Wearable Computing Systems (APWCS), pp. 44–46. IEEE (2010)
168. Reddy, S., Mun, M., Burke, J., Estrin, D., Hansen, M., Srivastava, M.: Using mobile phones to determine transportation modes. ACM Trans. Sens. Netw. (TOSN) **6**(2), 13 (2010)
169. Siirtola, P., Röning, J.: Recognizing human activities user-independently on smartphones based on accelerometer data. IJIMAI **1**(5), 38–45 (2012)
170. Hemminki, S., Nurmi, P., Tarkoma, S.: Accelerometer-based transportation mode detection on smartphones. In: Proceedings of the 11th ACM Conference on Embedded Networked Sensor Systems, p. 13. ACM (2013)
171. Zhang, Z., Poslad, S.: A new post correction algorithm (POCOA) for improved transportation mode recognition. In: 2013 IEEE International Conference on Systems, Man, and Cybernetics (SMC), pp. 1512–1518. IEEE (2013)
172. Xia, H., Qiao, Y., Jian, J., Chang, Y.: Using smart phone sensors to detect transportation modes. Sensors **14**(11), 20843–20865 (2014)
173. Widhalm, P., Nitsche, P., Brändie, N.: Transport mode detection with realistic smartphone sensor data. In: 2012 21st International Conference on Pattern Recognition (ICPR), pp. 573–576. IEEE (2012)
174. Jahangiri, A., Rakha, H.A.: Applying machine learning techniques to transportation mode recognition using mobile phone sensor data. IEEE Trans. Intell. Transp. Syst. **16**(5), 2406–2417 (2015)
175. Xing, S., Caceres, H., Tong, H., He, Q.: Online travel mode identification using smartphones with battery saving considerations. IEEE Trans. Intell. Transp. Syst. **17**(10), 2921–2934 (2016)
176. Yu, M.-C., Yu, T., Wang, S.-C., Lin, C.-J., Chang, E.Y.: Big data small footprint: the design of a low-power classifier for detecting transportation modes. Proc. VLDB Endow. **7**(13), 1429–1440 (2014)
177. Gjoreski, H., Ciliberto, M., Wang, L., Morales, F.J.O., Mekki, S., Valentin, S., Roggen, D.: The university of Sussex-Huawei locomotion and transportation dataset for multimodal analytics with mobile devices. IEEE Access **6**, 42592–42604 (2018)

178. Carpineti, C., Lomonaco, V., Bedogni, L., Di Felice, M., Bononi, L.: Custom dual transportation mode detection by smartphone devices exploiting sensor diversity. In: 2018 IEEE International Conference on Pervasive Computing and Communications Workshops (PerCom Workshops), pp. 367–372. IEEE (2018)
179. Islam, Z.Z., Tazwar, S.M., Islam, Md.Z., Serikawa, S., Ahad, M.A.R.: Automatic fall detection system of unsupervised elderly people using smartphone. In: Annual Conference on Artificial Intelligence. IEEE (2017)
180. Frank, K., Nadales, M.J.V., Robertson, P., Pfeifer, T.: Bayesian recognition of motion related activities with inertial sensors. In: Proceedings of the 12th ACM International Conference Adjunct Papers on Ubiquitous Computing-Adjunct, pp. 445–446. ACM (2010)
181. Vavoulas, G., Pediaditis, M., Spanakis, E.G., Tsiknakis, M.: The mobifall dataset: an initial evaluation of fall detection algorithms using smartphones. In: 2013 IEEE 13th International Conference on Bioinformatics and bioengineering (BIBE), pp. 1–4. IEEE (2013)
182. Kwolek, B., Kepski, M.: Human fall detection on embedded platform using depth maps and wireless accelerometer. Comput. Methods Programs Biomed. **117**(3), 489–501 (2014)
183. Gasparrini, S., Cippitelli, E., Spinsante, S., Gambi, E.: A depth-based fall detection system using a Kinect® sensor. Sensors **14**(2), 2756–2775 (2014)
184. Ojetola, O., Gaura, E., Brusey, J.: Data set for fall events and daily activities from inertial sensors. In: Proceedings of the 6th ACM Multimedia Systems Conference, pp. 243–248. ACM (2015)
185. Vilarinho, T., Farshchian, B., Bajer, D.G., Dahl, O.H., Egge, I., Hegdal, S.S., Lønes, A., Slettevold, J.N., Weggersen, S.M.: A combined smartphone and smartwatch fall detection system. In: 2015 IEEE International Conference on Computer and Information Technology; Ubiquitous Computing and Communications; Dependable, Autonomic and Secure Computing; Pervasive Intelligence and Computing (CIT/IUCC/DASC/PICOM), pp. 1443–1448. IEEE (2015)
186. Micucci, D., Mobilio, M., Napoletano, P.: UniMiB SHAR: a dataset for human activity recognition using acceleration data from smartphones. Appl. Sci. **7**(10), 1101 (2017)
187. Casilari, E., Santoyo-Ramón, J.A., Cano-García, J.M.: Analysis of a smartphone-based architecture with multiple mobility sensors for fall detection. PLoS ONE **11**(12), e0168069 (2016)
188. Vavoulas, G., Chatzaki, C., Malliotakis, T., Pediaditis, M., Tsiknakis, M.: The mobiact dataset: recognition of activities of daily living using smartphones. In: ICT4AgeingWell, pp. 143–151 (2016)
189. Sucerquia, A., López, J.D., Vargas-Bonilla, J.F.: SisFall: a fall and movement dataset. Sensors **17**(1), 198 (2017)
190. Zhang, M., Sawchuk, A.A.: USC-HAD: a daily activity dataset for ubiquitous activity recognition using wearable sensors. In: Proceedings of the 2012 ACM Conference on Ubiquitous Computing, pp. 1036–1043. ACM (2012)
191. Yang, A.Y., Jafari, R., Sastry, S.S., Bajcsy, R.: Distributed recognition of human actions using wearable motion sensor networks. J. Ambient Intell. Smart Environ. **1**(2), 103–115 (2009)
192. Stiefmeier, T., Roggen, D., Troster, G.: Fusion of string-matched templates for continuous activity recognition. In: 2007 11th IEEE International Symposium on Wearable Computers, pp. 41–44. IEEE (2007)
193. Wirz, M., Roggen, D., Troster, G.: Decentralized detection of group formations from wearable acceleration sensors. In: International Conference on Computational Science and Engineering, CSE 2009, vol. 4, pp. 952–959. IEEE (2009)
194. Saha, S.S., Rahman, S., Rasna, M.J., Mahfuzul Islam, A.K.M., Ahad, M.A.R.: DU-MD: an open-source human action dataset for ubiquitous wearable sensors. In: Joint 7th International Conference on Informatics, Electronics & Vision; 2nd International Conference on Imaging, Vision & Pattern Recognition (2018)
195. Chereshnev, R., Kertész-Farkas, A.: HuGaDB: human gait database for activity recognition from wearable inertial sensor networks. In: International Conference on Analysis of Images, Social Networks and Texts, pp. 131–141. Springer (2017)

196. Ngo, T.T., Makihara, Y., Nagahara, H., Mukaigawa, Y., Yagi, Y.: The largest inertial sensor-based gait database and performance evaluation of gait-based personal authentication. Pattern Recogn. **47**(1), 228–237 (2014)
197. Ahad, M.A.R., Ngo, T.T., Antar, A.D., Ahmed, M., Hossain, T., Muramatsu, D., Makihara, Y., Inoue, S., Yagi, Y.: Wearable sensor-based gait analysis for age and gender estimation (2020)
198. Chen, C., Jafari, R., Kehtarnavaz, N.: UTD-MHAD: a multimodal dataset for human action recognition utilizing a depth camera and a wearable inertial sensor. In: 2015 IEEE International Conference on Image Processing (ICIP), pp. 168–172. IEEE (2015)
199. Ngo, T.T., Ahad, M.A.R., Antar, A.D., Ahmed, M., Muramatsu, D., Makihara, Y., Yagi, Y., Inoue, S., Hossain, T., Hattori, Y.: OU-ISIR wearable sensor-based gait challenge: age and gender. In: Proceedings of the 12th IAPR International Conference on Biometrics, ICB (2019)
200. Kwapisz, J.R., Weiss, G.M., Moore, S.A.: Activity recognition using cell phone accelerometers. ACM SigKDD Explor. Newsl. **12**(2), 74–82 (2011)
201. Faye, S., Louveton, N., Jafarnejad, S., Kryvchenko, R., Engel, T.: An open dataset for human activity analysis using smart devices (2017)
202. Nickerson, R.S.: Binary-classification reaction time: a review of some studies of human information-processing capabilities. Psychonomic Monograph Supplements (1972)
203. Unler, A., Murat, A.: A discrete particle swarm optimization method for feature selection in binary classification problems. Eur. J. Oper. Res. **206**(3), 528–539 (2010)
204. Zhu, W., Lan, C., Xing, J., Zeng, W., Li, Y., Shen, L., Xie, X.: Co-occurrence feature learning for skeleton based action recognition using regularized deep LSTM networks. In: Thirtieth AAAI Conference on Artificial Intelligence (2016)
205. Li, S., Jiang, T., Huang, T., Tian, Y.: Global co-occurrence feature learning and active coordinate system conversion for skeleton-based action recognition. In: The IEEE Winter Conference on Applications of Computer Vision, pp. 586–594 (2020)
206. Zhao, R., Wang, K., Su, H., Ji, Q.: Bayesian graph convolution LSTM for skeleton based action recognition. In: Proceedings of the IEEE International Conference on Computer Vision, pp. 6882–6892 (2019)
207. Si, C., Chen, W., Wang, W., Wang, L., Tan, T.: An attention enhanced graph convolutional LSTM network for skeleton-based action recognition. In: Proceedings of the IEEE Conference on Computer Vision and Pattern Recognition, pp. 1227–1236 (2019)
208. Liu, J., Shahroudy, A., Xu, D., Kot, A.C., Wang, G.: Skeleton-based action recognition using spatio-temporal LSTM network with trust gates. IEEE Trans. Pattern Anal. Mach. Intell. **40**(12), 3007–3021 (2017)
209. Si, C., Jing, Y., Wang, W., Wang, L., Tan, T.: Skeleton-based action recognition with spatial reasoning and temporal stack learning. In: Proceedings of the European Conference on Computer Vision (ECCV), pp. 103–118 (2018)
210. Huang, J., Xiang, X., Gong, X., Zhang, B., et al.: Long-short graph memory network for skeleton-based action recognition. In: The IEEE Winter Conference on Applications of Computer Vision, pp. 645–652 (2020)
211. Ke, Q., Bennamoun, M., An, S., Sohel, F., Boussaid, F.: Learning clip representations for skeleton-based 3D action recognition. IEEE Trans. Image Process. **27**(6), 2842–2855 (2018)
212. Luvizon, D., Picard, D., Tabia, H.: Multi-task deep learning for real-time 3D human pose estimation and action recognition. IEEE Trans. Pattern Anal. Mach. Intell. (2020)
213. Morais, R., Le, V., Tran, T., Saha, B., Mansour, M., Venkatesh, S.: Learning regularity in skeleton trajectories for anomaly detection in videos. In: Proceedings of the IEEE Conference on Computer Vision and Pattern Recognition, pp. 11996–12004 (2019)
214. Zhang, P., Lan, C., Xing, J., Zeng, W., Xue, J., Zheng, N.: View adaptive recurrent neural networks for high performance human action recognition from skeleton data. In: Proceedings of the IEEE International Conference on Computer Vision, pp. 2117–2126 (2017)
215. Lee, I., Kim, D., Kang, S., Lee, S.: Ensemble deep learning for skeleton-based action recognition using temporal sliding LSTM networks. In: Proceedings of the IEEE International Conference on Computer Vision, pp. 1012–1020 (2017)

216. Tang, Y., Tian, Y., Lu, J., Li, P., Zhou, J.: Deep progressive reinforcement learning for skeleton-based action recognition. In: Proceedings of the IEEE Conference on Computer Vision and Pattern Recognition, pp. 5323–5332 (2018)
217. Shi, L., Zhang, Y., Cheng, J., Lu, H.: Skeleton-based action recognition with directed graph neural networks. In: Proceedings of the IEEE Conference on Computer Vision and Pattern Recognition, pp. 7912–7921 (2019)
218. Cho, S., Maqbool, M., Liu, F., Foroosh, H.: Self-attention network for skeleton-based human action recognition. In: The IEEE Winter Conference on Applications of Computer Vision, pp. 635–644 (2020)
219. Baek, S., Kim, K.I., Kim, T.-K.: Augmented skeleton space transfer for depth-based hand pose estimation. In: Proceedings of the IEEE Conference on Computer Vision and Pattern Recognition, pp. 8330–8339 (2018)
220. Cai, J., Jiang, N., Han, X., Jia, K., Lu, J.: JOLO-GCN: mining joint-centered light-weight information for skeleton-based action recognition. In: Proceedings of the IEEE/CVF Winter Conference on Applications of Computer Vision, pp. 2735–2744 (2021)
221. Ahad, M.A.R., Ngo, T.T., Antar, A.D., Ahmed, M., Hossain, T., Muramatsu, D., Makihara, Y., Inoue, S., Yagi, Y.: Wearable sensor-based gait analysis for age and gender estimation. Sensors **20**(8) (2020)
222. Wang, L., Gjoreski, H., Ciliberto, M., Lago, P., Murao, K., Okita, T., Roggen, D.: Summary of the Sussex-Huawei locomotion-transportation recognition challenge 2020. In: Tentori, M., Weibel, N., Van Laerhoven, K., Abowd, G.D., Salim, F.D. (eds.) UbiComp/ISWC 2020: 2020 ACM International Joint Conference on Pervasive and Ubiquitous Computing and 2020 ACM International Symposium on Wearable Computers, Virtual Event, Mexico, 12–17 September 2020, pp. 351–358. ACM (2020)
223. Alia, S.S., Lago, P., Takeda, S., Adachi, K., Benaissa, B., Ahad, M.A.R., Inoue, S.: Summary of the cooking activity recognition challenge. In: Human Activity Recognition Challenge, pp. 1–13. Springer (2021)
224. Alia, S.S., Lago, P., Adachi, K., Hossain, T., Goto, H., Okita, T., Inoue, S.: Summary of the 2nd nurse care activity recognition challenge using lab and field data. In: Adjunct Proceedings of the 2020 ACM International Joint Conference on Pervasive and Ubiquitous Computing and Proceedings of the 2020 ACM International Symposium on Wearable Computers, pp. 378–383 (2020)
225. Komukai, K., Ohmura, R.: Optimizing of the number and placements of wearable IMUs for automatic rehabilitation recording. In: Kawaguchi, N., Nishio, N., Roggen, D., Inoue, S., Pirttikangas, S., van Laerhoven, K. (eds.) Human Activity Sensing. Springer Series in Adaptive Environments. Springer, Cham (2019)
226. Scholl, P.M., van Laerhoven, K.: Identifying sensors via statistical analysis of body-worn inertial sensor data. In: Kawaguchi, N., Nishio, N., Roggen, D., Inoue, S., Pirttikangas, S., van Laerhoven, K. (eds.) Human Activity Sensing. Springer Series in Adaptive Environments. Springer, Cham (2019)

Chapter 7
Radar-Based Non-Contact Physiological Sensing

Shekh Md Mahmudul Islam

Abstract Radar remote respiration sensing technology is gaining attention and over the last four decades, there has been significant advancement to utilize this technology in different biomedical application areas. Microwave Radar sensing technologies and systems are playing an important role in different healthcare applications such as remote breathing monitoring, identity authentication, occupancy sensing, sleep monitoring, security/surveillance, and so on. Radar is an attractive approach for remote breathing monitoring as it is a non-contact and privacy-invasive way to monitor. This chapter focuses on discussing the basics of radar remote respiration sensing technology and its associated advantages over other traditional wearable sensing methodologies. It also highlights the different potential new emerging application areas of this non-contact sensing technique. Additionally, it also discusses the research direction for young researchers to explore this new sensing methodology and continue their efforts in this area. Several future research challenges have been highlighted to solve these challenges to bring this sensor technology into the real world of implementation for ubiquitous deployment.

7.1 Introduction

Non-contact cardiopulmonary monitoring is gaining attention as a potential technological solution for vital signs monitoring during sleep, home healthcare sensing, infant monitoring, post-surgery monitoring and search and rescue applications. Without attaching any sensor to the body or human chest surfaces, this contactless sensing methodology could provide potential remote vital sign sensing to the chronically ill patients in an ambulance, enable sleep monitoring outside expensive sleep laboratories, and, detect survivors under the rubble. Over the last four decades, there has been a significant improvement in radar-based non-contact physiological sensing especially converting the laboratory experimental demonstration in real-world

---

S. M. M. Islam (✉)
Department of Electrical and Electronic Engineering, University of Dhaka, Dhaka, Bangladesh
e-mail: mahmud@du.ac.bd

M. A. R. Ahad and A. Inoue (eds.), *Vision, Sensing and Analytics: Integrative Approaches*, Intelligent Systems Reference Library 207, https://doi.org/10.1007/978-3-030-75490-7_7

settings for numerous applications [40], which could also potentially lead to the development of a non-contact, unobtrusive and long term respiration sensing device.

Cardiopulmonary monitoring is mostly carried out in a hospital setting with a sensor that needs to be attached to the body such as an electrocardiogram (ECG) electrodes. The attachment of the sensor to the body always creates an extra burden, skin irritations, and not applicable for long-term monitoring applications such as post-surgery patients, chronic and elderly patients, and, sleep disorder patients. Similarly, non-contact respiration sensing technology can also reduce the burden of premature infants and burn victims. Continuous monitoring of respiratory patterns without attaching any sensor to the body can also prevent subsequent cardiopulmonary arrests and that can also reduce the mortality rate. Early identification and recognition of respiratory abnormalities by continuous monitoring is one of the key steps of reducing the mortality rate. During this COVID-19 pandemic, the importance of non-contact respiratory sensing is also realized significantly by the scientific community, and also numerous publications have been reported [47]. Due to the contact nature of ECG, many patients are not continuously monitored. Thus, there is a growing need to bring non-contact physiological sensing technologies into the real world of implementation.

Non-contact respiration sensing technology has several potential impacts beyond healthcare applications. Such as occupancy sensing in a room environment for energy savings [71], recognizing people from their breathing diversity [35], finding individuals for disaster and emergency responses [37], and recognizing patients from their irregular breathing pattern for medical data integrity [29].

7.2 Basics of Radar

Radar stands for Radio Detection And Ranging described a transceiver that transmits radio frequency (RF) signal and senses the reflected echoes from reflecting targets, from the echoes it can gain different information of the target. In general, there is a time difference between the transmitted and reflected signal which is used to measure the distance between the target, the frequency deviation of the reflected echoes helps to extract the velocity information, and the reflected signal strength is also used to provide information about the target's shape, orientation, and composition. The transmitter emits the electromagnetic waveform toward a target which is usually in the microwave and millimeter-wave band. The receiver is in general very sensitive device which can extract very weak echoes from the target to extract information about the position, size, velocity of the target. To extract useful information, the received signal is amplified and then conditioned for further processing. One of the potential advantages of radar is it can penetrate through some objects such as walls where light cannot. Besides, the radar signal is not affected by weather conditions such as fog, mist, or variations in sunlight. Based on applications radar frequencies may vary ranges from megahertz to optical frequencies. Radar was originally developed for military applications and the word "Radar" always reminds us about the battlefield.

The development of the radar was related to providing warning the hostile aircraft and for directing aircraft weapons as well. The United States Army Signal Corps did the pioneering work during the second world war for the development of this modern radar concept. The term radar became popular to the scientific community at the end of the second world war. Before that, it was named RDF (radio detection and finding) in the UK [60]. In France, radar was also known as DEM (detection electromagnetic). Modern radar has been developed sophistically to cover a wide range of applications such as classification or identification of the target, imaging of the target (satellite ground mapping), and tracking wildlife (biological studies). Radar has now numerous civilian applications, including weather forecasting, position and location determination of aircraft's, ships, and spacecraft, and law enforcement.

7.2.1 Radar Principle of Operation and Waveforms

As shown in Fig. 7.1, in a simplistic way, a radar typically comprised of a transmitter and a receiver. In general, a transmitter generate radio-frequency (RF) signal and amplifies it, and then transmit it [59]. The antenna pattern used in the transmitter may be omnidirectional or directional. A directional antenna has a fixed radiation pattern and within the view of that pattern direction, it can determine the position of the target. In general, radar transmitters transmit a signal at a higher frequency towards the target. When the signal is being reflected and received by the receiver, it converts the high-frequency signal into a lower frequency signal which is also known as an intermediate frequency or baseband signal. The receiver also consists of signal processing chains where noise suppression is performed to reject clutter or other noises coming from the stationary targets. Radar operating frequency also varies in different ranges although the frequency depends on application purposes and also the consideration of the antenna size, transmitter power, range to the target. For most common applications, radio-frequency ranges from 3 KHz–300 GHz. At present, most of the radar system works on operating frequencies range from 3 MHz–300 GHz which is also defined as microwave and millimeter wave by the IEEE Standard for specific letter designation for frequency bands [60].

The basic principle of radar for determining the target's size, velocity, and shape is the underlying principle of the Doppler effect. The Doppler shift is the change in the frequency shift of the returning signal from the transmitted signal when a target is moving in the vicinity of the radar. Doppler effect was first observed and explained by the German physicist Christian Doppler and this same phenomenon is applied to all wave motions, including sound, and light [60]. Figure 7.1 illustrates that a target is moving at a fixed velocity, V_r and the associated frequency shift of the transmitted signal that represented as, $f_d = 2 \times \frac{V_r}{\lambda}$; where V_r is the radial velocity. From the frequency shift the velocity of the car can be determined.

Based on the transmitted antenna waveforms radar can be classified into three major categories. The major three categories are:

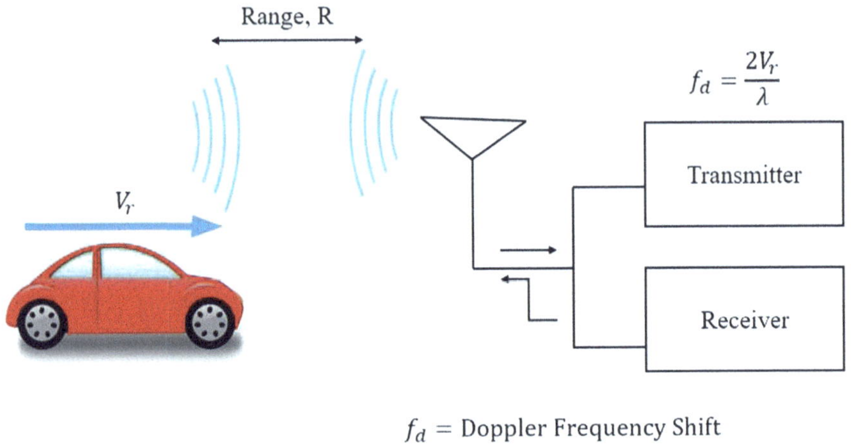

Fig. 7.1 Illustration of Doppler effect of a transceiver when a car is moving towards a radar with a constant radial velocity

1. Continuous-wave (CW) radar
2. Frequency-modulated continuous wave (FMCW) radar
3. Pulsed radar

In the next subsections we will describe in details of the classification of radar.

7.2.2 Continuous-Wave (CW) Radar

A CW radar continuously sends an electromagnetic signal toward the target and receives a signal which is very narrow band [59]. It has a very simple architecture. A CW radar transceiver continuously transmits and receives a signal so there is no need to control the switching between two different channels. However, a duplexer sometimes is also integrated with the transmission antenna for isolating the transmitted and received signals. For switching between two different channels separate antenna is also employed in CW radar. Figure 7.2 shows the simple architecture of CW radar. Due to the transmission and reception of the narrowband signal in CW radar, simple filters can be utilized in each stage of the receiver. Another potential advantage of transmission and reception of the narrowband signal helps to measure the Doppler shift in CW accurately to ambiguously determine the velocity of the target.

One of the bottlenecks of using the same CW radar antenna is the leakage from the transmitter to the receiver. This particular leakage signal adds to the received signal and frequency deviation takes place that is not because of the reflection from the target. Besides, if the signal strikes any stationary object known as the clutter also adds noise to the received signal. This unwanted signal is also known as DC

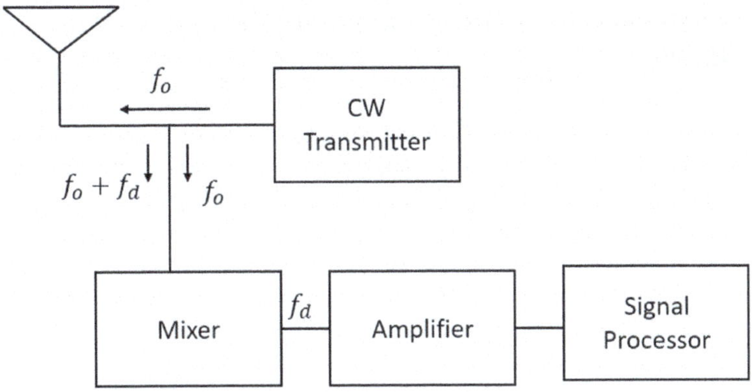

Fig. 7.2 Block-diagram of continuous wave (CW) radar

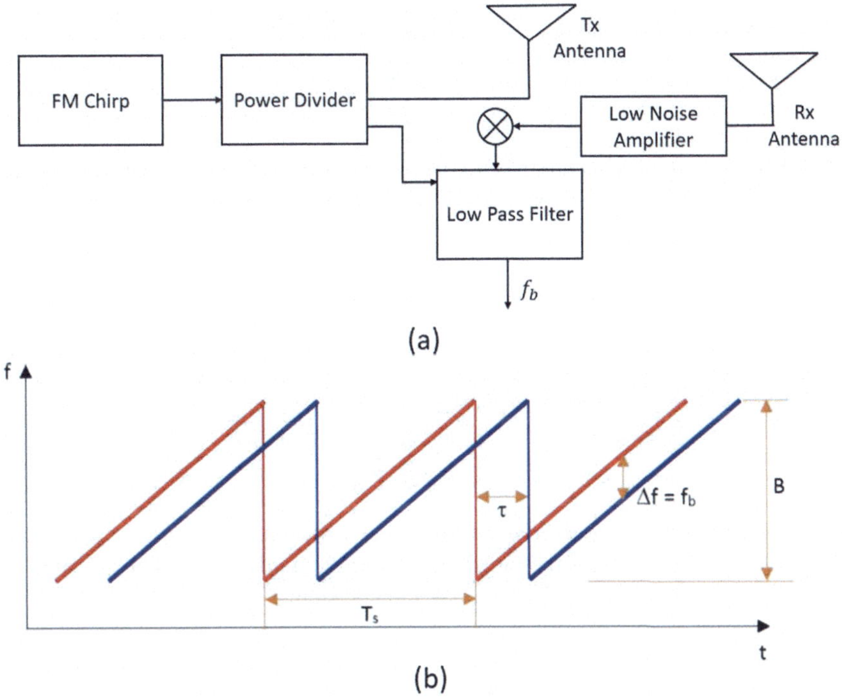

Fig. 7.3 **a** Block-diagram of FMCW radar system **b** Typical waveform of transmitted and reflected signal of FMCW radar system with the triangular modulation

offset. For removing the DC offset, post-processing of the received signal is required. However, the potential advantage of CW radar architecture is simple circuitry and low-cost.

7.2.3 Frequency Modulated Continuous Wave (FMCW) Radar

From the reflected signal frequency shift, CW radar can determine the velocity of the target unambiguously. But CW radar cannot determine the range of the target as the transmitted signal does not utilize any timing marker or the frequency sweep is also not considered during the transmission of the signal. For determining the velocity and range of the target, CW signal need to be modulated which is known as frequency modulation (FM) [59]. The frequency modulation is performed in FMCW radar system by transmitting the triangular waveform. Figure 7.3(a) illustrates the diagram of the FMCW radar system and Fig. 7.3(b) represents the transmitted and received FMCW triangular waveform. The transmitted signal is represented is illustrated in Fig. 7.3(b) by the red line and the blue line represents the received echo from the target. The frequency deviation, δf is the bandwidth of the FMCW signal. The transmitted signal is returned at the radar after a delay time, τ is represented as:

$$\tau = \frac{2R}{c} \tag{1.1}$$

where R is the target range and c is the free-space speed of light. In the receiver the received signal frequency is determined by the below equation:

$$f_b = \frac{B}{T_s}\tau \tag{1.2}$$

where B is the total frequency deviation of the chirp signal and T_s is the sweep time (chirp period). The target range can be determined by the below equation:

$$R = \frac{cT_s}{2B} f_b \tag{1.3}$$

One of the potential advantages of an FMCW radar system is the detection capability of the range and velocity of the target. However, the hardware circuitry of FMCW radar is much more complex than the CW radar system.

7.2.4 Pulse Doppler Radar

In general, a pulse radar system transmits a signal with narrow pulses having large peak power after a certain interval of time which is known as pulse repetition frequency (PRF), and then analyzes the reflected signal. In contrast to CW radar, the transmitter is turned off before receiving the pulses [60]. Pulse radar also uses the same principle of Doppler radar which is known as the Doppler principle. In other words, pulse radar is known as a moving target indicator (MTI) or Pulse-Doppler

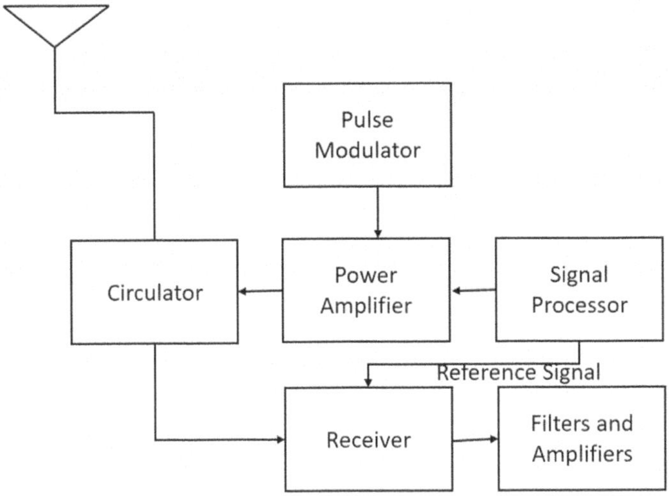

Fig. 7.4 Functional block diagram of the pulse-Doppler radar system

radar. One of the potential benefits of the pulse radar is it can estimate the velocity and range of the target simultaneously. One of the potential benefits of the pulsed radar system is its delay between the transmission and reception of the signal from the target. Due to the isolation of the transmitter to the receiver, there is no leakage in between the transmitter and receiver which also helps to mitigate the effect of clutters. To detect the Doppler shift of the reflected signal the pulse width of the transmitted signal should be higher and at the same time velocity of the target should be high enough for tracking.

A pulsed Doppler transmit signal is used which sends a pulse modulated RF carrier. Time delay between the transmission and reception of the signal helps to extract the range information accurately. Figure 7.4 shows a simple illustration of a block diagram pulsed radar system. The transmit antenna transmits a pulsed microwave signal. This pulse is reflected by the object and a weak reflection is received at the receive antenna. In simple terms the time delay of the two pulses is used to determine the distance between targets [59] using below equation:

$$R = t/2c \tag{1.4}$$

where t is the round-trip time of the pulse, c is speed of light and R is the range of the object. The technique is known as pulsed radar due to the nature of the transmission signal. The velocity of the moving target can also be determined using the Doppler effect. The radial speed of the object will alter the frequency of the received echo pulse by:

$$\nabla f = -V/c \tag{1.5}$$

where V is the speed (negative when the object is moving toward the radar) and ∇f is the frequency shift in the echo signal. There have been many techniques introduced to isolate the noise and interference of the received signal and the accuracy of speed and range estimation using phase and/or frequency modulation of the transmit signal. Strength of the transmit and received signals vary depending on the application. For extracting the Doppler frequency shift from the pulsed radar, pulse width of the transmitted signal, T is denoted as the time period of the signal where one cycle of Doppler frequency shift can exist. In general, Doppler frequency shift, f_d should be within the pulse to extract the target's information within a single pulse, $f_d T > 1$.

7.3 Traditional Methods of Respiratory Monitoring

The traditional method of assessing cardiopulmonary functions is mostly contacted sensor-based where sensors need to be attached to the human subject. In hospital settings, ECG is the gold standard to monitor the heart rate of patients. Although there is no equivalent gold standard for respiratory monitoring. Respiratory rate is one of the critical vital signs but not measured frequently. But based on respiratory rate physiological instability of human subjects can be assessed. In hospital settings, different vital signs such as blood pressure, pulse rate, and arterial oxygen saturation can be monitored frequently by healthcare professionals (nurses) with different sensors. Although the respiratory rate is measured by visual inspection with a nurse counting the chest movement.

The traditional method to monitor respiration measurement in a clinical setting can be divided into three major categories such as oxygen saturation measurement, airflow measurement, and breathing frequency measurement [67]. Pulse oximetry is the most popular technique where the calculation of the percentage of hemoglobin (Hb) is saturated within the oxygen. The basic mechanism of pulse oximetry is the transmission of a light source that is generated within a wavelength (650 and 805 nm) passed through blood and then absorption of hemoglobin takes place based on the concentration of blood flow. Another traditional way of measuring respiration flow rate is the spirometer with a mouthpiece or a face-mask. Spirometer utilizes the pressure transducer for measuring the flow rate. It also measures the volume and the amount of air inhaled or exhaled within a fixed time. Nowadays, this device has not been using hospital settings due to the large volume and high resistances. A thermocouple or capnograph is also traditionally used to measure the blood flow rate which has a less adverse effect than a spirometer. Additionally, it also requires close contact with the nose or mouth which is also uncomfortable for humans. There are numerous indirect ways to measure respiratory rate. All respiratory measurement has its advantages and drawbacks as well. Pulse oximetry is a quite popular technique to measure respiratory disturbances but it can not measure the respiratory rate. The whole body plethysmograph is more accurate but it requires sedation of the patients. In ambulatory conditions, the most popular way of respiratory monitoring of the patient is the chest belt which is a transducer but its performance degrades over time with the random body movement.

ECG is the traditional gold standard method to measure the cardiac activity of the patients. Electrical activity of the heart is recorded by ECG. Electrical wave causes the heart muscle to contract and expand. Electrodes are attached with the chest surface to determine the voltage differences between the pair of these electrodes. From the voltage differences of the electrodes, the overall rhythm of the heart can be measured. Another popular technique is photoplethysmograph (PPG) where optical sensor is employed to estimate the pulses of blood that is also caused by heartbeats. PPG is a little bit simpler in architecture than ECG and can be attached to the body comfortably. Although all these methods described above require close contact with the body to monitor vital signs.

7.3.1 Need for Non-contact Remote Respiration Monitoring and Potential Application Areas

There are several potential advantages of non-contact physiological sensing and it has several potential areas. The advantages of non-contact physiological sensing are listed below:

1. Remote respiration monitoring without attaching any sensor to the human body would help to reduce skin irritation.
2. Continuous monitoring is possible with the unobtrusive radar-based respiration sensing.
3. Electrode attachment sometimes obstructs the regular breathing pattern and is also not appropriate for long term monitoring applications (e.g. sleep).
4. Another potential benefit of this technique is the privacy-invasive capabilities. Vision-based sensing technology is already a well-established area. However, one of the fundamental limitations of vision sensing is privacy as intrusive video imaging is required.
5. Beyond healthcare potential areas, the method can also be beneficial for human identification, human presence detection, life sensing behind the wall, and post-disaster search and rescue applications.

7.3.2 Patients with Burn Injury

Non-contact and unobtrusive form of the respiratory monitoring devices can potentially play an important role in modern healthcare. For example, in an intensive care unit (ICU) neonates suffer skin damage due to the adhesive tape and ECG electrodes as there is no reliable non-contact respiratory monitoring device. Similarly, monitoring the vital signs of the burn victim patients is challenging as they may not have enough skin on which electrodes can be placed. In many cases, electrodes need to be stapled to the skin for monitoring the respiratory pattern. A non-contact, wireless heart, and breathing monitoring system could potentially help to reduce the sufferings of the neonates and burn victims.

7.3.3 Sleep Monitoring

Sleep is a biological imperative and a sufficient amount of daily sleep is required for better physical and mental health. Cardiovascular health, metabolism, weight, mood, and immune system function also depends on the quantity and quality of sleep. Obstructive sleep apnea (OSA) is the most common respiratory disorder where stoppage of airflow occurs more than 10 s and during OSA it happens multiple times for the patients during their sleeping time. There are several negative effects of OSA including excessive daytime sleepiness, hypertension, psychological distress, and so on. A current gold standard method for detecting OSA is polysomnography (PSG) where patients need to go to dedicated sleep laboratories. PSG consists of simultaneous recording of electrophysiological signals (ECG, EEG) during overnight monitoring. Attaching all those sensors to the human body always creates an extra burden and may also hamper the regular pattern of sleep. A Doppler radar-based non-contact, unobtrusive respiratory monitoring system has already provided its efficacy to recognize different sleep apnea events in a clinical study [6]. Thus, Doppler radar could be combined with the PSG system to make it non-contact and unobtrusive so that the suffering of the patients can be minimized [5]. Additionally, a radar-based portable sleep monitoring device could also help to diagnose OSA in an early stage [5].

7.3.4 Unobtrusive Identity Authentication

One of the most popular and commonly used authentication system is password and facial recognition where users need to go for one pass validation system which creates data falsification and information can be lost as well. Identity theft is one of the most common problem in day-to-day activities and the conventional identity authentication system is just a one-pass validation system [42]. In 2018, the number of consumers who were the victim of identity fraud was approximately 14.4 million in the United States. Recognizing people from radar-captured breathing pattern is gaining attention due to its non-contact and unobtrusive characteristics. It has also been proven that there is significant diversity in the breathing pattern of different people in terms of their airflow profile, breathing depth, and energy, and so on. From the radar captured respiratory traces, prior researches demonstrated the initial feasibility of recognizing people which opens a new door for non-contact identity authentication [24, 42, 57]. Additionally, Doppler radar has also been investigated to recognize people from abnormal breathing patterns such as obstructive sleep apnea (OSA) patients [29]. Authenticating people unobtrusively from their respiratory pattern could also reduce the shortcoming of the traditional identity authentication system (password, and facial recognition) [35]. Doppler radar could be a potential solution for overcoming the shortcoming of the existing identity authentication approaches. The proposed unobtrusive identity authentication technology is still not mature which requires further investigations and experimentation to bring this potential technology into the real world environment.

7.3.5 Elderly Monitoring

In-home healthcare monitoring of the elderly is one of the active areas of research and investigations as the percentage of the elderly people is increasing. Elderly people have reduced mobility in their home environment so long-term monitoring is one of the important tasks. Besides, due to the high risk of gait or balance disorder of seniors, there is a higher chance of falls in daily activities which can also lead to morbidity and mortality. For measuring vital signs in a home environment there are not that many off-body sensors in the market that can collect all this valuable information and track their health status.

The Doppler already proved its efficacy for human gait monitoring [41]. Additionally, Doppler radar has also been utilized for fall detection with very high accuracy to discriminate between normal activities and fall incidences [48]. Doppler radar system could potentially play an important role for elderly healthcare monitoring without attaching any sensor to the body.

7.3.6 Occupancy Sensing

Occupancy sensing is gaining attention due to its efficacy of energy savings. For building automation and energy savings, different commercial sensors such as passive infrared (PIR) and ultrasound is commonly used [68]. The commercial building can increase their energy savings by turning on/off on time based on detecting the presence of the human [38]. Camera-based occupancy sensing is also popular. However, due to long term monitoring, people feel uncomfortable with having intrusive video imaging for such a long time [72]. PIR and ultrasonic sensors create a false alarm when people are in a static position [71]. Due to the limitations of the traditional sensors, radar has also been investigated to test the feasibility of human presence detection based on respiratory motions. When the presence of the subject in an environment is increased the reflected echo becomes an interference of respiration patterns or the effective radar-cross section changes due to the presence of multiple subjects [72]. As the ERCS of the received signal changes, it also affects the RMS output voltage of the received signal [71, 72]. As the ERCS of the received signal changes, it also affects the RMS output voltage of the received signal. Radar is an attractive approach than a camera because of its unobtrusiveness and privacy-invasive properties.

7.3.7 Unmanned Aerial Vehicle (UAV)-Based Life Sensing Radar

Search and rescue operation during post-disaster and war need to be conducted unobtrusively so that it can minimize the loss [45]. Technology for detecting the presence of humans through barriers from vital signs can be promising for search and rescue applications. Additionally, there is a significant advancement of unmanned aerial vehicle (UAV) based technology for remote vital signs monitoring [49, 50]. Unmanned Aerial Vehicles (UAVs) with on board Doppler radar sensors can be used for health reconnaissance including the remote detection of respiratory patterns. Commercially available UAVs can travel anywhere a person can and beyond, only limited by communications range and battery life, and can relay triage relevant information to safely-isolated human responders via optical imagery. Prior research also demonstrated the feasibility of extracting respiratory signatures using microwave Doppler radar mounted with drone has also been gaining attention and proved its efficacy. It can also potentially used for search and rescue operations in the battlefield and after post disaster scenarios [27, 34]. UAV-borne radar sensors can provide a broad range of information about victims which can be useful for remote assessment by rescuers [31]. Non-contact sensing of physiological motion using stationary microwave Doppler radar has shown efficacy with proof of concepts demonstrated for vital signs sensing in various applications over the past four decades [40], and could be a powerful triage assessment tool if implemented on a mobile platform like a UAV. However, mobile platform motion will induce extraneous noise which will add an undesired phase component to the radar baseband signal making the extraction of respiratory information much more challenging and it is an active area of research and investigations for solving this issue [50]. Recent research efforts also demonstrated that adaptive filter technique has the potential to reduce the interference of the platform motion [31]. Further investigation is required to overcome other associated challenges for the improvement of the reliability and robustness of the system.

7.4 Basics of Doppler Radar for Non-contact Cardiopulmonary Monitoring

Vital signs monitoring using microwave Doppler radar is gaining attention and has been demonstrated its feasibility for about four decades with different potential application areas [14]. In this method, typically a radar transmits electromagnetic signal toward human subject under measurement. In general electromagnetic waves incident upon the human body since human body acts an in-homogeneity in the air due to the differences in permittivity and conductivity [44]. A live human body will have cardiopulmonary activities which results in physiological motion such as tiny movement of the chest surfaces [15]. These small variations of tiny movement of the chest surfaces is responsible for the time delay of the reflected echo and transmitted signal. The received echo is compared with the transmitted signal in a mixer which outputs the phase difference between the two signals. This idea was first demonstrated by J. Lin in 1970's [43] for measurement of respiration of human subject. Since then non-contact respiration sensing technology has improved significantly by integrating new miniaturized hardware systems, signal processing approaches, and intelligent algorithms [40].

In the below subsections, we will describe the basics of radar signal processing for estimating the vital signs such as breathing rate and heart rate of humans using CW and FMCW radar.

7.4.1 Theory of CW Doppler Radar for Cardiopulmonary Monitoring

A Continuous-wave (CW) radar topology consists of simpler architecture than FMCW radar for two reasons [59]. First, a similar oscillator can be utilized for the transmission or reception of the signal which helps to avoid interference, and at the same time, the computational cost of the filter is reduced. Secondly, CW transmits a narrow-band signal which requires simple and cheaper circuitry [26]. CW radar system can not measure the range of the target due to the continuous trasnmisison of the signal. However, the fundamental goal of physiological sensing is to track the tiny chest displacements due to cardio-respiratory activities. With having very simple circuitry, CW radar can measure the velocity of the signal accurately. On the other hand. FMCW or pulsed radar has limited resolution for velocity measurement [25]. Thus, all these features make CW radar a potential candidate for respiration sensing.

The basic of physiological sensing using microwave CW and FMCW radar is the Doppler effect. According to Doppler theory, the reflected echoes phase changes due to the tiny vibration of the chest surfaces and it is directly proportional to the displacement of the chest surfaces [14]. For example, consider that the target is moving at radial velocity v(t) in m/s, the reflected echoes frequency changes which is known as the Doppler shift [15]. The Doppler shift can be expressed as:

$$f_d\ (t) = 2f/c\ v(t) = 2v(t)/\lambda \tag{1.6}$$

where, f_d is denoted as change in frequency, f is the transmitted signal in Hz and λ is the wavelength of the transmitted signal in meter. This modulation is known as phase modulation. The phase modulation $\theta(t)$ can be described as:

$$\theta(t) = 2f/c\ (2\pi x(t)) = 4\pi x(t)/\lambda \tag{1.7}$$

where, x(t) is the periodic tiny displacement of the chest surfaces. When person's phase is target the phase change of the reflected signal occurs.To extract the breathing rate and heart rate of the signal demodulation is performed on two different channel signals for finding maximum displacement. A CW radar transmits a signal with fixed frequency. The signal is generally represented as:

$$T(t) - A\cos(2\pi f(t) + \phi(t)) \tag{1.8}$$

here, A is denoted amplitude of the transmitted signal, f is the single tone frequency of CW radar, $\phi(t)$ is the random phase noise of the oscillator. The reflected signal at the receiver can be modeled as:

$$R(t) = A_R \cos[2\pi f(t - t_d) + \phi(t - t_d) + \theta_0] \tag{1.9}$$

here, A_R is the amplitude of the reflected echoes, θ_0 is the phase shift at reflection surface and t_d is the time difference of the transmitted and reflected signal. The time delayed version of the signal can be expressed as:

$$t_d = 2d(t - d(t)/c)/c \tag{1.10}$$

Time difference between the transmitted and reflected signal is denoted as d(t), divided the velocity of the waveform, c. When signal strikes the chest surfaces, the distance between the antenna and the time of striking and reflecting back is $d(t - d(t)/c$. This tiny movement of the chest related information lies in the phase of the reflected echoes which need to be extracted by the detector.

7.4.2 Receiver Architecture

For detecting the phase information in the reflected echoes, the received signal is also multiplied with the same copy of the transmitted signal. By multiplying the received signal with the transmitted signal radio-frequency signal is converted into a baseband signal [39]. In general, receiver architecture can be divided into two categories. One of them is homodyne architecture and another one is heterodyne architecture [58]. In general, in a homodyne receiver, the transmitted signal is mixed with a similar transmitted local oscillator signal. On the other hand, the heterodyne architecture uses a different frequency band of the transmitted signal to multiply with the reflected signal. Therefore, the reflected signal is converted into the intermediate frequency rather than converting directly into the baseband signal.

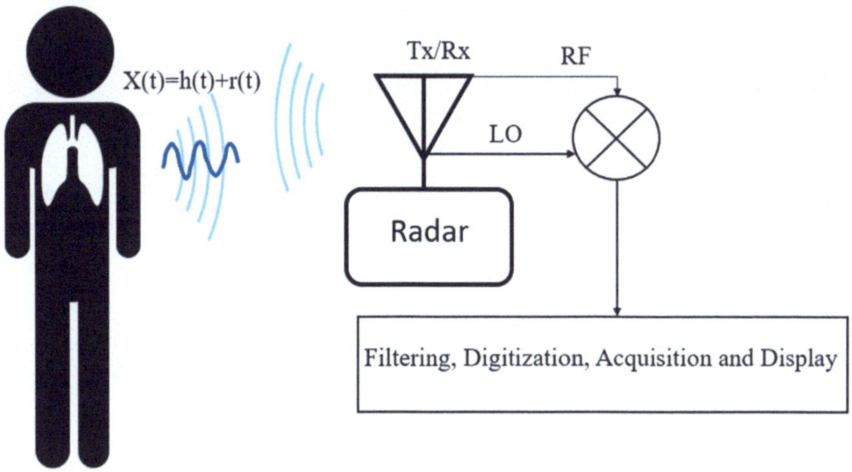

Fig. 7.5 Basics of Doppler radar physiological sensing. After capturing the signal it is down-converted to baseband signal and then post-processing is performed for extracting vital signs of humans

In a homodyne receiver, a band-pass filter is used to isolate noise and interference of the reflected signal and then mixed with the same local oscillator transmission signal. Although in a heterodyne receiver the received RF signal is converted into intermediate frequency (IF) signal by multiplying with the transmitted signal having different frequencies and then band-pass filtration and amplification is also performed for extracting vital signs [51].

A homodyne receiver architecture with frequency of 2.4 and 24 GHz is mostly employed in the cardiopulmonary monitoring research as it has simple circuitry. In a homodyne architecture, the transmitted signal is represented as:

$$L(t) = \cos(2\pi f t + \phi(t)) \tag{1.11}$$

Then the received signal is multiplied with the transmitted signal and it can be represented as:

$$B(t) = A_B \cos[\theta + 4\pi x(t)/\lambda + \delta\phi(t)] \tag{1.12}$$

where

$$A_B = A_R \sqrt{(G_R\ G_C)} \tag{1.13}$$

is the baseband amplitude with G_R is the receiver gain and G_C is the mixer conversion gain.

$$\delta\phi(t) = \phi(t) - \phi(t - (2d_0)/c) \tag{1.14}$$

is the residual phase noise and

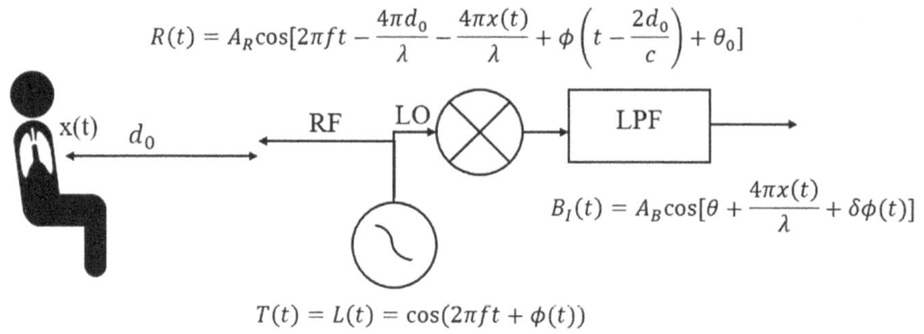

Fig. 7.6 Simplified Doppler radar signal processing flow for monitoring vital signs. The baseband signal output is directly proportional to the phase shift which is also determined by the distance of the target

$$\theta = (4\pi d_0)/\lambda - \theta_0 \tag{1.15}$$

is the constant phase shift depends on the nominal distance to the target, d_0. Figure 7.6 illustrates the signal processing flow to extract vital signs such as breathing rate and heart rate. If $x(t) \ll \lambda$ and θ in the above equation is an odd multiple of $\pi/2$, then the phase angle is very small where small approximation is valid and the output can me modeled as:

$$B(t) \approx A_B\ (4\pi x(t)/\lambda + \delta\phi(t)) \tag{1.16}$$

In this case, the optimum phase demodulation selectivity is achieved, the baseband output is proportional to the periodic chest displacement, x(t). When θ is an integer multiple of π then the output is approximately.

$$B(t) \approx A_B\ (1 - [4\pi x(t)/\lambda + \delta\phi(t)]^2) \tag{1.17}$$

If the distance of the target is an integer multiple of π then there is no direct relationship between the tiny movement of the chest surface. That particular distance of the target is known as null point and it mostly happens when the phase of the local oscillator (LO) of the transmitted signal are either in phase or 180° out of phase. Thus, there is a direct relation between the variable θ and distance of the target which introduces a null point at every quarter wavelength of the transceiver. For a frequency of 2.4-GHz transmitter, this null point occurs at every quarter wavelength which makes the receiver position sensitive device. This problem can be avoided using quadrature receivers which will be described in the next subsections.

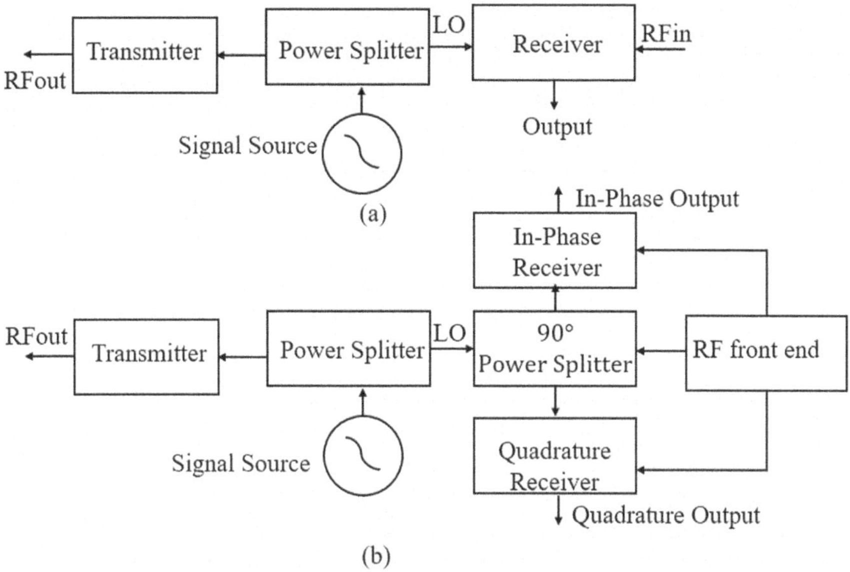

Fig. 7.7 Functional block diagram of the single channel **a** and quadrature channel transceiver **b**. Based on [71]

7.4.3 *Single Channel and Quadrature Receiver*

In general, single channel transceiver is simple but it has position sensitive issues for particular ranges of the target it creates null point which creates ambiguous result for physiological signal extraction [58]. The quadrature receiver can solve this issue by integrating two channels which also increases power consumption [54]. Figure 7.7 illustrates the block diagram of single channel and quadrature channel transceiver. The main challenges associated with single channel transceiver for phase demodulation accuracy varies with the range to the target. From the previous section, it is clear that there is direct relationship between the phase change of the reflected echoes and the tiny movement of the chest surfaces. Similarly, at the phase demodulation points the baseband signal depends on the square of the tiny movement of the chest surfaces due to cardio-respiratory activities [72].

To remove the position sensitivity issue, quadrature receiver employs two different channel signal. One is in-phase channel signal (I) and another one is quadrature-phase channel signal (Q).

A quadrature receiver block diagram is shown in Fig. 7.7 which consists of two receiver chains where one receiver chain is connected with the phase difference of 90° of the local oscillator (LO) signal to make sure if one channel is in null point and then another one will be maximum. The phase difference between the two channel signal is around $\pi/2$:

$$L_I\,(t) = \cos(2\pi f t + \delta\phi(t) + \pi/4) \tag{1.18}$$

$$L_Q\,(t) = \cos\left(2\pi f t + \delta\phi\,(t) - \frac{\pi}{4}\right) = \sin(2\pi f t + \delta\phi\,(t) + \frac{\pi}{4}) \tag{1.19}$$

Therefore, the quadrature receiver output can be represented as:

$$B_I\,(t) = A_B \cos[\theta + \pi/4 + 4\pi x(t)/\lambda + \delta\phi(t)] \tag{1.20}$$

$$B_Q\,(t) = A_B \cos[\theta - \pi/4 + 4\pi x(t)/\lambda + \delta\phi(t)] = A_B \sin[\theta + \pi/4 + 4\pi x(t)/\lambda + \delta\phi(t)] \tag{1.21}$$

From the above equations, if $\theta + \pi/4$ is an integer multiple of π then the in-phase channel signal will be at null point where the received signal output will be zero. On the other hand, if $\theta - \pi/4$ is an odd multiple of $\pi/2$, then Q signal will be at the maximum demodulation points. The basic criteria for extracting the tiny movement of the chest surface is $x(t) \ll \lambda$ where, the wavelength of the transmitted signal should be higher than the tiny displacement of the chest surfaces.

7.4.4 Data Acquisition and Signal Processing

After signal conditioning, generally, received signal is captured by data acquisition system (DAQ). In general, the bandwidth requirement for DAQ in physiological sensing is not that high because the vital sign information remains within the bandwidth of 0.2–0.8 Hz for respiratory-related information and 0.6–2 Hz for cardiac-related information. In general typical bandwidth of the signal is within 0–8 Hz [67]. According to the Nyquist theorem, for reconstructing the original signal the sampling frequency needs to be twice of the local oscillator transmission signal. So, the bandwidth of the DAQ signal should be 16 Hz [16]. Most of the physiological radar system utilizes the bandwidth 100 Hz for reconstructing the original signal [55].

The obtained baseband signal is further processed utilizing demodulation technique to extract respiratory rate and heart rate related information. Two popular methods that have been used for combining I/Q channel signal include linear demodulation [10] and arc-tangent demodulation [51]. Figure 7.5 illustrates the flow diagram of CW radar signal processing to extract vital signs. Linear demodulation is mostly known as principal component analysis (PCA). PCA is mostly employed for dimension reduction of the data by maximizing their variances and suppressing any redundant pieces of information [10]. After capturing the I/Q channel signal the eigenvalue and eigenvector are calculated using the eigenvalue decomposition method. Then the data is multiplied with the eigenvector for finding the maximum displacement direction [29]. In another method, non-linear demodulation considers that I/Q channel forms an arc in the complex plane and from the arc of the phase information of the signal can be extracted by extracting the angle in between the two-channel signals [51]. In

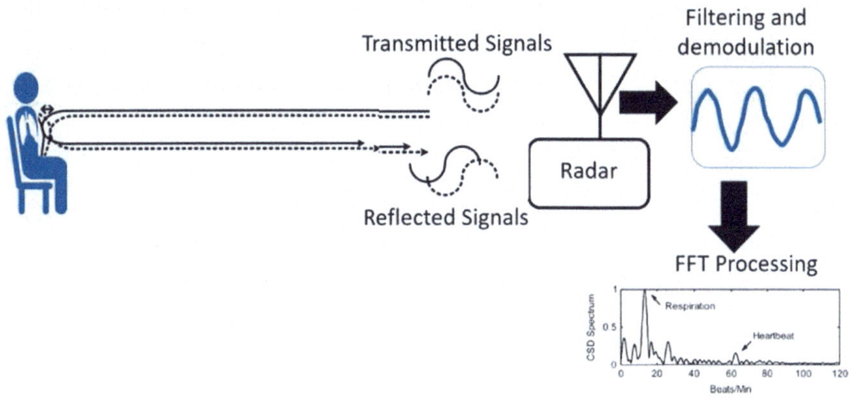

Fig. 7.8 Basics of Doppler CW radar physiological sensing. There is a direct relationship between the phase of the reflected echoes and chest displacement. Filtering and demodulation take place at the receiver end for further processing. After FFT analysis we can see two different dominant peak frequencies one is breathing rate and another one is heartbeat. Breathing rate and heartbeat is generally harmonics of each other

the next step, fast Fourier transform (FFT) is employed on the demodulated signal for finding the maximum vibrational frequency information. In general, the classic FFT operation converts the time domain signal into frequency domain [24]. From the FFT spectrum of the signal, we can see the highest peak which is known as breathing rate. In general, breathing is superimposed over heartbeat. By utilizing the bandpass filter with cutoff frequencies in between .2 to .5 Hz is used to find the breathing rate and within .6 2 Hz is used to extract the heart rate related information. In general, heartbeat is the harmonics of the respiration signal (Fig. 7.8).

7.4.5 FMCW Signal Processing for Vital Signs Sensing

In FMCW, a frequency modulated waveform such as triangular signal is generated by the local oscillator signal and transmitted towards the target [11]. In another term, transmitted signal is known as chirp. The transmitted frequency modulated signal can be modeled as [3]:

$$x_T(t) = A_T \cos(2\pi f_c t + \pi B/T_c\ t^2 + \phi(t)) \tag{1.22}$$

where, A_T, f_c, and B denotes amplitude of the chirp, starting frequency of the transmitted signal and bandwidth of the signal respectively. The received signal is then represented as [11]:

$$x_R(t) = \alpha A_T \cos(2\pi f_c (t - t_d) + \pi B/T_c\ (t - t_d)^2 + \phi(t - t_d)) \tag{1.23}$$

where, t_d is the time difference between the transmitted and reflected signal and it is represented as:

$$t_d = 2R(t)/c \tag{1.24}$$

R represents the range of the target and c is the velocity of the electromagnetic wave. In general, the received signal is multiplied with the same copy of the local oscillator signal known as in-phase (I) channel signal and the signal which is multiplied with the quadrature phase shifted of the transmitted signal is known as Q-channel signal. The in-phase channel signal can be represented as:

$$y(t) = A_R \cos\left(2\pi f_b\, t + \phi_b\,(t) + \delta\phi(t)\right) \tag{1.25}$$

where, A_R is the received signal amplitude and f_b is the beat frequency which can be represented as:

$$f_b = 2BR(t)/(cT_c\,) \tag{1.26}$$

The phase of the beat signal ϕ_b can be represented as:

$$\phi_b = 2\pi f_c\, t_d + \pi\, B/T_c\;\; t_d^2 \tag{1.27}$$

Residual phase noise can be removed for short range applications. The beat signal after the nth ADC sample for the mth chirp can be modeled as:

$$y[n\; m] = A_R \cos(2\pi f_b n T_f + 4\pi/\lambda R(nT_f + mT_s)) \tag{1.28}$$

There are two axis to represent the FMCW signal one is first time axis and another one is slow time axis. In general, first time axis is represented in the above equation as T_f and slow time axis as T_s. The tiny movement of the chest surfaces has very small amplitude (>10 mm) and low vibrational frequency (<4 Hz). To extract the vital signs related information generally processing the signal at slow time axis is sufficient.

The signal processing technique for FMCW radar is quite different than CW radar. After chirp design in a specified manner for extracting physiological information, FFT is performed on the captured signal which provides the range information of the target. After that, a second FFT is performed on the range FFT phase information to extract the breathing rate and heart rate [28].

7.5 Future Challenges

Over the four decades, there has been significant improvement and advancement of biomedical Doppler radar both in the hardware architectures and integrated signal processing algorithms [40]. The efficacy of the biomedical Doppler radar has also been proved in different application areas. In recent years, Google LLC (Mountain

View, CA, USA) has been developing Project Soli [66], which is focused on hand gesture recognition using miniaturized radar. Additionally, the successful project Soli is mounted in the Google Pixel 4 smartphone and attracted a lot of interest to the consumers, entrepreneurs, and researchers as well. A growing interest in engineering and life science communities towards utilizing radio frequency (RF) for healthcare sensing helped to continue the advancement of the biomedical Doppler radar. However, still there remain significant technical challenges that need to be addressed to bring this sensor technology into the real world of implementation for the benefit of humanity.

7.5.1 Multiple Subjects Respiration Monitoring

Isolating independent respiratory patterns from the combined mixture of breathing patterns of two subjects in front of the radar view has always been a difficult problem [40]. Multiple sources of motions from the presence of two or more subjects in front of the radar system create interference of respiration patterns which is hard to isolate the independent respiratory patterns from combined mixtures [9, 23]. For the CW radar system, it is much more challenging to isolate the respiratory patterns from the combined mixtures [33]. FMCW radar system can isolate the respiratory pattern based on range bin selection of the target [3]. Although isolating equidistant subjects from the radar especially within the same beamwidth is challenging both for CW and FMCW radar system [26]. However, some progress has been made to solve these critical challenges of isolating the motions of interference by having some prior knowledge of the clean signal [56]. Some of the common techniques are direction of arrival (DOA) of the target by utilizing MIMO radar [52] and blind source separation techniques [64]. Additionally, in the most recent attempt integrating two different approaches such as blind source separation and DOA with CW radar has been demonstrated to isolate the subjects in a view of the radar with a resolution limit less than 1 m [26]. Moreover, beamforming and beam switching technique is also gaining attention to separate the respiratory pattern from multiple subjects [28]. In one of the most recent attempts, the low-complexity multiple signal classification (MUSIC) based algorithm has successfully integrated with the FMCW radar system for concurrent respiration monitoring [36]. In another study, the adaptive digital beamforming technique is integrated with the single input multiple output radar for multi-subject respiration monitoring [69]. Recently, 76–81 GHz automotive radar integrated with beamforming and beam-steering technique for multiple subjects respiration measurement demonstrated its efficacy [4, 28]. Although most of this designed system can successfully separate motion artifacts in a controlled environment and cancellation of random body movement with isolating the breathing pattern is still a challenging problem.

7.5.2 Random Body Movement Cancellation

Life signs sensing using miniaturized Doppler radar has achieved significant progress although random body movement remains a significant technological challenge [40]. When people randomly move their hands in front of the radar system, extracting vital signs become challenging as it becomes a combined mixture of interference of respiration patterns. Although the successful demonstration of extraction of vital signs remotely during random body movement has been demonstrated in prior reported researches [12, 20]. In prior research, Doppler radar was also integrated with camera to cancel the random body movement [18, 20]. Other researches showed the efficacy of three different strategies for cancellation of random body movement such as (1) RF front hand can handle the phase distortion [12], (2) phase compensation at the baseband signal [18], and (3) cancellation during demodulation techniques [19]. Additionally, after demodulation techniques, a matched filter was integrated to extract vital signs when there was large scale noise caused by intense body movements [46]. A recent study also showed that adaptive noise cancellation technique with polynomial fitting is very effective for extracting the weak spectral components such as vital signs (respiration and heartbeat) that were submerged under various fast random body movement interference [70, 76]. Ongoing research investigations need to also consider different random body movement interference and its effect on vital sign sensing under this interference scenarios.

7.5.3 Identity Authentication of Multiple Subject

One of the fundamental limitations of CW radar is the detection of all motions in the field of view [24]. When there is a presence of two subjects in front of the radar system, basically radar gets an interference of respiration patterns [23]. Additionally, FMCW radar with the complex hardware system can only isolate the subjects when they are in different range bins [3]. If the subjects are within the beamwidth and at the same distances then isolation of subjects is not possible with FMCW radar [4]. Figure 7.9(a) illustrates the concurrent respiration sensing of two subjects when they are sharing bed and Fig. 7.9(b) represents the challenges if multiple subject identity authentication. The presence of multiple subjects will create interference of respiration patterns and at the same time extraction of unique features from the combined mixture signal would be difficult [26]. Recent advancement on isolating independent respiratory signatures from a combined mixture has been performed both on the hardware architecture of the radar system and integration of signal processing techniques also. For example, multiple inputs and multiple-output (MIMO) radar is mostly employed to estimate the spatial position of the target and switch the beam in certain directions to isolate respiratory signatures from combined mixtures [56]. Signal processing algorithms such as independent component analysis (ICA) is also employed to isolate the respiratory pattern from combined interference of

Fig. 7.9 **a** Source of undesired motion by the presence of two subjects who are sharing bed with partner during sleep study, and **b** continuous identity authentication of multiple subjects in the radar filed of view

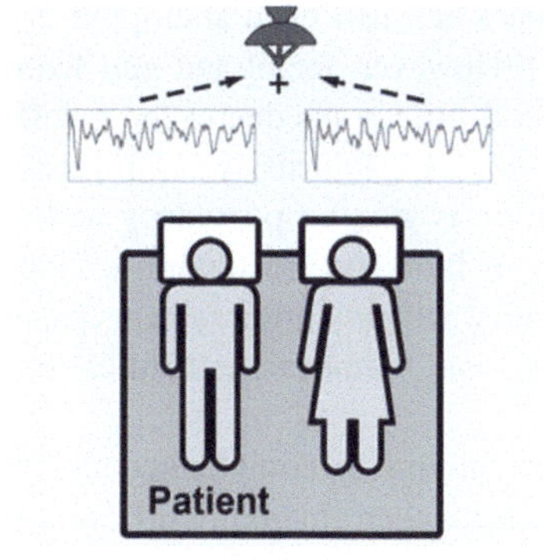

(a)

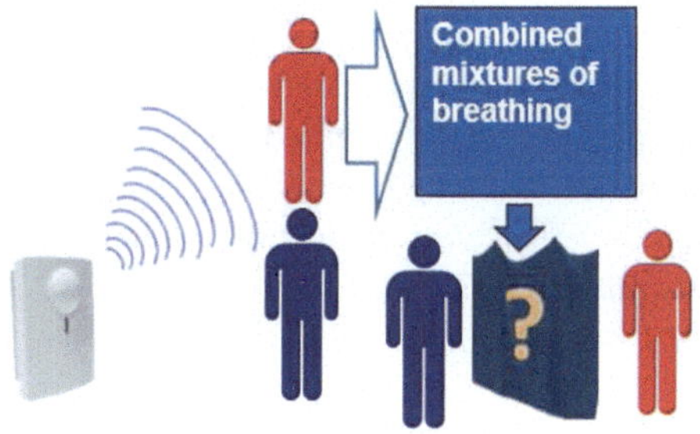

(b)

respiration pattern [23]. Additionally, a new algorithm with ICA and DOA has been developed to isolate the respiratory pattern both the subjects when they are closely-spaced within the beam and outside the beam as well [26]. Thus, considering all this prior research, a new method for ubiquitous monitoring must be sought to overcome the existing challenges of multi-subject scenarios for reliable identity authentication. Further investigation and continued experimentation on different experimental scenarios might help to solve this research challenges to bring this sensor technology into the real world of applications.

7.5.4 Breathing Diversity in Radar Authentication

Some of the research efforts are also attempted in recognizing people where vital signs and behavioral changes are captured with the off-body sensor. Optical, thermal cameras are also utilized for extracting features of face and gait pattern [65]. Face identification systems is also very popular and implemented into the real world to recognize and verify identity of human [7]. Recognizing people from gait using camera has also been attempted in literature and received a great deal of attention [61]. However, facial and gait feature based recognition system using camera has several limitations due to the insufficient resolution [62, 63]. Considering all these limitations, unobtrusive radar measurement of cardiac motion-based identity authentication is gaining popularity and many research efforts are exploring on this new way of biometric recognition. Doppler radar based non-contact cardio-respiratory identity authentication systems can open a new door for this technology by replacing traditional biometric authentication systems this new unobtrusive method [13, 17, 54, 75].

Breathing is a complicated anatomical process which is controlled by central neural mechanism involving different neurons and respiratory muscles as well [8, 30, 73]. Breathing pattern changes with emotional stress, physiological activities. So, recognizing people from different breathing diversity is important as breathing pattern diversity is dependent on emotional stress and physiological activities as well [30]. In addition, most of the prior research focused on recognizing people from sedentary breathing pattern [24, 42, 55]. Recent attempts focused on recognizing people from radar captured respiration pattern after performing post physiological activities (walking stairs) [21, 30]. Further investigations are required for evaluating the efficacy of the identity authentication system by bringing the variability of respiratory variations.

7.5.5 Radar Authentication Data Security

Radar based identity authentication system is comprised of extracting unique features and then integrating machine learning classifiers with the feature set [42]. One of the potential application of this system is to utilize this technology for medical applications such as in-home sleep apnea diagnostics [2]. For recognition of patients using microwave Doppler radar is inherently difficult as RF-based approach is vulnerable to noise [1]. Additionally, for integrating this proposed RF-based recognition system with a wireless communication system also make the system vulnerable to attack over the air [32]. The security problem can be subdivided into four categories [32].

1. Mix-up: In case of in-home unobtrusive sleep apnea screening, there is a high chance of sharing a bed or a mother sleeping with an infants. So, the research challenge is how to isolate independent respiratory pattern from combined mixture to extract unique features.
2. Surrogate: During in-home sleep apnea testing, the patient may substitute another human or a robotic phantom mover [34], perhaps to game a test for a job requirement. So, the system needs to e intelligent one to identify a person swap.
3. Eavesdrop: The patient may eavesdrop in between the wireless communication channel and OSA screening system so that he/she may manipulate the recorded data before the doctor. There is a probability of high chance of significant dis-

crepancy which may help to establish false excuse and may convince the doctor for accepting the test results [32].

4. Spoof: The patient may inject false data and manipulate the identity verification outcome. During identity authentication modality switching between the tracker and the OSA screening system is required. For instance, the patient may hijack data and falsify the test results.

Considering all these data authentication discrepancies, secure protocols need to be integrated with the system to make sure discrepancies and data falsification can not occur.

7.6 Radar Physiological Sensing for New Researchers

With the advent of a low-cost computing platform for radar chips, many young researchers can also contribute this field especially in the way we learn, experiment, and create imaginative solutions for the benefit of humanity to bring this technology into real-world applications. This particular subsection will focus on addressing how to initiate this non-contact physiological sensing technology-related research. We will mostly concentrate on preprocessing of the radar signal for vital signs extraction as different radar modules can be bought from online shops. For initiating, this particular research by the new aspirants, 2.4 to 24-GHz radar module can be bought online, and then it can also be connected with the computer for data collection. Most of the radar modules are designed for automotive applications where the target with higher velocity needs to be detected. In this particular research-related task the main goal is to extract different vital signs from the tiny movement of the chest surfaces. Additionally, CW radar is mostly preferred for this type of application as its related to motion sensing. Preprocessing of radar signal for extracting vital sign in formations is different than target with high velocity related in formations. So, we will discuss mostly how to pre-process the CW radar signal for utilizing it in remote physiological sensing. Figure 7.10 below illustrates the preprocessing steps of radar signal processing for breathing rate and heart rate extraction.

7.6.1 Step 1: Radar Data Channel Imbalance Minimization

The I and Q channels of a radar module data can never be exactly similar and it creates non-linearity and nonidealities in the data-set. This non-linearity create amplitude imbalance and phase imbalance of the received signal. This imbalance is caused by different values in different channels. This imbalance values not only depend on the hardware of the module but also changes with temperatures. Thus, a calibration step needs to be performed to measure the imbalance of the channels shown below [74]:

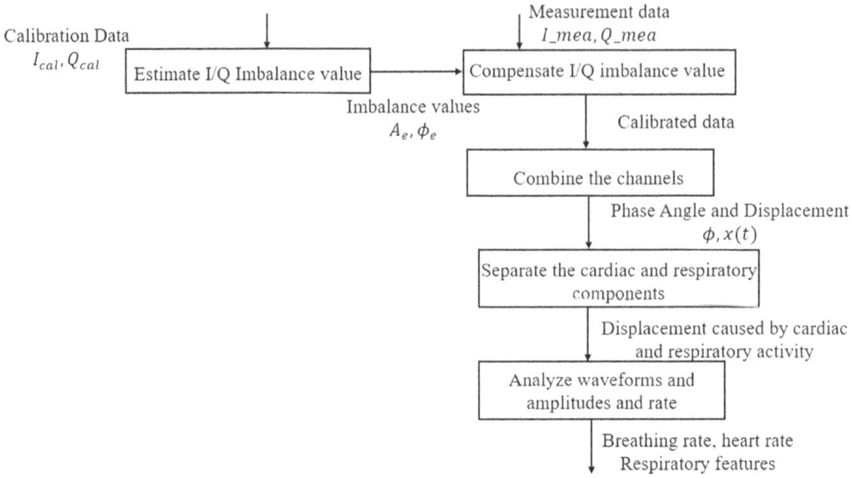

Fig. 7.10 Radar signal processing steps for extracting vital signs (breathing rate and heart rate). Based on [74]

$$\begin{bmatrix} B_I ort \\ B_Q ort \end{bmatrix} = \begin{bmatrix} 1 & 0 \\ -tan\phi_E & 1/(A_E cos\phi_E) \end{bmatrix} \times \begin{bmatrix} B_I \\ B_Q \end{bmatrix} \tag{1.29}$$

where, $B_I ort$ and $B_Q ort$ is the imbalance corrected baseband channel signal, and A_E, ϕ_E is the amplitude and phase imbalance respectively. The imbalance compensation steps must need to be performed for accurate vital signs extractions. Additionally, phase imbalance affects more in result analysis as the respiratory related information lies in the phase of the received signal.

7.6.2 Step 2: Calculating Imbalance Values

Amplitude and phase imbalance can be calculated in two different parts. At first calibration signal need to be generated and then calculation of the imbalance values of the generated calibration signal. The imbalance value of the generated signal can be calculated in four different ways. The first one is generation of the two signals with different frequencies using a local oscillator (LO) and then connecting it to the RF input port. Bypassing the local oscillator (LO) signal and removing the antenna for calculating the phase and amplitude imbalance is not feasible especially it requires laboratory settings. Another approach is to use the robotic phantom mover in front of the radar antenna to a fixed distance [34]. The robotic phantom mover can generate signals as a phase shifter of the moving target with a fixed velocity [34].

After generating the calibration signals in different ways, imbalance values can be calculated utilizing the time domain of the generated signal. The time-domain method is focused on extracting the extrema the minima and maxima points of the calibration signals that are detected both from I and Q channel data. The ratio of the peak to peak amplitude of two channel (I and Q) signal defines the amplitude error or imbalance values [74]. Similarly, the phase deviation from 90° in the phase delay between two channel peak points define the phase error or imbalance values. This time domain method is simpler as signals maxima and minima points need to be analyzed only. However, the time-domain method is sensitive to noise. Another way to find the imbalance values is the frequency domain method. In the frequency domain method, phase imbalance is calculated after performing FFT of the two channel signals. The deviation in FFT phase response of two channel signals are considered. The signal utilized for this method need to be sinusoidal signal with a constant velocity. For calculating the amplitude imbalance the data signal is shifted in the same phase signal of both channels to find the differences in the amplitude with having same phase. Another easiest way to find the phase imbalance is performing the cross-correlation operation in between I and Q channel signal. The cross-correlation operation is similar to the traditional FFT method.

7.6.3 Step 3: Channel Combining Method

Quadrature radar channel signals need to be combined together for making it one dimensional data space. One of the most popular technique is the principal component analysis (PCA) or eigenvalue decomposition method for dimensional reduction of the radar captured patterns. PCA tires to extract the direction of the variance of the data-set which is larger along the tangent of the arc in complex plane. For performing PCA, dc offset is removed by removing the mean from the signal and the data is rotated along with this principle component to find the maximum variance direction. This method is also known as linear demodulation.

The arc-tangent demodulation is also called non-linear demodulation and taking the arc-tangent of two-channel signals (I and Q). In general arc-tangent demodulation shows more accurate results for small-angle approximation. Figure 7.11 shows the graphical representation of the channel combining process for quadrature radar. Two main important factors need to be taken into account for channel combining. For quadrature radar, the radar carrier frequency and information that need to be extracted is important. For healthcare sensing applications and a small radar carrier frequency, the PCA is the most popular technique. The main reason for the popularity is less computational cost. For estimating accurate chest displacement information generally higher carrier frequency of radar is optimal as it is highly sensitive to the movement. Also, the arc-tangent demodulation technique provides better results because it actually find the phase information from the I-Q complex plane.

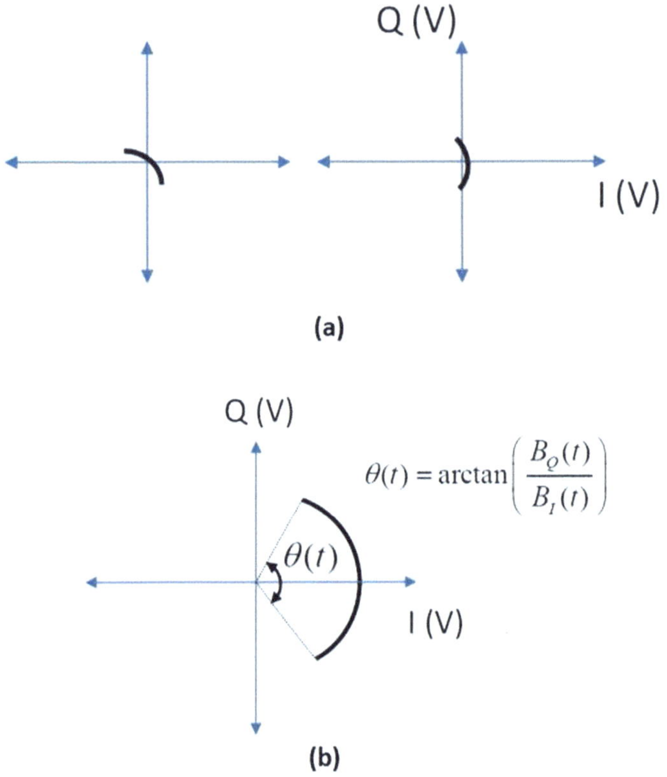

Fig. 7.11 Graphical representation of **a** principal component analysis **b** arc-tangent demodulation of I/Q plot

7.6.4 Step 4: Separation of Cardiac and Respiratory Component

In a general view, respiration signal is modeled as perfectly sinusoidal waveform where one particular period of the signal provided inhale and exhale information. Although in real life it does not look exactly sinusoidal. Especially when people take deep breathe it has seen that the waveform is either less than the one period of the sinusoid signal or it is larger than the time-period of the sinusoid signal [8]. Therefore, it is clearly elucidate that, the respiration signal has fundamental frequency as well as harmonics where in between the harmonics heartbeat related information exists [74].

Filtering is mostly employed to isolate the respiratory pattern and cardiac related information. Human in their sedentary breathing pattern typically respiratory rate lies between 0.13–0.4 Hz (8–24 breaths per minute) and heartbeat related vibrational frequency lies in between 0.8–1.5 Hz (50–90 beats per minute) [14]. So, from the frequency band, it is quite clear that the cardiac and respiratory components of the radar captured signal can be separated in the post-processing of the signal by integrating digital filters. Although there is a significant difference in the fundamental frequencies of the cardiac and the respiration signal, it may happen that respiratory related harmonics lies in the frequency range of the cardiac components. The transition band of the highest component of the respiratory and cardiac signal is in the interval of 0.4 and 0.8 Hz. The signal can be low pass filtered with a cut off frequency .5 Hz can provide the respiratory components and bandpass filtering from .6 2 Hz can also provide the heartbeat related information [16]. MATLAB is also popular platform for post-processing the demodulated signal.

7.6.5 Step 5: Analyze Waveform for Breathing and Heart Rate Extraction

After isolating the respiratory and cardiac phenomena utilizing a digital filter, a fast Fourier transform (FFT) is employed to extract the highest frequency components of the signal. FFT is a common signal processing approach to convert the time domain signal into a frequency-domain signal. In the frequency domain signal, the peak of the signal represents the fundamental frequency of the signal which also illustrates the breathing rate and heart rate-related information. FFT window size also defines the resolution limit for the collected sample size which is captured from the radar. In general, if the FFT window size is 256, for a sampling frequency 100 Hz the number of samples required to extract the breathing rate is around 25600 samples. If the collected signal is less than the particular sample number then the FFT window size needs to be reduced. Nowadays software (MATLAB) can optimize the window size based on the number of samples as well. Another important criterion for extracting reliable breathing rate and heart rate is the number of breathing cycle within the sample should be above two so that the fundamental frequency which is extracted resonates with the vital signs.

This section of the chapter briefly discusses the flow diagram for processing the CW radar signal processing steps for the beginners those who want to start their research in this exciting area. There are lot of commercial radar modules that need to be bought and then all these processes need to be followed for producing some meaningful results.

7.7 Conclusion

This chapter provides a brief introduction of radar-based non-contact physiological sensing technology and its emerging areas. It also discusses the basic theory of radar-based remote respiration sensing technique. Additionally, it also highlights the future challenges of this potential sensing technology for different applications. Finally, the chapter concludes with a brief introductory discussion for the new researchers to initiate their effort for conducting research on this area. Over the four decades due to the significant advancement of the performance and availability of the radio-frequency integrated circuits, traditional battlefield radar has been modified into miniaturized biomedical radar, rescue radar, surveillance radar, and so on. The chapter also discusses the fundamental future challenges of this remote respiration sensing technology. With the recent progress of machine learning and artificial intelligence, the next generation of ubiquitous remote sensing will open new doors to provide new biomedical information tools to help humanity. With the advancement of a new wireless communication system [22, 53], the radar-based remote cardiopulmonary monitoring system will become a common device for daily activities monitoring, search and rescue applications, and identity authentication as well. Moreover, further exploration and continued efforts need to be carried on to bring this potential non-contact remote respiration sensing technology into the diverse real-world application for the benefit of humanity and mankind.

References

1. https://patents.google.com/patent/US8679012B1/en
2. https://www.nhlbi.nih.gov/health-topics/sleep-deprivation-and-deficiency
3. Adib, F., Kabelac, Z., Mao, H., Katabi, D., Miller, R.C.: Demo: real-time breath monitoring using wireless signals. In: Proceedings of the 20th Annual International Conference on Mobile Computing and Networking—MobiCom 2014, Maui, Hawaii, USA, pp. 261–262. ACM Press (2014). ISBN 978-1-4503-2783-1. https://doi.org/10.1145/2639108.2641756. http://dl.acm.org/citation.cfm?doid=2639108.2641756
4. Ahmad, A., Roh, J.C., Wang, D., Dubey, A.: Vital signs monitoring of multiple people using a FMCW millimeter-wave sensor. In: 2018 IEEE Radar Conference (RadarConf18), Oklahoma City, OK, pp. 1450–1455. IEEE, April 2018. ISBN 978-1-5386-4167-5. https://doi.org/10.1109/RADAR.2018.8378778. https://ieeexplore.ieee.org/document/8378778/
5. Baboli, M., Singh, A., Soll, B., Boric-Lubecke, O., Lubecke, V.M.: Good night: sleep monitoring using a physiological radar monitoring system integrated with a polysomnography system. IEEE Microwave Mag. **16**(6), 34–41 (2015). ISSN 1527-3342. https://doi.org/10.1109/MMM.2015.2419771. http://ieeexplore.ieee.org/document/7118337/
6. Baboli, M., Singh, A., Soll, B., Boric-Lubecke, O., Lubecke, V.M.: Wireless sleep apnea detection using continuous wave quadrature doppler radar. IEEE Sens. J. **20**(1)538–545 (2020). ISSN 1530-437X, 1558-1748, 2379-9153. https://doi.org/10.1109/JSEN.2019.2941198. https://ieeexplore.ieee.org/document/8836512/
7. Balazia, M., Plataniotis, K.N.: Human gait recognition from motion capture data in signature poses. IET Biometr. **6**(2), 129–137 (2017). ISSN 2047-4938, 2047-4946. https://doi.org/10.1049/iet-bmt.2015.0072. https://digital-library.theiet.org/content/journals/10.1049/iet-bmt.2015.0072
8. Benchetrit, G.: Breathing pattern in humans: diversity and individuality. Respir. Physiol. **122**(2–3), 123–129 (2000). ISSN 00345687. https://doi.org/10.1016/S0034-5687(00)00154-7. https://linkinghub.elsevier.com/retrieve/pii/S0034568700001547

9. Boric-Lubecke, O., Lubecke, V.M., Host-Madsen, A., Samardzija, D., Cheung, K.: Doppler radar sensing of multiple subjects in single and multiple antenna systems. In: TELSIKS 2005—2005 uth International Conference on Telecommunication in ModernSatellite, Cable and Broadcasting Services, Nis, Serbia and Montenegro, vol. 1, pp. 7–11. IEEE (2005). ISBN 978-0-7803-9164-2. https://doi.org/10.1109/TELSKS.2005.1572052. http://ieeexplore.ieee.org/document/1572052/
10. Boric-Lubecke, O., Massagram ,W., Lubecke, V.M., Host-Madsen, A., Jokanovic, B.: Heart rate variability assessment using doppler radar with linear demodulation. In: 2008 38th European Microwave Conference, Amsterdam, Netherlands, pp. 420–423. IEEE, October 2008. ISBN 978-2-87487-006-4. https://doi.org/10.1109/EUMC.2008.4751478. http://ieeexplore.ieee.org/document/4751478/
11. Brooker, G.M.: Understanding millimetre wave FMCW radars. New Zealand 7 (2005)
12. Li, C., Lin, J.: Random body movement cancellation in doppler radar vital sign detection. IEEE Trans. Microwave Theory Tech. **56**(12), 3143–3152 (2008). ISSN 0018-9480, 1557-9670. https://doi.org/10.1109/TMTT.2008.2007139. http://ieeexplore.ieee.org/document/4682605/
13. Chen, V.C.: Doppler signatures of radar backscattering from objects with micro-motions. IET Signal Process. **2**(3), 291 (2008). ISSN 17519675. https://doi.org/10.1049/iet-spr:20070137. https://digital-library.theiet.org/content/journals/10.1049/iet-spr20070137
14. Droitcour, A., Lubecke, V., Lin, J., Boric-Lubecke, O.: A microwave radio for Doppler radar sensing of vital signs. In: 2001 IEEE MTT-S International Microwave Sympsoium Digest (Cat. No.01CH37157), vol. 1, Phoenix, AZ, USA, pp. 175–178. IEEE (2001). ISBN 978-0-7803-6538-4. https://doi.org/10.1109/MWSYM.2001.966866. http://ieeexplore.ieee.org/document/966866/
15. Droitcour, A.D.: Non-contact measurement of heart and respiration rates with a single-chip microwave doppler radar. Ph.D. thesis Dissertation, Stanford University, CA, USA (2006)
16. Droitcour, A.D., Seto, T.B., Park, B.-K., Yamada, S., Vergara, A., El Hourani, C., Shing, T., Yuen, A., Lubecke, V.M., Boric-Lubecke, O.: Non-contact respiratory rate measurement validation for hospitalized patients. In: 2009 Annual International Conference of the IEEE Engineering in Medicine and Biology Society, Minneapolis, MN, pp. 4812–4815. IEEE, September 2009. https://doi.org/10.1109/IEMBS.2009.5332635. http://ieeexplore.ieee.org/document/5332635/
17. Geisheimer, J.L., Marshall, W.S., Greneker, E.: A continuous-wave (CW) radar for gait analysis. In: Conference Record of Thirty-Fifth Asilomar Conference on Signals, Systems and Computers (Cat.No.01CH37256), Pacific Grove, CA, USA, vol.1, pp. 834–838. IEEE (2001). ISBN 978-0-7803-7147-7. https://doi.org/10.1109/ACSSC.2001.987041. http://ieeexplore.ieee.org/document/987041/
18. Gu, C., Li, R., Jiang, S.B., Li, C.: A multi-radar wireless system for respiratory gating and accurate tumor tracking in lung cancer radiotherapy. In: 2011 Annual International Conference of the IEEE Engineering in Medicine and Biology Society, Boston, MA, pp. 417–420. IEEE, August 2011a. ISBN 978-1-4577-1589-1 978-1-4244-4121-1 978-1-4244-4122-8. https://doi.org/10.1109/IEMBS.2011.6090054. http://ieeexplore.ieee.org/document/6090054/
19. Gu, C., Li, R., Li, C., Jiang, S.B.: Doppler radar respiration measurement for gated lung cancer radiotherapy. In: 2011 IEEE Topical Conference on Biomedical Wireless Technologies, Networks, and Sensing Systems, Phoenix, AZ, USA, pp. 91–94. IEEE, January 2011b. ISBN 978-1-4244-8316-7. https://doi.org/10.1109/BIOWIRELESS.2011.5724348. http://ieeexplore.ieee.org/document/5724348/
20. Gu, C., Wang, G., Li, Y., Inoue, T., Li, C.: A hybrid radar-camera sensing system with phase compensation for random body movement cancellation in doppler vital sign detection. IEEE Trans. Microwave Theory Tech. **61**(12), 4678–4688 (2013). ISSN 0018-9480, 1557-9670. https://doi.org/10.1109/TMTT.2013.2288226. http://ieeexplore.ieee.org/document/6662489/

21. Islam, S.M.M., Lubecke, V.M.: Extracting individual respiratory signatures from combined multi-subject mixtures with varied breathing pattern using independent component analysis with the JADE algorithm. In: IEEE Asia-Pacific Microwave Conference (APMC 2020), Hong Kong, China, December 2020, pp. 734–736, 8–11 December 2020. https://doi.org/10.1109/APMC47863.2020.9331715
22. Islam, S.M.M., Qazi, F.A., Iskander, M.F., Yun, Z., Sasaki, G.: Advanced directional networking: LTE vs WiFi radios. In: 2017 IEEE International Symposium on Antennas and Propagation & USNC/URSI National Radio Science Meeting, San Diego, CA, USA, pp. 189–190. IEEE, July 2017. ISBN 978-1-5386-3284-0. https://doi.org/10.1109/APUSNCURSINRSM.2017.8072137. http://ieeexplore.ieee.org/document/8072137/
23. Islam, S.M.M., Yavari, E., Rahman, A., Lubecke, V.M., Boric-Lubecke, O.: Separation of respiratory signatures for multiple subjects using independent component analysis with the JADE algorithm. In: 2018 40th Annual International Conference of the IEEE Engineering in Medicine and Biology Society (EMBC), Honolulu, HI, pp. 1234–1237. IEEE, July 2018a. ISBN 978-1-5386-3646-6. https://doi.org/10.1109/EMBC.2018.8512583. https://ieeexplore.ieee.org/document/8512583/
24. Islam, S.M.M., Rahman, A., Prasad, N., Boric-Lubecke, O., Lubecke, V.M.: Identity authentication system using a support vector machine (SVM) on radar respiration measurements. In: 2019 93rd ARFTG Microwave Measurement Conference (ARFTG), Boston, MA, USA, pp. 1–5. IEEE, June 2019a. ISBN 978-1-72810-506-2. https://doi.org/10.1109/ARFTG.2019.8739240. https://ieeexplore.ieee.org/document/8739240/
25. Islam, S.M.M., Yavari, E., Rahman, A., Lubecke, V.M., Boric-Lubecke, O.: Multiple Subject Respiratory Pattern Recognition and Estimation of Direction of Arrival using Phase-Comparison Monopulse Radar. In: 2019 IEEE Radio and Wireless Symposium (RWS), Orlando, FL, USA, pp. 1–4. IEEE, January 2019b. ISBN 978-1-5386-5944-1. https://doi.org/10.1109/RWS.2019.8714272. https://ieeexplore.ieee.org/document/8714272/
26. Islam, S.M.M., Boric-Lubecke, O., Lubekce, V.M.: Concurrent respiration monitoring of multiple subjects by phase-comparison monopulse radar using independent component analysis (ICA) with JADE algorithm and direction of arrival (DOA). IEEE Access **8**, 3558–73569 (2020a). ISSN 2169-3536. https://doi.org/10.1109/ACCESS.2020.2988038. https://ieeexplore.ieee.org/document/9067841/
27. Islam, S.M.M., Grado, C., Lubecke, V.: UAV radar sensing of respiratory variations for COVID-type disorders. In: IEEE Asia-Pacific Microwave Conference, pp. 734–736, 8–11 December 2020. https://doi.org/10.1109/APMC47863.2020.9331613
28. Islam, S.M.M., Motoyama, N., Pacheco, S., Lubecke, V.M.: Non-contact vital signs monitoring for multiple subjects using a millimeter wave FMCW automotive radar. In: 2020 IEEE/MTT-S International Microwave Symposium (IMS), Los Angeles, CA, USA, pp. 783–786. IEEE, August 2020c. ISBN 978-1-72816-815-9. https://doi.org/10.1109/IMS30576.2020.9223838. https://ieeexplore.ieee.org/document/9223838/
29. Islam, S.M.M., Rahman, A., Yavari, E., Baboli, M., Boric-Lubecke, O., Lubecke, V.M.: Identity authentication of OSA patients using microwave doppler radar and machine learning classifiers. In: 2020 IEEE Radio and Wireless Symposium (RWS), San Antonio, TX, USA, pp. 251–254. IEEE, January 2020d. ISBN 978-1-72811-120-9. https://doi.org/10.1109/RWS45077.2020.9049983. https://ieeexplore.ieee.org/document/9049983/
30. Islam, S.M.M., Sylvester, A., Orpilla, G., Lubecke, V.M.: Respiratory feature extraction for radar-based continuous identity authentication. In: 2020 IEEE Radio and Wireless Symposium (RWS), San Antonio, TX, USA, pp. 119–122. IEEE, January 2020e. ISBN 978-1-72811-120-9. https://doi.org/10.1109/RWS45077.2020.9050013. https://ieeexplore.ieee.org/document/9050013/
31. Islam, S.M.M., Lubecke, L.C., Grado, C., Lubecke, V.M.: An adaptive filter technique for platform motion compensation in unmanned aerial vehicle based remote life sensing radar. In: 50th European Microwave Week (EuMW 2021), Utrechet, Netherland, 12–14 January 2021, p. 4 (2021). https://doi.org/10.1109/EuMC48046.2021.9338011

32. Islam, S.M.: Non-contact and secure radar-based continuous identity authentication in multiple subject environments. Ph.D. thesis, Department of Electrical Engineering, University of Hawaii, p. 243, December 2020. http://hdl.handle.net/10125/73307
33. Islam, S.M.M., Yavari, E., Rahman, A., Lubecke, V.M., Boric-Lubecke, O.: Direction of arrival estimation of physiological signals of multiple subjects using phase comparison monopulse radar. In: 2018 Asia-Pacific Microwave Conference (APMC), Kyoto, pp. 411–413. IEEE, November 2018b. ISBN 978-4-902339-45-1. https://doi.org/10.23919/APMC.2018.8617539. https://ieeexplore.ieee.org/document/8617539/
34. Islam, S.M.M., Tomota, B., Sylvester, A., Lubecke, V.M.: A programmable robotic phantom to simulate the dynamic respiratory motions of humans for continuous identity authentication. In: 2019 IEEE Asia-Pacific Microwave Conference (APMC), Singapore, Singapore, pp. 1408–1410. IEEE, December 2019c. ISBN 978-1-72813-517-5. https://doi.org/10.1109/APMC46564.2019.9038224. https://ieeexplore.ieee.org/document/9038224/
35. Islam, S.M.M., Borić-Lubecke, O., Zheng, Y., Lubecke, V.M. : Radar-based non-contact continuous identity authentication. Remote Sens. **12**(14), 2279 (2020f). ISSN 2072-4292. https://doi.org/10.3390/rs12142279. https://www.mdpi.com/2072-4292/12/14/2279
36. Kim, B.-S., Jin, Y., Lee, J., Kim, S.: Low-complexity MUSIC-based direction-of-arrival detection algorithm for frequency-modulated continuous-wave vital radar. Sensors **20**(15), 4295 (2020). ISSN 1424-8220. https://doi.org/10.3390/s20154295. https://www.mdpi.com/1424-8220/20/15/4295
37. Chen, K.-M., Huang, Y., Zhang, J., Norman, A.: Microwave life-detection systems for searching human subjects under earthquake rubble or behind barrier. IEEE Trans. Biomed. Engi. **47**(1), 105–114 (2000). ISSN 0018-9294, 1558-2531. https://doi.org/10.1109/10.817625. http://ieeexplore.ieee.org/document/817625/
38. Levy, M.: Low-cost occupancy sensor saves energy, p. 2 (2010)
39. Li, C., Gu, C., Li, R., Jiang, S.B.: Radar motion sensing for accurate tumor tracking in radiation therapy. In: WAMICON 2011 Conference Proceedings, Clearwater Beach, FL, USA, pp. 1–6. IEEE, April 2011. ISBN 978-1-61284-081-9. https://doi.org/10.1109/WAMICON.2011.5872871. http://ieeexplore.ieee.org/document/5872871/
40. Li, C., Lubecke, V.M., Boric-Lubecke, O., Lin, J.: A review on recent advances in doppler radar sensors for noncontact healthcare monitoring. IEEE Trans. Microwave Theory Tech. **61**(5), 2046–2060 (2013). ISSN 0018-9480, 1557-9670. https://doi.org/10.1109/TMTT.2013.2256924. http://ieeexplore.ieee.org/document/6504804/
41. Liang, L., Popescu, M., Skubic, M., Rantz, M., Yardibi, T., Cuddihy, P.: Automatic fall detection based on doppler radar motion signature. In: Proceedings of the 5th International ICST Conference on Pervasive Computing Technologies for Healthcare, Dublin, Republic of Ireland. IEEE (2011). ISBN 978-1-936968-15-2. https://doi.org/10.4108/icst.pervasivehealth.2011.245993. http://eudl.eu/doi/10.4108/icst.pervasivehealth.2011.245993
42. Lin, F., Song, C., Zhuang, Y., Xu, W., Li, C., Ren, K.: Cardiac scan: a non-contact and continuous heart-based user authentication system. In: Proceedings of the 23rd Annual International Conference on Mobile Computing and Networking—MobiCom 2017, Snowbird, Utah, USA, pp. 315–328. ACM Press (2017). ISBN 978-1-4503-4916-1. https://doi.org/10.1145/3117811.3117839. http://dl.acm.org/citation.cfm?doid=3117811.3117839
43. Lin, J.C.: Noninvasive microwave measurement of respiration. Proc. IEEE **63**(10), 1530 (1975). ISSN 0018-9219. https://doi.org/10.1109/PROC.1975.9992. http://ieeexplore.ieee.org/document/1451922/
44. Lin, J., Li, C.: Wireless non-contact detection of heartbeat and respiration using low-power microwave radar sensor. In: 2007 Asia-Pacific Microwave Conference, Bangkok, Thailand, pp. 1–4. IEEE, December 2007. ISBN 978-1-4244-0748-4. https://doi.org/10.1109/APMC.2007.4554533. http://ieeexplore.ieee.org/document/4554533/
45. Lubecke, V.M., Boric-Lubecke, O., Host-Madsen, A., Fathy, A.E.: Through-the-wall radar life detection and monitoring. In: 2007 IEEE/MTT-S International Microwave Symposium, Honolulu, HI, USA, pp. 769–772. IEEE, June 2007. ISBN 978-1-4244-0687-6 978-1-4244-0688-3. https://doi.org/10.1109/MWSYM.2007.380053. http://ieeexplore.ieee.org/document/4263932/

46. Lv, Q., Chen, L., An, K., Wang, J., Li, H., Ye, D., Huangfu, J., Li, C., Ran, L.: Doppler vital signs detection in the presence of large-scale random body movements. IEEE Trans. Microwave Theory Tech. **66**(9), 4261–4270 (2018). ISSN 0018-9480, 1557-9670. https://doi.org/10.1109/TMTT.2018.2852625. https://ieeexplore.ieee.org/document/8415774/
47. Massaroni, C., Nicolò, A., Schena, E., Sacchetti, M.: Remote respiratory monitoring in the time of COVID-19. Front. Physiol. **11**, 635 (2020). ISSN 1664-042X. https://doi.org/10.3389/fphys.2020.00635. https://www.frontiersin.org/article/10.3389/fphys.2020.00635/full
48. Mercuri, M., Soh, P.J., Pandey, G., Karsmakers, P., Vandenbosch, G.A.E., Leroux, P., Schreurs, D.: Analysis of an indoor biomedical radar-based system for health monitoring. IEEE Trans. Microwave Theory Tech. **61**(5), 2061–2068 (2013). ISSN 0018-9480, 1557-9670. https://doi.org/10.1109/TMTT.2013.2247619. http://ieeexplore.ieee.org/document/6473854/
49. Nakata, R., Clemens, S., Martin, D., Jaquiro, C., Lubecke, V.: Unmanned aerial vehicle platform stabilization for remote radar life sensing. In: 2016 IEEE/ACES International Conference on Wireless Information Technology and Systems (ICWITS) and Applied Computational Electromagnetics (ACES), Honolulu, HI, USA, pp. 1–2. IEEE, March 2016. ISBN 978-1-5090-1259-6. https://doi.org/10.1109/ROPACES.2016.7465387. http://ieeexplore.ieee.org/document/7465387/
50. Nakata, R.H., Haruna, B., Yamaguchi, T., Lubecke, V.M., Takayama, S., Takaba, K.: Motion compensation for an unmanned aerial vehicle remote radar life sensor. IEEE J. Emergi. Sele. Top. Circu. Syst. **8**(2), 329–337 (2018). ISSN 2156-3357, 2156-3365. https://doi.org/10.1109/JETCAS.2018.2821624. https://ieeexplore.ieee.org/document/8328824/
51. Park, B.-K., Boric-Lubecke, O., Lubecke, V.M.: Arctangent demodulation with dc offset compensation in quadrature doppler radar receiver systems. IEEE Trans. Microwave Theory Tech. **55**(5), 1073–1079 (2007). ISSN 0018-9480. https://doi.org/10.1109/TMTT.2007.895653. http://ieeexplore.ieee.org/document/4195675/
52. Petrochilos, N., Rezk, M., Host-Madsen, A., Lubecke, V., Boric-Lubecke, O.: Blind separation of human heartbeats and breathing by the use of a doppler radar remote sensing. In: 2007 IEEE International Conference on Acoustics, Speech and Signal Processing - ICASSP 2007, Honolulu, HI, pp. I–333–I–336. IEEE, April 2007. ISBN 978-1-4244-0727-9. https://doi.org/10.1109/ICASSP.2007.366684. https://ieeexplore.ieee.org/document/4217084/
53. Qazi, F.A., Iskander, M.F., Yun, Z., Sasaki, G., Islam, S.M.M.: Smart physical layer based directional communication networking. In: 2017 XXXIInd General Assembly and Scientific Symposium of the International Union of Radio Science (URSI GASS), Montreal, QC, pp. 1–2. IEEE, August 2017. ISBN 978-90-825987-0-4. https://doi.org/10.23919/URSIGASS.2017.8104982. http://ieeexplore.ieee.org/document/8104982/
54. Rahman, A., Yavari, E., Lubecke, V.M., Lubecke, O.-B.: Noncontact doppler radar unique identification system using neural network classifier on life signs. In: 2016 IEEE Topical Conference on Biomedical Wireless Technologies, Networks, and Sensing Systems (BioWireleSS), Austin, TX, USA, pp. 46–48. IEEE, January 2016. ISBN 978-1-5090-1694-5. https://doi.org/10.1109/BIOWIRELESS.2016.7445558. http://ieeexplore.ieee.org/document/7445558/
55. Rahman, A., Lubecke, V.M., Boric-Lubecke, O., Prins, J.H., Sakamoto, T.: Doppler radar techniques for accurate respiration characterization and subject identification. IEEE J. Emergi. Sel. Circ. Syst. 8(2), 350–359 (2018). ISSN 2156-3357, 2156-3365. https://doi.org/10.1109/JETCAS.2018.2818181. https://ieeexplore.ieee.org/document/8322143/
56. Samardzija, D., Park, B.-K., Boric-Lubecke, O., Lubecke, V., Host-Madsen, A., Sizer, T.: Experimental evaluation of multiple antenna techniques for remote sensing of physiological motion. In: 2007 IEEE/MTT-S International Microwave Symposium, Honolulu, HI, USA, pp. 1735–1738. IEEE, June 2007. ISBN 978-1-4244-0687-6 978-1-4244-0688-3. https://doi.org/10.1109/MWSYM.2007.380063. http://ieeexplore.ieee.org/document/4264188/
57. Shi, K., Will, C., Weigel, R., Koelpin, A.: Contactless person identification using cardiac radar signals. In: IEEE International Instrumentation and Measurement Technology Conference (I2MTC), Houston, TX, pp. 1–6. IEEE, May 2018. ISBN 978-1-5386-2222-3. https://doi.org/10.1109/I2MTC.2018.8409645. https://ieeexplore.ieee.org/document/8409645/

58. Singh, A., Gao, X., Yavari, E., Zakrzewski, M., Cao, X.H., Lubecke, V.M., Boric-Lubecke, O.: Data-based quadrature imbalance compensation for a CW doppler radar system. IEEE Trans. Microwave Theory Tech. **61**(4), 1718–1724 (2013). ISSN 0018-9480, 1557-9670. https://doi.org/10.1109/TMTT.2013.2249525. http://ieeexplore.ieee.org/document/6479350/
59. Skolnik, M.I. (ed.): Radar Handbook, 2nd edn. McGraw-Hill, New York (1990). ISBN 978-0-07-057913-2
60. Skolnik, M.I.: Introduction to Radar Systems, 2nd edn. McGraw-Hill, New York (1980). ISBN 978-0-07-057909-5
61. Springer, S., Seligmann, G.Y.: Validity of the kinect for gait assessment: a focused review. Sensors **16**(2), 194 (2016). ISSN 1424-8220. https://doi.org/10.3390/s16020194. http://www.mdpi.com/1424-8220/16/2/194
62. Tang, J., Luo, J., Tjahjadi, T., Gao, Y.: 2.5D multi-view gait recognition based on point cloud registration. Sensors **14**(4), 6124–6143 (2014). ISSN 1424-8220. https://doi.org/10.3390/s140406124. http://www.mdpi.com/1424-8220/14/4/6124
63. van Dorp, P., Groen, F.C.A.: Feature-based human motion parameter estimation with radar. IET Radar Sonar Navigat. **2**(2), 135–145 (2008). ISSN 1751-8784, 1751-8792. https://doi.org/10.1049/iet-rsn:20070086. https://digital-library.theiet.org/content/journals/10.1049/iet-rsn_20070086
64. Vergara, A., Petrochilos, N., Boric-Lubecke, O., Host-Madsen, A., Lubecke, V.: Blind source separation of human body motion using direct conversion doppler radar. In: 2008 IEEE MTT-S International Microwave Symposium Digest, Atlanta, GA, USA, pp. 1321–1324. IEEE, June 2008. ISBN 978-1-4244-1780-3. https://doi.org/10.1109/MWSYM.2008.4633306. https://ieeexplore.ieee.org/document/4633306/
65. Vivet, D., Derbel, A., Emile, B.: Access control based on gait analysis and face recognition. Electron. Lett. **51**(10), 751–752 (2015). ISSN 0013-5194, 1350-911X. https://doi.org/10.1049/el.2015.0767. https://digital-library.theiet.org/content/journals/10.1049/el.2015.0767
66. Wang, S., Song, J., Lien, J., Poupyrev, I., Hilliges, O.: Interacting with soli: exploring fine-grained dynamic gesture recognition in the radio-frequency spectrum. In: Proceedings of the 29th Annual Symposium on User Interface Software and Technology, Tokyo Japan, pp. 851–860. ACM, October 2016. ISBN 978-1-4503-4189-9. https://doi.org/10.1145/2984511.2984565. https://dl.acm.org/doi/10.1145/2984511.2984565
67. Clark Webster, J.G.: Medical Instrumentation: Application and Design, 4th edn. Wiley, Hoboken (2010)
68. Woyach, K., Puccinelli, D., Haenggi, M.: Sensorless sensing in wireless networks: implementation and measurements. In: 2006 4th International Symposium on Modeling and Optimization in Mobile, Ad Hoc and Wireless Networks, Boston, MA, USA, pp. 1–8. IEEE (2006). ISBN 978-0-7803-9549-7. https://doi.org/10.1109/WIOPT.2006.1666495. http://ieeexplore.ieee.org/document/1666495/
69. Xiong, J., Hong, H., Zhang, H., Wang, N., Chu, H., Zhu, X.: Multitarget respiration detection with adaptive digital beamforming technique based on SIMO radar. IEEE Trans. Microwave Theory Tech. **68**(11), 4814–4824 (2020). ISSN 0018-9480, 1557-9670. https://doi.org/10.1109/TMTT.2020.3020082. https://ieeexplore.ieee.org/document/9198909/
70. Yang, Z.-K., Shi, H., Zhao, S., Huang, X.-D.: Vital sign detection during large-scale and fast body movements based on an adaptive noise cancellation algorithm using a single doppler radar sensor. Sensors **20**(15), 4183 (2020). ISSN 1424-8220. https://doi.org/10.3390/s20154183. https://www.mdpi.com/1424-8220/20/15/4183
71. Yavari, E., Jou, H., Lubecke, V., Boric-Lubecke, O.: Doppler radar sensor for occupancy monitoring. In: 2013 IEEE Topical Conference on Power Amplifiers for Wireless and Radio Applications, Austin, TX, USA, pp. 145–147. IEEE, January 2013. ISBN 978-1-4673-2918-7 978-1-4673-2915-6. https://doi.org/10.1109/PAWR.2013.6490217. http://ieeexplore.ieee.org/document/6490217/
72. Yavari, E., Song, C., Lubecke, V., Boric-Lubecke, O.: Is there anybody in there?: intelligent radar occupancy sensors. IEEE Microwave Mag. **15**(2), 57–64 (2014). ISSN 1527-3342. https://doi.org/10.1109/MMM.2013.2296210. http://ieeexplore.ieee.org/document/6763174/

73. Yue, S., He, H., Wang, H., Rahul, H., Katabi, D.: Extracting multi-person respiration from entangled RF signals. In: Proceedings of the ACM on Interactive, Mobile, Wearable and Ubiquitous Technologies, vol. 2, no. 2, pp. 1–22, July 2018. ISSN 24749567. https://doi.org/10.1145/3214289. http://dl.acm.org/citation.cfm?doid=3236498.3214289
74. Zakrzewski, M.: Methods for doppler radar monitoring of physiological signals. Ph.D. thesis, Tampere University of Technology, September 2015
75. Zhang, J., Wei, B., Hu, W., Kanhere, S.S.: WiFi-ID: human identification using wifi signal. In: 2016 International Conference on Distributed Computing in Sensor Systems (DCOSS), Washington, DC, USA, pp. 75–82. IEEE, May 2016. ISBN 978-1-5090-1460-6. https://doi.org/10.1109/DCOSS.2016.30. http://ieeexplore.ieee.org/document/7536315/
76. Zhang, L., Ding, C., Zhou, X., Hong, H., Li, C., Zhu, X.: Body movement cancellation using adaptive filtering technology for radar-based vital sign monitoring. In: 2020 IEEE Radar Conference (RadarConf20), Florence, Italy, pp. 1–5. IEEE, September 2020. ISBN 978-1-72818-942-0. https://doi.org/10.1109/RadarConf2043947.2020.9266671. https://ieeexplore.ieee.org/document/9266671/

Chapter 8
Biomedical Radar and Antenna Systems for Contactless Human Activity Analysis

Aniqa Tabassum and Md Atiqur Rahman Ahad

Abstract The elderly population of the world is increasing, and along with that, the need for assisted living and health monitoring. Biomedical radars can be used to monitor the health and safety of patients, both at home and in a clinical setup. This chapter gives a comprehensive overview of biomedical radar and antenna systems for contactless human activity analysis. Recent advancements in this topic are discussed, including original research work for particular applications such as posture recognition, search and rescue, sleep monitoring, activity recognition, identification of individuals, monitoring vital signs, occupancy monitoring, and fall detection. The main challenges and opportunities in this direction are also discussed in detail.

8.1 Introduction

Over the last century, the proportion of the global population aged 65 and above has been increasing gradually. The eventually growing aging population usually suffers from various health conditions and chronic illnesses, like lung disorders, cardiovascular conditions, diabetes, cancer, stroke, and sleep disorders [1]. Such challenges require a patient-specific health monitoring system where such patients can be monitored in their homes remotely and continuously, using modern technology that is reliable and affordable.

Accurate and continuous observation of patients is required while allowing them to live undisturbed and independently in their comfortable environment. The monitoring system must identify falls or any such disease-specific phenomena and instantly inform family and hospital authority. Of particular interest is monitoring the health of the elderly who live by themselves, patients who have just returned from the hospital, who have recently had surgery, or critical individuals.

A. Tabassum · M. A. R. Ahad (✉)
Department of Electrical and Electronic Engineering, University of Dhaka, Dhaka, Bangladesh
e-mail: atiqahad@du.ac.bd

Department of Intelligent Media, Osaka University, Suita, Japan

M. A. R. Ahad and A. Inoue (eds.), *Vision, Sensing and Analytics: Integrative Approaches*, Intelligent Systems Reference Library 207, https://doi.org/10.1007/978-3-030-75490-7_8

Table 8.1 Advantages and Disadvantages of some of the existing healthcare monitoring technologies

Category	Example	Benefit	Limitation
Vision-based	RGB Video Camera, IR Camera, Depth Camera (e.g., Kinect Sensor), Optical Tracking System (e.g., Vicon)	- Multiple subject monitoring is possible - Non-contact monitoring - Immune to electromagnetic interference	- May interfere with privacy - Highly dependent on light - Computational cost is high
Sound-based	Acoustic Sensors, Ultrasonic Acoustic Audio, Mobile Phones	- No privacy concerns - Non-contact monitoring	Easily affected by noise or other signals
Inertial sensor-based	Accelerometers, Gyroscopes, Body-worn devices (embedded in vests, shoes, wrist bands and watches)	- No privacy concerns - Outdoor monitoring is possible - Less expensive - Some wearable sensors can monitor physiological parameters	Physical discomfort issues for patients

There are many existing technologies today for continuous health monitoring of patients. The most common type of sensor used is the camera-based sensor. These are dependent on light variation and fail to trace the patient if the subject goes out of the line-of-sight or visibility area of the sensor [2]. These also result in less accurate results outside of the laboratory environment. Another main concern is the privacy issue (e.g., in bathrooms and bedrooms). Ambient sensors like pressure pads, infrared technology, and acoustic receivers detect the environment of a subject. However, their operation can be disturbed by ambient interference (noise in the case of acoustic or other heat sources in case of thermal infrared). For correct functioning, however, the system has to be set up all over the house, increasing the overall expense. Wearable devices are relatively low in cost and consist of accelerometers, gyroscopes, and magnetic sensors, which can accurately and elaborately monitor health parameters as well as daily routine activities of subjects. The disadvantage is that these have to be worn or carried all the time by the patient being monitored, arising concerns of comfort, agreement, and compliance. A few of the benefits and disadvantages of the mentioned technologies are discussed in Table 8.1 [3, 4].

8.2 Contactless Monitoring for Healthcare

Modern healthcare monitoring devices can revolutionize the future of medicine because they can generate extremely elaborate individual physiological data [5].

Having access to their physiological data and respective analyses, patients can contribute much more actively in their health management. This evolution must happen in the regular lives of people, not just in research labs and hospitals. Contact-based sensors need users to wear or carry bulky sensors, straps, or electrodes all the time, which discourages continuous monitoring because such devices may cause discomfort and restrict normal movement. Contactless monitoring is the answer to all these concerns. Some of the most common contactless technologies for healthcare and wellness monitoring include Radio-Frequency (RF) sensing, such as using Radar, WiFi, and Radio-frequency identification (RFID) based systems [4].

Continuous monitoring of a patient is beneficial for doctors and researchers [6]. An example of a dangerous sleep disorder is obstructive sleep apnea (OSA). In this case, the breathing of a person continuously stops and starts, because of full or limited blockage of the upper airway of the lungs while sleeping. Sudden infant death syndrome (SIDS) is defined as the undiscovered and sudden death of a child and has been inferred to be related to sleep apnea. Biomedical radar-based systems can implement contactless, non-invasive, and monitoring subjects over long periods, and help detect OSA and SIDS, along with many other such medical conditions. Another major reason for using non-contact sensors is the reduction of cost and also to increase the reliability and longevity of the sensors. A few advantages of contactless monitoring include:

Long-term observation of rarely occurring medical conditions Certain medical conditions that cannot be detected during a short monitoring period require long-term monitoring. Contactless monitoring is best for such conditions [7].

Potential of long-term monitoring of gradual health changes at reasonable expense Observing the exercise recovery rate of a professional athlete, the health recovery monitoring of a surgery patient, lifestyle and habit changes while carrying out a diet, the influence of certain actions (like physical stress, sports, coffee, and sugar intake) on an individual's health or sleep are some examples.

No requirement for prior preparation ECG and EMG electrodes usually need thorough preparation before carrying out measurements and can cause unease and discomfort for newborn children, bedridden patients, and burn victims. Contactless monitoring raises no such concerns.

8.3 Radar-Based Sensing

RADAR stands for Radio Detection And Ranging. A complete radar-based system is made up of a transmitter that generates and transmits electromagnetic (EM) signals, a receiver to detect reflected echo signals, antennas to transmit and receive signals, and a processing unit that can analyze the collected data to obtain the desired characteristics of the target [4]. Transmitted radio waves reflect off the target, the echoes come back to the receiver, and characteristics such as the range, trajectory, velocity, and even

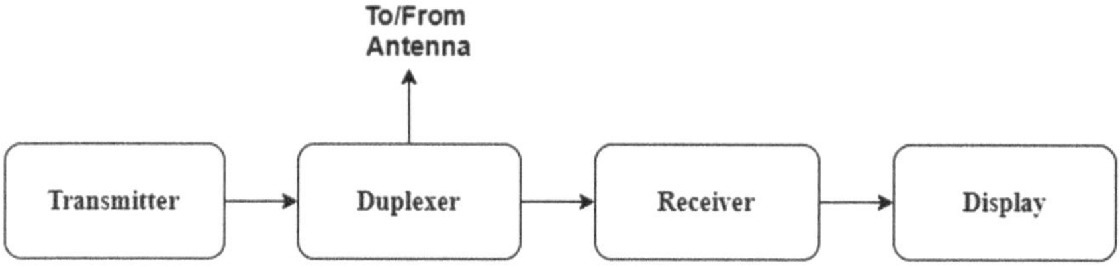

Fig. 8.1 Radar basic block diagram

recognition of the target can be obtained by correct processing of the echoes. Target velocity can be achieved using the Doppler Effect, i.e., by looking at the final shift of the radar signal frequency after reflection from the target. The Doppler Effect can be observed when a change in frequency occurs of either the transmitted signal or reflected signal, as a result of the relative motion between the transmitter and the receiver. Figure 8.1 shows the basic block diagram for a radar system.

Along with the change in Doppler frequency due to the overall motion of the object of interest of a radar, the oscillating or rotating parts of the target undergo micro-motion dynamics, i.e., mechanical rotations or vibrations. These fluctuations add Doppler changes to the received echo. This effect is known as the micro-Doppler effect [8]. It is possible to identify targets by studying their specific micro-Doppler signature, which is the pattern of micro modulations on the received echo. When monitoring human beings, the resulting micro-Doppler signatures are associated with oscillation patterns of the human body, the limbs, and the head of a person in movement. These signatures are very specific to each individual and also activity-specific. As a result, it is a useful feature for individual monitoring and classification.

Radar systems are classified into two categories. Active radars have an assigned transmitter and receiver, which can be merged within a single system. This has the benefit of complete control of the signal which is transmitted. The existence and the position of targets can be derived by studying the total time taken by the transmitted signal from the transmitter to the target to the receiver. A passive radar has no dedicated transmitter and depends on a signal which has been transmitted from a different source. The benefit, in such cases, is simpler hardware and no requirement of the allocated spectrum. The limitation is no control over the transmitted signal. The existence and the position of objects are derived by looking at the time-lapse between receiving the signal straight from the source and the signal after being reflected off a target.

Many different radar architectures have been used for different applications. Radars can be classified depending on the operating frequency, principle of operation, phase, range, antenna type, application area, and arrival time changes. A radar that functions using a pulse signal for the detection of stationary targets is called a Pulse Radar. A single antenna is used for both transmitting and receiving signals. A Continuous Wave (CW) radar, or Doppler radar, sends one frequency, continuous-wave, RF signal, which then gets reflected from a target and is then received. From the Doppler effect, the target's velocity can be obtained accurately. When the radar target has a periodic motion, such as the heart or the chest, a phase shift can be obtained which is proportional to the surface displacement. A patient's chest wall position can thus be determined, which provides further information about chest movements because of heartbeat and respiration, making the CW radar a feasible option for vital signs monitoring. The major limitation, however, is that it cannot determine absolute distances. In contrast to the CW radar, a Frequency-Modulated Continuous Wave (FMCW) radar is capable of altering its operating frequency, i.e., the transmitted signal can be frequency modulated. It can measure not only the speed but also the position of the target. These properties allow FMCW radar to be more suitable for vital sign detection, particularly for multiple subjects. FMCW radar power consumption is also low with relatively small compatible sizes [9]. Ultra-wideband (UWB) radars use transmitting signals over a comparatively wider frequency range than traditional systems and hence cannot be detected easily. The very short pulses required for generating a UWB signal offer several advantages like lower power, better output, better resistance to jamming, and the ability to coexist with existing radio services. The narrow pulse also allows better spatial resolution and very short-range capability compared to other systems. A UWB radar can obtain the velocity as well as the absolute distance of the target and improve the signal-to-clutter ratio for more accurate detection of targets, above the clutter and noise. Millimeter-wave (mm-Wave) radars use transmission signals which have a wavelength in the millimeter range. This relatively short wavelength has a particular advantage. The size of the system components like the antennas needed for mm-Wave signals is also comparatively smaller. The high accuracy of mm-Wave radars is a second advantage. A system that functions at 76–81 GHz (resultant wavelength is about 4 mm) are capable of perceiving movements comparative to a fraction of a millimeter. The sensing principle of radar almost remains the same, though, no matter which radar type is used.

Radar systems transmit electromagnetic or radio waves. Radars can generally be categorized depending on the operating frequency, which is the frequency band used. The operating frequency of the transmitted waves depends on the particular radar application. Radar systems use frequencies ranging from 5 megahertz to 130 gigahertz. Specific frequencies, however, are more suitable for certain applications. Radar systems for very long range surveillance generally use frequencies in the lower range. The minimum antenna size is proportional to the wavelength and inversely proportional to the frequency used [10]. Radar systems are often named by the frequency band they use for operation, using the band designations shown in Table 8.2.

The Industrial, Scientific, and Medical (ISM) frequency band contains radio and microwave frequencies grouped around 2.4GHz. These frequencies are booked and assigned for industrial, scientific, and medical devices or equipment using radio frequency, as defined by the ITU Radio Regulations to ensure safety and avoid interference. Radars allow undisturbed home-based healthcare and monitoring facilities. Even though radars can obtain motion or distance-based parameters, they can also monitor many important biomedical parameters and recognize risky medical emergencies such as falls, apnea, sudden infant death syndrome [11], and irregular heart rate [12]. Radar allows such monitoring remotely and unobtrusively, within the

Table 8.2 Radar Frequency Bands (Source: American Institute of Aeronautics and Astronautics) [https://www.aewa.org/Library/]

Band Classification	Frequency Range	Application
VHF	50–330 MHz	Used for monitoring over a very long range
UHF	300–1000 MHz	Used for monitoring over a very long range
L	1–2 GHz	Used for monitoring over a long range, en-route air traffic
S	2–4 GHz	Used for monitoring over a moderate range, terminal air traffic control, and long-range weather monitoring
C	4–8 GHz	Used for tracking over a long range, airborne weather surveillance
X	8–12.5 GHz	Used for tracking over a short range, guidance of missiles, mapping, marine radar, airborne interception
Ku	12.5–18 GHz	High resolution mapping, satellite altimetry
K	18–26.5 GHz	Rarely used
Ka	26.5–40 GHz	Very high resolution mapping, surveillance at the airport
mm	40–100 GHz	Used in experiments

comfortable environment of a patient's home, allowing healthcare monitoring over extended periods.

8.3.1 Data Processing

The basic signal processing stage of a radar-based healthcare monitoring system begins by using an experimental data set; this may consist of practically collected data, data produced using a numerical model, or a combination of both to enhance the diversity of the data set. Such data are then processed through various stages to extract useful parameters, i.e., range, amplitude, phase, or even micro-Doppler information. Micro-Doppler information, which is a distinctive signature of the radar target, can be derived using time-frequency transformation, e.g., Short-Time Fourier Transform (STFT) [13], Wavelet Transform (WT) [14, 15], Extended Modified B-Distribution [16], etc. Next, parameters or properties called features must be extracted from the processed data. This can be useful for data classification, i.e., in distinguishing between more similar activities. For better accuracy and to identify false alarms, data are prescreened, and feature selection is then carried out. Feature extraction can be carried out in time as well as frequency domains [17]. Some common time-domain features include the mean value, standard deviation, median absolute deviation, maximum and minimum values of samples in a frame, and signal entropy. Fast Fourier Transform (FFT) can be employed for the derivation of features in the frequency domain. Features can be used for the classification of activities. These can be selected using the importance factor, which is derived using the Random Forest (RF) algorithm. Some other feature selection steps can also be applied, such as repetitive feature removal, univariate feature selection, elimination of those features that have higher correlation, and lower variance.

8.3.2 Data Classification

The final stage requires using these extracted feature samples as inputs in classifiers, which are machine learning-based algorithms. A portion of the samples is used in training the classifier, and the remainder in testing its performance. Multiple classification algorithms are being looked into, with different computational complexities. Classification methods can be applied judging by the data type, amount of data, homogeneity between the activities, how many activities are being studied, and the overall count of classes. To classify dynamic and static activities, the most common classifiers, i.e., Support Vector Machine (SVM) [18–20], Random Forest (RF) [19], Linear Discriminant Analysis (LDA), Logistic Regression (LR), and K-Nearest Neighbour (KNN) [19, 20] can be used. However, kernel choice for Support Vector Machine, K-value (number of neighbors), and the method to calculate the distance for KNN, how many trees to use for RF, and tolerance value for LDA are important in determining the performance of these classifiers and must be carefully set. Deep learning networks are also applicable to classifying human activities. Since deep neural networks require more processing time, there must be a compromise between the processing time and classifier performance. The performance of such algorithms is usually observed in terms of higher sensitivity, accuracy, specificity, or their combinations. Many such algorithms assume identical distribution of the input features. This may not hold in the case of human activities, where certain actions may be related to past actions and also influence future activities.

8.4 Radar-Based Human Activity Analysis

Radar has several applications in the biomedical sector [1]. It has been used in combination with Wi-Fi to help in assisted living and health monitoring of the elderly [4]. Figure 8.2 illustrates some radar-based health applications.

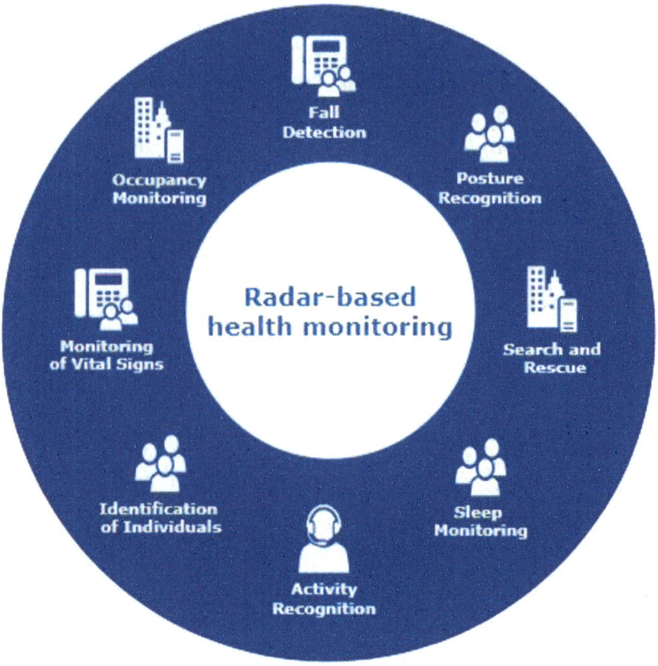

Fig. 8.2 Radar-based health monitoring

8.4.1 Posture Recognition

When a radar target is stationary, consecutive pulses reflected from the target travel the same distance, and hence there is no resulting phase shift between the reflected waves. However, when the subject has a velocity towards or in the direction opposite to the radar, successive waves return to the radar with different phases. The net phase change is a result of the target motion, with larger velocities causing greater phase changes. Phase change information derived from the Doppler radar can be associated with the subject chest motions [4]. The total signal back-scattered by the subject's body can be defined using the Doppler radar cross-section (RCS). It shows how easily a subject can be detected by the radar. The RCS is proportionate to the area the subject's chest covers, along with the surface reflectivity, and the radar directivity at the operating frequency.

Identifying movements while sleeping is important for monitoring patients having sleep disorders. Radars have already been used to identify a patient's posture on the bed. This is extremely beneficial in monitoring certain sleep conditions and disorders, which may be more recognizable in certain sleep positions. Kiriazi et al. [21] have introduced a convenient and effective method to identify a patient's sleep posture on the bed. This study examined the sleep postures of a patient by studying the RCS. The cardiopulmonary activity of a subject is reflected in the movements of the chest. The echo from the subject is phase modulated by the chest movements of the patients as a result of the Doppler effect. The RCS for the backside of a subject was found to be about nine times more compared to his front side. This rise in RCS is a result of the difference in the geometrical shapes of the front and back sides of the human body. Sleep posture transition (SPT) recognition has been implemented using ultra-wideband (UWB) radar [22]. The Multi-View Multi-Task Learning, called SleepPoseNet or SPN, has classified four standard SPTs. This application can extract features in both time and frequency domains, as well as movement and directions of sleeping positions. Data obtained from 26 subjects showed potential in performance.

Millimeter-Wave radars have also been used for capturing and analysis of human posture [23]. Because a single radar has a limited beamwidth and can only detect part of a person, to obtain the complete picture, a vertical radar array has been used in this work. This increases the coverage showing a 55% improvement on average. Mm-Wave radars can obtain detailed data with a rich set of features for human posture recognition.

Sengupta et al. [24] have introduced mm-Pose, a novel real-time approach for estimating skeletal pose employing the mm-Wave radar. This method is capable of detecting greater than about 15 different skeletal joints using radar reflections. A forked CNN architecture predicted the position of the skeletal joints in the body in 3-D by utilizing a novel high-resolution radar-to-image representation. The current implementation was tested with the data obtained using four different basic motions, i.e., walking, swinging left arm, swinging right arm, and swinging both arms to test accuracy in range, azimuth, and elevation. Even a greater number of motions can be included by carrying out data collection and identification for a bigger range of spatial motions.

Classifying and identification of hand gesture activities of combatants can also be done using the analysis of multi-domain micro-Doppler data obtained using radar systems [25]. Several common combat gestures are studied by utilizing a through-the-wall radar system, to obtain information on human pose and movement. Three different human subjects were used to form a database of radar data obtained in real-time with more than 2000 recordings for a group of experiments. Using the micro-Doppler signatures obtained, 6 major features are extracted by analyzing data in the time domain, frequency domain, and STFT time-frequency domain respectively. Results show that the classification accuracy is more than 90%. Figure 8.3 shows the block diagram for posture recognition.

8.4.2 Search and Rescue Purposes

Radar-based systems are being developed that can sense and help rescue alive people under wreckage and debris in case of an earthquake or an avalanche [26]. Doppler radar sensors can locate the exact position of survivors under collapsed buildings and other obstacles. When a transmitted wave is targeted towards a pile of collapsed

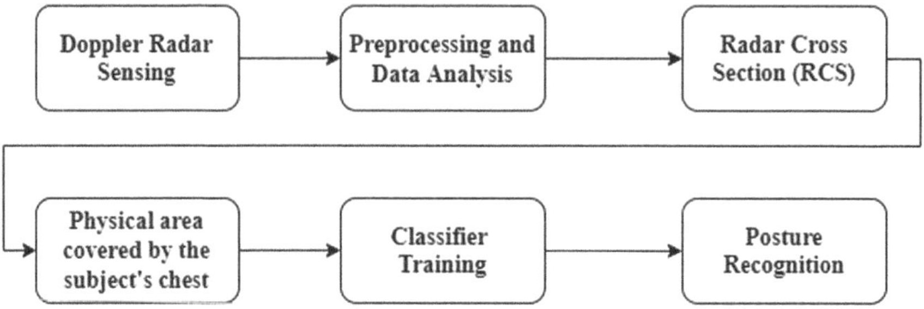

Fig. 8.3 The multiple stages in human posture recognition using radar-based systems

debris, covering an alive human survivor or through a barrier blocking a human, the signal can pass through the rubble or the obstacle and reach the survivor. The echo reflected on the survivor is modified by the person's physiological motion, like respiratory rate as well as the heartbeat. If the clutter due to the reflected wave of the surrounding static objects and the background is correctly nullified, the actual respiratory and heartbeat signals of the survivor can be obtained. This would provide concrete proof of the existence of a living person underneath the remains [27].

Studies have proved that breathing as a major and basic vital sign is possible to detect conveniently and with high accuracy and confidence. Thus, this can be the spotlight feature for the development of radar-based systems which can be used for search and rescue of an earthquake, as well as avalanche survivors [28]. The difficult task is monitoring the breathing of an unconscious person. Trapped people who are moving can cause an imprint in the reflected radar signals. Unconscious victims are incapable of such action. The only possible physiological variations for them are due to their breathing motions and heartbeat. Heartbeat is a weaker signal, and it is difficult to extract information from it, so the detection of breathing rate is the main focus for such systems [29]. Vital sign detection using radar sensors is highly disrupted due to motion artifact, and interference caused by the presence of multiple victims. Multi-person respiration sensing is difficult to detect and label, especially due to the background noise and clutter present. Filtering and advanced signal processing schemes may help improve system performance in such cases. Another major challenge is the effect of different building materials and obstacle layer thicknesses on the received echo [28]. Considering the distance and blockages between the survivor and the system, the radar penetration capability is the most important factor. This depends mainly on the chosen operating frequency and the transmitting power. Many researchers have chosen the L and S-band as the operating frequency ranges for their system as a result [30]. Radar systems based on various new signal processing techniques are being proposed, to identify survivors during an earthquake, tsunami, or avalanche [31]. The focus is on detecting as well as separating the vital signs obtained from multiple trapped survivors.

After natural disasters like earthquakes, avalanches, tsunamis, or explosions, the affected may find shelter in certain voids that form naturally in collapsed structures.

The rescue team needs to discover and locate these voids for quick and efficient search and rescue of victims. However, because of the complicated and unstructured blockages in such affected areas, conventional searches take up time and can be hampered by human errors. Hu et al. [32] introduced a novel technique capable of automatically sensing, locating, and identifying voids within the debris by employing ground-penetrating radar (GPR). After preprocessing the obtained radar data, borders of possible voids are segmented, using patterns of radar signals, and 95% confidence intervals are formed surrounding the segmented borders for representing uncertainties. For improved accuracy of detection, the geometric relationships between the obtained borders and their corresponding signal characteristics are studied, to determine whether a void is present. The void size and location are then approximated using the calibrated velocity of the radar signal, and the time it takes for travelling. The possibility of using GPR to identify, determine the location, and characterize voids in demolished structures is tested experimentally. This proposed approach was examined in simulations, in labs as well as in field experiments, to determine performance. 60 scenarios were tested in total, comprising of 30 simulated and 30 real radargrams. This approach showed an accuracy of 93% on average, 95% precision, and a recall of 95% for radar data in real-time and in simulation. The findings prove the potential of the introduced method, which will provide the first responders with knowledge of the situation for search and rescue of survivors trapped and fighting for their lives.

A novel multiscale residual attention network was proposed to recognize stationary human subjects and some animals via a through-the-wall condition using UWB radar. Human subjects, cats, rabbits, dogs, and empty data are acquired and evaluated [33]. Deep learning was first utilized in this particular application in this work. Comparison with some other common approaches indicates the suitability of the approach for search and rescue work. The requirement of advanced professional knowledge was also not reduced here due to automatic feature extraction.

A framework, AiRobSim, was developed to simulate an aerial robot to obtain above the ground as well as underground information with the aim of search and rescue work [34]. This is done to improve the existing robots that cannot detect the voids in collapsed buildings that may contain survivors. The aerial robot is given a holistic view of the disaster areas by integrating an unmanned aerial vehicle (UAV) platform with a ground-penetrating radar (GPR) and some common sensors, like a global navigation satellite system (GNSS), inertial measurement unit (IMU), and cameras. The obtained data can be used to identify critical spaces under the rubble to rescue trapped survivors. Results show that the robot can assist in search and rescue with an accuracy of 79.2%. Figure 8.4 shows the block diagram for a radar system used for search and rescue purposes.

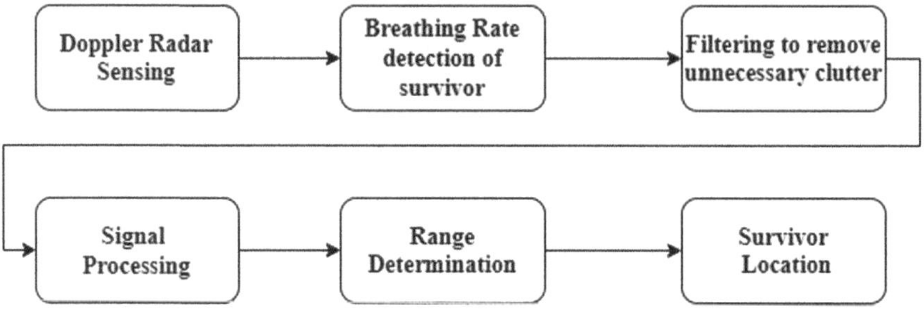

Fig. 8.4 Radar-based search and rescue of survivors during disasters like earthquakes and avalanches

8.4.3 Sleep Monitoring

About half of the elderly population of the world, whose number is ever increasing, experience disturbances in sleep. Sleep monitoring studies today, hence, have become a considerable health monitoring factor for medical professionals [35]. Patients suffering from sleep disorders usually have abnormal sleep habits, capable of leading to mental and physical stress. Studies have discovered several categories of sleep disorders. Obstructive sleep apnea (OSA) is a familiar sleep problem for the elderly. Biomedical radar-based sensing systems are being developed to monitor a patient's sleep pattern remotely [36]. The quality and period of sleep heavily influence a person's physical and mental health, metabolism, weight, safety, cardiovascular health, and body immunity [37].

The standard method to diagnose OSA is an overnight sleep monitoring technique based on polysomnography (PSG). For this technique, patients have to be monitored overnight in a specialized lab. The associated parameters can be achieved via wearable sensors, which turn the entire process difficult, complicated, and very expensive. Despite giving reliable results, this system is not appropriate for continuous monitoring for extended periods, as it may restrict movement as well as cause discomfort.

These shortcomings have led to the development of contactless sleep monitoring systems. Singh et al. [38] have introduced a noninvasive radar system for measuring physiological parameters, which is known as a physiological radar monitoring system (PRMS). This system is reasonable in cost, easy to use, and can correctly observe the sleep dynamics and habits of a person. The system capable of monitoring sleep apnea was based on the CW radar. Biomedical radar-based systems have several advantages over other sensors in case of sleep-related applications. Contactless sleep monitoring frees the subject from the electrodes required in existing technologies. The high sensitivity and accuracy of microwaves allow capturing little variations in vital signs and body movements while sleeping [39].

The Circadia Contactless Breathing Monitor is a novel approach that employs pulsed UWB radar with a target of monitoring the breathing function and body movements [40]. The Circadia Sleep Analysis Algorithm can estimate different stages in a subject's sleep, bed occupancy, and can determine standardized sleep metrics, and makes monitoring of sleep feasible in a patient's home. A radar sensor can perform occupancy monitoring in bedrooms and also detect falls. The sleep monitoring findings of the approach were compared against the standard for sleep monitoring, PSG. The C100 device was first utilized to obtain sleep findings over 17 nights from 9 subjects along with PSG, with a following 24 nights of PSG data to validate findings. Body movement and respiration features were obtained using data from sensors, and a machine learning algorithm was used for the prediction of the sleep stage. This approach has outperformed a commercial as well as a medical-grade monitoring device worn on the wrist (Fitbit Alta HR and Philips Respironics Actiwatch) when comparing performance. The obtained respiration data was also high in accuracy. The findings indicate that this approach can provide a good replacement to conventional wrist-worn sensors, and also that monitoring the same subject repeatedly for extended periods, of breathing and sleep monitoring is feasible at homes.

Many systems for contactless vital sign monitoring have been developed, but most use "static" and "direct-facing" patient findings, which does not apply to the complicated sleep situation. Some concerns on accuracy for these systems are because of background clutter, DC offsets, null point, artifacts due to motion, and interferences from various sources. Particularly for continuous sleep monitoring, the complex sleep environment, the uncertainty of body movements and positions, sudden changes in position, cancellation of multiple people present, and identification of the desired subject are added challenges.

8.4.4 Human Activity Recognition

Human activity recognition (HAR) based research and development have gained attention and there has been significant progress in this sector in recent times. HAR based on radars has many advantages. Radar does not depend on the lighting or the weather, hence the technology is applicable in harsh environments. There are no privacy concerns regarding radar. Instead of working with the physical form of the target, the radar echoes modulated by the target contain plenty of time–varying range and velocity data regarding the activities. Radars are also capable of detecting subjects through walls, which makes HAR based on radars feasible in many difficult scenarios. Also, such systems do not need any tags to be attached to the subject's body, which makes it more user-friendly. When a subject is in motion, the Doppler frequencies of the subject's body parts vary in proportion to the subject's motion. The subjects' range, speed, and angle obtained by the radar can be observed to identify activities. Because of its advantages like simplicity in design, comparatively lower cost, and good penetration property, radar represents a practical option to monitor human activities. Types of radars used for this purpose are continuous-wave radar, ultra-wideband radar, and noise radar. Deep learning provides a feasible option to obtain high-level features automatically and has displayed efficient results in the case of HAR. Experimental findings show the practicality of using deep learning approaches for human activity recognition based on radar in 1D, 2D, and 3D echo [41].

Recognition of daily basic activities, and monitoring gait, which refers to a person's way of walking or moving on foot, are the building blocks of a contactless health monitoring system. Such systems have the potential to allow patient-specific treatment and therapy. These continuously collect individual medical data to allow doctors to assess an individual's health and the effect of treatment on the patient, reduce the cost of healthcare, and also ensure an immediate response in case of medical emergencies or accidents [42].

A biomedical radar-based system can recognize different human activities by obtaining the amplitude, range, phase, and other information from the reflected echo [43]. Such systems can be used to identify daily routine activities, as well as classify different human walking and moving styles [44], e.g., walking, standing, sitting, squatting, crawling, bending low to pick something, walking while holding or lifting something, and risky incidents such as fall [45, 46]. These basic individual actions can be merged to identify regular household activities like cooking, laundry, house cleaning, etc., which can be health indicators of an individual. When a subject is in good health, these activities would be performed regularly, and break in such habits may imply worsening health conditions.

Li et al. [20] introduced a multi sensor-based approach to monitor everyday regular activities using radar-based systems as well as wearable sensors to get better results. The radar sensor was frequency modulated and used an operating frequency of 5.8 GHz. The experiment was carried out in a laboratory/office room at the University of Glasgow, whose environment has been described in the paper. Nine volunteers (aged between 23 and 31) participated in the experiment and performed ten activities, repeating every activity three times. These activities were walking empty-handed, walking holding a box, sitting, standing back up, bending down, bending down to tie shoelaces, drinking water, talking on the phone, emulate a frontal fall, and bending down before coming back up [47]. The study included similar events on purpose to examine the validity of the classification technique. Information fusion methods were adopted by using both radar-based as well as wearable sensors. This overcame the limitations of using a single sensor, and the final accuracy of the system rose from 80 to 98% when this multisensory approach was applied.

Yang et al. [48] provided a different approach for radar-based sensing for human activity recognition. This paper identified three different types of human gaits, i.e., walking empty-handed, walking with a handbag in a hand, and walking holding a handbag using both hands. This work utilized an X-band (9.8 GHz) continuous-wave radar. The classification is useful for safety as well as security applications, e.g., to identify thieves, etc. This work extracted micro-Doppler features, which are distinctive signatures of subjects, and used machine learning algorithms like Support vector machines (SVM) for classifying the obtained sensor data. The data were analyzed to test the accuracy of the proposed approach, which was relatively good, considering three different human gaits.

Deep neural networks (DNNs), particularly convolutional neural networks (CNNs), have shown great potential in HAR using radar. The limitation of these networks is a scarcity of enough RF data to allow the construction of deep neural networks that have more than a few layers. A large training data set is difficult to obtain because of the cost and the resources needed for collecting real-time data. Erol et al. [49] have proposed using a generative model, auxiliary classifier generative adversarial networks (ACGANs) to produce a large quantity of training data which can consider environmental effects, like clutter because of through-the-wall scattering. The classification performance obtained by training using real data compared to training on the synthetic data produced by ACGAN is obtained. Using ACGAN-generated synthetic motion signatures lead to a 14% betterment in the accuracy of classification.

8.4.5 Identification of Individuals

Contactless monitoring of vital signs utilizing a Doppler radar-based system has a lot of scope in continuous identity authentication, which may counteract the vulnerability of the existing one-pass validation authentication systems [50]. Doppler radar presents an exciting choice for identity authentication, since it is non-contact and does not need line-of-sight, and there is no performance degradation in absence of ambient light. This can provide low cost and faster identification without violating privacy like camera-based identifiers.

Each individual has a unique respiratory pattern, which differs from others in parameters like expiratory direction, movements of the lungs due to air intake, tidal volume, and airflow pattern. Besides this, breathing patterns have been obtained in ventilation when a subject is at rest, in a normal breathing state, and during abnormal breathing. A unique motif is obtained in each case. Rahman et al. [51] introduced a unique human identification system that can recognize subjects using their breathing patterns, utilizing a continuous-wave (CW) radar-based system. This proposed system can identify individuals with more than 90% accuracy and help to diagnose physiological conditions and illnesses. Radar-based identification of people has a whole lot of possibilities, including IoT applications, health monitoring, security, and virtual reality. A CW Doppler radar-based system for characterizing respiration correctly has been proposed to obtain a better dynamic range [52]. Minor component analysis of the respiration data is done, which helps identify individuals. Non-contact individual identification is appealing due to being biometric that can be sensed beyond the wall. The count of total subjects, however, was low and more work and research is needed to obtain authentic results.

The possibility of obtaining features that are useful for identification using radar respiratory patterns was studied for subjects in resting state and after physical activities [53]. Features associated with respiratory function, e.g., respiration rate, breathing depth, spectral entropy, inhale/exhale area ratio, and the mean and standard deviation of the peaks were obtained from the respiratory data obtained using radar, and differences in these features after physical activities were also observed. Findings show that after short exertions, the breathing pattern exhale area and breathing depth increased by more than 1.4 times for the subjects, showing the individuality of surplus heart volume after expiration for identifying a person after activities. This method is combined with a Support Vector Machine (SVM) and gives a success rate of identification of about 98.55% for rest state and about 92% for combined (rest and after exercise).

A study introduced a continual identification system called “Cardiac Scan” [54]. This comprised of a 2.4 GHz Doppler radar transceiver with 2 antennas (a transmitting and a receiving one). From the experimental findings, it was established that every subject had a different heart, tissue, and blood circulation system, due to the presence of noticeable variations in points in their cardiac cycle evaluated using the data set obtained from radar. A contactless biometric authentication approach was introduced using heartbeat related features obtained by utilizing a microwave Doppler sensor [55]. This work employed a 24-GHz radar system to obtain features associated with a heartbeat to identify 11 subjects. The proposed method demonstrates the success of the heart-based power spectral density (PSD) feature for individual identification. But if there is a change in position between the radar and subject, or there is a lot of fluctuation in heart rate, then this approach may cause false classification. Micro-Doppler heart-related characteristics are patient-specific and not easy to replicate, and hence are useful for identification purposes. A study demonstrated radar-based identity authentication based on short-time Fourier Transform (STFT) [56].

A challenge in radar-based identity authentication is due to the noise generated as a result of random body movements when studying breathing and due to multiple subjects being present. Unlike non-critical applications, incorrect detection in identity authentication could result in permanent loss of access privilege. Hence a sophisticated system has to be developed. Also, for subjects with cardiovascular diseases, the cardiac cycle would be affected, and unique identification may be problematic using heart rate. Figure 8.5 shows the block diagram for the personal identification of individuals.

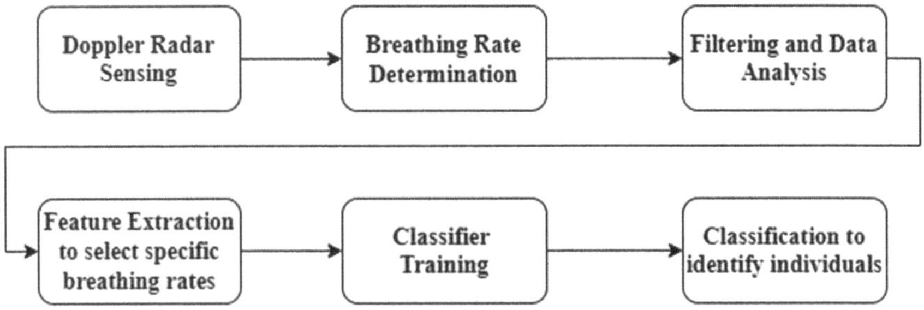

Fig. 8.5 Multiple stages in the identification of individuals using radar-based systems

8.4.6 Monitoring of Vital Signs

Vital signs are measurements of the human body's vital (life-sustaining) functions. The four standard vital signs which are regularly monitored by medical professionals are:

1. Breathing rate
2. Pulse rate (Heart rate)
3. Body temperature
4. Blood pressure

Vital signs can reflect the health of a subject, detect illnesses or medical problems, and display a patient's path to recovery. The standard ranges for a subject's vital signs differ with gender, age, weight, and overall health and wellness. One of the most common applications of biomedical radar-based systems is vital sign monitoring. Radars have been considered suitable for this purpose, as they can work through obstacles and harsh environmental conditions [9]. For physiological parameters and vital signs monitoring, the radar target is normally taken to be the patient's chest or the abdominal region. The reflected signal can be used to extract information about the patient's chest or abdominal movements. Chest movements of patients can traditionally determine the respiratory and heart rates. Abdominal movements are known, for stronger Doppler shifts, because of higher amplitude and deeper displacement than chest movements.

Several recent studies have focused on correctly monitoring the breathing rates of patients [15, 57], which are very clear indicators of patient health conditions. FMCW radar has been considered applicable in this field due to its ability to detect both the target position and chest micro-motions as a result of breathing [9]. Zito et al. [58] have proposed a UWB system-on-chip radar sensor for monitoring the breathing rate. This system was designed in 90 nm CMOS technology, and tested on adults and babies in various situations, proving that the sensor can detect the respiration activity as a result of sub-centimeter subject chest movements. Mercuri et al. [1] have shown a radar-based sensor that can be employed for multi-people tracking and to observe the vital signs (heartbeat and respiration) of more than one subject in a practical setup. This non-contact approach is strong against mild random body movements and is capable of monitoring individuals during vigorous movement. Kuutti et al. [59] have presented an evaluation of a non-contact Doppler radar-based system for detecting the vital signs, the presence, and activities of an individual in an RF shielded room, i.e., a test room with radar signal reflecting off aluminium-coated surfaces. The results obtained were promising and the system seems capable of being utilized in position-independent activity and health monitoring, e.g., for aged patients or intoxicated people in police custody.

The mechanical cardiac activity of a human subject can be monitored unobtrusively and non-invasively using a biomedical radar. However, there is an intermodulation phenomenon that happens between cardiac and respiratory signals. Hence, signal separation and filtering steps are needed before analysis can be carried out.

Yao et al. [60] have used a non-linear filter for differentiating between breathing and heartbeat signal components obtained from radar data. Monitoring physiological functions of the human body contactlessly through walls is extremely convenient in surveillance, security, and healthcare, and radar technology is extremely suitable here. A new approach for the detection of the human heartbeat using ultra-wideband (UWB) impulse radar, which consumes low power, and is harmless to the human body, has been introduced [61]. The heart rate has been detected with high accuracy, by processing the reflected time-domain radar signal, instead of the conventionally used frequency domain.

In most cases, radar-based human vital signs monitoring has been done under ideal conditions, in an indoor setup environment [62], and has mostly been examined without any disturbances or external objects present [63]. In practice though, in clinical environments, as well as homes, there are many other factors, such as stationary clutter, reflections from unwanted objects, the presence of humans apart from the subject, external noise, and interference. These may cause problems in the correct monitoring and tracking of the subject vital signs. Alternate approaches are being developed now to solve these problems and improve system accuracy. One such device called the WiTrack [64] has been developed recently. This can trace the three-dimensional movement of a subject using the radio signals which are reflected off the subject's body. WiTrack functions properly even when the subject is obstructed from the device or is placed in a different room altogether. The system has 4 antennas (one for transmitting and three for receiving) and can localize multiple subjects in 2D, carry out vital sign monitoring, as well as gesture recognition. Adib et al. [65] have also introduced the Vital-Radio, a wireless system capable of contactless monitoring of the respiratory and heart rates. Vital-Radio takes advantage of the feature that wireless signals are affected by movements of the human chest while breathing in and out, and also by skin vibrations as a result of the heartbeats. It has been proved through experimental study, that the device can correctly monitor users' heart and respiratory rates with a high median accuracy, despite subjects being at 8 m distance from the device, or in another room. Multiple subjects' vital signs can be observed simultaneously. This Vital-Radio may be used in the future to build smart homes that can observe the vital signs of their residents without the need for body contact, and help in their healthcare monitoring.

In the presence of multiple subjects, such radar systems get a combined mixture of breathing patterns, which is the interference of the respiration pattern of individuals. Separating a particular individual's respiratory pattern from the overall mixture is challenging. However, such methods work well for dedicated single subjects, particularly when the subjects are maintaining social distance. Attempts focused on using a separate antenna for each dedicated subject were not scalable in realistic scenarios. An intelligent decision algorithm has been proposed to integrate two different approaches to separate individual respiratory patterns from combined mixtures [66]. Independent component analysis (ICA) is a signal processing technique to extract individual signals from a mixture of breathing patterns. The direction of arrival (DOA) technique was also used for approximating the spatial position of the subject. Using the location of the subject, the antenna beam can be switched in different directions to separate respiratory signatures.

During the COVID-19 pandemic, hospitals are facing a rush of patients, so monitoring vital signs based on conventional contact-based sensors may increase the chances of infection of medical professionals because of close contact. Radar-based breathing monitoring can be useful in overcoming this problem because of its unobtrusive and non-contact sensing methodology. Radar-based systems can also generate an alarm or warning if a patient's breathing rate abruptly changes, indicating an emergency. The fatality rate and the spread of the virus may be minimized by using this monitoring technique.

A lot of issues associated with traditional contact-based monitoring of blood pressure could be overcome by measuring blood pressure using a video of the subject's face using a smartphone [67]. Such widely available new tools can allow data collection in an increased number of people and with more regularity than ever before and are likely to have significant diagnostic value. Luo et al. [68] recently showed blood pressure values obtained using videos of the face with accuracy close to existing automated blood pressure monitors. Their method merges two major advancements in this field. The first is a form of variation of video photoplethysmography known as transdermal optical imaging (TOI), which assists in obtaining usable photoplethysmography signals based on a video. The other is a set of original blood flow features that have been merged using advanced machine learning techniques to determine blood pressure value.

Figure 8.6 shows the block diagram for vital sign monitoring, i.e., breathing rate or heart rate. The different stages in this case include:

- Vital sign selection: Breathing rate, heart rate, etc.
- Sensing of the chosen vital sign via a suitable radar system.
- Radar Signal Processing: Time-frequency analysis like Short-Time Fourier Transform (STFT), Wavelet Transform (WT), Numerical Analysis, etc.
- Feature Extraction: Required features are selected to extract the most relevant data.
- Classifier: SVM, K-nearest neighbor, Random forest (RF), Deep neural networks, etc.
- Classification: Data are classified, and then the system performance is evaluated.

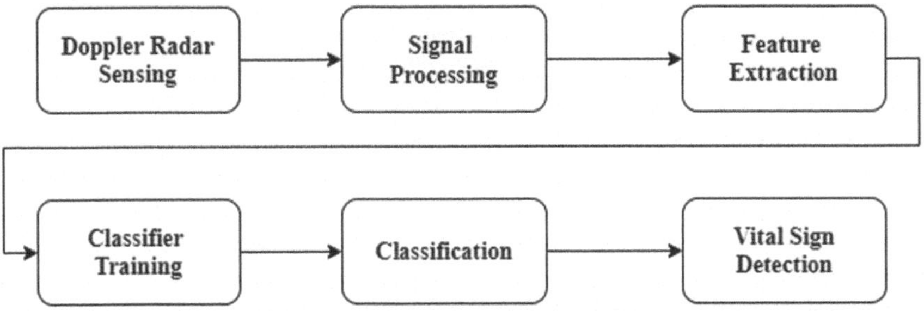

Fig. 8.6 The different stages in the monitoring of vital signs, breathing rate or heart rate, using radar-based systems

8.4.7 Occupancy Monitoring

The major approach for achieving energy efficiency in commercial spaces is to provide the basic services, which are lighting, heating, ventilating, and air conditioning (HVAC), in the exact amount, place, and time that they are needed. For this purpose, what is necessary is to correctly evaluate the real-time occupancy as an input to optimal control [69]. Occupancy detection can conserve about 80% of the energy required for lighting and HVAC systems, which will result in huge monetary savings [70]. The already existing occupancy sensors show high false alarm rates and have also failed to detect static subjects on instances. Such sensors, thus, effectively become motion sensors, not exactly presence sensors. In addition to the detection of overall human motion, Doppler radar-based sensors can detect weak vital signs, which presents a useful approach to counteract the limitations of conventional sensors. The detection of true presence is achieved using Doppler radar occupancy systems to counteract the shortcomings of traditional occupancy sensors.

Utilizing microwave Doppler radar for the detection of human cardiopulmonary motion can be a potential method to identify the occupants in a room and determine the total number [66]. Sensing human cardiopulmonary motion based on Doppler radar can present an apt approach to counteract the limitations of false detection and dead spots of existing occupancy sensors when a subject is stationary. Radars sense uniquely modulated back-scattered electromagnetic signals. As the number of people in a room increases, it has an impact on the received signal strength (RSS) of the received signal. The RSS method can determine the number of occupants in a room, as it is directly affected by the number of people. This method has several advantages. It is contactless, unobtrusive, and is efficient in tracking even when the subject is stationary. During this COVID-19 pandemic, governments are struggling to implement complete lockdown in their countries. Occupancy detection using radar can assist in monitoring the number of occupants in a home, and identify any new entry as well. Thus social isolation and proper quarantine can be ensured by using a radar-based smart physiological sensing system.

Doorpler [71] is a system based on the radar which is aware of space, time, and power, and can detect occupancy at zone transition points owing to its ability to sense crossings and direction of the crossings. Crossings are detected using the Doppler Principle. A subject moving in the direction of the radar leads to an increase in the frequency of the RF signal transmitted. The crossing direction is derived by measuring the angle-of-arrival of the echo. The performance was evaluated using 2 in-situ studies and a scripted study. The findings indicated that this system could obtain accuracy over 99% and 95% for all metrics evaluated in the scripted and in-situ studies, respectively. Occupancy sensing based on such systems, however, can be both expensive and bulky [71]. Such systems may be falsely triggered, by the presence of pets, because they can also cause a Doppler effect. Pets can be differentiated based on their gait which is different from human subjects.

A low-power millimeter-wave FMCW radar was used with the aim of occupancy monitoring in vehicles, which was of relatively low-cost [19]. In this paper, FMCW radars were used to obtain the count of the total passengers in an automobile and to recognize which seat each subject was sitting on. Machine learning was used to correctly recognize the existence and position of a person within the vehicle. The low angular resolution of the radar restricts the visual perception of the occupant position. Detection accuracy was high for various scenarios because of using a machine learning classifier.

Many of the radar-based occupancy detection solutions use motion signatures. The distinctive individual pattern of each subject in the micro-Doppler signature can be analyzed to recognize subjects [72]. Deep neural networks (DNNs) have been employed to allow micro-Doppler signature classification sensors based on the radar to total the subjects in a setting, identify an individual in a group of 4 to 20 people, as well as recognize trespassers using already obtained signatures of the members of the household. How someone carries a weapon, like a gun, has been proved to translate distinctive impact in the micro-Doppler signature, hence allowing uses like recognition of shooters and personnel, whether they are armed or unarmed.

The practicality of occupancy sensor based on the Doppler radar to obtain both the respiratory and heart signals when the subject is static and in motion has also been explored [73]. The system detects human presence by obtaining the respiratory as well as the heart signals when a person is static, and while in motion at various levels of activity. The obtained findings prove that the true presence of a subject can be correctly detected while differentiating between non-human periodic motions and recovered respiratory patterns, which may have triggered the false positive. For static subjects, false alarms can be eliminated by monitoring heart and breathing rates not possible with existing sensors. In cases where someone deliberately tries to maintain a very constant periodic respiration, the presence of a human can be detected by looking at fluctuations in heart rate.

8.4.8 Fall Detection

Falling is an extreme health concern, especially for the elderly, that may cause hurt or even death. Hence discovering effective and robust methods for fall detection is important. In the past, several systems have been developed for fall detection, using different approaches based on wearable sensors and video cameras. Radar-based contactless monitoring of the elderly is a rapidly growing research field due to being non-invasive, and also there are no privacy concerns compared to traditional techniques [74].

A method for detecting falls has been introduced by combining time-series data obtained using UWB radar echoes, and a deep neural network [75]. This was implemented employing a deep convolutional neural network for extracting multi-level features, using time series data obtained from radar. Echoes from the subject are summed up over all the range bins in fast-time, and a time series is obtained in slow time. This is the input to the deep convolutional neural network with the target of automatic feature extraction. In comparison to other existing methods, this method depends on multi-level feature learning directly using time-series signals obtained through radar. Particularly, this approach uses a deep convolutional neural network for automated feature extraction and a global maximum pooling technique for improving model differentiability. Data were obtained in a room by maintaining the at-home healthcare setup. The performance of this approach was compared with that of the state-of-the-art like multilayer perceptron, recurrent neural network, and dynamic time warping techniques. Findings show that this method performs better than the other compared techniques in terms of precision, accuracy, specificity values, and sensitivity.

A UWB radar is employed to observe regular subject activities and fall detection. A transfer learning approach is used based on a pre-trained model using the ImageNet data set for robust feature extraction using data obtained from radar [74]. The model architecture and depth are adjusted according to radar time-frequency representations. Findings show that this method can achieve a fall detection accuracy higher than others. This approach has the potential to extract features using radar echoes and classifying them, despite the size of the data set being comparatively small.

A new framework with the aim of information fusion obtained using data from radar and wearable sensors, based on a bidirectional Long Short-Term Memory (bi-LSTM) neural network, has been proposed for human activity recognition (HAR) and to detect falls [76]. Data used are continuous activity streams obtained using FMCW radar and 3 wearable sensors. Every activity has a fluctuating time duration within the data stream, which is why transitions between activities can occur at any random time, and not the traditional fixed-duration snapshots. Radar and wearable sensor data are processed within the recurrent LSTM neural network as a continuous-time series. The sequence of data was processed instead of individual images like done normally in convolutional neural networks. Instead of considering radar data or data from wearable sensors separately, the proposed bi-LSTM approach performs soft feature fusion between radar data and wearable sensors as well as two robust hard-fusion methods using the confusion matrices of both sensors. An original hybrid fusion scheme was introduced to merge both soft and hard fusion to increase the classification accuracy to about 96% in the identification of activities and falls. The developed hybrid-fusion approach can stabilize classification performance for distinct subjects by lowering accuracy variance of up to 18.1% and raising the lowest, worst-case accuracy to 16.2%.

8.5 Future Challenges

Biomedical radar-based healthcare monitoring is a research sector with tremendous potential. Such radar-based sensing, unlike contact-based sensors, does not disrupt the regular activities of patients, does not cause physical discomfort or privacy issues. However, even with the developing hardware and algorithms, there are still limitations and challenges that have to be overcome for the complete potential use of this technology. Some important limitations and their possible solutions include:

8.5.1 Development of Suitable Approaches

For the broader application of radar-based sensing, techniques, algorithms, and approaches must be developed, which are feasible and mature enough to be suitable for a wide variety of applications. Also, the developed methods must allow their growth into technical standards for commercialization [2].

8.5.2 Development of Readily Available Data Sets

There is a lack of well-designed clinical trials and readily available data sets. Diversity in terms of gender (male and female subjects), body shapes and sizes, and age groups (young and elderly) need to be included in the data set so that it is more representative of practical situations [20]. Larger data sets are required to improve the accuracy and overall performance of the systems, so test groups should include a higher number of subjects. If the number of data samples is low, it would be difficult to evaluate the performance, accuracy, and reliability of the method.

8.5.3 Well-Designed Clinical Trials

Testing must be done in appropriate settings for the results to be valid and authentic. Most radar-based studies obtain data from stationary people. Testing must be done in environments with furniture, obstacles, where many people are in motion, representing normal household conditions [77]. In the case of sleep monitoring, for example, the complexity of the sleep environment must be included in the study. These include noises related to sudden body movements, changes in posture and position while sleeping, cancellation of multiple subjects, and interference. Improvements can massively add to the realization of contactless sleep monitoring technology [36].

8.5.4 Compensation of Noise and Artifacts

In the case of monitoring vital signs remotely, innovative, and improved signal processing algorithms are needed to identify and separate the cardiorespiratory function of a subject even in the presence of noise [78]. A major limitation of contactless radar-based monitoring is its sensitivity to non-physiological activities, like background noise or movement, change in body orientations, and motion in case of multiple targets. Such interferences happening within the same frequency band as physiological motion are very complicated to nullify using basic filters. These can degrade the performance of the system.

8.5.5 Elimination of Clutter

In practical scenarios, i.e., in both clinical environments and homes, there are certain factors, such as stationary clutter, obstacles, reflections from unwanted objects, presence of humans apart from the subject, and interference. These may cause problems in the monitoring and tracking of the target. Suitable filtering and preprocessing may help eliminate such background clutter.

8.5.6 Multi-person Monitoring

Another major concern is the existence of clutter in indoor environments that can cause multipath effects and false target identification. People other than the subject can be present and remain in the monitoring zone, hence adding noise to the obtained radar signatures [4]. Detection of complicated events like falls regardless of the sequence of actions is a tough task. Particular issues such as physical impairment or any other limitation associated with a subject must also be considered.

8.5.7 Generalization of Motion

Radar micro-Doppler signatures are affected not only by an action being carried out, but because of the velocity, style of walking, and physique of the subject as well. The potential to generalize movement of subjects for all people is imperative to proving the reliability of the radar classifier. This means that the targeted results must be maintained for a "generic" subject [72]. There are many possible human activities, and newly developed approaches must be developed for addressing this open-set problem.

8.5.8 Subject-Specific Classification

Non-contact health and wellness monitoring applications need to characterize human gait with high sensitivity in order to identify abnormalities in gait. Problems like imbalance have a role in the evaluation of the risk of falls, and also in recovery after stroke. Detection of parameters related to gait over long periods is highly important in neuromuscular disorder treatment. In these cases, minor changes can occur, which are only possible to be identified by a highly sensitive, customized radar system, adjusted according to movement patterns of the specific subject.

8.5.9 Recognition of Complex Human Activities

Human activities that are complex, like drinking coffee, cooking, or working on a computer, are a culmination of many simple activities that occur simultaneously. In comparison with simple activities like sitting, walking, and running, complex activities that are more representative of someone's intentions, are worth studying [41]. Because of the difficult semantic and context information, complex human activities are more difficult to identify than simple activities. Effective approaches to study such complex activities should be developed for more accurate monitoring of health and wellness.

8.5.10 Search and Rescue Operation

For search and rescue purposes, a major concern is the device cost, along with its dimensions and weight, which should be convenient, as this is necessary for proper management while doing search and rescue work. Static clutter is much stronger than the desired survivor echo. Its elimination needs stable antenna positioning. As a result, many sensors are required to cover a reasonable search area, increasing cost [29]. Another concern is the detection and separation of the weak cardiopulmonary signal of an unconscious subject from the noises made by moving objects or people present in the search area [62]. To counteract the effect of different building materials and thicknesses of the obstacle layer on the received echo [28], suitable operating frequency and transmitting power have to be chosen [30]. For in-depth research and testing, more realistic scenarios of such disasters should be researched for search and rescue purposes, and further testing is needed to observe the effect on system performance of changing locations, objects, and clutter [33].

8.6 Conclusion

Radar-based sensing has the benefit of high range resolution and improved sensitivity. Such systems are thus well suited for contactless health monitoring of the elderly and diseased. The chapter includes a review of the recent developments of radar-based systems in the healthcare sector. Such systems provide effective, low cost, and convenient techniques to continuously monitor patients in their own homes without disturbing their everyday lives. Radar applications in healthcare, though challenging, is a research track with potential.

References

1. Mercuri, M., Lorato, I.R., Liu, Y.-H., Wieringa, F., Van Hoof, C., Torfs, T.: Vital-sign monitoring and spatial tracking of multiple people using a contactless radar-based sensor. Nat. Electron. **2**(6), 252–262 (2019)
2. Angelov, G.V., Nikolakov, D.P., Ruskova, I.N., Gieva, E.E., Spasova, M.L.: Healthcare sensing and monitoring. In: Enhanced Living Environments, pp. 226–262. Springer (2019)
3. Ahad, M.A.R., Antar, A.D., Shahid, O.: Vision-based action understanding for assistive healthcare: a short review. In: CVPR Workshops, pp. 1–11 (2019)
4. Shah, S.A., Fioranelli, F.: RF sensing technologies for assisted daily living in healthcare: a comprehensive review. IEEE Aerosp. Electron. Syst. Mag. **34**(11), 26–44 (2019)
5. Poh, M.-Z., McDuff, D., Picard, R.: A medical mirror for non-contact health monitoring. In: ACM SIGGRAPH 2011 Emerging Technologies, p. 1 (2011)
6. Hall, T., Lie, D.Y., Nguyen, T.Q., Mayeda, J.C., Lie, P.E., Lopez, J., Banister, R.E.: Non-contact sensor for long-term continuous vital signs monitoring: a review on intelligent phased-array doppler sensor design. Sensors **17**(11), 2632 (2017)
7. Zakrzewski, M.: Methods for doppler radar monitoring of physiological signals (2015)
8. Chen, V.C., Li, F., Ho, S.-S., Wechsler, H.: Micro-doppler effect in radar: phenomenon, model, and simulation study. IEEE Trans. Aerosp. Electron. Syst. **42**(1), 2–21 (2006)
9. Abdulatif, S., Aziz, F., Altiner, P., Kleiner, B., Schneider, U.: Power-based real-time respiration monitoring using FMCW radar. arXiv preprint arXiv:1711.09198 (2017)
10. Parker, M.: Digital Signal Processing 101: Everything you need to know to get started. Newnes (2017)
11. Li, C., Lin, J., Xiao, Y.: Robust overnight monitoring of human vital signs by a non-contact respiration and heartbeat detector. In: 2006 International Conference of the IEEE Engineering in Medicine and Biology Society, pp. 2235–2238. IEEE (2006)
12. Nosrati, M., Tavassolian, N.: Experimental evaluation of the effects of antenna radiation characteristics on heart rate monitoring radar systems. arXiv preprint arXiv:1801.00484 (2018)
13. Birsan, N., Munteanu, D.-P., Iubu, G., Niculescu, T.: Time-frequency analysis in doppler radar for noncontact cardiopulmonary monitoring. In: 2011 E-Health and Bioengineering Conference (EHB), pp. 1–4. IEEE (2011)
14. Tariq, A., Shiraz, H.: Doppler radar vital signs monitoring using wavelet transform. In: 2010 Loughborough Antennas & Propagation Conference, pp. 293–296. IEEE (2010)
15. Iyer, B., Garg, M., Pathak, N.P., Ghosh, D.: Contactless detection and analysis of human vital signs using concurrent dual-band RF system. Procedia Eng. **64**, 185–194 (2013)
16. Amin, M.G., Zhang, Y.D., Ahmad, F., Ho, K.D.: Radar signal processing for elderly fall detection: the future for in-home monitoring. IEEE Signal Process. Mag. **33**(2), 71–80 (2016)
17. Antar, A.D., Ahmed, M., Ahad, M.A.R.: Challenges in sensor-based human activity recognition and a comparative analysis of benchmark datasets: a review. In: 2019 Joint 8th International Conference on Informatics, Electronics & Vision (ICIEV) and 2019 3rd International Conference on Imaging, Vision & Pattern Recognition (icIVPR), pp. 134–139. IEEE (2019)
18. Inui, S., Okusa, K., Maeno, K., Kanakura, T.: Recognizing aspiration presence using model parameter classification from microwave doppler signals. In: 2012 World Congress on Engineering and Computer Science, WCECS 2012, pp. 509–512. Newswood Limited (2012)
19. Abedi, H., Luo, S., Shaker, G.: On the use of low-cost radars and machine learning for in-vehicle passenger monitoring. In: 2020 IEEE 20th Topical Meeting on Silicon Monolithic Integrated Circuits in RF Systems (SiRF), pp. 63–65. IEEE (2020)
20. Li, H., Shrestha, A., Heidari, H., Le Kernec, J., Fioranelli, F.: A multisensory approach for remote health monitoring of older people. IEEE J. Electromagnet. RF Microwaves Med. Biol. **2**(2), 102–108 (2018)
21. Kiriazi, J.E., Boric-Lubecke, O., Lubecke, V.M.: Radar cross section of human cardiopulmonary activity for recumbent subject. In: 2009 Annual International Conference of the IEEE Engineering in Medicine and Biology Society, pp. 4808–4811. IEEE (2009)

22. Piriyajitakonkij, M., Warin, P., Lakhan, P., Leelaarporn, P., Pianpanit, T., Niparnan, N., Mukhopadhyay, S.C., Wilaiprasitporn, T.: SleepPoseNet: multi-view multi-task learning for sleep postural transition recognition using UWB, arXiv preprint arXiv:2005.02176 (2020)
23. Cui, H., Dahnoun, N.: Human posture capturing with millimetre wave radars. In: 2020 9th Mediterranean Conference on Embedded Computing (MECO), pp. 1–4. IEEE (2020)
24. Sengupta, A., Jin, F., Zhang, R., Cao, S.: mm-Pose: real-time human skeletal posture estimation using mmWave radars and CNNs. IEEE Sens. J. (2020)
25. Ma, B., Chen, B., Zhang, Z., Ma, J., Kong, F.: Combat gesture classification using through-the-wall radar based on multi-domain features association. In: 2020 IEEE Radar Conference (RadarConf20), pp. 1–5. IEEE (2020)
26. Thi Phuoc Van, N., Tang, L., Demir, V., Hasan, S.F., Duc Minh, N., Mukhopadhyay, S.: Microwave radar sensing systems for search and rescue purposes. Sensors **19**(13), 2879 (2019)
27. Chen, K.-M., Huang, Y., Zhang, J., Norman, A.: Microwave life-detection systems for searching human subjects under earthquake rubble or behind barrier. IEEE Trans. Biomed. Eng. **47**(1), 105–114 (2000)
28. Liu, L., Liu, S.: Remote detection of human vital sign with stepped-frequency continuous wave radar. IEEE J. Sel. Top. Appl. Earth Obs. Remote Sens. **7**(3), 775–782 (2014)
29. Sachs, J., Helbig, M., Herrmann, R., Kmec, M., Schilling, K., Zaikov, E.: Remote vital sign detection for rescue, security, and medical care by ultra-wideband pseudo-noise radar. Ad Hoc Netw. **13**, 42–53 (2014)
30. JalaliBidgoli, F., Moghadami, S., Ardalan, S.: A compact portable microwave life-detection device for finding survivors. IEEE Embed. Syst. Lett. **8**(1), 10–13 (2015)
31. Nakanishi, T., Hirose, A.: Proposal of adaptive search-and-rescue radar system with online complex-valued frequency-domain independent component analysis. In: 2019 IEEE International Geoscience and Remote Sensing Symposium, IGARSS 2019, pp. 9431–9434. IEEE (2019)
32. Hu, D., Li, S., Chen, J., Kamat, V.R.: Detecting, locating, and characterizing voids in disaster rubble for search and rescue. Adv. Eng. Inform. **42**, 100974 (2019)
33. Ma, Y., Qi, F., Wang, P., Liang, F., Lv, H., Yu, X., Li, Z., Xue, H., Wang, J., Zhang, Y.: Multiscale residual attention network for distinguishing stationary humans and common animals under through-wall condition using ultra-wideband radar. IEEE Access **8**, 121 572–121 583 (2020)
34. Chen, J., Li, S., Liu, D., Li, X.: AiRobSim: simulating a multisensor aerial robot for urban search and rescue operation and training. Sensors **20**(18), 5223 (2020)
35. Matar, G., Lina, J.-M., Carrier, J., Kaddoum, G.: Unobtrusive sleep monitoring using cardiac, breathing and movements activities: an exhaustive review. IEEE Access **6**, 45 129–45 152 (2018)
36. Tran, V.P., Al-Jumaily, A.A., Islam, S.M.S.: Doppler radar-based non-contact health monitoring for obstructive sleep apnea diagnosis: a comprehensive review. Big Data Cogn. Comput. **3**(1), 3 (2019)
37. Sadek, I., Seet, E., Biswas, J., Abdulrazak, B., Mokhtari, M.: Nonintrusive vital signs monitoring for sleep apnea patients: a preliminary study. IEEE Access **6**, 2506–2514 (2017)
38. Singh, A., Baboli, M., Gao, X., Yavari, E., Padasdao, B., Soll, B., Boric-Lubecke, O., Lubecke, V.: Considerations for integration of a physiological radar monitoring system with gold standard clinical sleep monitoring systems. In: 2013 35th Annual International Conference of the IEEE Engineering in Medicine and Biology Society (EMBC), pp. 2120–2123. IEEE (2013)
39. Hong, H., Zhang, L., Zhao, H., Chu, H., Gu, C., Brown, M., Zhu, X., Li, C.: Microwave sensing and sleep: noncontact sleep-monitoring technology with microwave biomedical radar. IEEE Microwave Mag. **20**(8), 18–29 (2019)
40. Lauteslager, T., Kampakis, S., Williams, A.J., Maslik, M., Siddiqui, F.: Performance evaluation of the circadia contactless breathing monitor and sleep analysis algorithm for sleep stage classification. In: 42nd Annual International Conference of the IEEE Engineering in Medicine & Biology Society (EMBC), pp. 5150–5153. IEEE (2020)
41. Li, X., He, Y., Jing, X.: A survey of deep learning-based human activity recognition in radar. Remote Sens. **11**(9), 1068 (2019)

42. Seyfioğlu, M.S., Serinöz, A., Özbayoğlu, M., Gürbüz, S.Z.: Feature diverse hierarchical classification of human gait with CW radar for assisted living (2017)
43. Çağlıyan, B., Karabacak, C., Gürbüz, S.Z.: Indoor human activity recognition using bumblebee radar. In: 2014 22nd Signal Processing and Communications Applications Conference (SIU), pp. 1055–1058. IEEE (2014)
44. Seifert, A.-K., Schäfer, L., Amin, M.G., Zoubir, A.M.: Subspace classification of human gait using radar micro-doppler signatures. In: 2018 26th European Signal Processing Conference (EUSIPCO), pp. 311–315. IEEE (2018)
45. Jokanović, B., Amin, M.: Fall detection using deep learning in range-doppler radars. IEEE Trans. Aerosp. Electron. Syst. **54**(1), 180–189 (2017)
46. Erol, B., Amin, M.G.: Radar data cube analysis for fall detection. In: 2018 IEEE International Conference on Acoustics, Speech and Signal Processing (ICASSP), pp. 2446–2450. IEEE (2018)
47. Li, H., Shrestha, A., Fioranelli, F., Le Kernec, J., Heidari, H., Pepa, M., Cippitelli, E., Gambi, E., Spinsante, S.: Multisensor data fusion for human activities classification and fall detection. In: 2017 IEEE SENSORS, pp. 1–3 . IEEE (2017)
48. Yang, L., Li, G., Ritchie, M., Fioranelli, F., Griffiths, H.: Gait classification based on micro-doppler features. In: 2016 CIE International Conference on Radar (RADAR), pp. 1–4. IEEE (2016)
49. Erol, B., Gurbuz, S.Z., Amin, M.G.: GAN-based synthetic radar micro-doppler augmentations for improved human activity recognition. In: 2019 IEEE Radar Conference (RadarConf), pp. 1–5. IEEE (2019)
50. Islam, S.M.M., Borić-Lubecke, O., Zheng, Y., Lubecke, V.M.: Radar-based non-contact continuous identity authentication. Remote Sens. **12**(14), 2279 (2020)
51. Rahman, A., Yavari, E., Lubecke, V.M., Lubecke, O.-B.: Noncontact doppler radar unique identification system using neural network classifier on life signs. In: 2016 IEEE Topical Conference on Biomedical Wireless Technologies, Networks, and Sensing Systems (BioWireleSS), pp. 46–48. IEEE (2016)
52. Rahman, A., Lubecke, V.M., Boric-Lubecke, O., Prins, J.H., Sakamoto, T.: Doppler radar techniques for accurate respiration characterization and subject identification. IEEE J. Emerg. Sel. Top. Circuits Syst. **8**(2), 350–359 (2018)
53. Islam, S.M., Sylvester, A., Orpilla, G., Lubecke, V.M.: Respiratory feature extraction for radar-based continuous identity authentication. In: 2020 IEEE Radio and Wireless Symposium (RWS), pp. 119–122. IEEE (2020)
54. Lin, F., Song, C., Zhuang, Y., Xu, W., Li, C., Ren, K.: Cardiac scan: a non-contact and continuous heart-based user authentication system. In: Proceedings of the 23rd Annual International Conference on Mobile Computing and Networking, pp. 315–328 (2017)
55. Okano, T., Izumi, S., Kawaguchi, H., Yoshimoto, M.: Non-contact biometric identification and authentication using microwave doppler sensor. In: 2017 IEEE Biomedical Circuits and Systems Conference (BioCAS), pp. 1–4. IEEE (2017)
56. Cao, P., Xia, W., Li, Y.: Heart id: human identification based on radar micro-doppler signatures of the heart using deep learning. Remote Sens. **11**(10), 1220 (2019)
57. Huang, M.-C., Liu, J.J., Xu, W., Gu, C., Li, C., Sarrafzadeh, M.: A self-calibrating radar sensor system for measuring vital signs. IEEE Trans. Biomed. Circuits Syst. **10**(2), 352–363 (2015)
58. Zito, D., Pepe, D., Mincica, M., Zito, F., Tognetti, A., Lanata, A., De Rossi, D.: SoC CMOS UWB pulse radar sensor for contactless respiratory rate monitoring. IEEE Trans. Biomed. Circuits Syst. **5**(6), 503–510 (2011)
59. Kuutti, J., Paukkunen, M., Aalto, M., Eskelinen, P., Sepponen, R.E.: Evaluation of a doppler radar sensor system for vital signs detection and activity monitoring in a radio-frequency shielded room. Measurement **68**, 135–142 (2015)
60. Yao, Y., Sun, G., Kirimoto, T., Schiek, M.: Extracting cardiac information from medical radar using locally projective adaptive signal separation. Front. Physiol. **10**, 568 (2019)
61. Cho, H.-S., Park, Y.-J.: Detection of heart rate through a wall using UWB impulse radar. J. Healthcare Eng. **2018** (2018)

62. Schreurs, D., Mercuri, M., Soh, P.J., Vandenbosch, G.: Radar-based health monitoring. In: 2013 IEEE MTT-S International Microwave Workshop Series on RF and Wireless Technologies for Biomedical and Healthcare Applications (IMWS-BIO), pp. 1–3. IEEE (2013)
63. Mercuri, M., Schreurs, D., Leroux, P.: SFCW microwave radar for in-door fall detection. In: 2012 IEEE Topical Conference on Biomedical Wireless Technologies, Networks, and Sensing Systems (BioWireleSS), pp. 53–56. IEEE (2012)
64. Adib, F., Kabelac, Z., Katabi, D., Miller, R.C.: 3D tracking via body radio reflections. In: 11th USENIX Symposium on Networked Systems Design and Implementation (NSDI 2014), pp. 317–329 (2014)
65. Adib, F., Mao, H., Kabelac, Z., Katabi, D., Miller, R.C.: Smart homes that monitor breathing and heart rate. In: Proceedings of the 33rd Annual ACM Conference on Human Factors in Computing Systems, pp. 837–846 (2015)
66. Islam, S.M.M.: Can radar remote life sensing technology help to combat Covid-19? (2020)
67. Barszczyk, A., Lee, K.: Measuring blood pressure: from cuff to smartphone. Curr. Hypertens. Rep. **21**(11), 84 (2019)
68. Luo, H., Yang, D., Barszczyk, A., Vempala, N., Wei, J., Wu, S.J., Zheng, P.P., Fu, G., Lee, K., Feng, Z.-P.: Smartphone-based blood pressure measurement using transdermal optical imaging technology. Circ.: Cardiovasc. Imaging **12**(8), e008857 (2019)
69. Shen, W., Newsham, G., Gunay, B.: Leveraging existing occupancy-related data for optimal control of commercial office buildings: a review. Adv. Eng. Inform. **33**, 230–242 (2017)
70. Gu, C.: Short-range noncontact sensors for healthcare and other emerging applications: a review. Sensors **16**(8), 1169 (2016)
71. Kalyanaraman, A., Soltanaghaei, E., Whitehouse, K.: Doorpler: a radar-based system for real-time, low power zone occupancy sensing. In: 2019 IEEE Real-Time and Embedded Technology and Applications Symposium (RTAS), pp. 42–53. IEEE (2019)
72. Gurbuz, S.Z., Amin, M.G.: Radar-based human-motion recognition with deep learning: promising applications for indoor monitoring. IEEE Signal Process. Mag. **36**(4), 16–28 (2019)
73. Yavari, E., Jou, H., Lubecke, V., Boric-Lubecke, O.: Doppler radar sensor for occupancy monitoring. In: 2013 IEEE Topical Conference on Power Amplifiers for Wireless and Radio Applications, pp. 145–147. IEEE (2013)
74. Sadreazami, H., Bolic, M., Rajan, S.: TL-fall: contactless indoor fall detection using transfer learning from a pretrained model. In: 2019 IEEE International Symposium on Medical Measurements and Applications (MeMeA), pp. 1–5. IEEE (2019)
75. Sadreazami, H., Bolic, M., Rajan, S.: Fall detection using standoff radar-based sensing and deep convolutional neural network. IEEE Trans. Circuits Syst. II Express Briefs **67**(1), 197–201 (2019)
76. Li, H., Shrestha, A., Heidari, H., Le Kernec, J., Fioranelli, F.: Bi-LSTM network for multimodal continuous human activity recognition and fall detection. IEEE Sens. J. **20**(3), 1191–1201 (2019)
77. Girão, P.S., Postolache, O., Postolache, G., Ramos, P., Pereira, J.D.: Microwave doppler radar in unobtrusive health monitoring. In: Journal of Physics: Conference Series, vol. 588, no. 1, p. 012046. IOP Publishing (2015)
78. Pisa, S., Pittella, E., Piuzzi, E.: A survey of radar systems for medical applications. IEEE Aerosp. Electron. Syst. Mag. **31**(11), 64–81 (2016)

Chapter 9
Contactless Monitoring for Healthcare Applications

K. M. Talha Nahiyan and Md Atiqur Rahman Ahad

Abstract Contactless sensors have brought a new way of patient monitoring for healthcare applications. The patient is only required to be in the vicinity of the sensor for health monitoring. Also, there is opportunity for a contactless sensing system to monitor multiple patients simultaneously, which is not possible with wearable contact-based sensors. As a result, a contactless monitoring system would require fewer resources and can be cost effective. In this chapter, different contactless physiological monitoring is discussed. Contactless cardiac and blood related monitoring is discussed including Electrocardiogram (ECG), heart rate, blood pressure etc. Respiratory monitoring is another important area which is included. Monitoring of neurological diseases is reviewed according to different applications. A limited analysis of blood glucose monitoring in diabetes is also provided. Majority of these works involve video camera, motion sensors and Doppler radar as a contactless sensor. In this work, it is intended to analyze the pros and cons of different contactless monitoring and their feasibility to use in various healthcare requirements such as hospitals, nursing homes and homes.

9.1 Introduction

In a contactless monitoring method, the sensors and measurement instruments should not have any contact with a person. For different healthcare applications, the physiological measurements should be done in a way so that no sensors or electrodes

---

K. M. T. Nahiyan
Department of Biomedical Physics and Technology, University of Dhaka, Dhaka, Bangladesh
e-mail: tnahiyan@du.ac.bd

M. A. R. Ahad (✉)
Department of Electrical and Electronic Engineering, University of Dhaka, Dhaka, Bangladesh
e-mail: atiqahad@du.ac.bd

Department of Intelligent Media, Osaka University, Suita, Japan

M. A. R. Ahad and A. Inoue (eds.), *Vision, Sensing and Analytics: Integrative Approaches*, Intelligent Systems Reference Library 207, https://doi.org/10.1007/978-3-030-75490-7_9

are attached to the body of the patient. Contact-based non-invasive methods in healthcare applications are well established and widely used in hospitals. However, there are special cases where contactless methods might be required for patients. In general, some patients experience irritation or scarring of skin upon attachment of electrodes and sensors. Also, such attachment of contact sensors can be impossible for patients with open wounds, injury or burn. Moreover, in case of children, any contact-based sensor placement is usually difficult. Though the accuracy or precision of contactless methods might be lesser than conventional gold standard contact-based methods but for these criteria, it can be appropriate. A trade-off between accuracy and patient well-being can be done here.

Contact-based wearable sensors allow continuous monitoring of patients. Contactless monitoring is even more ubiquitous. For any physiological measurement, the patient does not need to wear or carry any sensor rather they just need to be in the vicinity of a contactless monitoring system. Thus, a number of patients can be monitored in a shorter time as no time is taken to put on and off the contact sensors. Furthermore, there is possibility of monitoring multiple patients simultaneously. A large portion of contactless sensing systems is based on video camera and Doppler radar. In different contactless healthcare monitoring, a Region of Interest (ROI) is selected from the video. Depending on targeted application, the ROI can be face, hand or any other part of the body. For simultaneous multiple monitoring of patients, the ROIs can be selected simultaneously from multiple patients. If the patients cannot be captured in a single video, then a single camera can periodically capture different patients. But the duration of recording for each subject would be depended on the clinical requirements and ability of the systems algorithm to estimate the desired parameter. Radar-based systems might be operated in a similar manner by directing the transmitting wave at different patients within a specified time. Overall, these scopes could make contactless monitoring more cost effective and reduce load of a healthcare system.

In this chapter, cardiac and blood related monitoring is presented extensively. It has seen the most research among contactless physiological monitoring. Contactless monitoring of Electrocardiogram (ECG), heart rate, blood pressure, blood flow and oxygen saturation has been discussed. A brief description on Remote Photoplethysmography (rPPG) is also provided which forms the basis of many video-based contactless monitoring. Respiration rate and its application in sleep monitoring are reviewed in contactless respiratory monitoring. Further, contactless techniques applied for neurological disease monitoring are analyzed. Also, a recent development in contactless blood glucose measurement for diabetes patients is explained. Moreover, the efficacy and usability of contactless systems in different healthcare settings as compared to conventional contact-based methods are debated.

9.2 Cardiac and Blood Related Monitoring

Cardiac and blood related physiological measurements are most important in healthcare monitoring. Many physiological changes in body are usually detected from various cardiac and blood related parameters. In this section, discussion is provided on contactless techniques for the following measurements –

- ECG
- Heart rate and heart rate variability
- Blood pressure
- Blood flow
- Oxygen saturation

Apart from ECG, contactless techniques for other physiological parameters are mostly video or radar-based methods. Video-based techniques are generally based on Remote Photoplethysmography (rPPG), where Photoplethysmography (PPG) signals are extracted from video signals of human body (usually face or hand). So, a brief discussion on rPPG is also provided.

9.2.1 Contactless ECG

The ECG records electrical activity of the heart. Figure 9.1 shows an ECG signal denoting the P wave and T wave along with the QRS complex. Normally in ECG, electrodes are attached to the body surface. The principle concept behind contactless ECG is capacitive-coupled electrodes. A basic illustration of contactless ECG is shown in Fig. 9.2. The conductive body surface and a capacitive electrode separated by electrical insulation form a capacitor through which ECG signal is measured. Any clothing, air or other materials present in between provides the electrical insulation.

Capacitive Electrode-Based A contactless ECG measurement system is proposed in [1], where the capacitive electrodes are attached to a chair and ECG can be acquired when a person sits on it. The results are validated against a classical ECG. However, the quality of ECG acquired is depended on the insulation provided by the clothing; so an adequate distance is needed to be measured according to the clothing. Later, the system is also clinically tested on thirty patients successfully and it is recommended to be an economically realizable setting for screening of heart arrhythmia [2].

Chamadiya et al. [3] developed a textile-based contactless ECG monitoring system based on similar capacitive-coupled principle. The system is implemented in non-ICU (non-Intensive Care Unit) clinical arrangements on stretcher, hospital-bed and wheelchair. The efficacy of the system is tested against conventional ECG by comparing accurate detection of QRS complexes. The system has excellent accuracy (>96%) for stretcher and wheelchair but hospital-bed suffers from low accuracy of 70% [3]. Nevertheless, the study has presented a permanent efficient way of unobtrusively monitoring ECG.

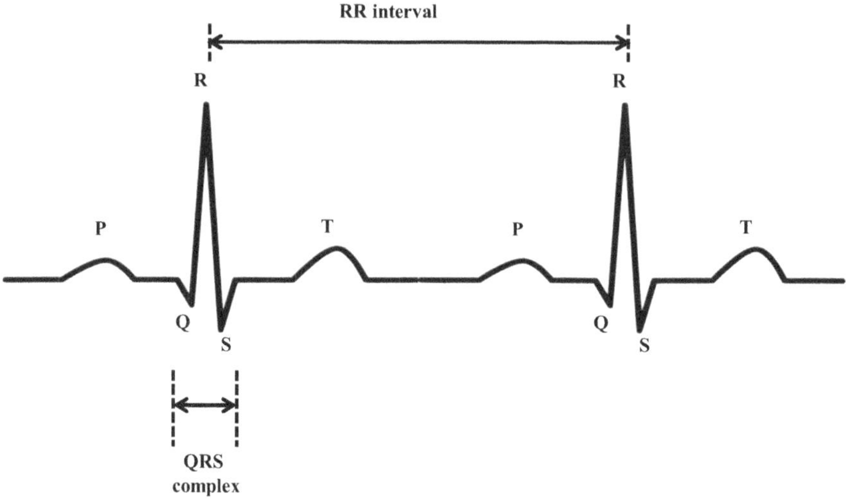

Fig. 9.1 An ECG signal denoting major waves and intervals

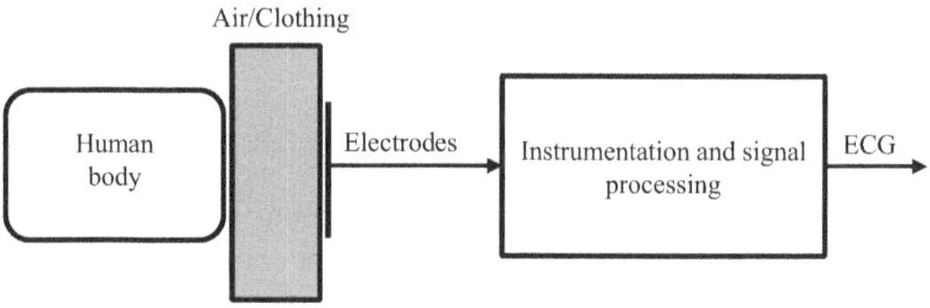

Fig. 9.2 A capacitive electrode-based contactless ECG measurement system. Air/clothing acts as insulator between human body and electrode to form capacitive effect

Conductive Textile-Based Lim et al. [4] implemented a conductive textile-based ECG monitoring system. The principle is similar to capacitive-coupled technique, where the textile is conductive rather than insulating. In this system, the electrode array is attached on a bed but without direct skin contact. This is a progress from the same group's work on non-contact conductive ECG on a chair [5]. The array of electrodes are placed on a mattress and covered with a conductive textile. In this way, the patient's body is no longer in direct contact with the electrodes and shows possibility of monitoring ECG during sleep in home environment. Recently, sensor array system with flexible printed electrodes [6, 7] and electro-conductive textile electrodes [8] are developed for contactless ECG measurement.

Challenges In general contactless ECG suffers from low signal amplitude. It is due to dependency on the textile clothing which is either insulating or conductive. Difference in textile requires varied separation between electrode and body. Thus there are works which focused on the circuitry improvement and signal processing to compensate these factors for enabling contactless ECG acquisition [9–12]. However, analysis of existing research indicates that it is possible to implement it for monitoring and screening purposes. But clinical level diagnosis is still not attempted. Further, once it is possible to record and process ECG, there is opportunity to derive other cardiac parameters such as heart rate from it. Long term recording would also enable Heart Rate Variability (HRV) analysis. Though contactless techniques solely dedicated towards heart rate and HRV measurement is discussed in the following section.

9.2.2 Remote Photoplethysmography (rPPG)

Remote Photoplethysmography (rPPG) is a video-based method to detect changes in blood volume pulse. This provides a contactless approach to the conventional Photoplethysmography (PPG). In conventional contact PPG, a light emitter is placed on the skin and returned light is detected by a light detector. The arterial pulse wave travels in the artery in accordance with cardiac cycle. When blood flow increases in the artery, the arteries expand and more light is absorbed. The remaining reflected light is detected (Fig. 9.3a). The same principle is applied in rPPG but in a contactless manner. Ambient light source is sufficient as a light emitter and a camera is used to detect the variance in red, green and blue channels of the video (Fig. 9.3b). The variation in absorption of light is in resemblance with the periodic information of Heart Rate (HR) and Respiration Rate (RR). All three channels show the variation but the green channel has the strongest signal of blood volume pulse [13, 14]. This is due to the fact that hemoglobin in blood absorbs the green light more than red or blue light. Further discussion on rPPG in the detection of HR, other blood related parameters and RR is presented in respective sections.

9.2.3 Heart Rate and Heart Rate Variability Measurement

Conventionally Heart Rate (HR) is measured from either ECG or Photoplethysmography (PPG) both of which require contact with human body. Contactless approaches for HR monitoring are either video-based or microwave radar-based systems. Further, possibility of Heart Rate Variability (HRV) measurement is also discussed.

Video-Based System Considerable researches are done on contactless HR monitoring using face video. Verkruysse et al. [13] first showed that ambient light is enough to record video capable of measuring heart rate. The technique is termed Remote PPG (rPPG). Later, the first real study on measuring heart rate from face video is done in [14]. Figure 9.4 depicts an outline of video-based HR detection. From captured video, Region of Interest (ROI) is detected which is usually the subject's face. It is tracked frame by frame and RGB information is extracted from the video. Any Blind Source Separation (BSS) technique is then required to analyze the contribution

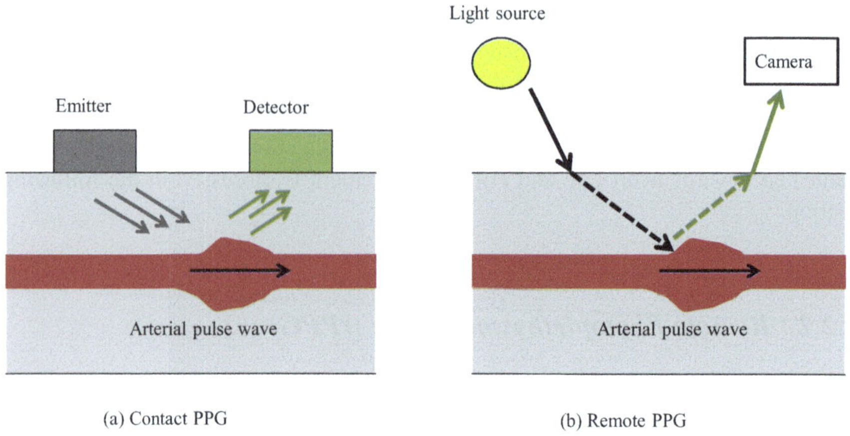

Fig. 9.3 Light absorption increases while artery expands with increase in blood flow (a) Contact PPG: Light emitter and detector are in contact with skin while reflected light is measured (b) Remote PPG: A light source and a camera is used to measure reflected light in contactless method

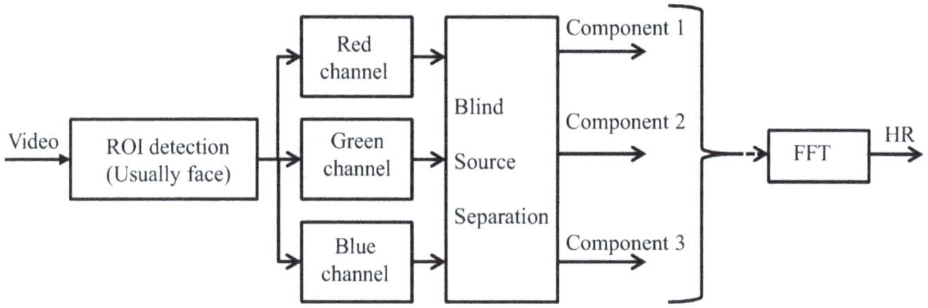

Fig. 9.4 HR detection method from RGB information of face video. BSS technique is applied to separate the HR signal and FFT leads to HR frequency

of different channels in the video. The most prominent channel is taken for further analysis, which is usually the green channel. The data is transformed into frequency domain and the peak frequency denotes the HR. This concept has more or less been the basis of HR detection from face video. [15]. The steps of this method have evolved. Face detection and tracking have improved from manual to automatic process. Also, researchers found that selecting cheeks [16] and carotid region [17] as ROI in face allow better estimation of HR. Different BSS techniques such as Independent Component Analysis (ICA) and Principal Component Analysis (PCA) were applied and use of machine learning was introduced in recent studies [15].

Microwave and Radar-Based System Microwave systems and radar are also applied in non-contact monitoring of HR. Small movements of the chest during each cardiac cycle as a result of volumetric change due to blood flow can be detected by the Doppler effect. The principle of radar-based detection is similar for Respiration Rate

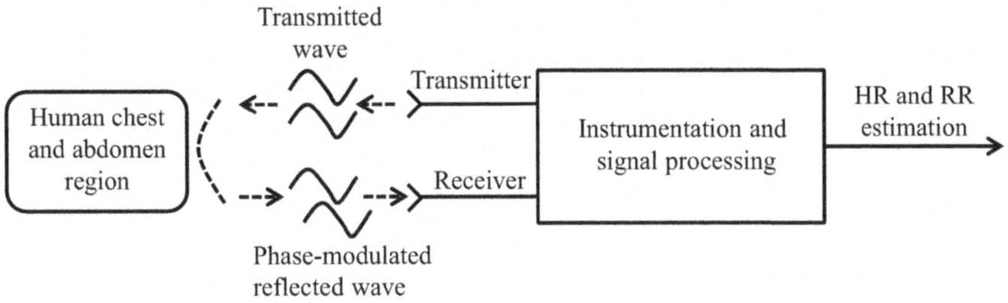

Fig. 9.5 Radar-based HR and RR detection from phase-modulated reflected wave due to movement of chest-abdomen region

(RR) also (see Sect. 9.2.6). The chest and abdomen area moves during respiration and that overrides the movement due to HR. Thus, it is difficult to measure HR using radar in normal breathing conditions. A radar-based HR and RR detection system is shown in Fig. 9.5. Due to movement of chest, the transmitted wave is reflected with the phase modulated. The phase modulation contains information about HR. The efficacy of the systems depends on frequency and number of antennas. The sensitivity of such system varies with frequency. High frequency brings high sensitivity, allowing detecting small displacements [18]. A two-antenna system performs slightly better under normal breathing conditions [19]. Frequency-Modulated Continuous Wave (FMCW) radar has also been used for HR estimation [20]

Towards Heart Rate Variability Measurement Long term recording is required for calculating several time domain and frequency domain parameters of Heart Rate Variability (HRV) which are early indicators of various underlying physiological conditions such as heart diseases and sleep disorders. It is possible to monitor HRV if there is long term non-contact HR data available. Rodriguez et al. [21] showed that it is possible to analyze HRV from video-based HR by measuring pulse-to-pulse intervals.

Challenges Heart rate detection from video-based techniques is quite accurate as found in literature. But the accuracy of the systems is depended on subject motion, illumination condition and subject's skin color. The algorithms need testing in practical scenarios before robust clinical application. Radar-based HR detection have several problems too. Under normal breathing condition, the effect of respiration rate makes HR detection complicated. Also, there are issues regarding employment of antennas in clinical environment.

9.2.4 Blood Pressure Monitoring

Blood Pressure (BP) is the pressure exerted by blood on the walls of blood vessels. Conventionally BP is measured non-invasively using a sphygmomanometer and a

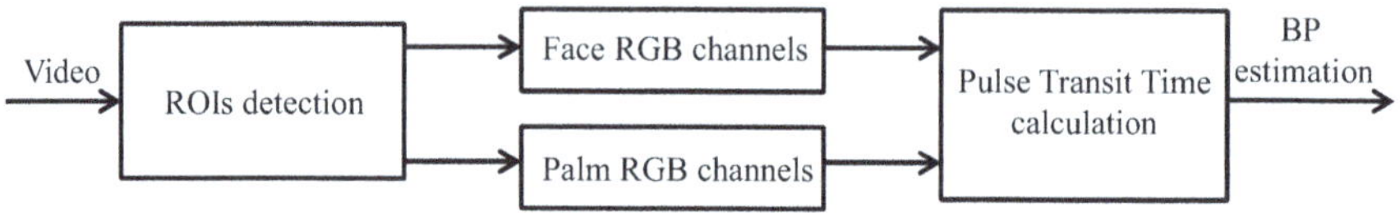

Fig. 9.6 Video-based BP estimation from PTT calculation between two arterial sites; face and palm

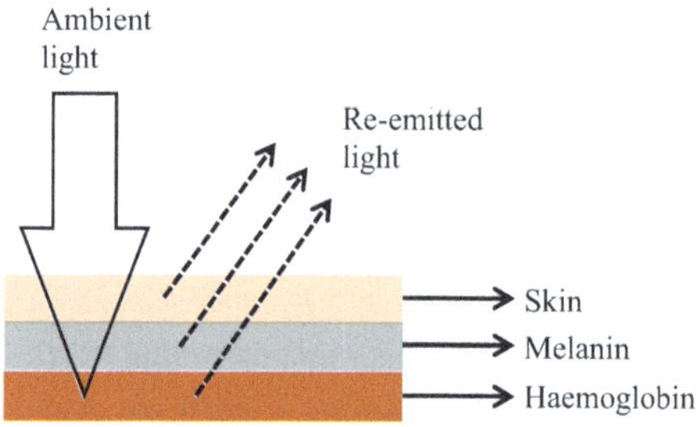

Fig. 9.7 Re-emitted light from beneath the skin. Light reflected by haemoglobin contain information related to BP

stethoscope. There are some recent advances to measure BP in contactless manner using video-based methods.

Video-Based Pulse Transit Time Calculation One of the first works in contactless BP monitoring is based on PPG signal extracted from rPPG signal and subsequent calculation of Pulse Transit Time (PTT) [22]. PTT is the time taken for a pulse to travel between two arterial sites. As in Fig. 9.6, from the video, face and palm areas are the two arterial sites detected as region of interests. PTT is calculated once rPPG signal is extracted from RGB channels of these two sites. PTT calculated in this way is found to be correlated strongly with BP for individual subjects.

In reference [22], high-speed camera is used for recording video which are expensive. Fan et al. [23] proposed an improved technique to measure BP from PTT using RGB video. They applied Gaussian curve fitting to account for missed peaks in rPPG, hence making PTT calculation more accurate than previous methods. The method is further improved by using adaptive Kalman filter to adjust the peaks of the rPPG signal [24] Finally, it is also established that PTT has strong correlation with BP. But none of these works are accurate enough to be used clinically.

Transdermal Optical Imaging Luo et al. [25] developed a novel contactless smartphone-based BP measurement system using Transdermal Optical Imaging (TOI). In TOI, light from visible spectrum which is re-emitted from beneath the skin is captured Fig. 9.7. In this work, the light reflected by the haemoglobin is of interest as blood flow is pulsating with change in BP. From this signal, 155 unique features are extracted and applied in a multilayer perception machine learning algorithm to build models to predict BP. In this study, there were a total of 1328 patients with normal BP and the method achieved 95% accuracy [25].

Challenges Contactless blood pressure monitoring is still far from conventional non-invasive method. PTT is strongly correlated with BP. However, the correlation

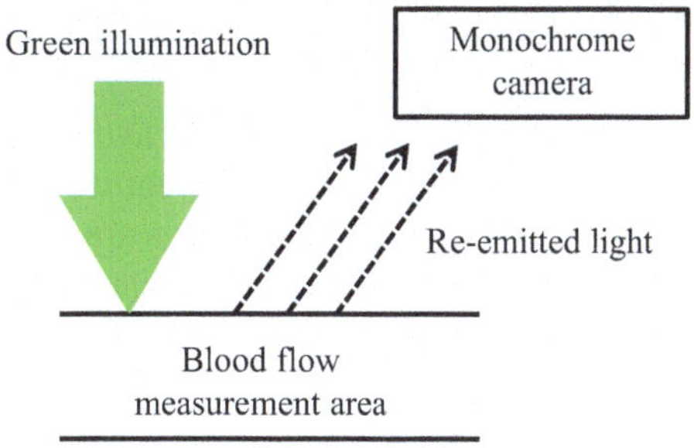

Fig. 9.8 Principle of blood flow measurement at green illumination. Monochrome camera captures re-emitted light

is not completely subject independent. Thus, it is difficult to calibrate. TOI based work shows promise in terms of globally measuring BP. But the study is conducted on normotensive population (people having normal BP). So, even for non-clinical purpose existing contactless BP monitoring methods are not complete.

9.2.5 Blood Flow Monitoring

Most commonly blood flow measurement is done by ultrasonic Doppler shift. It is a non-invasive method. An ultrasound probe is placed on the body surface, where the blood flow is measured. The red blood cells in the flowing blood induce Doppler shift on the reflected ultrasonic wave. It contains information about direction and velocity of the blood flow. There are not many advances in contactless monitoring of blood flow. Recently, there are some studies which introduced contactless methods based on video recording at green illumination. However, the works focus on blood flow changes in a region rather than precise measurements.

Video-Based Method Kamshilin et al. [26] proposed a blood flow measurement system from video recording at green illumination. The use of green light ensures low penetration in the skin and enables observing the changes in superficial capillaries. The video is recorded using a monochrome camera and the variation of blood flow is observed (Fig. 9.8). Thus, the system is able to perform blood flow changes in venous occlusion and the resultant measurement has a high correlation with conventional system. Further, they applied this technique to measure blood flow changes in upper limb [27]. Similar closely related contactless approach based on RGB video recording is found in [28]. However, the target of the work is to measure venous compliance, i.e., the stretchiness of the veins rather than direct blood flow monitoring.

Challenges Contactless methods discussed here are not exactly targeted towards general blood flow monitoring. These techniques are applicable for venous occlusion only. So, compared to conventional non-invasive method, the contactless advances in this area are in early stage and not suitable for use in practical scenarios.

9.2.6 Oxygen Saturation (SpO2) Monitoring

Typically a pulse oximeter is used to monitor the oxygen saturation in arterial blood. It consists of LEDs of two different wavelengths, generally of 660 nm and 940 nm. These wavelengths have maximum absorption by deoxygenated blood (660 nm) and oxygenated blood (940 nm). The reflected light intensity presents information about Photoplethysmography (PPG) waveform and through which oxygen saturation can be calculated [29].

Video-Based Methods Researchers focused on video recordings around the aforementioned wavelengths of a conventional pulse oximeter to work towards a camera-based contactless SpO2 monitoring. In Fig. 9.9, it is shown that, typically LEDs at two wavelengths (660 nm and 940 nm) are illuminated on a Region of Interest (ROI) of human body and reflected light information is captured by monochrome video camera. Further, PPG is produced from the videos and subsequent calculation of SpO2 is done. Wieringa et al. [30] described a CMOS camera which operates at three different wavelengths (660 nm, 810 nm and 940 nm) and is able to extract pulsatile PPG from the videos, indicating possibility of measuring SpO2. One of the first works to monitor SpO2 is presented by Humphreys et al. [31]. They capture video simultaneously at two wavelengths (760 nm and 880 nm) and are able to acquire good PPG from it. The work presents simulated relation of SpO2 with PPG derived from the system. Shao et al. [32] tested their system of two wavelength camera on subjects having SpO2 between 83%–98% and established high correlation with standard pulse oximeter. In these works, the variation of wavelength selection around 660 nm and 940 nm is done to find the optimum signal strength.

RGB video-based systems to monitor SpO2 has also been proposed, where signals from red and blue channels are used to estimate oxygen saturation [33, 34]. Verkruysse et al. [35] performed an extensive research by recording forehead video of 41 subjects in various controlled environment. The study is able to produce a single linear relation of SpO2 with video-based PPG with overall accuracy of 99% [35].

Challenges The biggest hindrance in contactless monitoring of SpO2 is the calibration of device for a wide range of SpO2 values. Normal human oxygen saturation level is between 95%-100%. Below this level, body functions gradually start to degrade. So, simulating a SpO2 level below 95% is challenging as it could be harmful for the subject. However, researchers produced hypoxic level of SpO2 in controlled clinical environment with necessary precautions and successfully calibrated the device between 83%–100% range [35]. Thus, application to clinical purposes is possible but more widespread trials may be required.

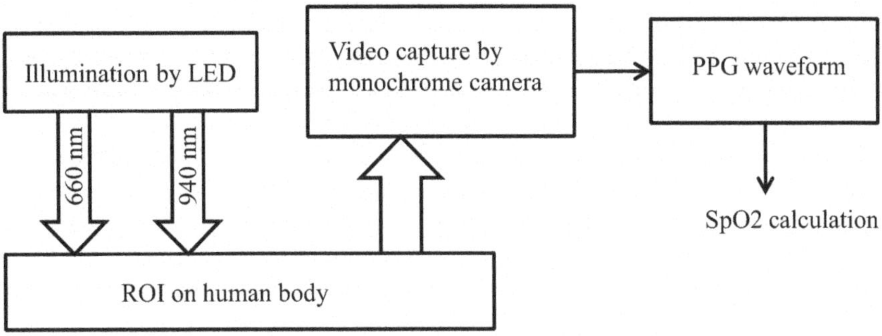

Fig. 9.9 Video-based SpO2 measurement with 660 nm and 940 nm LED as source. PPG waveform can be derived from reflected light intensity recorded by monochrome camera and hence SpO2 calculated

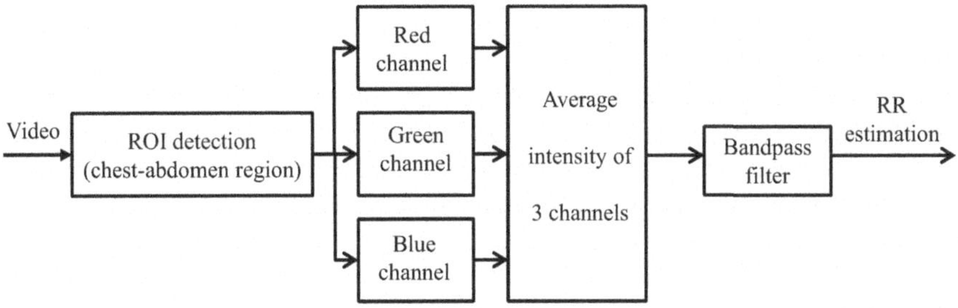

Fig. 9.10 RR detection from RGB video of chest-abdomen region. Respiration movement is indicated by periodic change in average pixel intensity

9.3 Respiratory Monitoring

Respiratory monitoring is another very important aspect of physiological monitoring. Contactless monitoring of Respiration Rate (RR) or breathing rate has been attempted by researchers. Moreover some works used the change in RR and breathing pattern to monitor or diagnose different diseases.

9.3.1 Respiration Rate Estimation

Respiration Rate (RR) is measured by counting the number of breaths per minute. Each time the chest rises it represents a breath. In contactless RR measurement, the same principle is applied in both video-based and Doppler radar-based methods.

Video-Based Methods An extensive amount of research has applied video-based techniques to measure RR. A general approach to video-based RR detection is shown in Fig. 9.10. The Region of Interest (ROI) is mostly abdomen-chest area (and rarely face). The RGB channel data from ROI is averaged frame by frame. Then a bandpass filter having lower and higher cutoff in accordance with lowest and highest RR measurement is used. Finally, detecting the frequency of the bandpass filtered pixel intensity series gives RR.

Tan et al. [36] proposed a video-based technique of detecting RR from the slight changes in abdomen and chest area due to respiration. Image subtraction technique is applied to detect the changes but the movements are very small making them undetectable if the subject's clothing does not have high contrast pattern. Also, the work is not suitable when the subject or the background is not static. But the method replicates measurement value of conventional methods in RR estimation. Similar camera-based technique to detect RR from healthy subjects and for various breathing patterns using a phantom is discussed in [37]. They also detected and eliminated any other motion artifacts, reducing the need for stationary background. Bernacchia et al. [38] applied Independent Component Analysis (ICA) on video signal acquired from abdomen region to extract RR information. The system is tested on 10 subjects but no patients were involved.

Lucas-Kanade optical flow technique has been used to measure RR from examining video of chest-abdomen region [39, 40]. The basic idea is that the pixel brightness remains constant even when there is motion. Tracking this brightness pattern enables detecting the motion. Lukac et al. [39] used this principle to estimate RR signal from the whole frames of a video. It does not select any ROI, thus prone to incorrect measurement. This problem is overcome in [40] by selecting the chest region as ROI and subsequently applying optical flow technique to detect the RR signal. The method is tested on 10 subjects with 84% accuracy [40].

Janssen et al. [41] introduced an efficient automatic region of interest detection algorithm and detected RR successfully. The uniqueness of the study is their developed benchmark dataset which consists of different scenarios of breathing pattern, non-respiratory motions and lighting conditions for adults and neonates. In reference [42], RR is detected accurately on a breath-by-breath approach. The system uses RGB video of chest-abdomen region and set a bandpass filter in the range of 0.05–2 Hz to extract the respiratory movement. The system is tested for different clothing conditions on both male and female and the error in all cases is below 1% [42]. A RR measurement method from hue saturation value (HSV) of RGB face video is described in [43]. A 20 s video of face region is recorded for each subject and the proposed method is more accurate in calculating RR than conventional green channel measurements.

Doppler Radar-Based Method Apart from video-based techniques, RR detection has also been done by Doppler radar methods. The basics are similar to HR detection as shown in Fig. 9.5 and explained in Sect. 9.2.2. The chest and abdomen region acts as a moving object during breathing; it expands during inspiration and contracts during expiration. The reflected wave undergoes a frequency shift which is proportional to the phase shift due to surface displacement of chest-abdomen region. These changes are of periodic nature resembling the respiration.

Min et al. [44] used an ultrasonic proximity sensor to measure time of flight between transmitted and received sound wave from abdominal wall motion due to respiration. RR is measured for different normal and rapid breathing conditions using Doppler radar in [45]. Sun et al. [46] applied auto-correlation model to measure RR from Doppler radar. The method suppresses body movement artifact and rapidly calculates RR within 10 s [46]. Different antennas were studied in [47] to estimate RR and antenna with least cross-polarization yielded lowest error rate. However, all these studies were performed in laboratory environment on healthy individuals. Goldfine et al. [48] proposed an ultra-wideband (UWB) impulse radar-based system to monitor respiration rate in laboratory setting and limitedly in clinical environment.

Challenges Both video-based and Doppler radar-based methods to detect RR in contactless manner have several drawbacks. Both the methods can be prone to subject motion and background movement. Some video-based methods do compensate for this but radar-based techniques are also prone to clothing variation of subject. None of the techniques has yet advanced towards application in hospitals completely but can be used for other healthcare monitoring purposes.

9.3.2 Sleep Monitoring from Breathing Pattern

Contactless respiratory monitoring has been targeted towards sleep monitoring purposes by analyzing breathing patterns. It could be used for classifying sleep stages and diagnosis of different sleep disorders.

Prochazka et al. [49] proposed a video-based technique to classify sleep stages. They acquired sleep video using MS Kinect depth sensor and created a depth map of the chest region. Variation of mean distances of chest during sleep are measured from the map and breathing pattern is developed. Then Bayesian classification is applied to distinguish between sleep and wake state from the breathing pattern. The study is conducted in a sleep laboratory with patients and healthy subjects involved. Siam et al. [50] developed an integral form of video frames to estimate respiration rate and monitor breathing pattern from sleep simulated videos.

Normal RR in adults are usually 12–20 breaths per minute. Tachypnea is defined as a RR which is more than normal and results in rapid breathing. A preliminary study to detect tachypnea from RGB video is explored in [42] by producing mimicked data from two subjects. They detected the increased RR that indicates tachypnea.

Hypopnea is a type of sleep disorder indicated by excessively low RR. It causes several disruptive episodes of breathing during sleep because the respiratory airway is partially blocked. Yang et al. [52] proposed a microwave non-contact sensing system for early detection of hypopnea from RR. The system consists of two S-band antennas for transmitting and receiving signals. Hypopnea episodes are distinguished from normal breathing pattern by detecting absence of inspiration peaks during breathing.

In Obstructive Sleep Apnea (OSA) the respiratory airway gets completely blocked and causes periods of breathing disruption during sleep. Tran et al. [51] provided an outline to detect OSA using Doppler radar. Various Doppler radar-based methods are discussed that were applied to analyze several respiratory parameters including RR, breathing pattern and sleep/wake stages. They suggested that the advances can be focused towards monitoring and diagnosis of OSA.

Challenges Contactless sleep monitoring from breathing pattern offers a more comfortable and flexible option than conventional polysomnography method. However, there are several areas to address to improve the diagnostic capability. In these works, researchers collected data in laboratory or clinical settings. In future, the target should be to make the systems usable in home environment. Subjects in these studies included patients but the sample size is very low (maximum 6 subjects in [52]) to assert the results statistically significant. None of the works attempted multi-patient monitoring, which can be explored in further studies.

9.4 Neurological Monitoring

In neurological diseases contactless techniques are used for different purposes. Figure 9.11 shows an outline of contactless techniques for different neurological monitoring. A patient can be monitored through a camera or motion sensor. Further, those data can be used in activity recognition which can be applied to detect symptoms, monitor patient's well-being or in rehabilitation process. In this section, different contactless methods for such purposes are discussed.

9.4.1 Symptoms Detection in Neuro-Degenerative Diseases

Neuro-degenerative disease is an umbrella term used for a wide range of medical condition where the neurons in brain get damaged and/or die affecting the normal func-

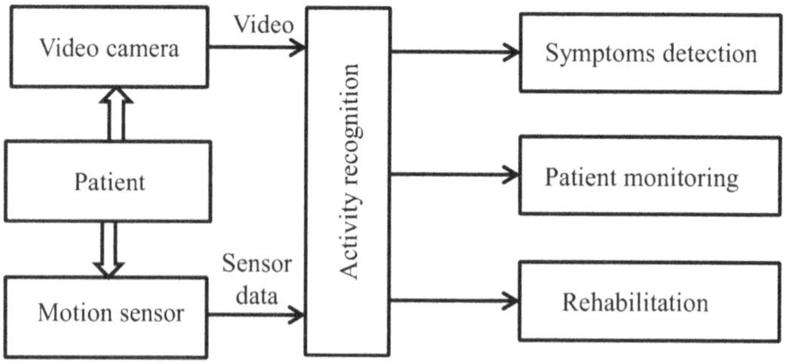

Fig. 9.11 An outline of contactless neurological monitoring. Activity recognition is done from video and motion data for symptoms detection, patient monitoring and rehabilitation

tion of brain, resulting in various hindrances in physical and mental functions. These kinds of diseases are progressive and incurable. Most common neuro-degenerative diseases are Parkinson's Disease (PD) and Alzheimer's Disease (AD). The symptoms of these diseases are not clearly evident until significant progression of the disease. Conventional diagnostic methodology involves visiting the doctor a few times yearly to check the symptoms and monitor progression of the disease. But it does not enable regular daily monitoring, which might helpful to slowdown the advancement of the disease to severe stages through medication. One of the common symptoms in such patients is difficulty in performing normal physical movements that can be monitored on a daily basis.

Abramiuc et al. [53] proposed a technique to detect two early motor symptoms of neuro-degenerative diseases by extracting silhouette from videos. Silhouette is delineated in HSV color space by applying background subtraction method. Then the silhouette is different body regions are distinguished by anatomical ratio. Data is taken for two scenarios for 23 step lengths and 10 arm swing angles; 1) walking into a scene 2) siting on a chair, then standing up an walking. After that step length and arm swing angle are calculated in pixels and degrees respectively. The results are tested against ground truth annotated by an human observer and Mean Absolute Error (MAE) in all cases are within 2.2–5.3% [53]. Continuous monitoring can detect any reduction in these two parameters which is an early sign of neuro-degenerative diseases. However, in this work data is collected from one normal subject only and no patients were involved in the study.

As an early symptom of PD, patients' exhibit hand tremor, i.e. uncontrolled shaking of hands. Shi et al. [54] applied an inductive sensor to detect hand tremors in a contactless manner. A sensing system was built consisting of a inductive coil. Whenever a human hand would shake close to the coil, there would be a change in the resonant frequency of the inductive system. Thus indicating the extent of tremor. The system is tested by moving a wooden hand 5 Hz for a maximum swing magnitude of 2 cm, which mimics hand tremor of a PD patient. Later, the inductive system is also verified against a wearable accelerometer.

Challenges Contactless monitoring can be helpful for symptoms detection for several neuro-degenerative diseases. But there is challenge in implementing it in real scenarios. Both [53] and [54] does not involve any patients in their studies. Also, the mimicked conditions does not include several situations that could be faced practically. Though the ideas are promising, they need to be experimented and implemented on real patients.

9.4.2 Home Monitoring for Alzheimer's Patients

In Alzheimer's Disease (AD), patients suffer from mild to severe memory loss at different stages of the disease. They tend to forget time, place and events as well as fail to recognize other people or remind their names. When the memory loss is

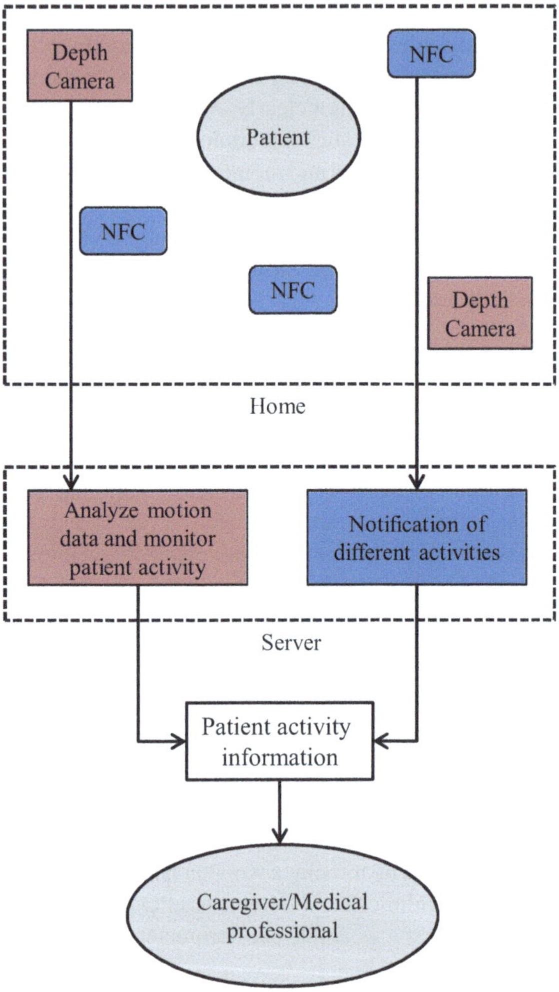

Fig. 9.12 An example of home monitoring system for Alzheimer's patients. Activities monitored by depth camera and near field communication (NFC) devices. Caregiver/Medical professional use the activity information to monitor patient remotely

severe, they become dependent on others for daily living activities. Either a family member or a caregiver is always needed. To overcome this problem contactless home monitoring could be solution.

An ambient sensing system in a house for Alzheimer's patient monitoring is described in [55]. A micro-controller-based alarm system is used to send message to caregiver on activity of the patient. The sensor automatically switches off or closes appliances if the patient forgets. Also it sets off alarm in case of fall out of bed or the main door remaining open. But the work presents a prototype rather than involving patients and caregiver.

Lam et al. [56] developed a system which is able to monitor daily activity of Alzheimer's patient in home environment. Images and motion posture data are collected using a Kinect device placed in a living room. Skeleton images consisting of 16 joints are extracted and used to detect basic postures of sitting, lying and standing. The classification is done using machine learning and Support Vector Machine (SVM) returns highest accuracy of 99% [56]. Also sequence of activities are correctly identified which opens possibility of monitoring more complex activities such as having food or doing exercise. It can give an overall estimation of normal activity of a patient compared to healthy people by analyzing time taken and spent on different activities.

An overview of a home monitoring system for AD patients is shown in Fig. 9.12. To monitor a patient in contactless manner, depth cameras and Near Field Communication (NFC) devices are placed in the home. Number of cameras may depend on the size and number of rooms in the house. The video captured would be sent to a server to analyze and monitor patient's motion activity. NFC devices can be attached to different objects to notify the usage by the patient. For example, NFC on a medicine box could monitor timely consumption of medicine. Data from NFC would also be sent to the server. Further, these patient activity information would be sent to the caregiver and medical professional. They would be able to continuously monitor the patient remotely and also take necessary action in case of any emergency.

Challenges Providing a complete solution to monitor AD patients in home environment is challenging. It is not possible to monitor all kinds of activities as evident from the researches. Also the studies do not involve any AD patient [55] or have very few participants (1 normal and 1 patient) [56].

9.4.3 Rehabilitation of Post-stroke Patients

Stroke patients suffer from motor movement impairment. They are not able to perform different limb movements and gestures as previously in normal condition. Rehabilitation attempts to restore these functions through training various exercises. Conventionally the patient is required to attend a doctor or trained personnel to get trained. However, this process is dependent on availability of doctors. Also continuous rehabilitation training might require the patient to remain admitted at a hospital incurring high cost.

Nikishina et al. [57] developed a video-based software for post-stroke rehabilitation of patients that can be conducted in home environment. A set of hand exercises are followed by the patient through video instructions. The patients' hand exercise videos are captured and images are extracted it. Contours of the patient hand is cut-out using image segmentation techniques. Then features are extracted and SVM is applied to recognize the images. Image recognition accuracy is reported 98% [57]. The speed and accuracy of the activities are analyzed then against reference images. A total of 57 patients took part in the study. Though their progress was non-uniform but an improvement in speed and accuracy of movements is observed in the patients after the rehabilitation program.

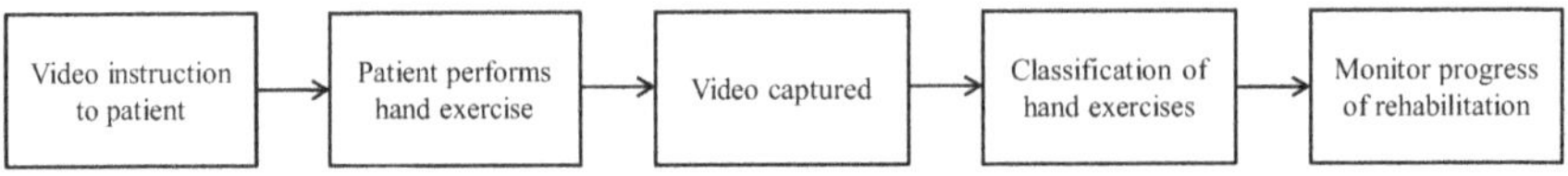

Fig. 9.13 Video-based training and rehabilitation of post-stroke patients

Figure 9.13 illustrates the concept of training and rehabilitation monitoring of post-stroke patients. A video instruction is provided to the patient to show different hand exercises. Then the patient performs those exercises by following the instructions. The exercise video is captured and hand exercises are classified accordingly. It would allow to analyze the accuracy of hand exercises by comparing against reference images and enable monitoring of rehabilitation in a contactless way.

Challenges Post-stroke patients also suffer from leg movements and speech problems along with difficulties in hand movements. So, future works could explore these areas. The hand exercise rehabilitation can be easily extended towards leg exercise. Also, speech processing training can be used to help recover from speech problems.

9.5 Blood Glucose Monitoring

Even noninvasive monitoring of blood glucose has been difficult. There are hardly any devices which are truly noninvasive, accurate and reliable. Several noninvasive techniques are explored, such as –

- Bioimpedance spectroscopy
- Electromagnetic sensing
- Near Infrared (NIR) spectroscopy
- Optical coherence tomography
- Ultrasound

However, none of these techniques led to reliable commercial or clinical glucose monitoring device [58]. Also, there has been focus on noninvasive epidermal blood glucose monitoring from interstitial fluids and sweat. Some devices are available but the reliability needs to be tested against gold standard clinical methods in widespread trial [59]. So, development of a contactless method is also rare.

Eardrum Temperature-Based Method Recently, Novikov [60] proposed a technique which measures blood glucose from eardrum and inner surface of head recorded by a contactless thermometer. Increase in temperature is associated with increase in blood glucose following food intake. The work develops an empirical equation to take these factors into consideration to measure blood glucose from eardrum temperature. The experiment is conducted on healthy and diabetes patients having a wide range of blood glucose level at different times (before and after food intake). Standard Clark error grid analysis shows that the technique has an average accuracy range of 87.5% for estimated blood glucose which are within 20% of a reference sensor [60].

Challenges Apparent from discussion of contactless techniques in various physiological monitoring is that these are either video-based or radar-based methods. However, change in blood glucose is not known to produce any significant change which can be detectable in a video-based method. Moreover, there is no evidence yet on similar measurability using radar. In this context, the eardrum temperature-based contactless blood glucose monitoring presents optimism. But before a large clinical trial is conducted, it cannot be said whether this method could be applicable in daily blood glucose monitoring.

9.6 Discussion

In this work, several contactless physiological monitoring techniques are analyzed. The most important factor is whether these techniques are comparable to existing contact-based techniques in terms of accuracy and diagnostic capability. Applicability of contactless monitoring in current state and direction of future researches in also discussed.

Diagnostic Capability Most researches did not attempt diagnosis of specific diseases using contactless techniques. Analysis of each method indicates that most of the research are on experimental level and conducted in laboratory environment [4, 19, 26, 27]. In some cases, number of subjects is very low [15, 42, 48]. So even with the methods with high accuracy, larger studies in real clinical scenario is necessary to assert those with confidence. Moreover, some studies conducted in clinical settings are derived from normal subjects [25, 35]. So, diagnostic capability of these methods is very limited compared to existing clinical contact-based methods.

Application in Screening and Monitoring Though contactless techniques are not widely tested in clinical settings but high efficacy of some of the researches show it is possible to use it for screening [2] and monitoring [4, 49, 55, 56]. Apart from diagnosis of diseases in symptomatic patients, a clinical test is also done for screening in large asymptomatic population for early detection of a disease. In screening applications, low accuracy or lack of clinical trial is not a big issue. Thus, contactless techniques present a low cost and time saving solution for screening diseases.

Lack of Datasets Availability of datasets for different contactless measurements is also an issue. Benchmark datasets are required for validating new datasets and assessing novel contactless advancements. There are very few works which attempted to create such datasets [23, 41].

Future Direction At this stage, contactless techniques are more suitable for screening and simple monitoring applications rather than clinical diagnosis. But these works present strong indication that these can be used for diagnosis in future. For that to happen, future researches should move towards testing and validation in clinical environment.

9.7 Conclusion

In this chapter, advances in contactless monitoring for most important physiological measurements are presented. Advantages and applicability of these methods compared to conventional contact-based methods are looked into. Also, future challenges for contactless techniques to be suitable for widespread healthcare applications are discussed.

In cardiac and blood related health monitoring several contactless monitoring methods are discussed. Contactless ECG is described which requires capacitive coupled electrodes for contactless acquisition of ECG signals. The proposed methods are tested in various scenarios and indicates applicability in screening and monitoring purposes. However, advancement towards clinical diagnosis is still not met. A brief background of rPPG is also presented which is the basic principle behind many video-based blood related monitoring and respiration rate. Heart rate monitoring is done using video-based rPPG methods or microwave radar-based methods. These methods are quite accurate in detecting HR but more testing in needed for application in clinical purposes. In case of blood pressure monitoring contactless methods are in very early stage. One type of video-based methods estimate BP from PTT calculated from rPPG signals. However, the correlation between PTT and BP is not yet well established and subject-dependent. Transdermal optical imaging measures BP from variation of light reflection in hemoglobin during pulsatile blood flow. The methods is highly accurate for normal BP range but not tested for high or low BP cases. For blood flow monitoring, there is no specific work that have precisely measured blood flow changes in a contactless manner but rather on studying venous compliance and occlusion. So, these methods cannot offer any screening or diagnostic opportunity in blood flow monitoring. Oxygen saturation monitoring done by rPPG video-based methods are also discussed. Light illumination and video capture around 660 nm and 940 nm are essential to record the changes of SpO2. Though, generating low level of SpO2 in human subjects are challenging and potentially hazardous, researchers have successfully calibrated a camera-based SpO2 measurement system for clinical application.

Contactless respiratory monitoring methods are also either video-based or Doppler radar-based. Respiration rate is measured using these systems. Along with it, respiration rate and breathing pattern are analyzed for different sleep monitoring applications.

Neurological monitoring in a contactless way has been done for different application including symptoms detection in diseases, home monitoring or rehabilitation of patients. All these methods use camera or motion sensors to detect several activities of patients. Identified activities then assist in further monitoring patients remotely or in home environment. In case of blood glucose monitoring, even non-invasive methods are not well established and not available for commercial or clinical use. However, a very interesting recent work measures blood glucose from eardrum temperature variations.

Finally, the chapter provides discussion on the drawbacks of the current contactless healthcare monitoring methods and suggests future directions. Diagnostic capability of the methods are discussed in comparison to conventional methods. Diagnostic capability is limited because most studies are done in experimental setup rather than clinical environment and generally normal subjects are involved instead of patients. Also, the availability of standard datasets is discussed and a lack of benchmark datasets is observed. So, it is concluded that the contactless methods are more suitable for screening and monitoring purposes in the current state and in future more clinical testing is required for these methods.

References

1. Aleksandrowicz, A., Leonhardt, S.: Wireless and non-contact ECG measurement system - the “Aachen SmartChair". Acta Polytechnica **47**(4), 68–71 (2007)
2. Czaplik, M., Eilebrecht, B., Walocha, R., Schauerte, P., Rossaint, R.: Clinical proof of practicability of a contactless ECG device. Eur. J. Anaesthesiol. (EJA) **27**, 65–66 (2010)
3. Chamadiya, B., Mankodiya, K., Wagner, M., et al.: Textile-based, contactless ECG monitoring for non-ICU clinical settings. J. Ambient Intell. Human Comput. **4**, 791–800 (2013)
4. Lim, Y.G., Kim, K.K., Park, K.S.: ECG recording on a bed during sleep without direct skin-contact. IEEE Trans. Biomed. Eng. **54**(4), 718–725 (2007)
5. Lim, Y.G., Kim, K.K., Park, K.S.: ECG measurement on a chair without conductive contact. IEEE Trans. Biomed. Eng. **53**(5), 956–959 (2006)
6. Weeks, J., Elsaadany, M., Lessard-Tremblay, M., et al.: A novel sensor-array system for contactless electrocardiogram acquisition. In: 42nd Annual International Conference of the IEEE Engineering in Medicine & Biology Society (EMBC), Montreal, Canada, pp. 4122–4125 (2020)
7. Lessard-Tremblay, M., Weeks, J., Morelli, L., et al.: Contactless capacitive electrocardiography using hybrid flexible printed electrodes. Sensors **20**(18), 5156 (2020)
8. Babušiak, B., Šmondrk, M., Balogová, L., Gála, M.: Mattress topper with textile ECG electrodes. Fibres Text. **27**(3), 25–28 (2020)
9. Hernández-Ortega, J., et al.: Morphological analysis on single lead contactless ECG monitoring based on a beat-template development. Comput. Cardiol. **41**, 369–372 (2014)
10. Parente, F.R., Santonico, M., Zompanti, A., et al.: An electronic system for the contactless reading of ECG signals. Sensors (Basel) **17**(11), 2474 (2017)
11. Bujnowski, A., Kaczmarek, M., Osiński, K., et al.: Capacitively coupled ECG measurements - a CMRR circuit improvement. In: Eskola, H., Väisänen, O., Viik, J., Hyttinen, J. (eds.) EMBEC & NBC, IFMBE Proceedings, vol. 65. Springer, Singapore (2018)
12. Wang, T.W., Lin, S.F.: Negative impedance capacitive electrode for ECG sensing through fabric layer. IEEE Trans. Instrum. Measur. **70**, 1–8 (2021)
13. Verkruysse, W., Svaasand, L.O., Nelson, J.S.: Remote plethysmographic imaging using ambient light. Opt. Exp. **16**(26), 21434–21445 (2008)
14. Poh, M.Z., McDuff, D.J., Picard, R.W.: Non-contact, automated cardiac pulse measurements using video imaging and blind source separation. Opt. Exp. **18**(10), 10762–10774 (2010)
15. Rouast, P.V., Adam, M.T.P., Chiong, R., et al.: Remote heart rate measurement using low-cost RGB face video: a technical literature review. Front. Comput. Sci. **12**, 858–872 (2018)
16. Lamba, P.S., Virmani, D.: Contactless heart rate estimation from face videos. J. Stat. Manag. Syst. **23**(7), 1275–1284 (2020)
17. Maji, S., Massaroni, C., Schena, E., Silvestri, S.: Contactless heart rate monitoring using a standard RGB Camera. In: 2020 IEEE International Workshop on Metrology for Industry 4.0 & IoT, Roma, Italy, pp. 729–733 (2020)

18. Obeid, D., Sadek, S., Zaharia, G., Zein, G.E.: Noncontact heartbeat detection at 2.4, 5.8, and 60 GHz. A comparative study. Microwave Opt. Technol. Lett. **51**(3), 666–669 (2009)
19. El-Samad, S., Obeid, D., Zaharia, G., et al.: Heartbeat rate measurement using microwave systems: single-antenna, two-antennas, and modeling a moving person. Analog Integr. Circ. Sig. Process **96**, 269–282 (2018)
20. Arsalan, M., Santra, A., Will, C.: Improved contactless heartbeat estimation in FMCW radar via Kalman filter tracking. IEEE Sens. Lett. **4**(5), 1–4 (2020)
21. Rodríguez, A.M., Ramos-Castro, J.: Video pulse rate variability analysis in stationary and motion conditions. Biomed. Eng. Online **17**(1), 11 (2018)
22. Jeong, I.C., Finkelstein, J.: Introducing contactless blood pressure assessment using a high speed video camera. J. Med. Syst. **40**(4), 77 (2016)
23. Fan, X., Ye, Q., Yang, X., et al.: Robust blood pressure estimation using an RGB camera. J. Ambient Intell. Human Comput. (2018). https://doi.org/10.1007/s12652-018-1026-6
24. Fan, X., Tjahjadi, T.: Robust contactless pulse transit time estimation based on signal quality metric. Pattern Recognit. Lett. **137**, 12–16 (2020)
25. Luo, H., Yang, D., Barszczyk, A., et al.: Smartphone-based blood pressure measurement using transdermal optical imaging technology. Circ. Cardiovasc. Imaging **12**(8), e008857 (2019)
26. Kamshilin, A.A., Zaytsev, V.V., Mamontov, O.V.: Novel contactless approach for assessment of venous occlusion plethysmography by video recordings at the green illumination. Sci. Rep. **7**(1), 464 (2017)
27. Zaytsev, V.V., Miridonov, S.V., Mamontov, O.V., Kamshilin, A.A.: Contactless monitoring of the blood-flow changes in upper limbs. Biomed. Opt. Exp. **9**(11), 5387–5399 (2018)
28. Nakano, K., Aoki, Y., Satoh, R., et al.: Visualization of venous compliance of superficial veins using non-contact plethysmography based on digital red-green-blue images. Sensors (Basel) **16**(12), 1996 (2016)
29. Webster, J.: Design of Pulse Oximeters. Institute of Physics, Bristol (1997)
30. Wieringa, F., Mastik, F., Steen, V.D.: Contactless multiple wavelength photoplethysmographic imaging: a first step toward "SpO2 Camera" technology. Ann. Biomed. Eng. **33**(8), 1034–1041 (2005)
31. Humphreys, K., Ward, T., Markham, C.: Noncontact simultaneous dual wavelength photoplethysmography: a further step toward noncontact pulse oximetry. Rev. Sci. Instrum. **78**, 044304–044306 (2007)
32. Shao, D., Liu, C., Tsow, F., et al.: Noncontact monitoring of blood oxygen saturation using camera and dual-wavelength imaging system. IEEE Trans. Biomed. Eng. **63**(6), 1091–1098 (2015)
33. Bal, U.: Non-contact estimation of heart rate and oxygen saturation using ambient light. Biomed. Opt. Exp. **6**, 86–97 (2015)
34. Guazzi, A.R., Villarroel, M., Jorge, J., et al.: Non-contact measurement of oxygen saturation with an RGB camera. Biomed. Opt. Exp. **6**, 3320–3338 (2015)
35. Verkruysse, W., Bartula, M., Bresch, E., et al.: Calibration of contactless pulse oximetry. Anesth. Analg. **124**(1), 136–145 (2017)
36. Tan, K.S., Saatchi, R., Elphick, H., et al.: Real-time vision based respiration monitoring system. In: 7th International Symposium on Communication Systems, Networks & Digital Signal Processing (CSNDSP), Newcastle upon Tyne, pp. 770–774 (2010)
37. Bartula, M., Tigges, T., Muehlsteff. J.: Camera-based system for contactless monitoring of respiration. In: 35th Annual International Conference of the IEEE Engineering in Medicine and Biology Society (EMBC), Osaka, pp. 2672–2675 (2013)
38. Bernacchia, N., Scalise, L., Casacanditella, L., et al.: Non contact measurement of heart and respiration rates based on Kinect™. In: IEEE International Symposium on Medical Measurements and Applications (MeMeA), Lisboa, pp1–5 (2014)
39. Lukac, T., Pucik, J., Chrenko, L.: Contactless recognition of respiration phases using web camera. In: IEEE RADIOELEKTRONIKA, 24th International Conference, pp. 1–4 (2014)
40. Ganfure, G.O.: Using video stream for continuous monitoring of breathing rate for general setting. SIViP **13**, 1395–1403 (2019)

41. Janssen, R., Wang, W., Moço, A., et al.: Video-based respiration monitoring with automatic region of interest detection. Physiol. Meas. **37**(1), 100–114 (2016)
42. Massaroni, C., Lo Presti, D., Formica, D., et al.: Non-contact monitoring of breathing pattern and respiratory rate via RGB signal measurement. Sensors (Basel) **19**(12), 2758 (2019)
43. Sanyal, S., Nundy, K.K.: Algorithms for monitoring heart rate and respiratory rate from the video of a user's face. IEEE J. Transl. Eng. Health Med. **6**, 1–11 (2018)
44. Min, S.D., Kim, J.K., Shin, H.S., et al.: Noncontact respiration rate measurement system using an ultrasonic proximity sensor. IEEE Sens. J. **10**, 1732–1739 (2010)
45. Lee, Y.S., Pathirana, P.N., Evans, R.J., et al.: Noncontact detection and analysis of respiratory function using microwave doppler radar. J. Sensors (2015). https://doi.org/10.1155/2015/548136
46. Sun, G., Matsui, T.: Rapid and stable measurement of respiratory rate from Doppler radar signals using time domain autocorrelation model. In: Conference Proceedings IEEE Engineering Medicine and Biology Society, pp. 5985–5988 (2015)
47. Alemaryeen, A., Noghanian, S., Fazel-Rezai, R.: Respiratory rate measurements via Doppler radar for health monitoring applications. In: 39th Annual International Conference of the IEEE Engineering in Medicine and Biology Society (EMBC), Seogwipo, pp. 829–832 (2017)
48. Goldfine, C.E., Oshim, F.T., Carreiro, S.P., et al.: Respiratory rate monitoring in clinical environments with a contactless ultra-wideband impulse radar-based sensor system. In: Proceedings of Annual Hawaii International Conference on System Sciences, pp. 3366–3375 (2020)
49. Procházka, A., Schätz, M., Centonze, F., et al.: Extraction of breathing features using MS kinect for sleep stage detection. SIViP **10**, 1279–1286 (2016)
50. Siam, A.I., El-Bahnasawy, N.A., El Banby, G.M., et al.: Efficient video-based breathing pattern and respiration rate monitoring for remote health monitoring. J. Opt. Soc. Am. A **37**(11), C118–C124 (2020)
51. Tran, V.P., Al-Jumaily, A.A., Islam, S.M.S.: Doppler radar-based non-contact health monitoring for obstructive sleep apnea diagnosis: a comprehensive review. Big Data Cogn. Comput. **3**(1), 3 (2019)
52. Yang, X., Fan, D., Ren, A., et al.: Diagnosis of the hypopnea syndrome in the early stage. Neural Comput. Appl. **32**, 855–866 (2020)
53. Abramiuc, B., Zinger, S., de With, P.H.N., et al.: Home video monitoring system for neurodegenerative diseases based on commercial HD cameras. In: IEEE 5th International Conference on Consumer Electronics—Berlin (ICCE-Berlin), pp. 489–492 (2015)
54. Shi, W.Y., Chiao, J.-C.: Contactless hand tremor detector based on an inductive sensor. In: IEEE Dallas Circuits and Systems Conference (DCAS), Arlington, TX, pp. 1–4 (2016)
55. Almagooshi, S., Hakami, M., Alsayyari, M., et al.: An assisted living home for Alzheimer's patient in Saudi Arabia, a prototype. In: Stephanidis, C. (eds) HCI International 2015 - Posters' Extended Abstracts. HCI: Communications in Computer and Information Science, p. 529. Springer, Cham (2015)
56. Lam, K., Tsang, N.W., Han, S., et al.: Activity tracking and monitoring of patients with Alzheimer's disease. Multimed. Tools Appl. (2015). https://doi.org/10.1007/s11042-015-3047-x
57. Nikishina, V.B., Petrash, E.A., Nikishin, I.I.: Application of a hardware and software system of computer vision for rehabilitation training of post-stroke patients. Biomed. Eng. **53**, 44–50 (2019)
58. So, C.F., Choi, K.S., Wong, T.K., et al.: Recent advances in noninvasive glucose monitoring. Med. Devices (Auckl.) **5**, 45–52 (2012)
59. Kim, J., Campbell, A.S., Wang, J.: Wearable non-invasive epidermal glucose sensors: a review. Talanta **177**, 163–170 (2018)
60. Novikov, I.A.: Noninvasive determination of blood glucose concentration by comparing the eardrum and head skin temperatures. Biomed. Eng. **51**, 341–345 (2018)

Chapter 10
Personalized Patient Safety Management: Sensors and Real-Time Data Analysis

Md. Jasim Uddin and Monika Nasrin Munni

Abstract In recent years, we have observed a technology revolution in healthcare monitoring through the development of intelligent systems for vital signal monitoring. Sensors attached to computer-aided software can be used to track the activity of the medical devices, especially for chronic patients by providing high-quality data. Designing of personalized monitoring system through an automated device comprises software, interpreting sensors, and online databases. Considering the advantages of easy fabrication, and selective and sensitive detection, such sensors would be widely applied in the fields of pharmaceutical, and environmental and clinical monitoring. The new wearable medical device with a sensor showed a great observational output for maintaining a personalized health care program. Explicit measurement data will be provided to customized clinical instruments with sensors while indicating high responsiveness, high accuracy because of possible entanglements, and great steadiness in biological systems. The key emphasis of this chapter is the design and analytical performance needed of the new customized wearable devices and machinery containing the sensors to hold the arrays with astonishing accuracy and precision that could provide controllable, minimally intrusive real-time tracking of drug release and plasma drug concentration to advance the drug dose regimen and a more proficient treatment.

Keywords Personalized health care system · Sensors · Sensor-based medical device · Minimally invasive monitoring · Biosensor

10.1 Introduction

Tailor-made medical devices which are designed according to the individual patient requirements, categorizes the personalized devices on a broad category [21]. They can

Md. J. Uddin (✉)
Department of Pharmacy, Brac University, 66 Mohakhali, Dhaka 1212, Bangladesh
e-mail: jasim.uddin@bracu.ac.bd

Md. J. Uddin · M. N. Munni
Drug Delivery and Therapeutics Lab, Dhaka, Bangladesh

M. A. R. Ahad and A. Inoue (eds.), *Vision, Sensing and Analytics: Integrative Approaches*, Intelligent Systems Reference Library 207, https://doi.org/10.1007/978-3-030-75490-7_10

currently be classified as custom-made medical instruments, constructed, or modified medical devices that are tailored to a patient [72]. Personalized sensor-based medical devices vary significantly in size and other parameters from the instruments produced with sophisticated manufacturing techniques and technology such as bones, steel, exoskeletons, valves, bones of the neck, tissues, and organs, etc. [79]. A large number of doctors are aiming to introduce an online-based web portal to provide high-quality health care service and thus making it more accessible to the patients [79]. Further life experiences are found more accountable for the healthcare system. Electronic data management will allow an increased interaction between the patients and the doctor ensuring a better relationship and trust between them [65]. A huge number of sources of probabilistic authenticity are available for patients. Many doctors are preparing to revolutionize the modern age by establishing this personalized health care system [43]. The massive advancements in material science and futuristic technologies over the last 25 years allowed improvements in the construction of innovative medical devices and the development is occurring at a staggering pace [60]. Innovations now facilitate the development of uniquely customized medical products, including designing software and advanced production methods such as 3D printing, 4D printing, etc. [15]. This is in contrast to the traditional health care system. Individualized methods are used to make an interactive tool in a laboratory by the technician [48]. The new technology would allow the personalized medical instrument model and the quality of uniquely designed medical equipment to be improved, specialized medical equipment prices can be minimized and the number of specialty medical instruments supplied to the market improved [94]. The chapter aims to focus on unveiling some of the innovations in the material design and improvements in the medical devices that can impact the patient life by allowing them access to better treatment options for multiple diseases [68, 69]. The area of personalized medicine presents pharmacists with a range of resources, as they hold specialized expertise and skills to facilitate the usage of personalized medicine as a resource [87].

A brief graphical representation of the anticipated process has been outlined in Fig. 10.1 [112]. In situations where frequent examinations and finally post-processing is required, the information is gathered from an underlying network and is processed accordingly [112].

10.1.1 Overview of a Personalized Health Care System

Personalized healthcare is a general healthcare framework that combines predictive technology and patients to deliver high-quality care with the main focus of improving the health of patients and providing effective treatment of complex disease conditions [11]. Digital Personalized Health System (PHS) records data on electronic devices used for the advancement of predictive analytics. In addition to that, there is the consideration of health and cost–benefit when the proposal goes above the possibility of services for widespread investigation, appraisal, and adaptation [71]. To provide a competitive environment and high-quality health coverage, electronic

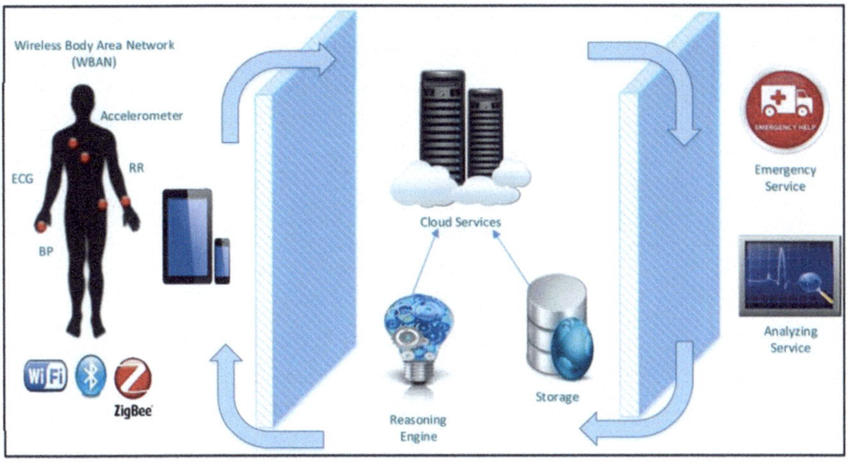

Fig. 10.1 Automated processes in healthcare delivery [1]

resources need to be offered to specialized medical sectors, such as pediatrics, general practitioners, and teledermatology for a small period [12]. Nevertheless, it enables useful data to be obtained for the potential production of PHS in the long term [30]. We have now reached an age of study and advancement through the way of large-scale evidence generated and mixed with clinical data obtained from patients such as psychological, cognitive, and medical outcomes [78]. The fundamental concepts of personal medicine (PCM), the methodological philosophy of medical science that positions an individual as the physical, psychological, and spiritual agent in the core of healthcare and treatment, are conventional, complementary, and integrative medical systems [91]. PCM aims to extend the bio-molecular reduced method to medical [23]. Diagnostic devices and engineering technologies are two fields that have a specific effect on digital apps [16]. An illustration of this is 3D printing where customized implantable devices, for example, a substitution hip would now be able to be made accessible to healthcare providers utilizing 3D printing innovation that is designed precisely to the requirements of a patient [5]. A custom-made device of this type would have been very costly and seldom used, although manufactured using more conventional techniques [81].

In Fig. 10.2, the expected IoT architecture is displayed [80]. This part of the section takes into account the literature for security observing, wearable systems, and sensor classifications, and utilizations of wearables in development and different enterprises [103]. Attributable to the hazardous working conditions at building destinations, staff likewise face conceivable health and security dangers during the entire process of construction [43]. Building safeguard has generally been evaluated and controlled responsively through endeavors to react to antagonistic injury sequences [60]. Dynamic following of the physiological information of representatives utilizing wearable gadgets will all things considered permit estimation of pulse, respiratory rate, and stance [87].

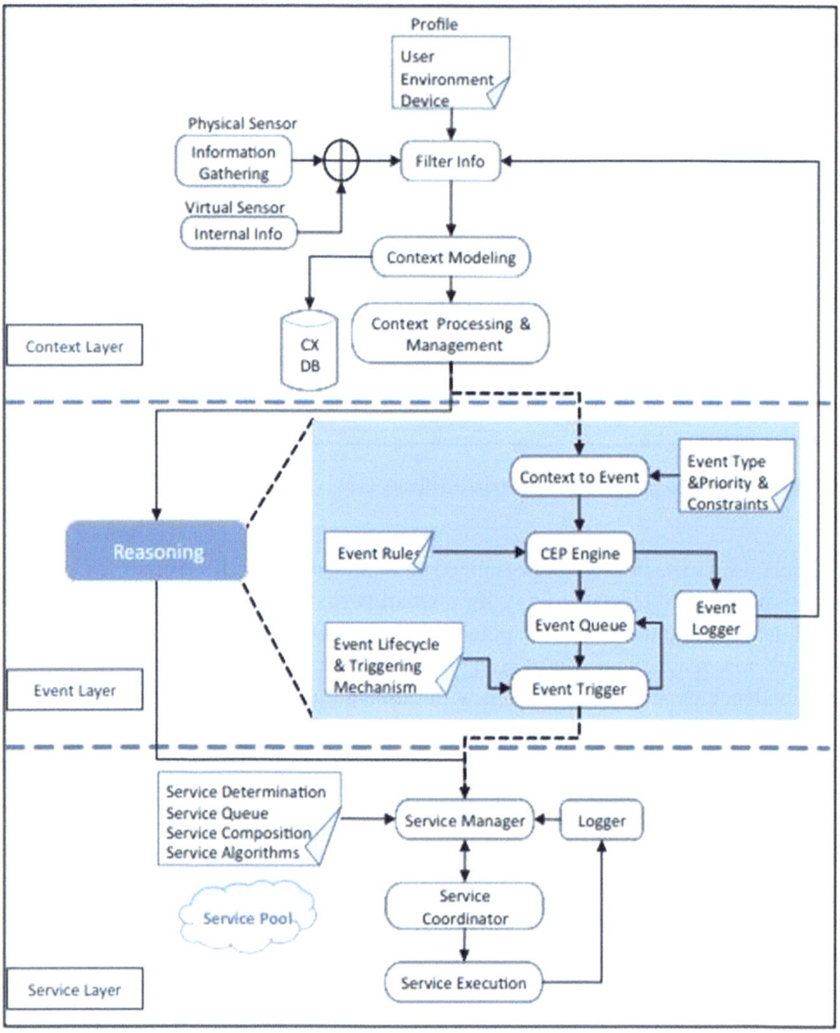

Fig. 10.2 Real-time reusable occasion has driven IoT architecture [80]

10.1.1.1 Construction Safety Measurement and Monitoring

Conventional methods to deal with the evaluation metrics of achieving safety are dominatingly manual and subject to individual decisions [26]. These strategies are centered around enormous human work to assemble information, and subsequently, information is accumulated at restricted frequencies and when accidents occur [113]. These methodologies are costly, helpless against data input blunders, and lead to data collections too little to even consider managing the undertaking effectively and precisely [82]. Computerized safety observation is perhaps the most encouraging methodologies for solid and reliable checking of building destinations to address the deficiencies of manual endeavors [117]. The programmed global positioning framework will gather information, transform it into coordinated information, and disseminate the information straightforwardly to project chiefs [118]. Many have expansive security executions among continuous tasks has the board working strategies [119]. The objective of security and wellbeing oversight is to guarantee that the wellbeing techniques of development laborers are viably surveyed and controlled adhering to current security plans and norms [120]. Sadly, customary modern observing gadgets for projects are not valuable because of the transient presence of development exercises and task associations [28]. Among other designing uses, it very well may be helpful for insurance, security, and cycle examination to follow the position and ways of individuals consequently [121]. Wearable devices can continually follow a huge assortment of vital signs [1].

10.1.1.2 Systems and Sensors for Wearable Technology

The machinery used to develop wearable devices depends on different systems including radiofrequency, magnetic field, radar, super-wide reach, ultrasonic, sonar, Bluetooth, worldwide setting, laser, film, static camera, and electromyography [94]. A body sensor network contains sensors, for example, galvanic skin responses, accelerometers, spinners, and magnetometers [122]. The use of cell phones has changed numerous features of the lives of individuals, with numerous models indicating how wearable innovation is really and hypothetically utilized in medical care [110]. Sensor innovation progress has been significant for body sensor systems administration and they have been combined with the improvements in short-range communication technology, for example, Bluetooth and super wideband radio frameworks that permit client estimated frameworks to be sent [101].

Table 10.1 summarizes the sensors based on different categories and enlists the measurements taken by each of them [1]. Physiological parameters such as the movement of the eyeball, heart rate, or blood pressure are precisely monitored by the sensors [64]. Furthermore, the movement of these physiological parameters is monitored by the use of inertial sensors, position sensors, and image sensors [56].

Table 10.1 Classification of sensors [1]

Category	Sensor	Measurements taken
Inertial Sensor	Magnetic field sensor	Higher spatial resolution's location
	Pressure sensor	The altitude of the object
	Gyroscope	Angular rotational velocity
	Accelerometer	Measure linear acceleration
Physiological Sensors	Galvanic Skin Response	The temperature of the skin surface
	Electrooculography	Movement of eye
	Spirometer	Lung parameters like volume, expiration, and flow rate
	Electrocardiogram	Heart Activity
	Blood Pressure cuff	Blood pressure
Image Sensors	SenseCam	Pictures of daily living activities
Location Sensors	GPS	Coordinates of outdoor location

Table 10.2 Safety performance metrics for construction safety and health hazards [36]

	Construction site hazards		Metrics
	Safety hazards	Health hazards	
Physiological Monitoring	Slips, Trips, and falls from height	Stress, heat, cold, strain injuries (carpal tunnel syndrome, Back injuries) Skin diseases (absorption), Cuts (injection), Breathing or respiratory diseases, Toxic gasses	Heart rate, heart rate variability, Respiratory rate, body posture, body speed, Body Acceleration, body rotation, and orientation, angular velocity and, blood oxygen, blood pressure, Body temperature, Activity level, Calories burned, and walking steps
Environmental Sensing	Slips, Trips, fire, and explosions	Chemicals (Paints, asbestos, solvents, chlorine), molds, noise, heat, cold, radiation, vibration, toxic gasses	Ambient temperature, ambient pressure, humidity, noise, level, light intensity, air quality
Proximity Detection	Caught in or between, Struck by moving vehicle or equipment electrocution	Chemicals (Paints, asbestos, solvents, chlorine), molds, noise, heat, cold, radiation, vibration, toxic gasses	Object detection, navigation, Distance measurements, and proximity detection
Location Tracking	Caught in or between, Struck by confined spaces, cave in electrocution	Hazardous chemicals (paints, asbestos, solvents, chlorine), molds, noise, heat, cold, radiation, Vibration	Worker location tracking, material tracking, and vehicle equipment location tracking

10.1.1.3 Wearable Technology in Other Industries

Different categories of wearable devices are utilized in areas, for example, clinics, construction, mines, and sports [120, 121]. A portion of these advancements has demonstrated promising advantages and the analysts and industry specialists are both putting forth attempts to build up these innovations and to gain from their underlying formation. Wearable technology has been increasingly more utilized in the investigation of wellbeing to support actual exercise by planning PC frameworks with low force utilization and minimal cost sensors [44]. The significant progression in figuring innovation, miniature sensors, and broadcast communications in a strong state has made conceivable the preparing and survey of human physiological measurements through individual wellbeing global positioning frameworks [122, 123]. There are now a few conservative wearable sensors [60]. Another age of GPS devices, which tracks the physiological conditions of individuals who complete regular work in both the homes and the rest of the world has been actualized in advancements in small sensors and remote innovations [125]. Moreover, online patient administration causes patients to screen their wellbeing while at the same time maintaining a strategic distance from unnecessary clinical visits [124] . A few enterprises were enlivened by the original work done by analysts at the Jet Propulsion Laboratory at the National Aeronautics and Space Administration and the improvement of framework bases of commercialized body sensors [79]. Devices like these offer health applications, wireless action recognition, and medical equipment, which track information, for example, pulse, work out, ventilation, breathing, and stance constantly, to bring down medical care expenses and result in an elevated efficiency.

These advancements focus on the developing exchange of data, productivity, and security in strategic approaches, including overseeing admittance, client relations, far-off checking, and stock tasks [103]. Wearable gadgets are usually utilized in games and wellness to screen execution through smooth, subtle estimations [42]. Used to gather real-time execution subtleties, wearable advances (for instance, GPS montres, pulse screens, pedometers, and so on). Wearable equipment is utilized in the scope of hardware utilized by specialists to follow their presentation and wellbeing. Sensors are utilized to analyze blackouts and keen pressure shirts, for example, wires for arms control, innovation and to evaluate throw quality in Major League baseball, for instance in the protective caps from the National Football League [126]. Moreover, remote wristband GPS sports watches are generally used to support swing mechanics during training meetings [127]. Further wearable innovations are utilized in the wellbeing and sports industry that are connected to dynamic life, including wellness checking, outside route, body cooling, and warming, intelligent instructing, and sports results. Frailty programming, cops, firemen, and paramedics test wearable devices to help and give contributions from distant interchanges and get data without taking care of basic assignments [128]. Also, enlightenment innovations and intelligent gear are utilized to improve visibility and individual security. A GPS-based close by the ready framework and shared systems administration were likewise inherent in the mining business to evade mishaps between mining hardware, little vehicles, and fixed structures [129]. The possibility of GPS-based mining gear nearness notice incorporates utilizing GPS differential recipients with the goal that hardware administrators know about encompassing vehicles and staff [130]. Indeed, even wearable innovations are forming the ordinary work of individuals as far as games and the instruments utilized for the activity of family gadgets or different devices [131]. This alludes to encounters with PC administrations, including admittance to information/media, vivid games, instinctive learning, and social experience [132].

10.1.1.4 Wearable Technology in Construction

The utilization of wearable technologies is increasing in construction at a rapid rate compared to other areas [133]. The utilization of wearable technologies in the structure business is not many and reported. An evaluation of an apparatus to test the closeness identification and warning systems to improve assurance at building locales was one of the not many executions [134]. Besides, sans hand gadgets are utilized to follow and improve representatives' information on conditions by an on-going assortment of business records, recognition of ecological perils, and the vicinity of laborers to territories of danger [3]. The powerlessness to consolidate comprehensively is halfway because of a lack of soundproof to help its future increases [48]. As of late, the savvy gadgets industry has begun to view and trade project information from distant areas [63]. Even though advancements in versatility and mechanization instruments and different developments that may build profitability may gradually be actualized in the development business, wearable innovation may be the key to uncover the potential for useful change [105]. There is accordingly a colossal chance in the structure business to consolidate wearable devices for altered security following [28].

10.1.1.5 Research Needs Statement

The construction business faces threat that, for various reasons firmly connected to the usage of work on the construction business, it's hard to evaluate [73]. The situation for each gathering of representatives shifts, however, any building site frequently extends with development, changing ordinary dangers for laborers [24]. Manuals are the principal hindrances to viable documentation, investigation, and unwavering quality of the most current ways to deal with information assortment [26]. Wearable innovation conveys a non-noisy methodology that gives dependable, solid, and positive dynamic information continuously [41]. Wearable innovation is well-being the executive's system idea that permits staff to follow and manage their wellbeing profile by constant contribution to distinguish and fix the most punctual indications of security issues [135]. Wearable devices may likewise furnish security of the board with prescient measures on building destinations to advance dynamic about the sufficiency of continuous intercession and, if fitting, quickly change the methodology. While these strategies may have helped, generally few are recorded in the writing, and composed endeavors to classify and investigate these methodologies still can't seem to be made [136]. This section adds to the comprehension of customized assembling security reconnaissance and the improvements by an intensive investigation of wearable innovation frameworks. This examination includes an appraisal of the qualities and future preferences of wearable devices to decrease mishaps and infections in building destinations and the security proof gathered [137].

10.1.2 Current Trends on the Development of the Personalized Medical Device

Molecular diagnostics have been merged with personalized medicine and molecular diagnostics have been developed with increasing importance [89]. The wide implementation of precision and customized medicine technology would require the integration of a range of variables from the advancement of innovation and medicine curriculum to government funding for innovative planned clinical trials, and scaling up the usage of Data management to facilitate funding for clinical decision-making [51]. The 3D printing technique is used for review of the electronic equipment sector and the pharmaceutical industry for their usage on different platforms of the medical industry [4]. An array of 3D printing methods have been developed for the production of novel robust dosages [99]. The primary objective of this analysis is to explain the different techniques used in pharmaceutical 3D printing [9]. Ongoing progressions in electronics and microelectronics have permitted the creation of minimal-cost devices that are regularly utilized by numerous individuals as medical services or preventive instruments [10]. Focused on the non-invasive and wearable devices, actuators, and progressed network and data innovation, this medical care foundation gives inventive technologies that permit patients to remain in their relaxed homes and to be protected in any capacity [100]. In a fact, as proactive measures at home are made, expensive hospital services can be used for the acute treatment of patients [55]. This scheme will track patients' very critical physiological parameters in real-time, conform with clinical problems and assess them, and offer input, most critically [14]. Sensors for diagnostic applications in appliances and diverse types of indications are translated into electrical signals [18]. In life-supporting devices, preventive strategies, long-term surveillance of people with handicaps, or illness indicators should be used [19].

In Fig. 10.3, WIoT advances can generally be classified into five classifications: physical activity control, electronic self-administration and tracking, automated diagnostics and handling clinical decisions making, crisis medical care, and live help for the older and others [137].

Fig. 10.3 Applications of the personalized healthcare system

10.1.3 Sensor-Based Medical Device

Devices to be used in medical systems for the inspection of humans on a broad scale, limited scope development and a few physiological signs have been proposed to be adaptable and wearable [5]. The hydrogel of nanocomposite is made of graphene oxide, polyvinyl liquor, and polydopamine and has fantastic mechanical and electrical properties in the treatment fields [81]. This diagram-based, auto adhesive, the conductive hydrogel can be amassed as wearable sensors to identify precisions and progressively signals of human movements and limited scope movements through cracking and recombining the diminished electric oxide pathway of the hydrogel organization's permeable structures [91]. It further serves to make the wearable hydrogel sensor available to monitor human body movement and detect physiological parameters in the long term and repetitively [16]. The modern and innovative remote monitoring device in real-time for telemedicine sensing-based mobile health security has substantially reduced and distributed connectivity components [23]. Characteristics included motivation and challenges associated with sensor-based authentication of researchers' smartphones and recommendations for strengthening this critical research area [78]. By using bio-sensors, the condition, the beginning of the disease, and its advancement can be assessed rapidly which can help to plan the treatment of many medical and nano-technological diseases [30]. The biocatalyst can identify a biological item and transducer that will transform the biocatalyst-biological entity mixing case into a recognizable component [71]. In recent years, bio-sensors for the diagnosis of diseases using immobilized cells and enzymes have entered the field [68, 69]. Ultrasmall sizes and special properties of nano-bio sensors for the production of diagnostic biosensors of the disease have also been reported.

10.1.4 Sensor-Based Personalized Healthcare Devices, Safety, and Well-Being

The introduction of advanced care tools dependent on sensors rendered the tracking of health problems simpler and the contact between physicians and patients better [11]. Current programs have helped to provide patients with sufficient preventive monitoring [87]. Such concerns are tackled mainly by the availability of inexpensive medical supplies for the low-cost health network, customized advice focused on patient clinical requirements, and emergency medical practitioner warnings [94]. The customized health management program meets clinical concerns in remote and rural communities by allowing the patient healthcare staff to track specific health issues to prevent incursions, direct guidance on sleep, exercise, lifestyle and send medical details to the community health center as emergency warnings to the patients and the physician [48]. Throughout the testing process, administration, and monitoring of Medicare recipients the program is assisted by the community health staff and works efficiently, with just sporadic connections to the Internet [60]. Predictive models also utilize EHR and management frameworks to classify at-risk individuals, provide prompt diagnosis and personalized treatment. Such methods have also increased treatment outcomes and prices [15].

10.2 Development of Sensors-Based Personalized Medical Devices

A huge part in the advancement of sensors is played by new ways to deal with the plan of sensors, for example, a lab-on-a-chip, printed (composed), and wearable sensors [65]. Utilizing the standards of power devices can lessen the energy utilization of estimating frameworks and give the likelihood to make small forms of them [43]. The electrochemical strategy for location remains the favored one. The consequences of these examinations will fill in as a reason for making sensor-based clinical gadgets that are required in a research facility, clinical preliminaries, yet in addition to the purpose of care diagnostics, the home center, and telemedicine [79].

10.2.1 Designing of Sensor-Based Medical Devices

Sensors-based medical devices are introduced by delicate, adaptable, and stretchable electronic devices expected for the checking of clinically significant files of human wellbeing [108, 109]. Compound wearable sensors, ideally with the electrochemical location of the reaction, are utilized for the assurance of individual synthetic substances and biomarkers [79]. Tattoo sensors-based medical devices are made utilizing exceptionally designed pressure suffering inks and biocompatible polymers. These are the most ergonomic and advantageous to-utilize sensors [106]. Patch sensors put together clinical gadgets made concerning a material premise are stronger, have a more evolved surface, and, along these lines, have more alternatives for immobilizing the receptor layer and coordinating the going with hardware [72]. Tattoo and fix sensors are minimal effort and not planned to be utilized for more than one time [82]. Band sensors-based clinical gadgets are created dependent on silicone, the most solid and solid of all the above gadgets, and there is the chance of substitution/recovery of the sensor layer just as multiplex recognition of analytes.

Possibilities for the advancement of wearable and composed sensors are broad, and the extent of their potential application is colossal [21]. Smaller than normal and remote sensors consider consistent checking of a patient's wellbeing both at the bedside and a long way past the medical clinic. Simultaneously, because of present-day worked in information trade frameworks, a specialist's proposals can be gotten a good way off [84]. Hence, the improvement of wearable and composed sensors is a significant advance towards customized medication [8]. Quantitative examinations of the hardware, mechanics, heat-move, and medication dissemination qualities approve the activity of individual parts, in this way empowering framework level multi-functionalities [88]. The incredible lion's share of sensor-based clinical gadgets is created by utilizing biosensors and synthetic sensors [77].

10.2.2 System Architecture and Application

Patients around the world have started to embrace wearable biosensors, new applications for individualized eHealth and mHealth advances have arisen [70]. The potential gains of these advances are clear: they are profoundly accessible, effectively open, and easy to customize; furthermore, they make it simple for suppliers to convey individualized substance cost-successfully, at scale [37]. To break down the top worries in IoT advances that relate to shrewd sensors for medical services applications, especially applications focused on individualized telehealth intercessions to empower better lifestyles [6]. These applications incorporate wearable and body sensors, progressed unavoidable medical services frameworks, and the Big Data investigation needed to educate these gadgets [7].

Medical services and technology have consistently been associated, yet that relationship because of the quick development of the Internet of Things and the fame of wearable gadgets has been altogether changed lately [104]. This prompts customized sensor-based medical care, expanding medical services access, and customization - any semblance of which we have never seen before [86]. These headways, while energizing, ought to be received cautiously, as there are as yet authentic concerns identified with consistency, security, cost-adequacy, and that's just the beginning [95].

10.2.3 Monitoring the Device Activity

Pervasive sensors have a special role to play in integrating health care by providing clinicians with an innovative and less biological view of their patients' habits [29]. Throughout hospital environments, the healthcare system routinely tracks the appropriate physiological parameters [102]. The usage of modern wireless sensor systems allows the tracking of higher yet still stable metrics, such as the degree of physical activity, the venue, the number of interactions, and the social background, beyond institutions [85].

Disruptive monitoring may contribute to enhanced health care for people with chronic conditions by supplying doctors with the ability to test their conformity with prescription schemes and behavior standards; to obtain a deeper understanding of variability in perceptions of patients in variables such as blood pressure, breathing rate, and weight; Such apps may be used by elderly patients not only to track medication therapies and other interventions identical to the ones above but also to control physical exercise and socialization [41].

Difficult and accurate details should be open to families to prioritize concerns to take maximum advantage of short hospital interactions [2]. Pervasive sensors have a special role to play in integrating health care by providing clinicians with an innovative and less biological view of their patients' habits [76]. Throughout hospital environments, the healthcare system routinely tracks the appropriate physiological parameters [49]. Current Wireless Sensor Platforms with generic sensors enable superior yet also scientifically relevant tracking, such as degree of physical activity, location, frequency of touch, and social circumstances beyond institutions [75].

The three-dimensional printing revolution is underway and reports frequently on printed reindeer and jaws, dolls and cars, food, and body armor [68, 69]. In the fourth dimension, the latest goal is to create 3D materials: time. Such "4D" materials can be flexible and durable to react to temperature, light, or even stress [39]. 4D implants are of special interest in pediatric medicine; the content will always develop as the individual ages [90]. Morrison et al. using secure, bioresorbable polymer mixture 3D printing technologies to create splices for three pediatric trachea-bronchomalacia (TBM) patients—an excessively common respiratory failure disorder [74]. Fixed-size implants currently available will change and often need to be resized [66]. The scientists used images and computer simulations to construct breaks for specific geometries of the TBM patient, configuration implants to conform to the nature of the airways, and resist outward strain, before being incorporated into the body for a while [67]. The 4D systems were inserted without question in all three patients (one with broken airways) [73]. After 1 month, all four implants were stable and working; one implant was in place and the airway was opened for more than three years [53]. This pilot trial exhibits that the fourth measurement is a reality for 3D-printed materials, and then proceeded with human examinations, 4D biomaterials guarantee to change how we imagine the up and coming age of regenerative medication [25].

10.2.4 Data Processing and Feature Extraction

The data analysis structures utilized for support systems for clinical decision-making are exponential in this area, as are more complex structures [58]. Although intelligent medical systems continue to be established in an intense area of study, some of this knowledge is being applied in low-resource computing platforms [50]. Driven by an explosion in the volume of data, it is essential to reduce data at source through efficient node processing to achieve the sustainability of overall sensing for the management of large populations [45].

To do that, analytical algorithms have been directly mapped to ultra-low-power and implantable sensors [97]. Musiani et al. reported that a Hilbert transform-based signal analysis performed in a Shimmer programmable sensor node needed over 100 million instructions [47]. Simple operations, including matrix inversions and decomposition, in machine learning and signal processing, have difficulty [52]. With just 100 samples a basic algebraic procedure will need approximately 1 million internal loop commands, without taking into consideration external operator directives such as floating points [83].

However, for basic processing elements in on-line Algorithms, such as noise filters, attribute extraction, and peak detectors, quick and light-weight processes have been introduced [34]. For example, in a continuous algorithm based on the wavelet transformation, the ECG heartbeats can be detected through an integrated ad-hoc low-cost framework [17]. Some studies have proposed reducing the on-node implementation to a pre-trained model deduction process in recognition of physiological activity [108, 109]. Concurring advancement in high-performance computing has allowed huge data volumes to be stored more effectively in broad databases [33].

Specifically, the software part has fulfilled a critical job in guiding decreased frameworks with modern format storing and in-memory handling [13]. In equal, advancements have permitted exceptionally escalated tasks and changes in equipment, as coprocessors and GPU quickening agents, including Nvidia Tesla [22]. Group learning is another intriguing system that consolidates deductions from different calculations that are prepared in information subsets utilizing a democratic procedure that can work as a synchronous activity [57].

Personalized healthcare programs provide e-health resources to address the safety and social demands of elderly adults [40]. The Internet of Things (IoT) marks a big change in the age of Big Data [93]. Analyzing IoT data streams has become a source of healthcare systems that use data to find information, predict early detection, and decide on the critical situation for quality of life improvement [54]. We carried out a detailed study on the new technologies in customized health care systems with a focus on cloud computing, fog computing, Big Data analysis, IoT, and mobile applications [111].

Researchers explored the complexities of designing a sustainable health care network to identify and treat diseases early and addressed possible approaches when delivering healthy e-health services [38]. In healthcare big data research has been used to forecast infections, reverse infectious disorders, avoid premature deaths, and enhance the quality of life [61].

Wearable sensor systems track consumers daily and capture both organized, unstructured data called Big Data [59]. Useful knowledge, which is part of big details, is a challenging method for decision-making [107]. More delay is induced by the cloud storage used to store and handle details on the Internet's central server [35]. Before the cloud, the information is stored, assessed, and prepared to diminish latency and to build the act of a fog layer [62]. At long last, the information is saved in the cloud progressively to make a swift judgment [31].

The utilization of important information in big data identifying with medical care is hard to store, oversee, and examine [68, 69]. To store and deal with sensor information for wellbeing applications, Manugaran et al. have proposed another design for IoT gadgets [20]. The Meta Fog Redirector (MC-R) has been created by Te creators with a gathering and assortment design (GC) for IoT and Large Data conditions to make sure about and hinder Big Data from interruption [18]. IoT systems capture the pulse rhythm, breathing pace, body temperature, blood pressure, and sugar in sensors and transfer them into the fog for further diagnosis [19].

The alarm message will be shipped off to doctors through fog computing during a crisis [14]. Enormous Data stockpiling is utilized on the cloud stage by Apache Pig and Apache HBase [55]. The administration of Big Data is a significant test in well-being dynamic progressively far off observing [12]. In huge information examination, the need of patients is an issue in telemedicine [100]. Khalid et al. associate patient need and constant far-off reconnaissance of clinical administrations utilizing Big Data [10]. Consistent far-off patient checking through sensors is vital for persistent illnesses [9].

Sensors can follow client information and inform doctors through ongoing information sending and telemedicine [4]. The expanded number of patients raises the adaptability issue [99]. They control populace maturing and debacle to give adequate telemedicine administrations [51]. Three regions: Procedure, Transplantation, and Treatment offer inclination to patients [89]. To address this issue, an emergency is utilized to evaluate the outrageous state of the case. The high-level medical care frameworks with the use of wearable gadgets and sensors and the need to give enormous information to give data on these gadgets assume an essential job in the arising scene [9].

The sensor-based organization has three parts [5]: (I) the Body Sensor Network used to accumulate patients' action subtleties and data on the atmosphere, for example, temperature, stickiness, area, time using body wear sensors, or implanted sensors. (ii) Advanced Gateway with Web access [91]. (iii) Cloud and Big Data administration for all sensor information assortment, survey, assessment, and joint effort with crisis guardians [78]. Electronic wellbeing records (EHR) include a huge audit of information zeroed in on Big Data Analysis to dodge the flighty advancement of clinical and clinical information [30].

Next, the emergency clinic information assortment is accumulated for outpatients, and additional doctors are required for determination [71]. The symptomatic model at the same time helped to build the work productivity of walking specialists and diminish their labor force [68, 69]. The built-up models' jobs are information assortment, information stockpiling, preprocessing of information, recovery of data, framework learnings, yield testing, and quick examination list determination for doctors. Data assortment catches meds, programs, care costs, impacts for the clinical-stage, assessment, and photo subtleties from patients outside [87].

PC reproduction utilizes the help vector machine and a neural organization calculation to recognize hyperlipidemia in information from past clinical informational indexes [94]. The job of large information investigation in the arising clinical universe of today is to handle a wide scope of information and to store it as an effective method to concentrate, update, and eliminate information [48]. The information is put away in a self-assertive multi-dimensional framework called a tensor focused on an idea progressive system [60]. A sensor-based way to deal with information mining to address three difficulties: multifaceted nature, vulnerability, and wellbeing [15].

Granularity Computing is the specialty of taking care of issues of registering at different granularity rates and gathering helpful information from the informational index accessible [65]. The framework recommended by Te is part of three phases to productively mine medical services information [43]. The information lattice is the fundamental advance wherein crude content, clinical records, sound, and video information are gathered in the tensor for wellbeing purposes. During the subsequent advance, the information grid is stretched out through different granularity rates utilizing the definition progression [79]. All through the third step, granules are utilized to deal with inquiries and to create the last tests [108, 109]. The yield examinations are assessed by the results for snappy assessment and lower costs comparative with PARAFAC2 and CANDELINC with tensor-based information portrayal [79]. Te's diabetic dataset is changed over for the proposed application in the tensor tool stash in MATLAB [106].

In Fig. 10.4, it is showing that smart footwear, fitness bands, smart watches are used in healthcare system many applications such as health care and personal assistance, wearable technology to ensure an improved healthcare system from the past decade and play a critical role. In certain applications, found a useful substitute for sensor-based devices IoT and wireless sensor networks provide multi-faceted solutions for expanding the range of wearable devices [75].

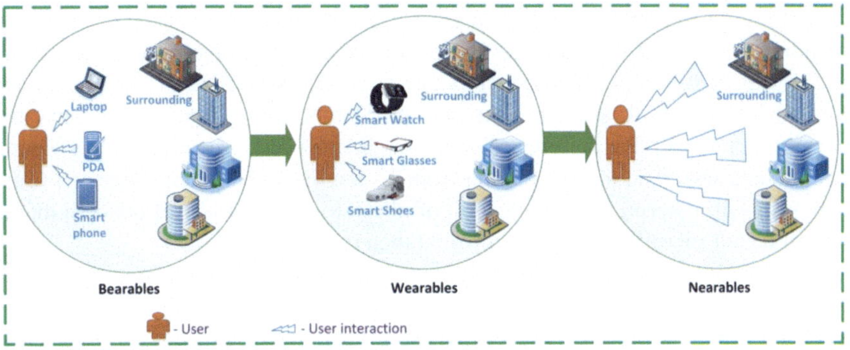

Fig. 10.4 Architecture of Amrita Jeevanam (Indian Organization) Health Awareness Platform [138]

10.2.5 Process of Data Analysis

The progression of sensing tools to add up to data stockpiling and heterogeneous information handling created huge deterrents for information security to approve the examination action [72]. Early frameworks included limited scope, review investigation, which frequently was handled disconnected [21]. In terms of the ability to capture data in real-time and the need for integration with a wide range of heterogeneous sources, sensor information has presented unique challenges [84]. Preeclampsia care, for example, maybe improved by the use of mobile home monitoring results, medical details, and disease indicators, such as severity and adverse medication effects [92]. In intensive care units, data from both bedside devices, tests, and electronic medical reports may be merged [70]. Researchers are progressively dealing with large databases, where patients care not only takes in the form of isolated data sources also the abstraction would start at a node stage, enabling on-node processing with silicone-based procedures [37]. The integration of various systems in combination with knowledge from general health devices offers a range of incentives and challenges [6]. Different criteria are required to effectively incorporate the various sensing data into current biomedical repositories [104]. There is an abundant yet underutilized online health database detailing procedures as well as patient outcomes [86]. Regional processing knowledge used with electronic health record-based data has also proved to be an efficient means of solving different health-care problems, such as promoting risk control programs, pharmacovigilance, and clinical trial recruiting support networks [95]. If quantitative health data is constantly replicated, it can be captured useful and rich time series to enable temporal data mining [29]. This function will assist in recognizing patient course trends across medications, diseases, and procedures [102]. Clinical testing repositories may be used to react easily to questions, such as potential medication reactions, risk factors, predictor levels, and signatures of disease [85]. Testing in several research sites is usually carried out in different laboratories can provide different datasets [41]. This problem makes a heterogeneous dataset more difficult [49].

Security on the board program might be acquainted with constantly acquiring the information needed in amazed and facilitated cycles, along these lines raising the impacts of different climate factors and seasons, and utilizing a few members simultaneously [75]. It smoothes out and accelerates applications for information preparation [68, 69]. For defining reliable biomarkers of a medication outcome with specific data sets, data mining of data from research data has been suggested [39]. Such tests will determine the collection of physiological markers of concern that a general health surveillance program would then rely on [90].

The combining of the effect of these new roles with overall health surveillance will enable the rapid gathering of the knowledge required to explain sequence variances in the human genome [74]. Genetic stratification of medical tests, in the case of vulnerable conditions such as high-risk Genetic abnormalities, milk and gluten allergy, and cystic fibrillation, EHR is the most efficient technique to avoid that [66]. Current practice takes weeks to screen genetic abnormalities and often requires a preliminary examination of the mutation [67].

According to the complications in disease conditions, recovery choices, and impacts on decision-making are also minimal [58]. In recent years, a significant interest is being felt in laboratory-on-a-chip approaches to DNA sensing [73]. The period is taken to sequence a genome that has decreased [53]. This sensing system attempts to identify nucleotide changes relevant to diseases, which may lead for example to the sensitivity to disease or the reaction to pathogens and medicines [45, 50].

At present, the findings could take longer than a month, triggering unnecessary gaps in diagnosis. The patients will apply for gene testing [47]. They may prefer unilateral operation and postpone tests [52]. The patients can subsequently found to be undergoing a high-risk mutation and a delayed contralateral mastectomy [83]. Sequencing in the future will require rapid identification of risk mutations at the same time as a diagnosis and the patients may undergo bilateral mastectomy and reconstruction simultaneously [34].

When an individual is diagnosed with a genetic mutation, that patient may be involved in finding a place in an expanded network of help networks and certain individuals [17]. The one given by data from the social network is a significant but new source of knowledge for the general healthcare system [108, 109]. Statistics from social networks also helped to explain the emergence of illnesses and good conduct from the place, actions, and time [33]. For example, social networking has analyzed the relational complexities of those who suffer from different diseases [22].

It is now used for disease and health emergencies, such as extreme acute respiratory syndrome, H1N1 influenza, hepatitis outbreaks, and most recently Ebola silenced [96]. The fast development of social network data will easily be paired with extensive health tracking, for instance assessing a patient's current health condition by their contacts with other individuals and the effects they have on their health status [57].

10.3 Implementation of Personalized Devices in the Physiological Systems

Digital models provide a significant tool for personalizing the treatment of patients in a specific state by utilizing clinical data and machine ID methods to generate so-called internet patients [40]. The secrets to interactive patients and model-based treatment are these sensitivities [93]. Establishing the interactive patient care system and model-based treatment is comprising of very difficult steps [54]. The special form of physiological deterministic models is being used [38, 39, 58, 70, 71, 83, 100, 111]. These techniques have an extended practice of biomedical sciences but in clinical trials, it has made it shorter for getting efficient data values [54]. Over the past 10 years, however, the successful design and implementation of model-based sensors or decision-making support systems have demonstrated the potential that this approach offers to ICU patients' solutions [61]. MPM data is integrated into interactive site systems, including an IMD personalization engine, to provide tailored information, training, and advice to patients with IMDs [46]. The data are processed by using online systems [38]. A system-operated questionnaire evaluates the patient's concerns by providing the most useful source of information, including references [59]. Through another element of the innovation, the patient is assessed for the intent of directing the patient to the right source of knowledge following the medical situation suggested by the IMD [35]. A personalization engine is also incorporated into IMD data for information, training, warning, or advice on an automatic assessment of the patient's data obtained from the IMD in yet another aspect of this invention [62]. IMD data is provided [31]. A network-centered multimedia framework for chronic patient care is also another dimension of innovation [27]. For a fact, comprehensive networks of different repositories of patients are incorporated into IMDs such that the interactions between patients, healthcare professionals, and other support groups and populations are smooth [68, 69].

10.3.1 Challenges During Implementation Procedures

The implementation procedures may cause cognitive dysfunctions like patients in depressive conditions may have a deficiency in verbal comprehension, memory, concentration, and management; whereas deficiencies typically are more frequent and evident in manic mode [20]. Tests should easily be directed and flexible towards depressive disorders ("state" markers), should be simple to administer, and should not take long [18]. They will need to be conveniently implemented on applications with the user interface [19]. PLT is a reactive, biased device, because anxious patients create more false alarms and perseverance, whereas stressed patients lack less and become far more sluggish [14]. Tests can be done remotely and require 7–10 min for patient evaluation [55]. Easily inserted the device into the body [12]. An analog visual sizing, the self-assigned measure is responsive to mood swings and emotional impacts. It can be used for an everyday attitude adjustment, together with other personalized features to promote diagnosis, biofeedback, and the intentions of the user, to be applied on a mobile network [10].

10.3.2 The Working Mechanism of the Device

A computer-implemented program which supports online a patient with an implanted medical device (IMD) that includes a framework consisting of a website resident on a server, an online user interface with the database, a browser that provides patient contact, a program that distinguishes the patient's client from the platform that enables the data to be retrieved [9]. The IMD is doing the recollection of data through the IMD app [37, 65, 76]. Incorporation of patient data with IMD data is being collected, and patient contact with the platform, which provides patient details based on the patient data entry and collection of IMD data from the plurality of IMDs of other patients [4]. Databases that contain psychological knowledge, information on school funding, transactional assistance, connections to certain websites, and historical records are given [99]. An IMD server is distinct from the other patient's server. It enables the processing of data gathered within IMD at least one external data access network gateway, consisting of a patient web interface to a database, to capture specific patient records, implied patient information, and transactional data [51] The server includes a personalization engine which provides patient-specific information, based on databases and interfaces [45, 47, 50, 51, 79, 82, 88, 89, 108, 109], where the databases store psychology, education, transactional support, web referrals, and historical data [89]. The personalized online patient support with IMD is a component diagnostic component for individual patient's diagnoses and treatment [100].

The process of delivering electronic assistance for an individual with an embedded medical instrument (IMD) includes: downloading documented details from an IMD via an input device; transmitting documented data from the input system to the repository through the network; capturing case-based tacit and explicit details; a customized system hypothesis focused on whether there is a concordance or disagreement between the explicit information and IMD data reported; and the patient's suggestion [89]. A personalization engine is designed to combine indirectly documented data and explicit details from the patient to shape a personalization engine assumption and to produce a suggestion for assistance [9]. The customization engine further collaborates the transactional information with the recorded data, implicit patient data, and explicit data for the conclusion of the customizing engine [4].

10.3.3 Monitoring System

Synergies in many fields such as biomedical innovations, micro and nanotechnology, materials processing, and ICT enable new solutions to personal safety and wellbeing to be encouraged [78]. Those provide, among other aspects, unobtrusive patient safety screening and preventive education monitoring for illness avoidance and early detection, and medication follow-up [30]. This report examines the emerging developments and obstacles in science which is in line with certain strategies in the area of wearable health systems [71]. The monitoring system of the personalized device is difficult sometimes [68, 69]. However, by using advanced technology we can easily overcome this situation. For example, Poor adherence to TB therapy impedes the recovery of a person and jeopardizes public health [11]. The norm in care now is explicitly applied (DOT) therapy; nevertheless, the high expense in maintenance restricts its effectiveness and fulfills the need for more efficient ways of verifying commitment [87]. Technologies including visual tracking and time-recording to open pill bottles cannot validate the actual intake of drugs [48, 94]. An ingestible sensor and a connected tracker on the body represent an innovative solution established by Proteus Digital Health, Inc.; they together electronically validate specific ingestions and monitor the date/time of the ingestion [60]. The device may accurately classify reasonably sensitive ingestible sensors that present a small risk to the patient and can be widely tolerated by patients [15]. The machine will validate dose-by-dose conformity with drug appropriate therapy [65]. The device will enable wirelessly monitored therapy (WOT) in conjunction with mobile technologies for tracking TB as an alternative to DOT [43].

10.4 Data Security for the Patient

An implantable tool that gathers and aggregates data from an external patient's body from unplanted medical equipment is identified. Data from medical equipment in the body may also be collected and aggregated [79]. The implantable device contains a remote handset for physiological information obtained from outer clinical gadgets and physiologic information stockpiling media [108, 109]. A processor gathers the physiological information and gives them to a distant administration framework [79]. The gadget may, for longer periods, gather and save physiological information from the different outside information hotspots for resulting chronicles [106]. The implantable tool may also gather diagnostic data from certain embedded medical instruments [72]. Throughout this way, the system represents a crucial point to capture and analyze patient physiological details [21]. Typically speaking, the technology-focused on an implantable data collection system that gathers and aggregates medical data from different sources for an individual [84]. The system extracts, for example, biology data from non-implanted diagnostic instruments beyond the patient's body [88]. A system captures, maybe over a lengthy time, physiological data from various external sources and retains the data [77]. The implantable system can gather physiological information from other implanted medical devices [88]. The app provides a crucial location for capturing and aggregating patient-related physiological details [92]. The technology is oriented towards a method that involves the collection, within a system implanted within the patient's body, of physiological data from a plurality of healthcare devices outside the patient's bodies [70]. The method consists of another way the innovation is geared into a process requiring the collection of clinical data from a diagnostic system outside the patient's body and the storing of data within a device inserted into the patient's body [37].

In another manner, the innovation is applied to an implantable instrument comprising of a remote handset to gather physiological information from an indicative framework outside a patient's body and a physiological information stockpiling medium [6]. The techniques may have one or more art advances [7]. An intentionally suitable IDAD may be used for continuously gathering and aggregating physiological data, irrespective of the patient's role, over a prolonged period [104]. The IDAD assembles details from certain clinical instruments, be that as it may, to give the patient a particular physiologic profile [86]. The IDAD may send the collected physiological information to a focal framework for patient care, the executives for clinician access [29]. To view the physiological data collected from numerous medical devices, external medical devices, additional implanted medical instruments the clinician may access patient management [102]. The techniques described here can thus be used to gather extensive physiological data for the patient more efficiently and to present that data to a clinician by a common system [85]. This can help physicians to better recognize the patient's current condition and to make the patient's diagnosis and care more effective [41].

10.4.1 Health Records in Electronic Forms and Health Information Systems

A personalized health information system includes a personalized source of health content, a source of entertainment information, a way of composing communication with entertainment sources and personalized health contents, and a means of communication with the means of composing [75]. The approach provides a mix of entertainment and customized health care materials [68, 69].

The source of entertainment content deals through an application for material, which enables the consumer to access particular information, which is to suggest, to "download" the material from the source of entertainment information [90]. In response to a patient's request for entertainment content, the entertainment content is provided to the composing device [74]. A health information collection allows that interacting with a range of sources becomes the basis of customized health material [67]. The selection of health content means that individual health content is generated according to inputs [73].

The collection of inputs comprises a specific output and inputs on the general safety knowledge base for the selection of material [53]. The content collection covers physical treatment, a wellness description, an instructional counseling program, a social profile, and a patient material history [25]. A care program means that a patient's medical scheme and well-being status should produce an informative intervention schedule [58]. The medical system specifies health parameters and the medical profile specifies the patient's actual health parameters [45]. The technique for treatment consists of aligning targets with real conditions [52]. The care program lays forth the criteria of the patient's academic target [83]. The criteria of the medical target calculate the significance of the educational goals for the individual patient [34]. The psychological profile includes data that characterize the patient's likes, dislikes, and motivators [17]. The process of construction generates a combination according to patient criticality [96]. The criticality of the condition tests the enforcement of the care program [57]. By contrasting the care plan and the safety record, criticality implies ensuring patient safety [38].

The composite is a structural component, ideally, and consists of a Hybrid website with both wellness and entertainment materials [61]. The Composite Page includes a section on entertainment and a section on health which includes health material [59]. The advertising generator provides an initial website with an entertainment segment and an unwelcome portion [35]. The layout implies the unhealthy section is substituted by a safe section and therefore a composite page is created [62]. The new web interface is essentially similar to the original website model [31]. A spatial compound includes ideally a hypertext markup (HTML) code [27]. The composite is temporary as an alternate method for television schooling [68, 69]. The hybrid show consists of a first and second image, with the one entertainment material in the first image and the other safety material in the second picture [20]. The first picture comprises primarily of the material and the second picture of the information is significant [18]. The framework consists of the view media application module, the server module, and a central network for communicating with a computer [19]. A Television set with a multimedia processor is ideally included in the setup [14]. The Web and cable tv transmission cables are acceptable satellite networks [55]. The section of the platform ideally involves writing means and origins of entertainment and safety material [10]. When dealing with several clientele subsystems, the application subsystem becomes a separate patient; for each client subsystem, the details are configured separately [9]. This innovation also offers a means of supplying an individual with clinical records [4]. The approach involves the measures to create and show the combination of individual healthcare and patient care material [99].

10.4.2 Security and Privacy

Developing, evolving healthcare structures around the world are facing changes in the institutional, legislation, and conceptual frameworks, and the mechanisms and responsibilities of stakeholders involved in providing healthcare services [5]. Within this sense, protection, privacy, and trust are key concerns to be handled properly [16, 30]. Fresh methods are necessary for the study and construction of future-proof health systems [68, 69]. A structure theory-based design model that describes every individual in the program within its particular sense explicitly reflects the current architecture of health systems. The innovative and tailored approach to health care provides strong rewards along with tremendous challenges for performers and patients in the first place [11]. Security, protection, and privacy, in particular, should be stated in this sense [87]. The configuration and actions of the procedure and, therefore, the laws which govern the procedures must be regulated by the key participant, the healthcare provider in the first place, and now the patient. The structure that would emerge will be very versatile and governed by legislation, in which patient discretion is paramount [94]. The clinical care provided under these cases describes the role of the individual, his or her views, desires, and preferences, as well as his personal and environmental background [48]. Therefore, the patient removes the accountable mechanism and circumstances provider from the field of treatment [60]. Concerning the current position of patients, the requisite specific procedures, structure, and tools must be introduced to allow the patient to perform that function [15].

10.4.3 Secure Transmission

Big data include a broader range of media data (organized, semi-structured, unstructured), unforeseeable (almost) moving through real-time data of several various channels (traditional source, Web server logs, and click-stream data, accounts of the social network operation, data from cell phone contact information, wearable technology, etc.) and systems (organized, partly structured, unstructured, unstructured) [65]. This also needs a comprehensive data management system, data importance commitment, and reliability, safety, and confidence issues solving [43]. Data review is an investigation method in which data are collected, lined up, translated, and interpreted to identify secret motifs, uneven connections, and other valuable details. As web systems are a source of cybercrime assault due to the absence of sufficient protection and privacy protections on the various pages, there is a huge amount of data for data analytics [79]. The accessibility and trust of implemented data and systems are needed in data and analytics [108, 109].

Failure to comply with privacy standards and policies undermines the trust that is important in the delivery of data [79]. Centered on context-sensitive governance, context-aware usage of personal information will rebuild confidence by creating an adequate environment of personal data [106]. The level of openness, responsibility, and representation of people in the current sense is also right [72].

Thus, uniquely specified, scalable, and complex policies and machine-processing policies are needed [21]. Specific regulations will be established and related to the regulatory and corporate policy of service suppliers and agencies to meet compliance criteria in an unbiased manner [84]. If access to information cannot be managed individually, data identification, and the proper management and authentication of

IDs must be given. In this respect, it is possible to implement user-centered or federated identity management arrangements, including single sign-on (SSO) capabilities [8]. For the safety of large data, the following protections are crucial: a) surveillance, b) review and auditing of illegal operations, c) the review and auditing of all confidential details, d) previously stated identity authentication, e) masking of data and f) encryption apps. Nevertheless, not only operational and technological challenges arise, but philosophical, psychological, and social concerns are often present [92]. Both confidentiality and data masking may be a legal and safety danger for digital data [70]. Users can be shielded from themselves because the dangers of inappropriate consumer activity in social networking are often overlooked [37]. Health environments based on the institution and procedure are controlled and linked to a particular context [6]. Security and privacy tests are established and function quite well in these environments [7]. The fundamental principles and guidelines implemented in healthcare system regulation include, for example, publicly identifying patient records, restrictions on access, data minimization and privacy approval, and its usage [95]. These will not operate in (partly) uncontrolled areas of potential health however and especially struggle inaccessible environments such as social networking [29]. Individual practices and legal as well as equal knowledge standards will account for the absence of regulations [102]. The Code of Ethics for Professionals in Health Care, developed by Kluge under the IMIA, is an illustration of (1) autonomy and reverence for the individual, (2) exclusion from the right of enforcement of the rights, (3) exclusion of specific differentials between right and performance (4) responsibility for the appropriate practice, (5) ensuring that the right to exercises is exercised. (4) The Equal Data Standards were also suggested, such as (1) transparency and visibility, (2) limitations on data collection, data dissemination, and the usage of information, (3) confidentiality, and (4) regulation of access [85]. A separate package called equal knowledge procedure, which included openness, person oversight, regard for the history, protection, access, quality, centered compilation, and accountability, was introduced by the United States Federal Trade Commission [41]. The increasingly complicated and transparent world on the one side and the growing lack of restrictions on procedures, actors and functions, etc. on the other, allow (1) several policy areas to be handled, (2) fluid policy management to be carried out, (3) tasks and obligations shift to involve growing agent and program control and (4) protection improvements to the product [76]. Last but not least, they will grow into professionals who handle protection through individuals, systems, and technology, to secure knowledge to the business goals of the enterprise, automate complex business processes and perform lasting risk management assessments in that sense [49]. Because data security, apps, and devices are unavoidable over their lifespan, remote services are often needed in mobile environments to remotely erase (wipe) data as well as inactivate and disruptive apps [75]. The latter model shifts contribute to (1) decentralization of production procedures, (2) decentralization and business alignment of organizational and defense decentralization, and privacy protection, and (3) digital decision-making assistance, (4) corporate intelligence, and (5) middleware systems [39]. As a consequence, protection and privacy solutions must be (1) structured safe, (2) distinguishing security and privacy programs from software, (3) security and privacy, and (4) supporting patients [90].

10.5 Regulatory Aspect for Commercialization

The problem of PM R&D and execution was often spoken about with regulatory instability [74]. Personalized medication (PM) targets bridling the surge of "omics" discoveries to advance investigation and improvement on custom-made diagnostics techniques just as to advance medical services offices' viability through recognizing and dealing with a disease or turmoil inclinations by and by [50]. While considerable speculations have been made, there have been little gains in promoting customized meds [45]. In the fields of PM development [12, 14, 66, 78, 106], interpretation, execution, and clinical treatment, we distinguish the primary administrative, licensed innovation, and repayment challenges apparent [58].

Related to different subjects, for example, discount [17], guideline of clinical preliminary; guideline of co-advancement; hazy proof-based prerequisites; unsuitable motivations for innovative work; contrary data frameworks; and a contrasting guideline of diagnostics appears to have been a cross-cutting issue [50]. The advantages of PM to health systems and patients, the problems in terms of legislative, intellectual property, and payment will be tackled following technological advancement [45].

The healthcare agencies and services are generating new PM legislation, policy materials, and relevant legislation [97]. There's a general perception that the legislation is now in advance both as regards the routine improvements during the medication life cycle or the growth of the diagnosis [31, 40, 57, 90, 99], and the versatility in which the unique PM products and services may be expanded [50]. Changes in the regulations became too quick for certain stakeholders while uncleanness in the regulations involved generated uncertainty for others [45].

Despite these issues, it has been primarily at the expense of public health to relax the proof standards for FDA clearance [97]. The Act further encourages a study of biomarkers and other tools that lead to the production of new pharmaceutical drugs [47]. This offers administrative guidance on the optimal way for hybrid drug approvals [108, 109]. To reduce the pressure on patients for licenses, the FDA has directed to issue guidelines on innovative or flexible prototypes for clinical trials [33]. The deteriorating effects on PM R&D and execution through the wider regulatory changes often contribute to instability by the more regulatory changes [13].

The writers discuss the value chain and the obstacles posed by nanotechnologies-based implantable biomedical products [22]. Through this way, researchers develop a map for the cycle from practical science to final product adoption and promotion, so they can measure the social cost of return from an experimental study [54].

10.5.1 Legislative Policies in the Sensor-Based Personalized Healthcare System

The Health Insurance Portability and Accountability Act

1) To ensure that the individual wellbeing records gathered, got, or handled clung to a progression of security orders, the Health Insurance Portability and Transparency Act (HIPAA) from 1996 had been passed [139]. These insurance rules characterize steps to all the more likely secure all electronic wellbeing records [140]. The laws, nonetheless, just apply to freely supported organizations and the security arrangements for businesses and medical services suppliers.
2) The Genetic Information Non-Discrimination Act (GINA), 2005, clarified straight forward boycott of the oppression and wellbeing backup plans on the grounds of their innate danger factors, in this manner filling a few weaknesses in HIPAA security assurance [141]. Under the Act of President George W, marked May 21, 2008 [142]. Shrubbery, GINA, has conquered critical difficulties to the improvement of customized prescriptions. The limitation of the utilization of hereditary data by wellbeing back up plans and businesses implies that protection issues concerning Health IT (HIT) are just as related worries about biobanking [143]. For partners in science, the GINA entry was additionally huge [144]. The probability of hereditary disparity debilitates headway in PM by hampering key examinations in the hereditary features of sickness: Employers and wellbeing suppliers have fears that the discoveries of hereditary testing may be available; this may deter people from moving toward the specialist for these tests and may neglect to recognize and foresee illnesses [145]. The worry was that hereditary information gathered from research study investment could be in the possession of bosses and medical care suppliers and might deter people from selecting hereditary exploration.
3) The Affordable Care Act (ACA) of 2010 sets up the ensured issue, implying that guarantors offering protection in either the gathering or individual market should give inclusion to all people who demand it. The law precludes backers of medical coverage from oppressing patients with hereditary illnesses by denying inclusion in light of "previous conditions." ACA offers extra assurances for patients with hereditary sicknesses by building up that specific medical coverage guarantors may just fluctuate charges dependent on a couple of determined elements, for example, age or geographic zone, subsequently disallowing the change of expenses as a result of ailments.
4) The American Disability Act (ADA) restricts segregation dependent on positions, public consideration, lodging, and incapacity correspondence [146]. In 1995, the Equal Employment Opportunity Commission (EEOC) delivered an agreement that ADA prohibits segregation dependent on hereditary data connected to incapacity, sickness, and different issues [147]. While hereditary protection strategy is changing to address patients' needs, the guideline will make assembling and auditing total clinical information harder for making new customized care and diagnostics [148] . For tolerant administrations to keep

on advancing, suspicions around the mystery and the need to advance science should be very much coordinated [149].

10.5.2 Government Agencies Shaping Personalized Medicine

Educating the future regarding customized medication, four associations in the USA and Europe will assume an essential part in 9 [150].

10.5.3 Habitats for Medicare and Medicaid Services (CMS)

CMS may have a significant impact as the biggest medical services supplier in the USA on the formation of focused findings and helpful items and the execution of a proactive medical care model that features medical care, health, and sickness avoidance. Most private guarantors have executed installment models set up by the CMS [151]. CMS will move towards an additional outcomes-based installment worldview that could spike the creation of custom-fitted demonstrative and helpful applications [152].

10.5.4 Food and Drug Administration (FDA)

As the government authority liable for medication and diagnostics leeway and control, the FDA faces a huge test in building up a basic and attainable course to new, custom-made diagnostics, therapeutics, and theragnostic frameworks [153]. Endeavors, for example, the Vital Road Project make ready for business organizations to address the issue and drive progression in individual medication [154]. FDA may likewise accelerate advancement with the assistance of contingent endorsements, empowering more modest and less expensive drugs, clinical preliminaries to be completed, utilizing individual cell phones (cell phones) to follow understanding consistency and results so that issues can be resolved and quality and wellbeing permitted.

10.5.5 Public Institutes of Health (NIH)

The US clinical examination office NIH has an NIH Roadmap for Medical Research, which incorporates research upholds for gadget science, genomics among proteomics, and different parts of customized medication. The NIH is the general wellbeing boss [155]. The NIH will grow its investigation into biomarkers of ailments, custom-made diagnostics, and treatments, supported by $10 billion from financial boosts. The NIH likewise delivers its clinical and translational exploration grants (CTSA) program, an interconnected organization of driving AMCs. These foundations are most likely at the cutting edge of the study of customized medication examines [156].

10.5.6 European Medicines Agency (EMA)

The European Medicines Agency (EMA) is the European Union's administering body that advances wellbeing and security in the public area and has administrative endorsement and oversight of new analyses and therapeutics [157]. It is around equivalent to the US FDA however it is a decentralized association [158]. Through its endorsement cycle, EMA can speed the spread of customized medication, which needs just one solicitation for the endorsement or analytic affirmation of medication across all EU nations. One of the vital objectives of the EMA is to furnish patients with protected and successful meds. Better medications should come available quickly to be tried with new procedures [159]. A vital point of the EMA Road Map 2010 is to advance drug exploration and development in the European Union [159] (Abugabah 2020). Therefore, an "Imaginative Drug Production EMA/CHMP thinking taken gathering" has been set up. The gathering comprises EMA staff and a few individuals from the agency's various logical/working gatherings filling in as an inner center gathering. These specialists target finding logical difficulties both in exploration and creation in the business and the scholarly world to making new drugs.

The long curve in clinical history has advanced from the otherworldly to the physiological, organic, and ultimately the atomic phases of indicative limit [160]. Presently, since infections can be sub-isolated into classes that anticipate the sickness course and its likely response to therapy utilizing information a long way past what is clear, there is a commitment to react to that information [161]. The advanced model for drug disclosure and the act of medication is Personalized Medicine [162]. While the upsides of individualized medication may incorporate the creation of protected and more suitable drugs for specific populaces of sicknesses, such advantages would be far-fetched before such hindrances have been wiped out to their usage [163]. Impediments in open strategy incorporate hazy administrative guidelines, lacking demonstrative test repayment, inadequate guideline protecting against hereditary separation, the nonattendance of current data innovation medical care, and of clinical schooling structure that has not prepared specialists to coordinate them into the medical care framework. Seeing all these large powers – from obstacles to benefits – is a stage in how instruments can be utilized to influence the turn of events and improvement of customized medication.

10.6 Conclusions and Future Perspectives

This part presents an investigation of wearable gadget answers for modified checking and improvements in undertaking the executives. A full audit of wearable gadget credits and the subsequent security intercessions are given that can figure the well-being productivity of current medical care rehearses. The examination found that in different areas a wide scope of wearable gadgets are utilized to improve security and productiveness, albeit little use in the structure business has been accounted for [80]. Realizing that the various enterprises with high utilization of wearable innovation are not a high-hazard modern field, for example, fabricating, the circumstance about the utilization of wearable innovation in development needs earnestly to improve. Building proprietors and specialists have taken on these new specialized developments to essentially expand the security effectiveness of medical care offices. Such countless examinations distinguished potential employments of wearable innovations in the catch and following of different measurements liable for building destinations for rehashed mishaps and passings. The investigation completed in different examinations shows that a wide scope of wellbeing metricians in the structure business can likewise be assessed.

References

1. Abdollahi, M., Ashouri, S., Abedi, M., Azadeh-Fard, N., Parnianpour, M., Khalaf, K., Rashedi, E.: Using a motion sensor to categorize nonspecific low back pain patients: a machine learning approach. Sensors (Switzerland) **20**(12), 1–16 (2020). https://doi.org/10.3390/s20123600
2. Acevedo, M., Varleta, P., Kramer, V., Quiroga, T., Prieto, C., Parada, J., Adasme, M., Briones, L., Navarrete, C.: Niveles de fosfolipasa A2 asociada a lipoproteína en sujetos sin enfermedad coronaria con riesgo cardiovascular variable. Rev. Med. Chil. **141**(11), 1382–1388 (2013). https://doi.org/10.4067/S0034-98872013001100003
3. Adly, A.S., Adly, A.S., Adly, M.S.: Approaches based on artificial intelligence and the internet of intelligent things to prevent the spread of COVID-19: scoping review. J. Med. Internet Res. **22**(8), 1–15 (2020). https://doi.org/10.2196/19104
4. Aguado, B.A., Grim, J.C., Rosales, A.M., Watson-Capps, J.J., Anseth, K.S.: Engineering precision biomaterials for personalized medicine. Sci. Transl. Med. **10**(424) (2018). https://doi.org/10.1126/scitranslmed.aam8645
5. Alam, M.M., Malik, H., Khan, M.I., Pardy, T., Kuusik, A., Le Moullec, Y.: A survey on the roles of communication technologies in IoT-based personalized healthcare applications. IEEE Access **6**, 36611–36631 (2018). https://doi.org/10.1109/ACCESS.2018.2853148
6. Alfian, G., Syafrudin, M., Ijaz, M.F., Syaekhoni, M.A., Fitriyani, N.L., Rhee, J.: A personalized healthcare monitoring system for diabetic patients by utilizing BLE-based sensors and real-time data processing. Sensors (Switzerland) **18**(7) (2018). https://doi.org/10.3390/s18072183
7. Amar, A.B., Kouki, A.B., Cao, H.: Power approaches for implantable medical devices. Sensors (Switzerland) **15**(11), 28889–28914 (2015). https://doi.org/10.3390/s151128889
8. Wasalathanthri, D., Rehmann, M., Song, Y., Gu, Y., Mi, L., Shao, C., Chemmalil, L., Lee, J., Ghose, S., Borys, M., Ding, J., Li, Z.: Technology outlook for real-time quality attribute and process parameter monitoring in biopharmaceutical development—A review. Biotechnol. Bioeng. **117**(10), 3182–3198 (2020)

9. Aquino, R.P., Barile, S., Grasso, A., Saviano, M.: Envisioning smart and sustainable healthcare: 3D printing technologies for personalized medication. Futures **103**, 35–50 (2018). https://doi.org/10.1016/j.futures.2018.03.002
10. Awolusi, I., Marks, E., Hallowell, M.: Wearable technology for personalized construction safety monitoring and trending: review of applicable devices. Autom. Constr. **85**(2016), 96–106 (2018). https://doi.org/10.1016/j.autcon.2017.10.010
11. Adams, S.J., Henderson, R.D.E., Yi, X., Babyn, P.: Artificial intelligence solutions for analysis of X-ray images. Can. Assoc. Radiol. J. **72**(1), 60–72 (2021). https://doi.org/10.1177/0846537120941671
12. Bariya, M., Shahpar, Z., Park, H., Sun, J., Jung, Y., Gao, W., Nyein, H.Y.Y., Liaw, T.S., Tai, L.C., Ngo, Q.P., Chao, M., Zhao, Y., Hettick, M., Cho, G., Javey, A.: Roll-to-roll gravure printed electrochemical sensors for wearable and medical devices. ACS Nano **12**(7), 6978–6987 (2018). https://doi.org/10.1021/acsnano.8b02505
13. Beckmann, S., Lahmer, S., Markgraf, M., Meindl, O., Rauscher, J., Regal, C., Gimpel, H., Bauer, B.: Generic sensor framework enabling personalized healthcare. In: 2017 IEEE Life Sciences Conference (LSC), pp. 83–86 (2017). https://doi.org/10.1109/LSC.2017.8268149
14. Blobel, B., Lopez, D.M., Gonzalez, C.: Patient privacy and security concerns on big data for personalized medicine. Health Technol. **6**(1), 75–81 (2016). https://doi.org/10.1007/s12553-016-0127-5
15. Campbell, M.R.: Update on molecular companion diagnostics - a future in personalized medicine beyond Sanger sequencing. Expert Rev. Mol. Diagn. **20**(6), 637–644 (2020). https://doi.org/10.1080/14737159.2020.1743177
16. Catherwood, P.A., Steele, D., Little, M., McComb, S., McLaughlin, J.: A community-based iot personalized wireless healthcare solution trial. IEEE J. Transl. Eng. Health Med. **6**(February), 1–13 (2018). https://doi.org/10.1109/JTEHM.2018.2822302
17. Chen, M., Yang, J., Zhou, J., Hao, Y., Zhang, J., Youn, C.: 5G-smart diabetes: toward personalized diabetes diagnosis with healthcare big data clouds. IEEE Commun. Mag. **56**(4), 16–23 (2018). https://doi.org/10.1109/MCOM.2018.1700788
18. Cheng, G.Z., Folch, E., Wilson, A., Brik, R., Garcia, N., Estepar, R.S.J., Onieva, J.O., Gangadharan, S., Majid, A.: 3D printing and personalized airway stents. Pulm. Ther. **3**(1), 59–66 (2017). https://doi.org/10.1007/s41030-016-0026 y
19. Choonara, Y.E., Du Toit, L.C., Kumar, P., Kondiah, P.P.D., Pillay, V.: 3D-printing and the effect on medical costs: a new era? Expert Rev. Pharmacoecon. Outcomes Res. **16**(1), 23–32 (2016). https://doi.org/10.1586/14737167.2016.1138860
20. Chung, K., Park, R.C.: Chatbot-based heathcare service with a knowledge base for cloud computing. Cluster Comput. **22**, 1925–1937 (2019). https://doi.org/10.1007/s10586-018-2334-5
21. Cirillo, D., Valencia, A.: Big data analytics for personalized medicine. Curr. Opin. Biotechnol. **58**, 161–167 (2019). https://doi.org/10.1016/j.copbio.2019.03.004
22. Clements, E.D., Roane, B.M., Alshabrawy, H., Gopalakrishnan, A., Balaji, S.: System for monitoring user engagement with personalized medical devices to improve use and health outcomes. In: 2019 41st Annual International Conference of the IEEE Engineering in Medicine and Biology Society (EMBC), pp. 4301–4305 (2019). https://doi.org/10.1109/EMBC.2019.8856859
23. Clifton, L., Clifton, D.A., Pimentel, M.A.F., Watkinson, P.J., Tarassenko, L.: Gaussian processes for personalized e-health monitoring with wearable sensors. IEEE Trans. Biomed. Eng. **60**(1), 193–197 (2013). https://doi.org/10.1109/TBME.2012.2208459
24. Coccia, M.: Deep learning technology for improving cancer care in society: new directions in cancer imaging driven by artificial intelligence. Technol. Soc. **60**, 101198 (2020). https://doi.org/10.1016/j.techsoc.2019.101198
25. Crosson, F. J.: An overview of the medical device industry. In: Report to Congress: Medicare and the Health Care Delivery System, pp. 207–242 (2017). https://www.medpac.gov/docs/default-source/reports/jun17_ch7.pdf?sfvrsn=0

26. Dahele, M., Tol, J.P., Vergeer, M.R., Jansen, F., Lissenberg-Witte, B.I., Leemans, C.R., Slotman, B.J., Verdonck-de Leeuw, I.M., Verbakel, W.F.A.R.: Is the introduction of more advanced radiotherapy techniques for locally-advanced head and neck cancer associated with improved quality of life and reduced symptom burden? Radiother. Oncol. **151**, 298–303 (2020). https://doi.org/10.1016/j.radonc.2020.08.026
27. Day, B.: Personalized blood flow restriction therapy: how, when and where can it accelerate rehabilitation after surgery? Arthroscopy - J. Arthrosc. Relat. Surg. **34**(8), 2511–2513 (2018). https://doi.org/10.1016/j.arthro.2018.06.022
28. Dev, A., Khanra, S., Shah, N. Advanced technologies in the modern era for augmented patient health care and drug delivery. J. Drug Deliv. Ther. **10**(1), 147–152 (2020). https://doi.org/10.22270/jddt.v10i1.3838
29. Di Prima, M., Coburn, J., Hwang, D., Kelly, J., Khairuzzaman, A., Ricles, L.: Additively manufactured medical products – the FDA perspective. 3D Print. Med. **2**(1), 4–9 (2016). https://doi.org/10.1186/s41205-016-0005-9
30. Di Sarsina, P.R., Tassinari, M.: Person-centred healthcare and medicine paradigm: it's time to clarify. EPMA J. **6**(1), 1–6 (2015). https://doi.org/10.1186/s13167-015-0033-3
31. Dodson, B.P., Levine, A.D.: Challenges in the translation and commercialization of cell therapies. BMC Biotechnol. **15**(1), 1–15 (2015). https://doi.org/10.1186/s12896-015-0190-4
32. Dong, Q., Li, B., Downen, R.S., Tran, N., Chorvinsky, E., Pillai, D.K., Zaghlou, M.E., Li, Z.: A cloud-connected NO2 and Ozone sensor system for personalized pediatric asthma research and management. IEEE Sens. J. **XX**(2), 1 (2020). https://doi.org/10.1109/jsen.2020.3009911
33. Dridi, A., Sassi, S., Faiz, S.: A smart IoT platform for personalized healthcare monitoring using semantic technologies. In: 2017 IEEE 29th International Conference on Tools with Artificial Intelligence (ICTAI), pp. 1198–1203 (2017). https://doi.org/10.1109/ICTAI.2017.00182
34. Feng, K., Leary, R.H.: Personalized medicine in digital innovation. Int. J. Pharmacokinet. **3**(4), 103–106 (2018). https://doi.org/10.4155/ipk-2018-0006
35. Firouzi, F., Rahmani, A.M., Mankodiya, K., Badaroglu, M., Merrett, G.V., Wong, P., Farahani, B.: Internet-of-things and big data for smarter healthcare: from device to architecture, applications and analytics. Future Gener. Comput. Syst. **78**, 583–586 (2018). https://doi.org/10.1016/j.future.2017.09.016
36. Garzón, V., Bustos, R.H., Pinacho, D.G.: Personalized medicine for antibiotics: the role of nanobiosensors in therapeutic drug monitoring. J. Pers. Med. **10**(4), 1–34 (2020). https://doi.org/10.3390/jpm10040147
37. Gu, J., Huang, R., Jiang, L., Qiao, G., Du, X., Guizani, M.: A fog computing solution for context-based privacy leakage detection for android healthcare devices. Sensors (Switzerland) **19**(5), 1–19 (2019). https://doi.org/10.3390/s19051184
38. Guan, A., Hamilton, P., Wang, Y., Gorbet, M., Li, Z., Phillips, K.S.: Medical devices on chips. Nat. Biomed. Eng. **1**(3), 1–10 (2017). https://doi.org/10.1038/s41551-017-0045
39. Guk, K., Han, G., Lim, J., Jeong, K., Kang, T., Lim, E.K., Jung, J.: Evolution of wearable devices with real-time disease monitoring for personalized healthcare. Nanomaterials **9**(6), 1–23 (2019). https://doi.org/10.3390/nano9060813
40. Guo, J.: Smartphone-powered electrochemical biosensing dongle for emerging medical IoTs application. IEEE Trans. Industr. Inf. **14**(6), 2592–2597 (2018). https://doi.org/10.1109/TII.2017.2777145
41. Gupta, S., Sharma, A., Verma, R.S.: Polymers in biosensor devices for cardiovascular applications. Curr. Opinion Biomed. Eng. **13**, 69–75 (2020). https://doi.org/10.1016/j.cobme.2019.10.002
42. Henman, P.: Improving public services using artificial intelligence: possibilities, pitfalls, governance. Asia Pac. J. Public Adm. **42**(4), 209–221 (2020). https://doi.org/10.1080/23276665.2020.1816188
43. Ho, D., Quake, S.R., McCabe, E.R.B., Chng, W.J., Chow, E.K., Ding, X., Gelb, B.D., Ginsburg, G.S., Hassenstab, J., Ho, C.M., Mobley, W.C., Nolan, G.P., Rosen, S.T., Tan, P., Yen, Y., Zarrinpar, A.: Enabling technologies for personalized and precision medicine. Trends Biotechnol. **38**(5), 497–518 (2020). https://doi.org/10.1016/j.tibtech.2019.12.021

44. Hu, M., Ge, X., Chen, X., Mao, W., Qian, X., Yuan, W.E.: Micro/nanorobot: a promising targeted drug delivery system. Pharmaceutics **12**(7), 1–18 (2020). https://doi.org/10.3390/pharmaceutics12070665
45. Huang, L., Wang, L., He, J., Zhao, J., Zhong, D., Yang, G., Guo, T., Yan, X., Zhang, L., Li, D., Cao, T., Li, X.: Tracheal suspension by using 3-dimensional printed personalized scaffold in a patient with tracheomalacia. J. Thoracic Disease **8**(11), 3323–3328 (2016). https://doi.org/10.21037/jtd.2016.10.53
46. Hussain, S., Kang, B.H., Lee, S.: A Wearable Device-Based Personalized Big Data Analysis Model. Lecture Notes in Computer Science (Including Subseries Lecture Notes in Artificial Intelligence and Lecture Notes in Bioinformatics), vol. 8867, pp. 236–242 (2014). https://doi.org/10.1007/978-3-319-13102-3_39
47. Jagadeeswari, V., Subramaniyaswamy, V., Logesh, R., Vijayakumar, V.: A study on medical Internet of Things and Big Data in personalized healthcare system. Health Inf. Sci. Syst. **6**(1), 1–20 (2018). https://doi.org/10.1007/s13755-018-0049-x
48. Jiang, H., Fu, J., Li, M., Wang, S., Zhuang, B., Sun, H., Ge, C., Feng, B., Jin, Y.: 3D-printed wearable personalized orthodontic retainers for sustained release of clonidine hydrochloride. AAPS PharmSciTech **20**(7), 260 (2019). https://doi.org/10.1208/s12249-019-1460-6
49. Jørgensen, J.T.: Twenty years with personalized medicine: past, present, and future of individualized pharmacotherapy. Oncologist **24**(7), 432–440 (2019). https://doi.org/10.1634/theoncologist.2019-0054
50. Kalogiannis, S., Deltouzos, K., Zacharaki, E.I., Vasilakis, A., Moustakas, K., Ellul, J., Megalooikonomou, V.: Integrating an openEHR-based personalized virtual model for the ageing population within HBase 08 Information and Computing Sciences 0806 Information Systems 11 Medical and Health Sciences 1117 Public Health and Health Services. BMC Med. Inform. Decis. Mak. **19**(1), 1–15 (2019). https://doi.org/10.1186/s12911-019-0745-8
51. Kaushik, A., Jayant, R.D., Nair, M.: Advances in personalized nanotherapeutics. In: Advances in Personalized Nanotherapeutics (2017). https://doi.org/10.1007/978-3-319-63633-7
52. Kennedy, M.J.: Personalized medicines–are pharmacists ready for the challenge? Integr. Pharmacy Res. and Practice **7**, 113–123 (2018). https://doi.org/10.2147/iprp.s133083
53. Knowles, L., Luth, W., Bubela, T.: Paving the road to personalized medicine: recommendations on regulatory, intellectual property and reimbursement challenges. J. Law Biosci. **4**(3), 453–506 (2017). https://doi.org/10.1093/jlb/lsx030
54. Korzun, D., Meigal, A.: Multi-source data sensing in mobile personalized healthcare systems: semantic linking and data mining. In: 2019 24th Conference of Open Innovations Association (FRUCT), pp. 187–192 (2019). https://doi.org/10.23919/FRUCT.2019.8711950
55. Kozitsina, A.N., Svalova, T.S., Malysheva, N.N., Okhokhonin, A.V., Vidrevich, M.B., Brainina, K.Z.: Sensors based on bio and biomimetic receptors in medical diagnostic, environment, and food analysis. Biosensors **8**(2), 1–34 (2018). https://doi.org/10.3390/bios8020035
56. Krittanawong, C., Rogers, A.J., Johnson, K.W., Wang, Z., Turakhia, M.P., Halperin, J.L., Narayan, S.M.: Integration of novel monitoring devices with machine learning technology for scalable cardiovascular management. Nat. Rev. Cardiol. (2020). https://doi.org/10.1038/s41569-020-00445-9
57. Kuhlmann, J., Halvorsen, T.: Precision medicine: integrating medical images, design tools and 3D printing to create personalized medical solutions. In: 2018 IEEE International Symposium on Medical Measurements and Applications (MeMeA), pp. 1–5 (2018). https://doi.org/10.1109/MeMeA.2018.8438798
58. Lee, Y., Lee, C.H.: Augmented reality for personalized nanomedicines. Biotechnol. Adv. **36**(1), 335–343 (2018). https://doi.org/10.1016/j.biotechadv.2017.12.008
59. Lewy, H., Barkan, R., Sela, T.: Personalized health systems—Past, present, and future of research development and implementation in real-life environment. Front. Med. **6**(July), 1–6 (2019). https://doi.org/10.3389/fmed.2019.00149
60. Li, G., Wen, D.: Wearable biochemical sensors for human health monitoring: sensing materials and manufacturing technologies. J. Mater. Chem. B **8**(16), 3423–3436 (2020). https://doi.org/10.1039/c9tb02474c

61. Li, P., Long, F., Chen, W., Chen, J., Chu, P.K., Wang, H.: Fundamentals and applications of surface-enhanced Raman spectroscopy–based biosensors. Curr. Opinion Biomed. Eng. **13**, 51–59 (2020). https://doi.org/10.1016/j.cobme.2019.08.008
62. Liang, K., Carmone, S., Brambilla, D., Leroux, J.C.: 3D printing of a wearable personalized oral delivery device: a first-in-human study. Sci. Adv. **4**(5), 1–12 (2018). https://doi.org/10.1126/sciadv.aat2544
63. Lopez-Jimenez, F., Attia, Z., Arruda-Olson, A.M., Carter, R., Chareonthaitawee, P., Jouni, H., Kapa, S., Lerman, A., Luong, C., Medina-Inojosa, J.R., Noseworthy, P.A., Pellikka, P.A., Redfield, M.M., Roger, V.L., Sandhu, G.S., Senecal, C., Friedman, P.A.: Artificial intelligence in cardiology: present and future. Mayo Clin. Proc. **95**(5), 1015–1039 (2020). https://doi.org/10.1016/j.mayocp.2020.01.038
64. Low, C.A.: Harnessing consumer smartphone and wearable sensors for clinical cancer research. NPJ Digit. Med. **3**(1) (2020). https://doi.org/10.1038/s41746-020-00351-x
65. Maturo, M.G., Soligo, M., Gibson, G., Manni, L., Nardini, C.: The greater inflammatory pathway—high clinical potential by innovative predictive, preventive, and personalized medical approach. EPMA J. **11**(1) (2020). https://doi.org/10.1007/s13167-019-00195-w
66. Melnykova, N., Shakhovska, N., Gregus, M., Melnykov, V., Zakharchuk, M., Vovk, O.: Data-driven analytics for personalized medical decision making. Mathematics **8**(8), 1211 (2020). https://doi.org/10.3390/math8081211
67. Metkar, S.K., Girigoswami, K.: Diagnostic biosensors in medicine – a review. Biocatal. Agric. Biotechnol. **17**, 271–283 (2019). https://doi.org/10.1016/j.bcab.2018.11.029
68. Morrison, R.J., Hollister, S.J., Niedner, M.F., Mahani, M.G., Park, A.H., Mehta, D.K., Ohye, R.G., Green, G.E.: Erratum: mitigation of tracheobronchomalacia with 3D-printed personalized medical devices in pediatric patients. Sci. Transl. Med. **7**(287), 1–12 (2015). https://doi.org/10.1126/scitranslmed.aac4749
69. Morrison, R.J., Kashlan, K.N., Flanangan, C.L., Wright, J.K., Green, G.E., Hollister, S.J., Weatherwax, K.J.: Regulatory considerations in the design and manufacturing of implantable 3D-printed medical devices. Clin. Transl. Sci. **8**(5), 594–600 (2015). https://doi.org/10.1111/cts.12315
70. Mphil, M.T., Alivia, M., Poma, L., Roberti, P., Sarsina, D., Tassinari, M.M.: The latest demographic surveys on traditional, complementary and alternative medicine commented by Italian scientific societies of the sector. Eur. J. Pers. Center. Healthc. **4**(4), 684–692 (2016)
71. Mule, S.T., Bhusnure, O.G., Waghmare, S.S., Mali, M.R.: Recent trends, opportunities and challenges in 3D printing technology for personalize medicine. J. Drug Deliv. Ther. **10**(4), 242–252 (2020). https://doi.org/10.22270/jddt.v10i4.4143
72. Münker, T.J.A.G., van de Vijfeijken, S.E.C.M., Mulder, C.S., Vespasiano, V., Becking, A.G., Kleverlaan, C.J., Becking, A.G., Dubois, L., Karssemakers, L.H.E., Milstein, D.M.J., van de Vijfeijken, S.E.C.M., Depauw, P.R.A.M., Hoefnagels, F.W.A., Vandertop, W.P., Kleverlaan, C.J., Münker, T.J.A.G., Maal, T.J.J., Nout, E., Riool, M., Zaat, S.A.J.: Effects of sterilization on the mechanical properties of poly(methyl methacrylate) based personalized medical devices. J. Mech. Behav. Biomed. Mater. **81**(January), 168–172 (2018). https://doi.org/10.1016/j.jmbbm.2018.01.033
73. Munoz-Guijosa, J.M., Martínez, R., Cendrero, A.M., Lantada, A.D.: Rapid prototyping of personalized articular orthoses by lamination of composite fibers upon 3D-printed molds. Materials **13**(4) (2020). https://doi.org/10.3390/ma13040939
74. Nagarajan, N., Dupret-Bories, A., Karabulut, E., Zorlutuna, P., Vrana, N.E.: Enabling personalized implant and controllable biosystem development through 3D printing. Biotechnol. Adv. **36**(2), 521–533 (2018). https://doi.org/10.1016/j.biotechadv.2018.02.004
75. Nedungadi, P., Jayakumar, A., Raman, R.: Personalized health monitoring system for managing well-being in rural areas. J. Med. Syst. **42**(1) (2018). https://doi.org/10.1007/s10916-017-0854-9
76. Palo, M., Holländer, J., Suominen, J., Yliruusi, J., Sandler, N.: 3D printed drug delivery devices: perspectives and technical challenges. Expert Rev. Med. Devices **14**(9), 685–696 (2017). https://doi.org/10.1080/17434440.2017.1363647

77. Penh, C.U.T.O.M., Leichner, R., Park, M., Beaulieu, P., Jose, S., Us, C.A., Arne, L., Alto, P., Us, C.A., Zdeblick, M.: Communicati u tom penh. 2 (2018)
78. Popescu, D., Ilie, C., Laptoiu, D., Hadar, A., Barbur, R.: Web-based collaborative platform for personalized orthopaedic applications. Stud. Inform. Control **25**(4), 517–526 (2016). https://doi.org/10.24846/v25i4y201613
79. Purohit, B., Kumar, A., Mahato, K., Chandra, P.: Smartphone-assisted personalized diagnostic devices and wearable sensors. Curr. Opinion Biomed. Eng. **13**, 42–50 (2020). https://doi.org/10.1016/j.cobme.2019.08.015
80. Rahmani, A.M., Babaei, Z., Souri, A.: Event-driven IoT architecture for data analysis of reliable healthcare application using complex event processing. Cluster Comput. (2020). https://doi.org/10.1007/s10586-020-03189-w
81. di Sarsina, P.R., Alivia, M., Guadagni, P.: The contribution of traditional, complementary and alternative medical systems to the development of person-centred medicine-the example of the charity association for person-centred medicine. OA Altern. Med. **1**(2) (2013). https://doi.org/10.13172/2052-7845-1-2-655
82. Roberti di Sarsina, P., Alivia, M.: Widening the paradigm in medicine and health: the memorandum of understanding between the European Association for Predictive, Preventive and Personalised Medicine EPMA and the Italian Charity? Association for Person Centred Medicine? Altern. Integr. Med. **01**(01) (2013). https://doi.org/10.4172/2327-5162.1000101
83. Roberti di Sarsina, P., Tassinari, M.: Inclusive healthcare, medicine (health care) focused on the person: a step beyond integrative medicine, complementary and alternative, non conventional medicine. Curr.tradit. Med. **2**(1), 18–21 (2016). https://doi.org/10.2174/2215083802999160722153006
84. Rodriguez, A., Smielewski, P., Rosenthal, E., Moberg, D.: Medical device connectivity challenges outline the technical requirements and standards for promoting big data research and personalized medicine in neurocritical care. Mil. Med. **183**, 99–104 (2018). https://doi.org/10.1093/milmed/usx146
85. Senthamizhan, A., Balusamy, B., Uyar, T.: Recent progress on designing electrospun nanofibers for colorimetric biosensing applications. Curr. Opinion Biomed. Eng. **13**, 1–8 (2020). https://doi.org/10.1016/j.cobme.2019.08.002
86. Shaban Nejad, A., Michalowski, M., Buckeridge, D.L.: Health intelligence: how artificial intelligence transforms population and personalized health. NPJ Digit. Med. **1**(1) (2018). https://doi.org/10.1038/s41746-018-0058-9
87. Shikama, M., Nakagami, G., Noguchi, H., Mori, T., Sanada, H.: Development of personalized fitting device with 3-dimensional solution for prevention of niv oronasal mask-related pressure ulcers. Respir. Care **63**(8), 1024–1032 (2018). https://doi.org/10.4187/respcare.05691
88. Sinu, I., Abraham, B., Ranch, P., Us, C.A., Mcmahon, C.M., Us, C.A., Agrawal, P., Us, C.A., Zhong, Y., Us, C.A., Huzefa, F., Examiner, P., Gray, P.A.: United States Patent (2019)
89. Sivaramakrishnan, M., Kothandan, R., Govindarajan, D.K., Meganathan, Y., Kandaswamy, K.: Active microfluidic systems for cell sorting and separation. Curr. Opinion Biomed. Eng. **13**, 60–68 (2020). https://doi.org/10.1016/j.cobme.2019.09.014
90. Solaimuthu, A., Vijayan, A.N., Murali, P., Korrapati, P.S.: Nano-biosensors and their relevance in tissue engineering. Curr. Opinion Biomed. Eng. **13**, 84–93 (2020). https://doi.org/10.1016/j.cobme.2019.12.005
91. Stanley, K.G., Osgood, N.D.: The potential of sensor-based monitoring as a tool for health care, health promotion, and research. Ann. Family Med. **9**(4), 296–298 (2011). https://doi.org/10.1370/afm.1292
92. Tashkandi, A., Wiese, I., Wiese, L.: Efficient In-database patient similarity analysis for personalized medical decision support systems. Big Data Res. **13**(May), 52–64 (2018). https://doi.org/10.1016/j.bdr.2018.05.001
93. Tasic, J., Gusev, M., Ristov, S.: A medical cloud. In: 2016 39th International Convention on Information and Communication Technology, Electronics and Microelectronics (MIPRO), pp. 400–405 (2016). https://doi.org/10.1109/MIPRO.2016.7522176

94. Tasnim, F., Sadraei, A., Datta, B., Khan, M., Choi, K.Y., Sahasrabudhe, A., Vega Gálvez, T.A., Wicaksono, I., Rosello, O., Nunez-Lopez, C., Dagdeviren, C.: Towards personalized medicine: the evolution of imperceptible health-care technologies. Foresight **20**(6), 589–601 (2018). https://doi.org/10.1108/FS-08-2018-0075
95. Therapeutic Goods Administration Department of Health. Consultation: Proposed regulatory scheme for personalised medical devices, including 3D-printed devices, 1–23 February 2019 (2019).https://www.tga.gov.au/sites/default/files/consultation-proposed-regulatory-scheme-personalised-medical-devices-including-3d-printed-devices.pdf
96. Timokhov, G.V, Semenova, E.A., Yuldashev, Z.M.: An intelligent system of remote personalized medical care with the possibility of a therapeutic impact. In: 2019 IEEE Conference of Russian Young Researchers in Electrical and Electronic Engineering (EIConRus), pp. 1336–1340 (2019). https://doi.org/10.1109/EIConRus.2019.8657242
97. Tong, Y., Kucukdeger, E., Halper, J., Cesewski, E., Karakozoff, E., Haring, A.P., McIlvain, D., Singh, M., Khandelwal, N., Meholic, A., Laheri, S., Sharma, A., Johnson, B.N.: Low-cost sensor-integrated 3D-printed personalized prosthetic hands for children with amniotic band syndrome: a case study in sensing pressure distribution on an anatomical human-machine interface (AHMI) using 3D-printed conformal electrode arrays. PLoS ONE **14**(3), 1–23 (2019). https://doi.org/10.1371/journal.pone.0214120
98. Ushigome, E., Yamazaki, M., Hamaguchi, M., Ito, T., Matsubara, S., Tsuchido, Y., Kasamatsu, Y., Nakanishi, M., Fujita, N., Fukui, M.: Usefulness and safety of remote continuous glucose monitoring for a severe COVID-19 patient with diabetes. Diab. Technol. Ther. **22**(9), 3–5 (2020). https://doi.org/10.1089/dia.2020.0237
99. van der Stelt, M., Verhulst, A.C., Vas Nunes, J.H., Koroma, T.A.R., Nolet, W.W.E., Slump, C.H., Grobusch, M.P., Maal, T.J.J., Brouwers, L.: Improving lives in three dimensions: the feasibility of 3D printing for creating personalized medical aids in a rural area of Sierra Leone. Am. J. Trop. Med. Hygiene **102**(4), 905–909 (2020). https://doi.org/10.4269/ajtmh.19-0359
100. Wangatia, L.M., Yang, S., Zabihi, F., Zhu, M., Ramakrishna, S.: Biomedical electronics powered by solar cells. Curr. Opinion Biomed. Eng. **13**, 25–31 (2020). https://doi.org/10.1016/j.cobme.2019.08.004
101. Yan, D., Chen, S., Krauss, D.J., Deraniyagala, R., Chen, P., Ye, H., Wilson, G.: Inter/intratumoral dose response variations assessed using FDG-PET/CT feedback images: impact on tumor control and treatment dose prescription. Radiother. Oncol. **154**, 235–242 (2021). https://doi.org/10.1016/j.radonc.2020.09.052
102. Yan, X., Yu, M., Ramakrishna, S., Russell, S.J., Long, Y.-Z.: Advances in portable electrospinning devices for in situ delivery of personalized wound care. Nanoscale **11**(41), 19166–19178 (2019). https://doi.org/10.1039/C9NR02802A
103. Yang, Y.J.: The future of capsule endoscopy: the role of artificial intelligence and other technical advancements. Clin. Endosc. **53**(4), 387–394 (2020). https://doi.org/10.5946/ce.2020.133
104. Yu, J., Hou, X., Cui, M., Zhang, S., He, J., Geng, W., Mu, J., Chou, X.: Highly skin-conformal wearable tactile sensor based on piezoelectric-enhanced triboelectric nanogenerator. Nano Energy **64**, 103923 (2019). https://doi.org/10.1016/j.nanoen.2019.103923
105. Yu, K.H., Beam, A.L., Kohane, I.S.: Artificial intelligence in healthcare. Nat. Biomed. Eng. **2**(10), 719–731 (2018). https://doi.org/10.1038/s41551-018-0305-z
106. Zema, L., Melocchi, A., Maroni, A., Gazzaniga, A.: Three-dimensional printing of medicinal products and the challenge of personalized therapy. J. Pharm. Sci. **106**(7), 1697–1705 (2017). https://doi.org/10.1016/j.xphs.2017.03.021
107. Zhan, X., Hu, R., Wang, X.: Multi-parameter systematic strategy opinion that predicts, prevents, and personalized treats a cancer. EPMA J. **5**(S1), A25 (2014). https://doi.org/10.1186/1878-5085-5-s1-a25
108. Zhang, J., Hu, Q., Wang, S., Tao, J., Gou, M.: Digital light processing based three-dimensional printing for medical applications. Int. J. Bioprint. **6**(1), 12–27 (2020). https://doi.org/10.18063/ijb.v6i1.242

109. Zhang, Y., Liang, B., Jiang, Q., Li, Y., Feng, Y., Zhang, L., Zhao, Y., Xiong, X.: Flexible and wearable sensor based on graphene nanocomposite hydrogels. Smart Mater. Struct. **29**(7) (2020). https://doi.org/10.1088/1361-665X/ab89ff
110. Zhao, Z., Ukidve, A., Kim, J., Mitragotri, S.: Targeting strategies for tissue-specific drug delivery. Cell **181**(1), 151–167 (2020). https://doi.org/10.1016/j.cell.2020.02.001
111. Zheng, J., Wu, T., Shen, Y., Zhang, G., Zhang, Z., Lu, H.: Emerging wearable medical devices towards personalized healthcare. In: BODYNETS 2013 - 8th International Conference on Body Area Networks, pp. 427–431, September 2013. https://doi.org/10.4108/icst.bodynets.2013.253725
112. Zhong, C.L., Li, Y.L.: Internet of things sensors assisted physical activity recognition and health monitoring of college students. Meas.: J. Int. Meas. Confed. **159**, 107774 (2020). https://doi.org/10.1016/j.measurement.2020.107774
113. Badman, R., Hills, T., Akaishi, R.: Navigating Uncertain Environments: Multiscale Computation in Biological and Artificial Intelligence March 2020. https://doi.org/10.31234/osf.io/ced3t
114. Currivan-Incorvia, J.A., Siddiqui, S., Dutta, S., Evarts, E.R., Ross, C.A., Baldo, M.A.: Spintronic logic circuit and device prototypes utilizing domain walls in ferromagnetic wires with tunnel junction readout. In: Tech. Dig. - International Electron Devices Meeting IEDM, vol. 2016, pp. 32.6.1–32.6.4. (2015). https://doi.org/10.1109/IEDM.2015.7409817
115. Basinger, K.L., Keough, C.B., Webster, C.E., Wysk, R.A., Martin, T.M., Harrysson, O.L.: Development of a modular computer-aided process planning (CAPP) system for additive-subtractive hybrid manufacturing of pockets, holes, and flat surfaces. Int. J. Adv. Manuf. Technol. **96**(5–8), 2407–2420 (2018). https://doi.org/10.1007/s00170-018-1674-x
116. Tortorella, G.L., Mac Cawley Vergara, A., Garza-Reyes, J.A., Sawhney, R.: Organizational learning paths based upon industry 4.0 adoption: an empirical study with Brazilian manufacturers. Int. J. Prod. Econ. **219**, 284–294 (2020) https://doi.org/10.1016/j.ijpe.2019.06.023
117. Usak, M., Kubiatko, M., Shabbir, M.S., Dudnik, O.V., Jermsittiparsert, K., Rajabion, L.: Health care service delivery based on the Internet of things: a systematic and comprehensive study. Int. J. Commun. Syst. **33**(2), 1–17 (2020). https://doi.org/10.1002/dac.4179
118. Kim, H., Hong, H., Ho Yoon, S.: Diagnostic performance of ct and reverse transcriptase polymerase chain reaction for coronavirus disease 2019: a meta-analysis. Radiology **296**(3), E145–E155 (2020). https://doi.org/10.1148/radiol.2020201343
119. Calvaresi, D., Schumacher, M., Calbimonte, J.P.: Personal data privacy semantics in multi-agent systems interactions. Lect. Notes Comput. Sci. (including Subser. Lect. Notes Artif. Intell. Lect. Notes Bioinformatics), LNAI, vol. 12092, pp. 55–67. January 2021 (2020). https://doi.org/10.1007/978-3-030-49778-1_5
120. Miller, T.: Explanation in artificial intelligence: insights from the social sciences. Artif. Intell. **267**, 1–38 (2019). https://doi.org/10.1016/j.artint.2018.07.007
121. Jill Hopkins, J., Keane, P.A., Balaskas, K.: Delivering personalized medicine in retinal care: from artificial intelligence algorithms to clinical application. Curr. Opin. Ophthalmol. **31**(5), 329–336 (2020). https://doi.org/10.1097/ICU.0000000000000677
122. Oniani, S., Marques, G., Barnovi, S., Pires, I.M., Bhoi, A.K.: Artificial intelligence for internet of things and enhanced medical systems. Stud. Comput. Intell. **903**, 43–59 (2021). https://doi.org/10.1007/978-981-15-5495-7_3
123. Parimi, S., Chakraborty, S.: Application of big data & iot on personalized healthcare services. Int. J. Sci. Technol. Res. **9**(3), 1107–1111 (2020)
124. Camacho-Cogollo, J.E., Bonet, I., Iadanza, E.: RFID technology in health care. Second edn., Elsevier Inc. (2019)
125. Radanliev, P., De Roure, D., Van Kleek, M., Santos, O., Ani, U.: Artificial intelligence in cyber physical systems. AI Soc. **0123456789**, (2020). https://doi.org/10.1007/s00146-020-01049-0
126. Barrow, N.J., Debnath, A., Sen, A.: Measurement of the effects of pH on phosphate availability. Plant Soil **454**(1–2), 217–224 (2020). https://doi.org/10.1007/s11104-020-04647-5

127. Mitchell, A.L., et al.: MGnify: The microbiome analysis resource in 2020. Nucleic Acids Res. **48**(D1), D570–D578 (2020). https://doi.org/10.1093/nar/gkz1035
128. Kamath, A., McDonough, C.E., Monk, J.D., Lambert, M.R., Giglio, E.: A. Kamath et al. reply. Nat. Ecol. Evol. **4**(6), 786–787 (2020). https://doi.org/10.1038/s41559-020-1188-4
129. Brandão-marques, L., Gelos, G.: Leaning against the wind: an empirical cost-benefit analysis
130. Moreira-Teixeira, L., et al.: Mouse transcriptome reveals potential signatures of protection and pathogenesis in human tuberculosis. Nat. Immunol. **21**(4), 464–476 (2020). https://doi.org/10.1038/s41590-020-0610-z
131. Moonla, C., et al.: An integrated microcatheter-based dual-analyte sensor system for simultaneous, real-time measurement of propofol and fentanyl. Talanta **218**, 121205 (2020). https://doi.org/10.1016/j.talanta.2020.121205
132. Bernal Monroy, E., Polo Rodríguez, A., Espinilla Estevez, M., Medina Quero, J.: Fuzzy monitoring of in-bed postural changes for the prevention of pressure ulcers using inertial sensors attached to clothing. J. Biomed Inform. **107**, 103476 (2020). https://doi.org/10.1016/j.jbi.2020.103476
133. Mohammed, M.N., Syamsudin, H., Al-Zubaidi, S., Sairah, A.K., Ramli, R., Yusuf, E.: Novel covid-19 detection and diagnosis system using iot based smart helmet. Int. J. Psychosoc. Rehabil. **24**(7), 2296–2303 (2020). https://doi.org/10.37200/IJPR/V24I7/PR270221
134. Zulfiqar, U., Sreeram, V., Du, X.: Frequency-limited pseudo-optimal rational Krylov algorithm for power system reduction. Int. J. Electr. Power Energy Syst. **118**, 1–12 (2020). https://doi.org/10.1016/j.ijepes.2019.105798
135. Fischer, A.M., et al.: Function Testing, pp. 1065–1071 (2020)
136. Yang, Y., Zhou, L., Shi, W., He, Z., Han, Y., Xiao, Y.: Interstage difference of pressure pulsation in a three-stage electrical submersible pump. J. Pet. Sci. Eng. **196**, 107653 (2021). https://doi.org/10.1016/j.petrol.2020.107653
137. Mishra, Z., Mishra, B., Aloosh, O.: Impact of artificial intelligence on the healthcare industry. Trends Appl. Sci. Res. **15**(2), 59–65 (2020). https://doi.org/10.3923/tasr.2020.59.65
138. Kumar, S., et al.: Ultrapure green light-emitting diodes using two-dimensional formamidinium perovskites: achieving recommendation 2020 color coordinates. Nano Lett. **17**(9), 5277–5284 (2017). https://doi.org/10.1021/acs.nanolett.7b01544
139. Rehman, U., et al.: Depression, anxiety and stress among Indians in times of covid-19 lockdown. Commun. Ment. Health J. **57**(1), 42–48 (2021). https://doi.org/10.1007/s10597-020-00664-x
140. Venkataraman, Y.R., et al.: General DNA methylation patterns and environmentally-induced differential methylation in the eastern oyster (Crassostrea virginica). Front. Mar. Sci. **7**, 1–14 (2020). https://doi.org/10.3389/fmars.2020.00225
141. Liu, H., et al.: 501Y.V2 and 501Y.V3 variants of SARS-CoV-2 lose binding to Bamlanivimab in vitro, bioRxiv, p. 2021.02.16.431305, (2021). https://www.biorxiv.org/content/10.1101/2021.02.16.431305v1
142. Ahmed, S.F., Quadeer, A.A., McKay, M.R.: Preliminary identification of potential vaccine targets for the COVID-19 coronavirus (SARS-CoV-2) based on SARS-CoV immunological studies. Viruses **12**(3) (2020). https://doi.org/10.3390/v12030254
143. Jang, A.I., Sharma, R., Drugowitsch, J.: Optimal policy for attention-modulated decisions explains human fixation behavior. Elife **10**, 1–31 (2021). https://doi.org/10.7554/eLife.63436
144. Cioffi, R., Travaglioni, M., Piscitelli, G., Petrillo, A., De Felice, F.: Artificial intelligence and machine learning applications in smart production: progress, trends, and directions. Sustain **12**(2), (2020). https://doi.org/10.3390/su12020492
145. Brainina, K.Z., Kazakov, Y.E.: Electrochemical hybrid methods and sensors for antioxidant/oxidant activity monitoring and their use as a diagnostic tool of oxidative stress: Future perspectives and challenges. Chemosensors **8**(4), 1–14 (2020). https://doi.org/10.3390/chemosensors8040090
146. Phillips-Wren, G., McKniff, S.: Overcoming resistance to big data and operational changes through interactive data visualization. Big Data **8**(6), 528–539 (2020). https://doi.org/10.1089/big.2020.0056

147. Beardslee, L.A., et al.: Ingestible sensors and sensing systems for minimally invasive diagnosis and monitoring: the next frontier in minimally invasive screening. ACS Sens. **5**(4), 891–910 (2020). https://doi.org/10.1021/acssensors.9b02263
148. Baumann, F., Lorenz-Spreen, P., Sokolov, I.M., Starnini, M.: Modeling echo chambers and polarization dynamics in social networks. Phys. Rev. Lett. **124**(4), 48301 (2020). https://doi.org/10.1103/PhysRevLett.124.048301
149. Ginsburg, O., et al.: Breast cancer early detection: a phased approach to implementation. Cancer **126**(S10), 2379–2393 (2020). https://doi.org/10.1002/cncr.32887
150. Uz-Zaman, K.A., Biswas, B., Rahman, M.M., Naidu, R.: Smectite-supported chain of iron nanoparticle beads for efficient clean-up of arsenate contaminated water. J. Hazard. Mater. **407**, 124396 (2021). https://doi.org/10.1016/j.jhazmat.2020.124396
151. Greene, T., Shmueli, G., Fell, J., Lin, C.F., Shope, M.L., Liu, H.W.: The hidden inconsistencies introduced by predictive algorithms in judicial decision making. arXiv, no, 101 (2020)
152. Koledova, E., Tornincasa, V., Van Dommelen, P.: Analysis of real-world data on growth hormone therapy adherence using a connected injection device. BMC Med. Inform. Decis. Mak. **20**(1), 1–8 (2020). https://doi.org/10.1186/s12911-020-01183-1
153. Khan, A.I., Shah, J.L., Bhat, M.M.: CoroNet: a deep neural network for detection and diagnosis of COVID-19 from chest x-ray images. Comput. Methods Programs Biomed. **196**, (2020). https://doi.org/110.1016/j.cmpb.2020.105581
154. De Sarkar, S., et al.: The leishmanicidal activity of artemisinin is mediated by cleavage of the endoperoxide bridge and mitochondrial dysfunction. Parasitology **146**(4), 511–520 (2019). https://doi.org/110.1017/S003118201800183X
155. Barry, L., Charpentier, A.: Personalization as a promise: can big data change the practice of insurance? Big Data Soc. **7**(1) (2020). https://doi.org/10.1177/2053951720935143
156. Cheung, A.I., Projan, S.J., Edelstein, R.E., Fischetti, V.A.: Cloning, expression, and nucleotide sequence of a staphylococcus aureus gene (fbpA) encoding a fibrinogen-binding protein. Infect. Immun. **63**(5), 1914–1920 (1995). https://doi.org/10.1128/iai.63.5.1914-1920.1995
157. Poppe, M., Veerkamp, R.F., van Pelt, M.L., Mulder, H.A.: Exploration of variance, autocorrelation, and skewness of deviations from lactation curves as resilience indicators for breeding. J. Dairy Sci. **103**(2), 1667–1684 (2020). https://doi.org/10.3168/jds.2019-17290
158. Gillmann, K., Wasilewicz, R., Hoskens, K., Simon Zoula, S., Mansouri, K.: Continuous 24-hour measurement of intraocular pressure in millimeters of mercury (mmHg) using a novel contact lens sensor: Comparison with pneumatonometry. PLoS One **16**(3), 1–13 (2021). https://doi.org/10.1371/journal.pone.0248211
159. Bhavsar, K.A., Abugabah, A., Singla, J., AlZubi, A.A., Bashir, A.K., Nikita: A comprehensive review on medical diagnosis using machine learning. Comput. Mater. Contin. **67**(2), 1997–2014 (2021). https://doi.org/10.32604/cmc.2021.014943. [68]
160. Teymourian, H., Barfidokht, A., Wang, J.: Electrochemical glucose sensors in diabetes management: an updated review (2010–2020). Chem. Soc. Rev. (2020). https://doi.org/10.1039/d0cs00304b
161. Nimri, R., et al.: Insulin dose optimization using an automated artificial intelligence-based decision support system in youths with type 1 diabetes. Nat. Med. **26**(9), 1380–1384 (2020). https://doi.org/10.1038/s41591-020-1045-7
162. Lamb, L.E., Bartolone, S.N., Ward, E., Chancellor, M.B.: Rapid detection of novel coronavirus/Severe Acute Respiratory Syndrome Coronavirus 2 (SARS-CoV-2) by reverse transcription-loop-mediated isothermal amplification. PLoS One **15**(6), 1–15 (2020). https://doi.org/10.1371/journal.pone.0234682
163. Fujita, H.: AI-based computer-aided diagnosis (AI-CAD): the latest review to read first. Radiol. Phys. Technol. **13**(1), 6–19 (2020). https://doi.org/10.1007/s12194-019-00552-4

Chapter 11
Electrical Impedance Tomography Based Lung Disease Monitoring

Aniqa Tabassum and Md Atiqur Rahman Ahad

Abstract Electrical impedance measurements can detect many diseases and disorders in the human body. Electrical Impedance Tomography (EIT) is a fast-developing medical imaging technique. In this chapter, we present some applications of EIT in lung disease detection. Existing literature in this subject has been investigated, including original research work on EIT based lung imaging, carried out in clinical settings for particular respiratory diseases: Acute respiratory distress syndrome (ARDS), Chronic obstructive pulmonary disease (COPD), Cystic fibrosis, Pneumonia, and Pleural effusion. Information about the purpose of the tests, studied subjects, the procedure followed, and results of analysis have been included as well as limitations of the studies. EIT is a promising technique in the direction of non-invasive diagnostic medicine because this can perform imaging of the human body, without posing any visible risk.

11.1 Introduction

Electrical Impedance Tomography (EIT) is a promising and noninvasive imaging technique of the human body. It generates cross-sectional images representing the electrical impedance or conductivity distributions in the body. Usefulness of EIT in diagnostic imaging has been observed in numerous studies with time. The basic principle of EIT is over thirty years old, yet the growing number of publications in this field suggest possible clinical applications of this method in recent times. In EIT, potentials measured through surface electrodes help derive the electrical impedance of a specific part of the body and form a corresponding tomographic image. EIT is particularly useful when a specific physiological event of interest results in contrasts in electrical characteristics of tissues. Examples include the presence of cancerous and diseased tissues [1].

A. Tabassum · M. A. R. Ahad (✉)
Department of Electrical and Electronic Engineering, University of Dhaka, Dhaka, Bangladesh
e-mail: atiqahad@du.ac.bd

M. A. R. Ahad
Department of Intelligent Media, Osaka University, Suita, Japan

M. A. R. Ahad and A. Inoue (eds.), *Vision, Sensing and Analytics: Integrative Approaches*, Intelligent Systems Reference Library 207, https://doi.org/10.1007/978-3-030-75490-7_11

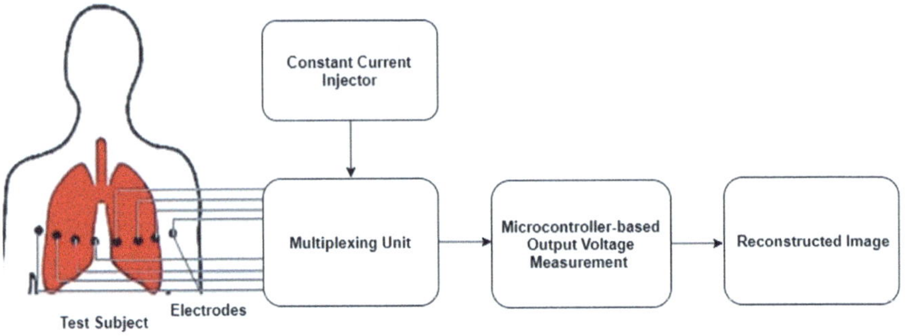

Fig. 11.1 Electrical Impedance Tomography (EIT) system basic block diagram [4]

The theoretical concept of EIT was defined over 30 years ago [2]. EIT hardware design, image reconstruction algorithm, and achievable, practical applications were introduced by Barber and Brown, Department of Medical Physics and Clinical Engineering, Sheffield, UK. They named this technique Applied Potential Tomography, which is called Electrical Impedance Tomography (EIT), today, studied absolute and functional EIT images, and suggested its ability to monitor lung ventilation dynamics. An extensive development of this method has taken place since then, to apply EIT in a clinical environment [3]. This method involves the continuous measurement of potentials, resulting from the cyclic injection of low-amplitude and high-frequency alternating currents, through surface electrodes placed on a test subject. The goal is to derive the electrical conductivity distributions within the subject through solving an inverse problem. Figure 11.1 shows a basic block diagram for an EIT system.

11.1.1 Applications of EIT

The idea behind EIT imaging was first used in geology, where this is named Electrical Resistivity Tomography (ERT) [5]. In the science of forestry, trees that are ready for harvesting can be identified by employing EIT [6]. EIT has also been used for pressure-sensitive artificial skin for applications in robotics [7]. Henderson and Webster were first to introduce images of the impedance of human tissues [8]. Also, anatomy and function of the brain were widely studied through slight variations of this method. A technique called magnetic resonance electrical impedance tomography (MREIT), which combines magnetic resonance with EIT [9], has been used to detect breast cancer [10].

The main focus of the imaging applications of EIT was to detect and locate pathologies, such as the presence of cancerous cells, e.g., for breast [11] and prostate [12] since malignant tissue has significantly different characteristics compared to benign tissue. One of the biggest advantages that EIT offers is the monitoring of lung function dynamics in ventilator aided ICU patients. Inspiration leads to a rise in impedance value of tissues, proportionate to the volume of air inhaled, and can be measured using EIT. Detection of the breathing rate and blood flow in the human body are some major monitoring applications of EIT [13]. In the case of mechanical ventilation, optimum support pressure is required, and EIT can provide a patient-specific solution to this problem. EIT can correctly show whether a recruitment maneuver is capable of opening up closed lung alveoli and keep them open by appropriate ventilation settings, all to lessen the possibility of VALI (Ventilator-Associated Lung Injury).

For obstructive lung diseases like COPD, emphysema, asthma, etc., lung inhomogeneity cannot be detected by existing methods. EIT shows improved disease detection in this case [14]. It allows noninvasive monitoring of the heart function and blood flow and can compute important measures such as cardiac output and pulmonary perfusion. Neural and brain activity imaging has been done using EIT [15]. Applications also include detecting edema in the lungs, brain, etc., determining the volume of fluid in the bladder and monitoring for internal bleeding. From dedicated research work on EIT, a prospective future can be expected in extensive medical applications.

11.1.2 Benefits of EIT

Imaging of the lungs is the basis of respiratory medicine. It provides undisputed information about the structure and function of the lungs. Considering Lung Function Imaging to be the aim, we look at existing imaging modalities. While many methods give better resolution than EIT, their major drawback is the expense of the procedure and exposure to harmful radiation. A lot of clinical research has been done to evaluate the applicability of Electrical Impedance Tomography (EIT) in monitoring lung function in healthy subjects as well as patients with ARDS, Cystic fibrosis, COPD, Pneumonia and many other respiratory diseases in both adults and infants. EIT is a promising, safe, non-hazardous and noninvasive monitoring technique [16]. Its portability and faster imaging capabilities give EIT significant benefit over other common imaging techniques. Though the images have poor spatial resolution, the high temporal resolution makes it capable of monitoring impedance changes over time. EIT is free from radiation exposure and allows continuous observation of lung function over extended periods. It allows testing of pulmonary function in children, in particular infants, which is considerably difficult for other tests due to lack of cooperation from this category of patients.

EIT measurement devices are comparatively inexpensive, and there have not been any such instances to expect harmful biological effects from the small amplitude currents applied in the process. Different measures can be obtained and processed faster using EIT to acquire tomographic images corresponding to impedance changes. Due to its portability, EIT can be used to continuously monitor critical and bedridden patients as well. This is mainly because of the constantly improving and easily available hardware and data processing modalities, and successful implementation will lead to a huge number of opportunities in the clinical sector in the future. Table 11.1 gives the specifications of some commonly used EIT devices.

Table 11.1 Some Commercial EIT Devices frequently used in examining human subjects [13]

EIT System	Manufacturer	Electrodes	Image Reconstruction Algorithm	Data Acquisition
CareFusion	Goe MF II	16 individual electrodes	Sheffield Back-projection Algorithm	Pair drive (adjacent), serial measurement
Dräger Medical	PulmoVista 500	16 electrode belt	FEM-based Newton-Raphson method	Pair drive (adjacent), serial measurement
Maltron International	Mark 1 Mark 3.5	16 individual electrodes 8 individual electrodes	Sheffield Back-projection Algorithm	Pair drive (adjacent), serial measurement
Swisstom AG	Swisstom BB	32 electrode belt	GREIT (Graz consensus Reconstruction algorithm for EIT)	Pair drive (adjustable skip), serial measurement
Timpel SA	Enlight 1800	32 electrode stripes	FEM-based Newton-Raphson method	Pair drive (3-electrode skip), parallel measurement

11.1.3 Fundamental Characteristics of EIT

The physical aspect of EIT lung imaging is built on the determination of impedance or conductivity distributions within the human thorax. Electrodes are fixed around a patient's chest. Alternating currents of high frequency and low amplitude are applied to the chest by successive electrode pairs and potential obtained from the remaining pairs. Commercial systems mostly have 16 or 32 electrodes. The greater the number of electrodes, the better the resolution but complexity and computational burden rises as a result. A full circle of all the electrodes produces a voltage profile, also called a frame, which is then used to reconstruct a tomographic image of the thorax. Sheffield Backprojection Algorithm was the basis of EIT image reconstruction, though several new algorithms with additional features and easier implementations have been developed or are at work. The obtained image reflects the impedance distribution within the lungs. Figure 11.2 presents the path by which significant clinical measures can be obtained using EIT. However, the most difficult challenge was probably extracting time-varying impedance information.

The different data sampling techniques and analysis methods are explained in detail in a wide number of publications. They can be divided into the following categories:

1. functional EIT (fEIT) [17]
2. absolute EIT (aEIT) [18, 19]
3. EIT spectroscopy or multi-frequency EIT (MF-EIT)

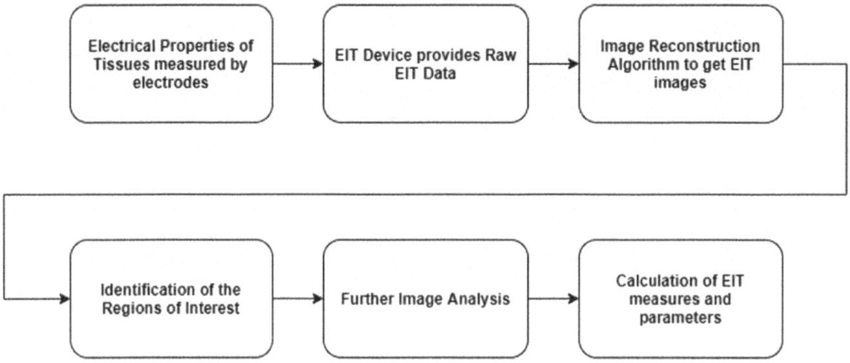

Fig. 11.2 Image Reconstruction Path by which clinically important measures and parameters are derived from raw EIT measurements [1]

The most common and effective use of Electrical Impedance Tomography in case of lung function monitoring is f-EIT. Taking into consideration the potential of EIT to continuously monitor lung function, a lot of work has been done to generate and study functional EIT images of the dynamics of lung function either using customized or commercially available EIT devices [20].

11.1.4 Drawbacks of EIT

EIT also has possible drawbacks. Firstly, its comparatively lower spatial resolution makes using EIT difficult for obtaining morphological information, as CT and MRI both give significantly higher resolutions. Specific observation of absolute impedance is extremely challenging and involves many considerations, like patient-specific anatomical information. So EIT is better suited for monitoring changes in impedance with time, in a particular individual, rather than for comparisons between many.

Another problem is the practicality of the placement of electrodes, especially in bedridden patients and newborn children. EIT images show relatively poor spatial and contrast resolution and only allow the observation of a slice of the lungs, with the assumption that the entire remaining lung space has the same characteristics. It can explain only a relative impedance change and cannot distinguish the physical borders between the lung and non-aerated tissues in case of lung imaging.

Several reviews have been published on the different aspects of Electrical Impedance Tomography (EIT). There have been overviews of the basic concept, focused surveys on the image reconstruction algorithms, and the basic clinical applications. This book chapter is focused on EIT Lungs Imaging, specifically on the detection of lung diseases. Such a review is timely, since the latest research in EIT has been extremely focused on disease diagnosis. With the developing hardware and easily accessible algorithms, clinical applications have richer prospects today. We have reviewed literature relevant to some common respiratory diseases, and the various applications of EIT in their study and detection. Recent studies have been included comparing lung healthy patients with those who have non-contagious lung diseases, e.g., ARDS, COPD, Cystic fibrosis, Pleural effusion, and also communicable diseases such as Emphysema and Pneumonia.

This chapter has been organized as follows. Section 11.2 reviews recent literature of EIT-based Lung Disease Monitoring for some common respiratory diseases. We also introduce several standard methods of disease detection. Section 11.3 presents a discussion and analysis of the existing work. We also include some future challenges and possible solutions. Finally, the chapter is concluded in Sect. 11.4.

11.2 Lung Disease Monitoring

Electrical Impedance Tomography (EIT) can be used for the detection of many lung diseases. This review focuses on some specific respiratory diseases and EIT findings for those cases. The included diseases are ARDS, COPD, Cystic fibrosis, Pneumonia, and Pleural effusion.

11.2.1 Acute Respiratory Distress Syndrome (ARDS)

Acute respiratory distress syndrome (ARDS) is an abrupt failure of the respiratory system due to increased inflammation in the lungs. It is a serious illness and shows a high mortality rate [21]. The main complication in ARDS is caused by fluid that builds up in the lungs making breathing difficult or impossible. ARDS is the most severe form of acute lung injury (ALI), which is defined as a significant decline in lung function due to characteristic pathological abnormalities in normal structure or architecture of the lungs. Pulmonary edema is a lung condition due to excess fluid accumulation in the lungs. Possibly the most common form of noncardiogenic pulmonary edema is ARDS.

Pulletz et al. [22] used EIT to observe the dynamics of regional lung aeration in mechanically-ventilated ARDS patients caused by an increase and decrease in airway pressure in steps and its effect on positive end-expiratory pressure (PEEP). PEEP corresponds to the lung (alveolar) pressure above atmospheric pressure that remains at the end of expiration. The study involving 12 lung healthy subjects and 20 ARDS patients differentiated between these two categories under all test conditions. Modified Sheffield Back-projection Algorithm was used to obtain EIT scans and inflation and deflation dynamics were observed by means of a custom-made MATLAB toolbox. Because the study was done in clinical settings, additional factors may have contributed, and age-dependent lung ventilation differences may have influenced the results as well.

Researchers from Boston Children's Hospital concluded from experimental findings with 9 pediatric ARDS patients that lung compliance or pulmonary compliance measured by EIT and mechanical ventilator settings showed a high correlation coefficient during a recruitment maneuver (RM) [23]. This implies that the anatomical locations used for placement of electrodes are suitable for monitoring lung ventilation mechanics in ALI/ARDS patients.

In a group of 10 intubated ARDS patients, Mauri et al. [24] compared a lower PEEP with a higher PEEP value, under similar support ventilation settings. The results showed that for ARDS patients who require support pressure mechanical ventilation, an increase in PEEP and a decrease in support pressure improve the fraction of lung aeration for the dependent lung regions. This leads to much better and homogeneous lung ventilation and improved ventilation/perfusion matching as a result.

Taking benefit from EIT's high temporal resolution, Yoshida et al. [25] monitored the effects of an arduous spontaneous breathing effort using a lung injury animal model and in ARDS patients. Despite the identical global tidal volumes, the contraction of the diaphragm leads to air in the lungs from the non-dependent to the dependent regions (pendelluft phenomenon). Spontaneous breathing effort in case of mechanical ventilation causes unpredicted overstretch of dependent lung regions during initial inflation (also leading to deflation of the non-dependent lung regions as a result). Though it does not increase tidal volume, this may potentially increase lung damage.

A study examined homogeneity of tidal ventilation of lungs using EIT measurements during two modes of ventilation, pressure support ventilation (PSV) and pressure-controlled ventilation (PCV) under similar conditions. Results exhibited that tidal volume will be more evenly distributed using PSV, in patients with and without ARDS [26]. For 20 cardiac postoperative ICU patients, PSV displayed better dependent lung region aeration because of more spontaneous diaphragm activity compared to PCV. This is much more prominent in case of lower support levels. So EIT can be applied to monitor and optimize a patient-specific ventilation guide to obtain better tidal ventilation in assisted patients in either ventilator mode.

Cinnella et al. [27] studied the effects of an open lung approach (OLA) strategy (based on recruitment maneuver and decremental PEEP) using Electrical Impedance Tomography (EIT). This method leads to better homogeneity of tidal ventilation. Improved oxygenation caused an increased volume of air reaching the dorsal lung regions.

In an observational study, 18 critically ill ARDS patients were studied. The results categorized them into 2 groups: 13 of the patients were responders and 5 were non-responders. For responders, decremental PEEP level leads to a reduction in recruited and overdistended pixels in dependent lung regions, and also fall in overdistended pixels in non-dependent lung regions. However, this was not the case for non-responders [28]. PEEP titration therefore remarkably influenced regional gas distribution in the lungs, which can be visualized. EIT can thus optimize the PEEP level during RMs.

In ARDS patients, a rise in ventilation heterogeneity can be caused by the continuous opening and closing of the alveoli in the lungs, and this increases the risk of ventilator-induced lung injury (VILI). Applying a Recruitment Maneuver (RM) and introducing PEEP decreases the chances of hypoxemia. An appropriate PEEP level and lower tidal volume are imperative for critically ill patients [29].

A promising development of EIT, still requiring extensive study and validation, is its ability to image lung regional perfusion or the cardiac related impedance changes because of the blood flow pulsatility in blood vessels [30].

Hsu et al. [31] carried out a study with 19 artificially-ventilated ARDS patients. EIT has been used appropriately to monitor recruitment and status of oxygenation with variation in levels of PEEP. There were long-term effects on the PaO2/FiO2 ratio followed by an increase in PEEP, which was associated with the rise in end-expiratory lung impedance (EELI) change. Observations showed a remarkable connection between (EELI) which was obtained by the standard PulmoVista 500 device, and the PaO2/FiO2 ratio after a recruitment maneuver (RM). Findings recommend the use of EIT in assisting artificial ventilation management and determination of the level of ARDS severity.

EIT can also measure changes in bioimpedance for the complete cardiac cycle and thus compute ventilation-perfusion mismatch that varied remarkably in the dependent (dorsal) lung regions preceding and succeeding a Recruitment Maneuver, in a study involving 20 ARDS patients done from January 2014 to December 2014 [32]. An inconsistency in lung tissue reopening and oxygenation improvement during RM was found. Hence it can be concluded that similar improvement in oxygenation was not exhibited by all subjects. EIT can thus determine the success of a RM by combining different measures of oxygenation.

The TRanslational EIT developmeNt stuDy group has published a consensus paper, which summarizes expert recommendations on data analysis, terminology, and applications of EIT lung imaging in the clinical field, by investigating and studying relevant research carried out during the last 30 years. EIT can provide two types of information, both on regional level: the amount of tidal ventilation for a particular region and differences in end-expiratory lung impedance (reflecting differences in end-expiratory lung volume) due to ventilatory setting changes. EIT has been widely applied in preclinical studies, demonstrating its ability to track alveolar recruitment and overdistension [33]. An increasing number of publications in this subject suggests the potential of EIT in clinical practice.

11.2.2 Chronic Obstructive Pulmonary Disease (COPD)

Chronic obstructive pulmonary disease (COPD) is a common lung disease which is characterized by the long-term obstruction of airflow within the lungs, which causes breathlessness. COPD is a collective term used to refer to progressive diseases of the lungs such as Chronic bronchitis, Emphysema, and Asthma.

Significant research work was also done relating to COPD using the Goe MF II commercial EIT system, which can be compared to the findings from the PulmoVista 500 device [34]. A prospective study compared the inhomogeneity of local lung ventilation in young and older healthy controls as well as COPD patients using EIT [35].

The findings show the ability of EIT to detect not just pathological but also the less pronounced age-related inhomogeneity in ventilation distribution in healthy subjects and COPD patients. The traditional filtered Back-projection Algorithm was used to obtain the resulting EIT scans, and the application of some different algorithms may provide improved results.

EIT measurements performed on 10 COPD patients (age: 63–83 years) with sudden worsening cases of COPD, and hypercapnic respiratory failure, and requiring intensive care treatment, showed better homogeneously distributed ventilation during high-frequency oscillatory ventilation (HFOV) than during the initial conventional mechanical ventilation (CMV) [36].

Benefiting from EIT's high temporal resolution, monitoring rapid regional lung volume changes determined the lung function in adult asthma patients. To investigate the impact of bronchodilator at the regional level, asthma and COPD patients were tested using EIT preceding and succeeding bronchodilator reversibility testing [37]. A possible limitation of the study may be a result of the subject's chest expansion or organ movement. Simulation studies and practical research indicates that artefacts in EIT lung signals can be introduced by such factors [38].

Schullcke et al. [39] proposed an approach for the detection and examination of obstructive lung diseases in a simulation study. A 3D Finite Element Method (FEM) based model of the human thorax (developed using the CT dataset of a human male subject) simulates the voltages of electrodes corresponding to specific lung conditions. Measurements were carried out in 2 different electrode planes using 16 electrodes, and differing levels of mucus accumulation and emphysema were simulated within the lungs. Patient-specific anatomical information has been used to reconstruct EIT images. Comparisons show that this type of starting information can contribute to improved results compared to traditional methods. However, further studies on patients in clinical settings are necessary to fully validate this research work.

Many EIT research studies have talked about obstructive lung diseases, however, the level of obstruction has not been thoroughly investigated. Obstructive ventilatory defect (OVD) is a respiratory abnormality common in diseases like bronchial asthma and COPD. Zhang et al. [40] evaluated the capabilities of EIT to determine the obstruction level in patients showing OVD on both the local and global levels.

11.2.3 Cystic Fibrosis (CF)

Cystic fibrosis (CF) is a common metabolic disease of the human body, affecting mostly the lungs and the digestive system. Thick mucus is produced by the body, which can lead to blocking of the lungs and obstruction of the pancreas. This disease may lead to potentially life-threatening consequences. The resulting change in lung anatomy and small airways causes reduced airflow in inspiration as well as expiration. Respiratory failure can cause death in cystic fibrosis patients. A standard method for detection of this disease is known as low-dose high resolution computed tomography

(HRCT). This, however, leads to patient exposure to a significant dose of ionizing radiation and so alternate detection techniques came into focus.

Zhao et al. [41] first used EIT clinically to monitor regional lung ventilation in CF patients. EIT-based global inhomogeneity (GI) index is an important measure to monitor the inhomogeneity in lung function. The obstruction of airways leads to slower distribution of air within the lungs. GI index showed ventilation inhomogeneity at the start of the forced inspiration, however, there was a gradual improvement in homogeneity. The same group carried out another study using 5 patients. This time, maximum expiratory flow ratios set at 25 and 75% of vital capacity corresponding to relative changes in bioimpedance were computed for specific regions in the obtained EIT images. The results have proved that EIT measures were compatible with well established CT based findings [42].

Wettstein et al. [43] explored the impact of various breathing aids on lung function dynamics in CF patients and suggested the use of EIT to personalize respiratory physiotherapy. In 9 patients with CF and 11 healthy controls, measurements were done while spontaneous breathing, positive expiratory pressure (PEP) and continuous positive airway pressure (CPAP) were applied in the upright and lateral positions in the subjects. Results prove that EIT may be useful in determining the most suitable breathing aid for an individual and applying customized patient-specific therapy.

EIT can observe the regional and global lung function dynamics of the lungs with very high temporal resolution. A lot of research comparing EIT with various other lung imaging and functional tests has been done. Lehman et al. [44] showed an important statistical correlation between existing spirometry tests and global EIT measures in a study with 11 pediatric patients and 11 lung healthy controls.

In a study with 10 adult CF patients, spirometry, as well as EIT were done simultaneously at 2 different thoracic levels, i.e., the third and the fifth intercostal space [45]. Airway obstruction and heterogeneity in lung aeration could be studied by obtaining pixel ratios corresponding to changes in impedance equivalent to forced expiratory volume in 1 s (FEV1) and forced vital capacity (FVC). Findings imply that results obtained at more cranial thoracic planes will help more in deciding the best form of treatment, for instance, targeted physiotherapy.

This work validates EIT by comparing EIT measures of lung function testing in CF patients with those of spirometry and proposes EIT as a possible replacement for spirometry. EIT measures displayed a remarkable ability to identify cystic fibrosis patients, and also classify the states in the disease. EIT has shown a positive response to the treatment of acute pulmonary exacerbations using intravenous antibiotic treatment [46].

Frerichs et al. [47] designed a novel multiple sensor-based EIT system, with continuous monitoring integrated in a wearable vest. Cystic fibrosis patients may particularly benefit from this system which has been merged with simultaneous collection of multiple bio-signals. Ventilation-related changes in impedance were detected using EIT in various everyday scenarios, with walking showing relatively poorer results. The obtained results show the potential for future monitoring of patients outside clinics.

11.2.4 Pneumonia

Pneumonia is defined as inflammation of the lungs and accumulation of fluid or pus within the lung alveoli. This is caused by a bacterial or viral infection and inflames the air sacs in one or both the lungs. Pneumonia may result in cough with pus, fever, chills, and difficulty in breathing.

A study involving 24 adult subjects who have community-acquired pneumonia (CAP) showed that Electrical Impedance Tomography (EIT) measurements can be used to detect lung function disorder caused by pneumonia in the right and left sides of the lungs, and the results were compatible with X-ray of the chest [48]. The regional ventilation distribution moved to those areas of the lungs that were unaffected by CAP. EIT can be used to monitor lung function continuously while ongoing therapy. Chest X-ray cannot determine the level of pulmonary infiltration and hence EIT can be an appropriate additional tool to assess lung function in CAP.

In a study with 11 patients supported by a mechanical ventilator, EIT as a method was assessed to determine regional lung density as an identifying parameter for lung disease detection [49]. Results showed that EIT measurements were capable of diagnosis of Pneumonia, Atelectasis and Pleural effusion. When the subject was moved from supine to left lateral position, regional lung density value did not change for pneumonia. This is explained by the fact that filling of the alveoli with fluid or pus has no dependency on the posture. The research findings show that a prospective measure for diagnosis of lung disease may be measurement of lung density.

To monitor how EIT can be used as an additional method for the diagnosis of pediatric community-acquired pneumonia, 19 pediatric subjects were chosen who had unilateral pneumonia. Findings show a remarkable agreement between chest radiography and EIT measurements in the identification of the lung which was affected [50]. EIT seems capable of monitoring the lung function for the follow up six-month time frame.

11.2.5 Pleural Effusion

Pleural effusion (PE) is a lung condition that results in excess fluid to build up in the pleura around the lungs. The pleura are thin membranes that outline the lungs and the inside of the chest cavity and work to lubricate and facilitate breathing.

EIT measurements with reduced spatial resolution in 11 patients with pleural effusion exhibited a high-level correlation between the resistivity values of the thorax and the drained pleural fluid [51]. Measurements of impedance preceding and succeeding the drainage of pleural fluids support the application of bioimpedance technique and the approach of parametric lung resistivity in reconstruction, in specific, for pleural effusion patients.

A research work introduces a novel 5 electrode-based parametric Electrical Impedance Tomography (pEIT) approach and investigates the ability of this technique in the identification and monitoring of pleural effusion. This technique uses a comparatively smaller configuration of electrodes in a 3D computerized model of the human thorax [52]. Results also propose that cross-sectional projections are more sensitive for the detection of pleural effusion and more important for determining the amount of fluid in pleura within the lungs.

Nebuya et al. [49] showed that EIT measurements were able to diagnose Pneumonia, Atelectasis and Pleural effusion. Since pleural effusion results in the accumulation of excessive fluid in the pleura, which has a comparatively higher density than healthy lungs. Findings show that obtained lung density value was the greatest for pleural effusion patients. Change in posture from supine to left lateral, however, caused this value to change. This was explained by the fact that changes in posture resulted in the movement of fluid in the lungs.

Changes in impedance can paradoxically fall during inspiration. These impedance changes which are out of phase have been associated with PE by Becher et al. [53]. The study findings, which involved 20 ARDS patients, imply that such out-of-phase changes are characteristic of pleural effusion patients, and decrease remarkably with its drainage. The sum of these changes in impedance can accurately identify patients who have pleural effusion and must be followed by further monitoring. Table 11.2 provides a summary of some of the publications included in this review.

11.3 Discussion and Future Challenges

Even with the recent progress of EIT based disease diagnosis, there are still limitations that have to be overcome. In this section, such challenges and limitations have been included, with some possible solutions. The most basic limitation of Electrical Impedance Tomography (EIT) is the low spatial resolution, in contrast to other existing imaging techniques.

11.3.1 Development of Readily Available Data Sets

Before carrying out any sort of clinical trial using EIT, patient-tailored hardware interfaces and image reconstruction algorithms are required. A collaboration between physicians and manufacturers can provide patient-specific and age-appropriate solutions [13]. Subject's chest shape and measurements are critical in obtaining correct image representations. Existing data sets of chest shapes and measurements by age, gender, weight, etc. can immensely help in this and more work towards developing such data sets should be encouraged. Electrode interfaces have to be personalized now to account for the diversity of lung patients, from neonates and children, obese patients to those who are critically ill.

Table 11.2 Summary of some of the publications investigated in this review; ARDS: Acute Respiratory Distress Syndrome; ALI: Acute Lung Injury; COPD: Chronic Obstructive Pulmonary Disease

Publication	Year	Disease	Electrode Configuration	Image Reconstruction Algorithm	Additional Tools	Data set	Subjects
Gómez-Laberge et al. [23]	2012	ARDS/ ALI	16 electrodes Geo-MF II system or Dräger EIT Evaluation KIT 2	The one-step linear Gauss–Newton solver implemented by EIDORS	MATLAB Signal Processing Toolbox	Self	9 pediatric ALI/ARDS patients
Pulletz et al. [22]	2012	ARDS	16 electrodes Geo-MF II system	Modified Sheffield Back-projection Algorithm	A customised MATLAB toolbox	Self	20 ARDS patients and 12 healthy controls
Blankman et al. [26]	2014	ALI	16 electrodes PulmoVista 500 EIT device	Data analysis using special software by Dräger Medical	NA	Self	20 mechanically ventilated post-cardiac surgery patients
Yun et al. [32]	2016	ARDS	16 electrodes PulmoVista 500 EIT device	FEM based linearized Newton-Raphson-Reconstruction Algorithm	Dräger EIT Data Analysis Tool	Self	20 ARDS patients subjected to mechanical ventilation
Vogt et al. [35]	2012	COPD	16 electrodes Geo MF II EIT system	Weighted Back-projection Algorithm	NA	Self	1) 14 young healthy adults 2) 12 elderly healthy adults 3) 33 COPD patients
Frerichs et al. [36]	2012	COPD	16 electrodes Geo MF II EIT system	NA	NA	Self	10 patients requiring ICU treatment who failed on noninvasive ventilation
Frerichs et al. [37]	2016	Asthma	16 electrodes Geo MF II EIT system	GREIT Reconstruction Algorithm	NA	Self	7 chronic asthma patients and 7 matched healthy controls
Schullcke et al. [39]	2017	COPD	3D simulation model was used for blocked lungs, with 16 rectangular equidistantly placed electrodes	FEM models were generated using NETGEN. The lungs were incorporated by allocating individual FEM elements as 'lung tissue'	Custom-made MATLAB function	CT Dataset of a human male subject	Simulation of electrode voltages using 3D FEM model
Zhang et al. [40]	2018	COPD	16 electrodes PEIT4 system	Image Reconstruction using GREIT software with EIDORS	NA	Self	1) Control group, 41 subjects 2) Mild group, 17 patients 3) Moderate group, 16 patients 4) Moderate group, 16 patients 5) Severe group, 8 patients
Karsten et al. [48]	2014	Pneumonia	16 electrode belt	Modified Newton-Raphson Reconstruction Algorithm	Data analyzed by DataReview software	Self	24 community-acquired pneumonia patients
Nebuya et al. [49]	2015	Pneumonia, Pleural Effusion and Atelectasis	8 electrodes Sheffield Mk 3.5 system	Sheffield Back-projection Algorithm and Sensitivity matrix	NA	Self	11 male patients, 207 times measurement
Mazzoni et al. [50]	2017	Pneumonia	16 electrodes PulmoVista 500 device	Back-projection Algorithm	Dräger EIT Data Analysis Tool	Self	19 Patients (13 male, 6 female)
Arad et al. [51]	2009	Pleural Effusion	8 electrode belt 'PulmoTracePro' system	Modified Newton-Raphson Optimization Technique	NA	Self	11 PE patients (8 male, 3 female)
Becher et al. [53]	2018	Pleural Effusion	16 electrode belt PulmoVista 500 device	Dräger EIT Data Analysis Tool	Excel	Self	20 PE patients, 10 controls without PE
Zhao et al. [41]	2012	Cystic Fibrosis	16 electrodes EIT Evaluation KIT 2	MATLAB	NA	Self	3 CF patients (1 male, 2 female) 2 healthy controls (all male)
Wettstein et al. [43]	2014	Cystic Fibrosis	16 electrodes Goettingen Geo MF II EIT Tomograph	Weighted Back-projection Algorithm	MATLAB	Self	9 CF patients (2 female, 7 male) 11 healthy physiotherapists (5 female, 6 male)
Krueger et al. [45]	2016	Cystic Fibrosis	16 electrodes PulmoVista 500 device	FEM based linearized Newton-Raphson Reconstruction Algorithm	1) Dräger EIT Data Analysis Tool 2) MATLAB	Self	10 CF patients (8 male, 2 female) 10 healthy controls (all male)
Lehmann et al. [44]	2016	Cystic Fibrosis	16 electrodes PulmoVista 500 device	Intregrated FEM based linearized Newton-Raphson-Reconstruction Algorithm	MATLAB	Self	11 CF patients (7 female, 4 male) 11 controls (7 female, 4 male)
Muller et al. [46]	2018	Cystic Fibrosis	32 electrodes ACEI (Active Complex Electrode) EIT system	2D D-Bar Mathod	NA	Self	21 CF patients 14 healthy controls

11.3.2 Well-Designed Clinical Trials

Different trials to evaluate EIT's potential to detect different diseases or lung parameters have been done. A basic improvement that is a must is balancing the test subject group and the control group in terms of number of subjects, gender, age, and weight. Also, testing must be done in appropriate clinical settings for the results to be valid and authentic. Obesity has been considered a factor in many studies, and hence specific testing based on weight should be included to evaluate results. Maximum work on lung diseases has been done for adult subjects. Pediatric respiratory diseases are a growing concern all over the world. Further studies extending these findings to children, especially neonates must be done as well.

Initially, clinical applications should target using EIT for the detection of adverse events. Well-designed and available training tools including possible applications,

EIT image parameters and patterns of different diseases and uses must be developed. This will help clinicians to understand this technique better and make correct diagnostic decisions. Clinical trials must be carried out to validate the procedures and establish such information, and this is probably one of the main future research directions of EIT.

11.3.3 Studying the Effect of Diverse Testing Conditions

EIT measures change values when the subject position is changed from supine to left lateral for specific diseases [49]. It is very important thus to carry out examinations in different positions to address the optimal position according to the disease and patient condition, to identify between healthy lungs and pulmonary diseases. Obstructive lung diseases like COPD and emphysema and their detection have been studied for a long time, but very little work has been done on their degree of obstruction [40]. This field of study needs much more work and practical research.

11.3.4 Compensation of Noise and Errors

Interference and noise in EIT systems may lead to incorrect interpretation. Accuracy in EIT is thus imperative for proper disease diagnosis. While testing, proper positioning of electrodes and getting the right skin contact are tough challenges because of the constant movement of the patient. Identification of this is particularly important, so counteractive measures can be employed to make up for electrodes that may be faulty and to reduce errors due to electrode positioning, by automatic compensation for example. Biological tissue properties must also be accounted for to enhance image quality.

11.3.5 3D Image Reconstruction

Due to the better sensitivity of impedance measurements, 3D image reconstruction is the most recommended. Collecting data in 3D is difficult since the measurement set of electrodes has to cover the whole body surface rather than a particular segment of interest. The development of techniques like finite element method (FEM) make image reconstruction in 3D possible in recent times. Deep learning methods may also be incorporated for modeling and comparisons, in the case of 3D EIT image reconstruction. A challenge, in this case, is that the human thorax model employed in FEM involves truncation of the domain to reduce the evaluation time [54]. Corresponding boundary conditions have to be set correctly and optimization is still under research.

11.3.6 Verification of Phantom-Based Findings

Some recent studies in EIT have been phantom-based. Various approaches with different electrode configurations and parameters have been used. The findings in these cases, though interesting, must be further validated by well-designed and appropriate clinical testing on human subjects and balanced controls.

11.3.7 Other Possibilities

EIT has been recommended for functional lung imaging in various prospects, but static or anatomical lung imaging has been explored much less. Commercially available EIT devices mostly use time-difference EIT for image generation. This method, however, cannot determine the correct volume of the domain of interest.

Observation of lung perfusion employing EIT could help in evaluating the ventilation-perfusion mismatch and understand its importance in the case of respiratory medicine. This specific area of EIT application needs to be focused on more, as lung perfusion has the potential to incorporate cardiac disease monitoring along with lung diseases.

Lung cancer develops in lung tissues, mostly in cells that line the air passages. It is that cancer which causes the highest number of deaths all over the world. However, not a lot of work has been done on EIT-based lung cancer detection. Research so far has mostly been simulation-based. Clinical trials are thus required, including human subjects and healthy controls, and real-time monitoring must be done in appropriate clinical settings. Lung cancer detection is probably one of the latest research sectors in clinical applications of EIT.

Due to being a low-resolution imaging technique, EIT is more suited to be a monitoring technique rather than an imaging technique. Because it has a very high temporal resolution and nonradiative nature, there are good chances of evaluating specific physiological information and parameters for disease identification from functional EIT scans. EIT may be an optimum monitoring technique to directly analyze therapy results in many lung diseases. For this purpose, more work has to be done on EIT software and reference data used for modeling, algorithm analysis, providing an instinctive interface for individual users with appropriate settings and calibrated units [1].

11.4 Conclusion

EIT offers versatile possibilities in medical diagnostic imaging and dynamic lung function monitoring. Applications of EIT in lung disease detection have been reviewed in this chapter, focusing on some specific respiratory diseases, i.e., ARDS,

COPD, Cystic fibrosis, Pneumonia, and Pleural effusion. EIT is moving towards an extremely important stage of its development. Having a relatively high temporal resolution and no requirement of exposure to harmful radiation, EIT is useful for continuous real-time monitoring applications. It can detect acute diseases, as well as monitor complex clinical conditions such as optimization of ventilation monitoring in intensive care units. EIT, however, still needs some standardized methods, terminology, guidelines, and reference data through clinical trials to make its use easier and widespread in clinical applications.

References

1. Adler, A., Boyle, A.: Electrical impedance tomography: tissue properties to image measures. IEEE Trans. Biomed. Eng. **64**(11), 2494–2504 (2017)
2. Brown, B., Barber, D., Seagar, A.: Applied potential tomography: possible clinical applications. Clin. Phys. Physiol. Meas. **6**(2), 109 (1985)
3. Frerichs, I.: Electrical impedance tomography (EIT) in applications related to lung and ventilation: a review of experimental and clinical activities. Physiol. Meas. **21**(2), R1 (2000)
4. Chitturi, V., Farrukh, N.: Spatial resolution in electrical impedance tomography: a topical review. J. Electr. Bioimpedance **8**(1), 66–78 (2019)
5. Allaud, L., Martin, M.: Schlumberger: The History of a Technique. Wiley, Hoboken (1977)
6. Bodenstein, M., David, M., Markstaller, K.: Principles of electrical impedance tomography and its clinical application. Crit. Care Med. **37**(2), 713–724 (2009)
7. Silvera-Tawil, D., Rye, D., Soleimani, M., Velonaki, M.: Electrical impedance tomography for artificial sensitive robotic skin: a review. IEEE Sens. J. **15**(4), 2001–2016 (2014)
8. Henderson, R.P., Webster, J.G.: An impedance camera for spatially specific measurements of the thorax. IEEE Trans. Biomed. Eng. **3**, 250–254 (1978)
9. Woo, E.J., Seo, J.K.: Magnetic resonance electrical impedance tomography (MREIT) for high-resolution conductivity imaging. Physiol. Meas. **29**(10), R1 (2008)
10. Muftuler, L.T., Hamamura, M., Birgul, O., Nalcioglu, O.: Resolution and contrast in magnetic resonance electrical impedance tomography (MREIT) and its application to cancer imaging. Technol. Cancer Res. Treat. **3**(6), 599–609 (2004)
11. Assenheimer, M., Laver-Moskovitz, O., Malonek, D., Manor, D., Nahaliel, U., Nitzan, R., Saad, A.: The T-scan TM technology: electrical impedance as a diagnostic tool for breast cancer detection. Physiol. Meas. **22**(1), 1 (2001)
12. Borsic, A., Halter, R., Wan, Y., Hartov, A., Paulsen, K.: Sensitivity study and optimization of a 3D electric impedance tomography prostate probe. Physiol. Meas. **30**(6), S1 (2009)
13. Frerichs, I., Amato, M.B., Van Kaam, A.H., Tingay, D.G., Zhao, Z., Grychtol, B., Bodenstein, M., Gagnon, H., Böhm, S.H., Teschner, E., et al.: Chest electrical impedance tomography examination, data analysis, terminology, clinical use and recommendations: consensus statement of the translational EIT development study group. Thorax **72**(1), 83–93 (2017)
14. Vogt, B., Ehlers, K., Hennig, V., Zhao, Z., Weiler, N., Frerichs, I.: Heterogeneity of regional ventilation in lung-healthy adults (2016)
15. Aristovich, K.Y., dos Santos, G.S., Packham, B.C., Holder, D.S.: A method for reconstructing tomographic images of evoked neural activity with electrical impedance tomography using intracranial planar arrays. Physiol. Meas. **35**(6), 1095 (2014)
16. Durlak, W., Kwinta, P.: Role of electrical impedance tomography in clinical practice in pediatric respiratory medicine. ISRN Pediatrics, vol. 2013 (2013)
17. Hahn, G., Sipinkova, I., Baisch, F., Hellige, G.: Changes in the thoracic impedance distribution under different ventilatory conditions. Physiol. Meas. **16**(3A), A161 (1995)

18. Fuks, L.F., Cheney, M., Isaacson, D., Gisser, D.G., Newell, J.: Detection and imaging of electric conductivity and permittivity at low frequency. IEEE Trans. Biomed. Eng. **38**(11), 1106–1110 (1991)
19. Hahn, G., Just, A., Dudykevych, T., Frerichs, I., Hinz, J., Quintel, M., Hellige, G.: Imaging pathologic pulmonary air and fluid accumulation by functional and absolute eit. Physiol. Meas. **27**(5), S187 (2006)
20. Jang, G.Y., Ayoub, G., Kim, Y.E., Oh, T.I., Chung, C.R., Suh, G.Y., Woo, E.J.: Integrated EIT system for functional lung ventilation imaging. Biomed. Eng. Online **18**(1), 83 (2019)
21. Barišin, S., Ostovic, H., Prazetina, M., Sojcic, N., Gospic, I., Bradic, N.: Electrical impedance tomography as ventilation monitoring in ICU patients. Signa Vitae: J. Intensive Care Emerg. Med. **14**(Supplement 1), 21–23 (2018)
22. Pulletz, S., Kott, M., Elke, G., Schädler, D., Vogt, B., Weiler, N., Frerichs, I.: Dynamics of regional lung aeration determined by electrical impedance tomography in patients with acute respiratory distress syndrome. Multidiscip. Respir. Med. **7**(1), 44 (2012)
23. Gomez-Laberge, C., Arnold, J.H., Wolf, G.K.: A unified approach for EIT imaging of regional overdistension and atelectasis in acute lung injury. IEEE Trans. Med. Imaging **31**(3), 834–842 (2012)
24. Mauri, T., Bellani, G., Confalonieri, A., Tagliabue, P., Turella, M., Coppadoro, A., Citerio, G., Pesenti, A., et al.: Topographic distribution of tidal ventilation in acute respiratory distress syndrome: effects of positive end-expiratory pressure and pressure support. Crit. Care Med. **41**(7), 1664–1673 (2013)
25. Yoshida, T., Torsani, V., Gomes, S., De Santis, R.R., Beraldo, M.A., Costa, E.L., Tucci, M.R., Zin, W.A., Kavanagh, B.P., Amato, M.B.: Spontaneous effort causes occult pendelluft during mechanical ventilation. Am. J. Respir. Crit. Care Med. **188**(12), 1420–1427 (2013)
26. Blankman, P., Van Der Kreeft, S., Gommers, D.: Tidal ventilation distribution during pressure-controlled ventilation and pressure support ventilation in post-cardiac surgery patients. Acta Anaesthesiol. Scand. **58**(8), 997–1006 (2014)
27. Cinnella, G., Grasso, S., Raimondo, P., D'Antini, D., Mirabella, L., Rauseo, M., Dambrosio, M.: Physiological effects of the open lung approach in patients with early, mild, diffuse acute respiratory distress syndrome: an electrical impedance tomography study. Anesthesiol. J. Am. Soc. Anesthesiol. **123**(5), 1113–1121 (2015)
28. Long, Y., Liu, D.-W., He, H.-W., Zhao, Z.-Q.: Positive end-expiratory pressure titration after alveolar recruitment directed by electrical impedance tomography. Chin. Med. J. **128**(11), 1421 (2015)
29. Gong, B., Krueger-Ziolek, S., Moeller, K., Schullcke, B., Zhao, Z.: Electrical impedance tomography: functional lung imaging on its way to clinical practice? Expert Rev. Respir. Med. **9**(6), 721–737 (2015)
30. Mauri, T., Eronia, N., Turrini, C., Battistini, M., Grasselli, G., Rona, R., Volta, C.A., Bellani, G., Pesenti, A.: Bedside assessment of the effects of positive end-expiratory pressure on lung inflation and recruitment by the helium dilution technique and electrical impedance tomography. Intensive Care Med. **42**(10), 1576–1587 (2016)
31. Hsu, C.-F., Cheng, J.-S., Lin, W.-C., Ko, Y.-F., Cheng, K.-S., Lin, S.-H., Chen, C.-W.: Electrical impedance tomography monitoring in acute respiratory distress syndrome patients with mechanical ventilation during prolonged positive end-expiratory pressure adjustments. J. Formos. Med. Assoc. **115**(3), 195–202 (2016)
32. Yun, L., He, H.-W., Möller, K., Frerichs, I., Liu, D., Zhao, Z.: Assessment of lung recruitment by electrical impedance tomography and oxygenation in ARDS patients. Medicine, **95**(22) (2016)
33. Bellani, G., Rouby, J.-J., Constantin, J.-M., Pesenti, A.: Looking closer at acute respiratory distress syndrome: the role of advanced imaging techniques. Curr. Opin. Crit. Care **23**(1), 30–37 (2017)
34. Putensen, C., Wrigge, H., Zinserling, J.: Electrical impedance tomography guided ventilation therapy. Curr. Opin. Crit. Care **13**(3), 344–350 (2007)

35. Vogt, B., Pulletz, S., Elke, G., Zhao, Z., Zabel, P., Weiler, N., Frerichs, I.: Spatial and temporal heterogeneity of regional lung ventilation determined by electrical impedance tomography during pulmonary function testing. J. Appl. Physiol. **113**(7), 1154–1161 (2012)
36. Frerichs, I., Achtzehn, U., Pechmann, A., Pulletz, S., Schmidt, E.W., Quintel, M., Weiler, N.: High-frequency oscillatory ventilation in patients with acute exacerbation of chronic obstructive pulmonary disease. J. Crit. Care **27**(2), 172–181 (2012)
37. Frerichs, I., Zhao, Z., Becher, T., Zabel, P., Weiler, N., Vogt, B.: Regional lung function determined by electrical impedance tomography during bronchodilator reversibility testing in patients with asthma. Physiol. Meas. **37**(6), 698 (2016)
38. Zhang, J., Qin, L., Allen, T., Patterson, R.P.: Human CT measurements of structure/electrode position changes during respiration with electrical impedance tomography. Open Biomed. Eng. J. **7**, 109 (2013)
39. Schullcke, B., Gong, B., Krueger-Ziolek, S., Moeller, K.: Reconstruction of conductivity change in lung lobes utilizing electrical impedance tomography. Curr. Dir. Biomed. Eng. **3**(2), 513–516 (2017)
40. Zhang, C., Dai, M., Liu, W., Bai, X., Wu, J., Xu, C., Xia, J., Fu, F., Shi, X., Dong, X., et al.: Global and regional degree of obstruction determined by electrical impedance tomography in patients with obstructive ventilatory defect. PloS One **13**(12), e0209473 (2018)
41. Zhao, Z., Fischer, R., Müller-Lisse, U., Moeller, K.: Ventilation inhomogeneity in patients with cystic fibrosis measured by electrical impedance tomography. Biomed. Eng./Biomedizinische Technik **57**(SI-1 Track-L), 382–385 (2012)
42. Zhao, Z., Müller-Lisse, U., Frerichs, I., Fischer, R., Möller, K.: Regional airway obstruction in cystic fibrosis determined by electrical impedance tomography in comparison with high resolution Ct. Physiol. Meas. **34**(11), N107 (2013)
43. Wettstein, M., Radlinger, L., Riedel, T.: Effect of different breathing aids on ventilation distribution in adults with cystic fibrosis. PLoS One **9**(9), e106591 (2014)
44. Lehmann, S., Leonhardt, S., Ngo, C., Bergmann, L., Ayed, I., Schrading, S., Tenbrock, K.: Global and regional lung function in cystic fibrosis measured by electrical impedance tomography. Pediatr. Pulmonol. **51**(11), 1191–1199 (2016)
45. Krueger-Ziolek, S., Schullcke, B., Zhao, Z., Gong, B., Naehrig, S., Müller-Lisse, U., Moeller, K.: Multi-layer ventilation inhomogeneity in cystic fibrosis. Respir. Physiol. Neurobiol. **233**, 25–32 (2016)
46. Muller, P.A., Mueller, J.L., Mellenthin, M., Murthy, R., Capps, M., Wagner, B.D., Alsaker, M., Deterding, R., Sagel, S.D., Hoppe, J.: Evaluation of surrogate measures of pulmonary function derived from electrical impedance tomography data in children with cystic fibrosis. Physiol. Measur. **39**(4), e106591 (2018)
47. Frerichs, I., Vogt, B., Wacker, J., Paradiso, R., Braun, F., Rapin, M., Chételat, O., Weiler, N.: Wearable chest electrical impedance tomography system–a validation study in healthy volunteers. In: Electrical Impedance Tomography, p. 49 (2018)
48. Karsten, J., Krabbe, K., Heinze, H., Dalhoff, K., Meier, T., Drömann, D.: Bedside monitoring of ventilation distribution and alveolar inflammation in community-acquired pneumonia. J. Clin. Monit. Comput. **28**(4), 403–408 (2014)
49. Nebuya, S., Koike, T., Imai, H., Iwashita, Y., Brown, B.H., Soma, K.: Feasibility of using 'lung density' values estimated from EIT images for clinical diagnosis of lung abnormalities in mechanically ventilated ICU patients. Physiol. Meas. **36**(6), 1261 (2015)
50. Mazzoni, M.B., Perri, A., Plebani, A.M., Ferrari, S., Amelio, G., Rocchi, A., Consonni, D., Milani, G.P., Fossali, E.F.: Electrical impedance tomography in children with community acquired pneumonia: preliminary data. Respir. Med. **130**, 9–12 (2017)
51. Arad, M., Zlochiver, S., Davidson, T., Shoenfeld, Y., Adunsky, A., Abboud, S.: The detection of pleural effusion using a parametric EIT technique. Physiol. Meas. **30**(4), 421 (2009)
52. Omer, N., Abboud, S., Arad, M.: Classifying lung congestion in congestive heart failure using electrical impedance-a 3D model. In: 2015 Computing in Cardiology Conference (CinC), pp. 369–372. IEEE (2015)

53. Becher, T., Bußmeyer, M., Lautenschläger, I., Schädler, D., Weiler, N., Frerichs, I.: Characteristic pattern of pleural effusion in electrical impedance tomography images of critically ill patients. Br. J. Anaesth. **120**(6), 1219–1228 (2018)
54. de Castro Martins, T., Sato, A.K., de Moura, F.S., de Camargo, E.D.L.B., Silva, O.L., Santos, T.B.R., Zhao, Z., Möeller, K., Amato, M.B.P., Müeller, J.L., et al.: A review of electrical impedance tomography in lung applications: theory and algorithms for absolute images. Ann. Rev. Control 48, 442–471 (2019)

Chapter 12
Image Analysis with Machine Learning Algorithms to Assist Breast Cancer Treatment

Abu Asaduzzaman, Fadi N. Sibai, Shigehiko Kanaya, Md. Altaf-Ul-Amin, Md. Jashim Uddin, Kishore K. Chidella, and Parthib Mitra

Abstract Real-time imaging technology has the potential to be applied to many complex surgical procedures such as those used in treating people with breast cancer. Key delaying factors for the successful development of real-time surgical imaging solutions include long execution time due to poor medical infrastructure and inaccuracy in processing mammogram images. In this work, we introduce a novel imaging technique that identifies malignant cells and supports breast cancer surgical procedures by analyzing mammograms in real-time with excellent accuracy. According to this method, hidden attributes of a target breast image are extracted and the extracted pixel values are analyzed using machine learning (ML) tools to determine if there are malignant cells. A malignant image is divided into contours and the rate of change in pixel value is calculated to pinpoint the regions of interest (ROIs) for a surgical procedure. Experimental results using 1500 known mammograms show that the imaging

A. Asaduzzaman (✉)
Wichita State University, Wichita, KS, USA
e-mail: Abu.Asaduzzaman@wichita.edu

F. N. Sibai
Prince Mohammad Bin Fahd University, Al Khobar, Saudi Arabia
e-mail: fsibai@pmu.edu.sa

S. Kanaya · Md. Altaf-Ul-Amin
Nara Institute of Science and Technology, Nara, Japan
e-mail: skanaya@is.naist.jp

Md. Altaf-Ul-Amin
e-mail: amin-m@is.naist.jp

Md. Jashim Uddin
Vanderbilt University, Nashville, TN, USA
e-mail: jashim.uddin@Vanderbilt.Edu

K. K. Chidella
University of Nevada at Las Vegas, Las Vegas, NV, USA
e-mail: kishore.chidella@unlv.edu

P. Mitra
HCA Healthcare, Nashville, TN, USA
e-mail: parthibmtr@gmail.com

M. A. R. Ahad and A. Inoue (eds.), *Vision, Sensing and Analytics: Integrative Approaches*, Intelligent Systems Reference Library 207, https://doi.org/10.1007/978-3-030-75490-7_12

mechanism has the potential to identify benign and malignant cells with more than 99% accuracy. Experimental results also show that the rate of change in pixel values can be used to determine the ROIs with more than 98% accuracy.

Keywords Biomedical engineering · Feature extraction · Image processing · Machine learning · Mammography · Surgical procedure

12.1 Introduction

Cancer is a mortal disease that may begin in any internal body part of a patient. After starting, it normally spreads in the same and/or other body parts of the patient [30, 31]. Cancer is the second leading reason of death, after heart attack, in the U.S. and is a major wellbeing issue worldwide. According to the estimates published by the American Cancer Society Cancer Statistics Center, there will be 1,806,950 new cancer cases, of which 279,100 cases are in breast cancer type, only in the U.S. in 2020 [32, 45]. According to the same reports, there will be 606,520 cancer deaths, of which 42,690 deaths are in breast cancer type, in the U.S. in 2020.

In a typical human body, new cells flourish when the body desires those and they substitute old or dead cells. Normal cell division is a must to all living human beings. But, cancerous cells spread without any necessity and never die. Before long these cells materialize a lump, known as tumors [33]. Tumors can be divided into two classes: benign and malignant. Mostly benign tumors are considered to be harmless, because after a specified time, benign tumors do not grow. Unlikely, malignant tumors are dangerous; they do not stops growing, they may even spread to other body parts through the bloodstream. Figure 12.1 shows how cancer cells are different from normal cells [48]. Nucleus of a cancerous cell is larger and darker than that of a normal cell. Many cancerous cells divide and grow together (more than one nucleus in a cytoplasm), which is not the case for normal cells. The sizes and shapes of cancerous cells vary. Abnormal number of chromosomes in a cancerous cell are arranged in a disorganized fashion. A cluster of cancerous cells may grow without a boundary.

Generally, a cancer is named according to the organ where it is discovered. For an example, if cancer is found in the breast of a patient, it is called breast cancer. Figure 12.2 illustrates two anatomical views of human breasts, one is a front view, and the other one is a side view. There are two common types of breast cancer: non-invasive and invasive breast cancer. Figure 12.3 demonstrates the cellular structure of milk ducts: healthy milk duct, non-invasive cancer milk duct, and invasive milk duct [29]. The most common type of breast cancer is invasive breast cancer,it breaks the wall of the tissue and spreads all over the breast. Non-invasive type does not spread or invade to other cells.

In an ordinary (i.e., good) milk gland, cells multiply in an equal and precise fashion. However, in a non-invasive breast, cells increase randomly but stay inside

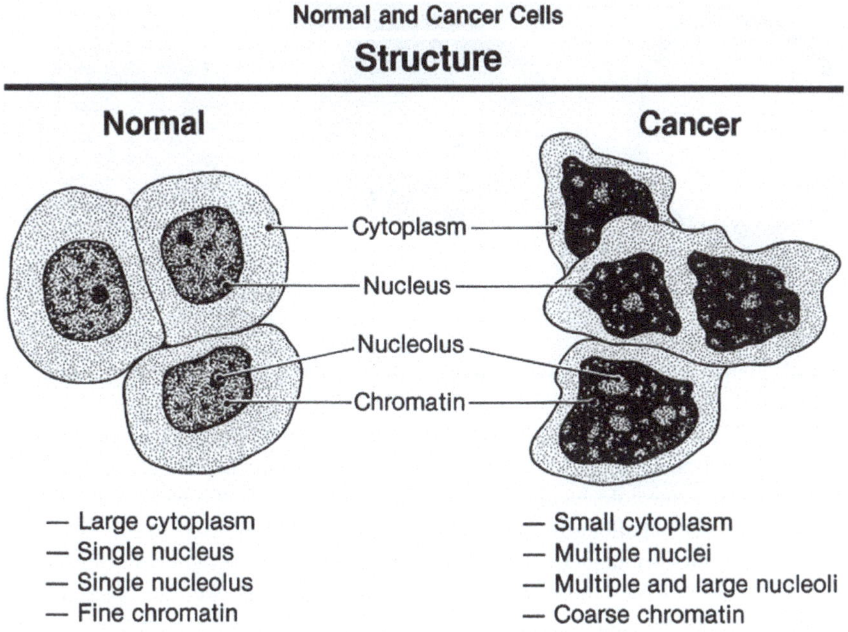

Fig. 12.1 Normal (left) and cancerous (right) cells (courtesy of [48])

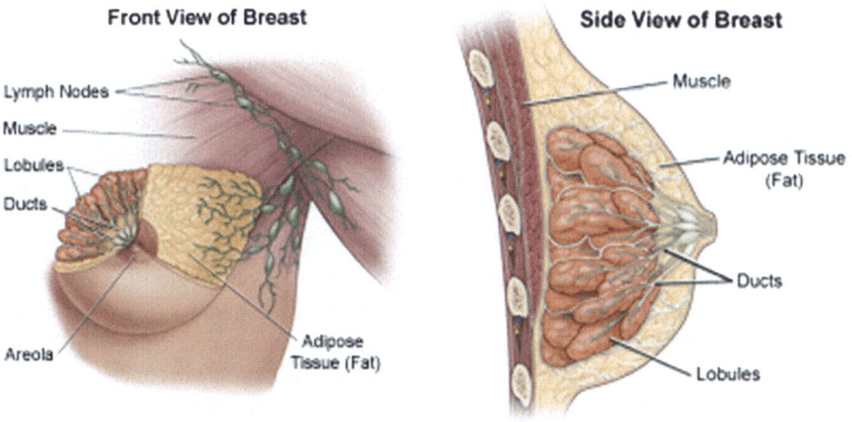

Fig. 12.2 Anatomical structure of women breasts (courtesy of [34])

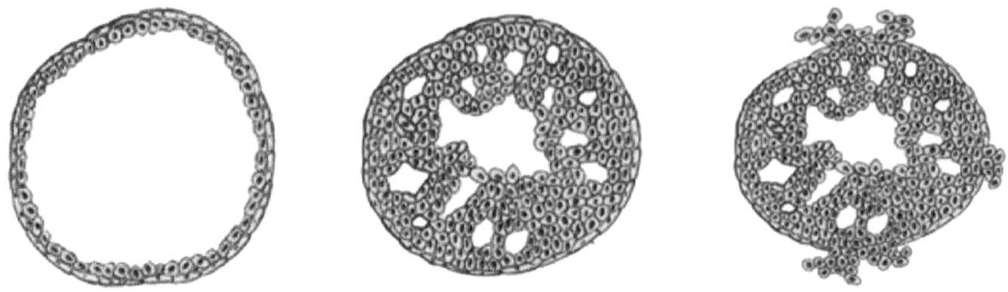

Fig. 12.3 Normal (left), non-invasive (middle), and invasive (right) milk ducts (courtesy of [29])

the milk gland; and in an invasive breast, cells produce randomly and broaden out from the milk gland [42].

Cancer infected breast tissues have more accumulated white spots as compared to normal breast tissues. Because both (normal and cancer infected) breast tissues have white spots and some colorless lines appear due to fatty tissue of breast, cancer detection becomes very challenging. Diagnostic imaging, widely used in breast cancer detection, uses electromagnetic radiation to get detailed images of the internal tissues and helps determine the illness and ensure accurate diagnosis. Well-accepted image-based approaches for cancer treatment include magnetic resonance imaging (MRI), mammography, and molecular breast imaging (MBI). When the imaging diagnosis shows better information on the abnormal cell growth, the physicians have a better tool to provide the most appropriate treatment that can save many lives.

Cancer is most treatable if determined before symptoms are experienced. American Cancer Society's guidelines suggests that the success rate of breast cancer treatment can be improved by detecting it early [30]. Low-dose X-rays in mammography are used to detect cancer early. The effectiveness of using mammogram images for treating breast cancer for younger women is not very clear. For women in the 40- to 49-year age group, mammography is found to be very helpful. For 50 years or older women, mammography is not very effective. Cancer Treatment Centers of America predict about the life expectation of breast cancer patients–about 30% of patients may be living after five years from their first identification of breast cancer [3]. Radiologists and internists uses various Computer Aided Diagnosis (CAD) systems and find those effective for examining breast cancer patients [23, 28]. However, CAD systems may be very complex and may produce inaccuracy in cancer analysis. Doctors commonly use their work experience and academic knowledge to examine breast cancer. A computer system that is able to analyze image data correctly can assist physicians improve accuracy and reduce errors in breast cancer treatment. Studies show that effective analysis of contrasts of mammogram images has potential to accurately predict if there are malignant cells on the images [2]. Contrast is a degree of difference between the elements that forms an image. Contrast depends on several features such as color, textures (e.g., background), and light intensity [43]. Contrast is one of the most important factors of an image because higher contrast gives an image a different feel than lower contrast. Physicians may become confused due to poor contrasts of images and give incorrect diagnosis for breast cancer treatment. Therefore, imaging techniques to address issues due to poor contrast are needed to improve the success rate of breast cancer treatment.

Recently scientists from various research institutions are trying to find computer-assisted solutions to diagnose breast and other types of cancer. The Algorithms, Machines, and People Laboratory scientists at the University of California at Berkeley (UCB) believe that computer scientists have the best skills to fight cancer in the next decade [22, 39]. Scientists from renowned universities (such as UCB, Brown University, University of Washington, and University of South Florida) anticipate that computer engineers and scientists will help treat breast cancer soon [28, 36, 39]. Lately computer-based techniques for analyzing images have turned out to be well-accepted for identifying breast cancer [1, 2, 9]. Studies show that various machine learning (ML) techniques using the Linear Discriminant Analysis (LDA) method [47] and Support Vector Machine (SVM) model [13] are used to analyze data for classification and to find a linear combination of features which characterizes two or more classes of objects [2]. In this work, we introduce a ML-based approach for processing images and accurately identifying malignant breast cells and guide surgical procedures with excellent accuracy.

The rest of the paper is organized as follows. Related published work is summarized in Sect. 12.2. The proposed image processing technique is presented and evaluated in Sects. 12.3 and 12.4, respectively. This work is concluded in Sect. 12.5. Appendix A, at the end, discusses some mammograms showing the images after preprocessing and Region of Interest (ROI) selection.

12.2 Related Work

In this section, we briefly discuss selected research articles that cover three related important areas: methods to treat breast cancer; mammography; and computer-assisted technology in breast cancer treatment.

12.2.1 Methods to Treat Breast Cancer

Most of the time, the breast cancer treatment is a complex and lengthy process. There are many exams at various points in the course of screening, identifying, and treating breast cancer patients. Important breast cancer exams include: physical exam, self-exam, chest X-rays, cancer index test, mammograms, MRI, biopsy, ultrasound, MBI, tumor genomic tests, digital tomosynthesis, computerized axial tomography (CAT) scan, computerized tomography (CT) scan, and positron emission tomography (PET) scans [25, 46]. Generally, the treatment starts by pinpointing the severity of cancer from the reports of MRI or mammogram tests. Image guided therapy plays a vital role in the treatment of breast cancer. The reports of diagnostic imaging serve as a better guide to physicians to have proper follow up for treating breast cancer.

Per recommendation, many people, especially elderly women, undergo thorough regular breast screening. However, the American Cancer Society's guidelines for early detection are not practiced properly. The screening should be considered every one to two years, irrespective of clinical breast examination [21, 27]. With the advances in breast cancer exams, incidences reported in cancer growth have increased [5, 17], fortunately, the mortality rate has been reduced significantly since 1990 [7, 19, 41].

Genetics testing is becoming popular in breast treatment. To avoid the high risk of BReast CAncer (BRCA) mutation, the U.S. Preventive Services Task Force (USPSTF) recommends women to undertake genetic testing from experts to assess the pros and cons [5]. As of now, usual test outcomes do not ensure spotting healthy genes. There are chances that some abnormal genes (that cause cancer) may appear as normal genes. Therefore, if the genes are not identified accurately, the risk of having cancer grows very high. If the exams are conducted to other family members, the genetic tests become time consuming, inconvenient, and quite expensive. A recent study suggests that the likelihood of inherited cancer is about 5%, where the highest risk of breast cancers is intermittent (up to 80%) [11]. It may be worthy to point out that this article does not focus on genetic research for breast cancer. In this work, we study computer-assisted ML-based image processing for breast cancer analysis.

Imaging techniques are effective for "looking" inside the target objects such as breast. Image processing has become a vital and crucial element in various fields such as medical, biomedical, and various clinical practices. In recent years, computer-based quantification and visualization techniques have become more useful in identifying breast cancer. A breast with cancer cells may not have any symptoms, especially at an early stage, which can be noticed from the outside. Among various imaging methods for breast cancer analysis, popular ones are breast tomography, MRI, ultrasound, biopsy, and mammography. In MRI method, a strong magnetic field is created around the breast, which penetrates the cells. As a result, it may produce blurred images, take longer time, and be prone to false positive (or false negative) readings. Breast tomography is a three-dimensional (3D) X-ray imaging technique; because it is moderately new, it may not be accessible in all hospitals and/or medical centers. More importantly, there is no evidence that breast tomography is the best method for breast cancer study [3, 14]. In ultrasound method, a computer run high frequency ultrasound machine is positioned over the breast and involves a mammogram for selecting the doubtful region [8, 15, 18, 50].

In order to screen/diagnose breast cancer and follow-up people with breast cancer, mammography is probably the most important tool that physicians use [20]. In 2002, the USPSTF recommended mammography screening as part of early detection for female over 40 years or older. The physicians may accurately estimate the size of cancer/tumor from mammogram images. Early finding of cancerous cells using digital image processing can advance the healing of cancer patients, thereby enhancing the survival rate.

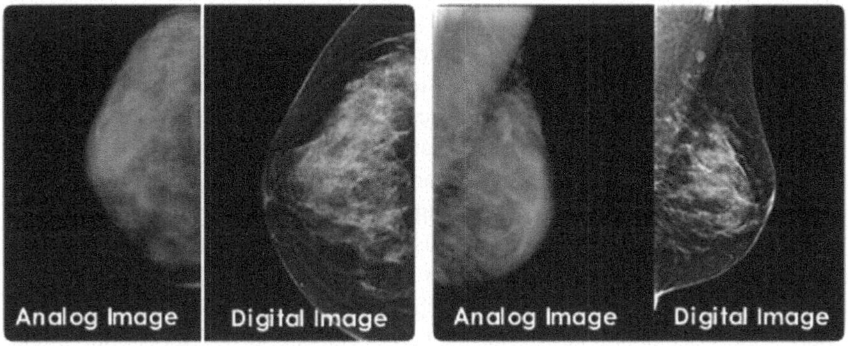

Fig. 12.4 Analog (left) and digital (right) images (courtesy of [35])

12.2.2 Mammography

Mammography is a special X-ray imaging technique that uses low radiation. The image generated by mammography technique is known as mammogram. There are two types of mammograms: analog and digital [20]. An analog mammogram image is like the film of an X-ray image. Analog mammography exploits low dose radiation and can reveal changes in tissue of one to two mm in size. In this technique, the X-ray beams are captured on film cassettes, the outcome becomes a film showing the breast from different angles. Figure 12.4 shows analog and digital images. A digital mammogram produces and stores digital images in a computer. Drawbacks of analog mammograms include a lower contrast than digital mammograms [4, 24]. Advantages of digital mammograms include low signal to noise ratio and enhanced image quality over analog mammograms [26].

The mammography technique is less sensitive to the denser breast; therefore, mammography is not advisable for women above 60 years' age. In addition, due to poor contrast, mammograms contribute to producing large number of false positive (i.e., a region in an image looks like a cancer but actually it is normal) and false negative (i.e., an area looks like benign but actually it is malignant) readings [3, 6]. Therefore, improved techniques are needed to reduce the contrast errors for accurate analysis of the mammogram images.

12.2.3 Computer-Assisted Technology in Breast Cancer Treatment

Other than the CAD systems, computer-assisted technologies such as computed tomography, machine learning, high performance computing (HPC), neural networks, and Android/Apple applications are being advanced to help physicians take care for breast cancer. Scientists from famous computer institutions are improving solutions that physicians use for treating cancer [28, 39]. Software such as LabVIEW, MATLAB [38], MeVisLab, Medical imaging, Bioimaging, and Photomania DX [37] are used to process, analyze, and/or view images. Programming languages such as Java, Python, and C/C++ are used to study cancer related data. Waikato Environment for Knowledge Analysis (Weka) [13], Statistical Package for the Social Sciences (SPSS) [47], and Microsoft Office software package are helpful to scrutinize and categorize the obtained pixel values of mammogram images.

HPC systems and parallel algorithms using graphics processing unit (GPU) are being improved for quicker evaluation of breast cancer data. [16] discuss classification models based on various ML techniques applied on different datasets. A WideResNet-based neural network is built by [49] using the Deep Learning Studio to categorize images to two classes one that contains breast cancer and other that does not. This approach takes a couple of hours and provides more than 85% accuracy. Furthermore, a computer aided detection approach for sorting breast cancer using machine learning and segmentation techniques and achieving up to 94% accuracy was proposed by [40]. In addition, a deep learning algorithm using an "end-to-end" educating approach was introduced to accurately detect breast cancer on screening mammograms. On an independent test set of full-field digital mammography images, the authors claim achieving about 96% accuracy [44]. [12] presented a pilot study about the use of convolutional neural networks for the quick finding of breast cancer via infrared thermography. Resnet34 and Resnet50 provide the best results,an analytical accuracy up to 100%. [10] apply genetic programming to select the best features and perfect parameter values of the ML classifiers. Based on sensitivity, specificity, precision, accuracy, and the roc curves, this study show that genetic programming can automatically find the best model by combining feature preprocessing methods and classifier algorithms.

The following tools are used in this work: MATLAB to process the images; Photomania DX to observe the images; and Microsoft Excel, SVM (via Weka), and LDA (via SPSS) to analyze the pixel values of the images.

12.3 Proposed Image Processing Mechanism

In this study, we introduce a ML-based approach to overcome errors due to poor contrasts of mammogram images by accurately detecting malignant microcalcifications (i.e., small marks inside the breast) through machine learning mechanisms. The methodology then can guide surgical procedures by accurately detecting the ROIs.

12.3.1 Detecting Malignant Cells

To identify if a mammogram has malignant cells, feature values of suspicious regions of an image are extracted. The extracted numerical values of some known ROIs

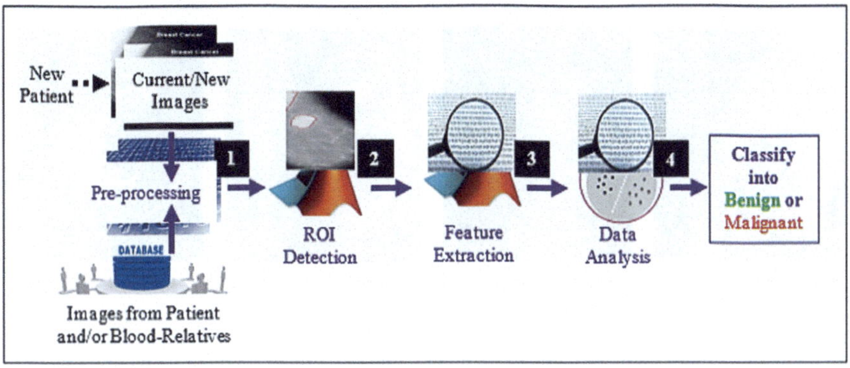

Fig. 12.5 Workflow of the proposed imaging methodology

are used to develop a training function to identify malignant and benign cells. The extracted numerical values of some other known ROIs are used to test the developed function. Figure 12.5 illustrates the essential steps required for the proposed method that include (i) pre-processing, (ii) ROI detection, (iii) feature extraction, and (iv) data analysis (to classify mammogram images as malignant or benign).

All target images are pre-processed so that the ROIs are precisely detected. After detecting ROIs, numerical pixel-values of the ROIs are extracted; the obtained numerical data is analyzed using SVM and LDA methods to classify images as benign or malignant. This process may be repeated as required.

12.3.1.1 Pre-processing

For image-based analyses, pre-processing is mandatory as most biological images are not in the correct format for further processing. In pre-processing, tasks such as removal of background, conversion of image type, contrast adjustment, unsharp mask, and filtering of images are involved. In this work, we use a portable grey map (pgm) image format, which is presently considered as a standard biological image format. Systematic activities in the pre-processing algorithm are shown in Fig. 12.6. Background removal is the first step after receiving a digital mammogram image. In this step, unwanted tissue and/or skin portions are separated from the

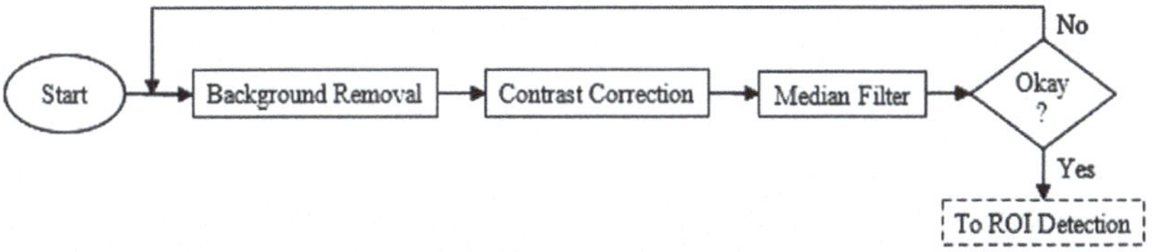

Fig. 12.6 Major steps of pre-processing

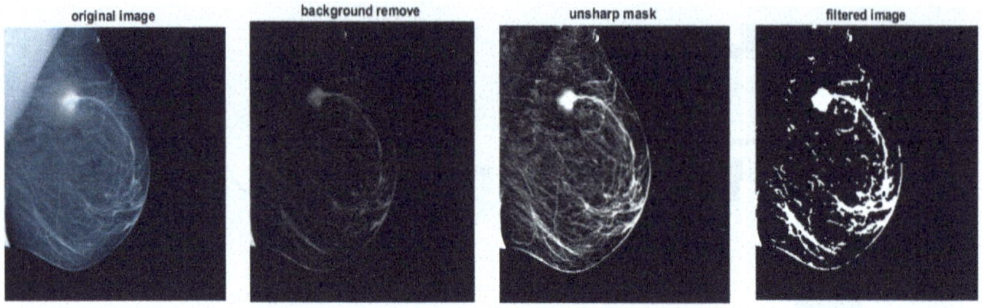

(a) Original image (b) background removal (c) contrast correction (d) median filter

Fig. 12.7 An image during pre-processing

mammogram. Morphological open image functions are used to execute this step. Morphological functions generally perform operations on a grayscale image with help of a structuring element. The pixel value range for images with 12-bit depth is 0 to 4095. As the pixel range is very high, the computation in GpuArray format is performed. GpuArray format uses the GPU, which generates results faster than conventional Central Processing Unit (CPU) computations.

The mammogram loses its brightness after removing the background. The built-in MATLAB functions *unsharp mask* and *imadjust* are used for contrast correction. Contrast correction is performed to improve brightness. *Unsharp mask* returns sharpened images with enriched edges. *Unsharp mask* has a built-in control element so that the quantity of sharpening can be regulated. The *imadjust* function maps the image strength to different values such a way that about 1% of data is saturated at high or low strengths. This process enhances contrast of the mammogram images. Median filtering is the next important step. Figure 12.7 illustrates an image at different major steps during pre-processing.

Throughout the process of background removal and improving contrast, the image may get some noise (i.e., unwanted data). In most experiments, "salt and pepper" type noise is found. The median filter should be selected after reviewing specific cases and types of noise to perform filtering operation. These steps should be repeated as many times as may needed. When the output image of the median filter is acceptable, it is ready for further computation and finding ROIs.

12.3.1.2 Region of Interest Detection

For cancer treatment, accurately identifying the area that contains malignant cells is extremely important. A doubtful vicinity where cancer source microcalcification may be realized is termed as a region of interest. The microcalcification may be produced in a body through a natural process due to the deposits of calcium. Clustered microcalcifications are considered bad because that may cause breast cancer in near future. Therefore, it is crucial to detect microcalcification in order to decide if the detected area is benign or cancerous. The fundamental hypothesis behind the algorithm (known as microcalcification) is that tumors in a breast are composed of calcium oxalate and calcium phosphate. A chemical property of calcium is that it may mitigate more X-rays than other cells. As a result, brighter marks in the mammogram images can be found. This algorithm finds the brighter marks (i.e., the ROIs) in the mammogram images.

After pre-processing, conversion of images to grey is the next step. It is done using a median filter. Conversion of images to grey is required for Red, Green, Blue (RGB) formatted images. For grayscale.pgm images, the conversion is optional. Thresholding is next. In thresholding process, the image is assigned a cut-off value. The cut-off value is determined by counting the frequency distribution in a histogram. In order to scrutinize the distribution of image concentration, histograms are plotted. Pixels with higher than the cut-off value are changed to 1's and the remaining pixels are converted to 0's. This process is known as binary image conversion. The next step records edges of breast cells. An edge is the boundary that separates two regions. To identify the edges of a ROI (i.e., perimeter), we use fuzzy logic. Generally, edge detection methods are created depending on the difference of intensity between two adjacent pixels. Occasionally this difference may be very small to separate the edges. In this process, fuzzy logic is used so that it authorizes control and specifies the criterion to discovery the edge. The image matrix or array used in fuzzy logic should be in double precision format. *GpuArray* is used in this work to process matrices with substantial computations. Using the fuzzy logic, the gradients along the x and y axes are calculated. The gradient is outlined as the slope. To unmask the edge along the x-axis, the x-axis gradient is convolved with the original image. Likewise, the edge along the y-axis can be uncovered by convoluting the y-axis gradient and the original image. The edge can be obtained by reconstructing both axis images. The edge is passed through a filter grid next. The steps described are iterated as many times as may require. The filter grid identifies different ROIs. Figure 12.8 illustrates an image at different major steps during ROI detection. After ROIs are identified, various feature values are extracted for ML-based image analysis.

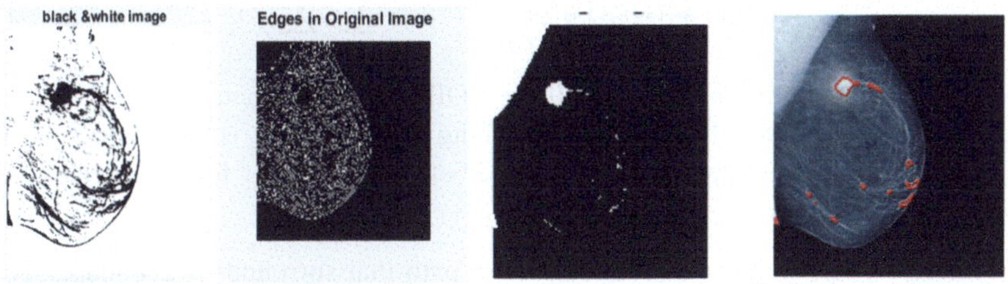

(a) Binary image (b) edge detection (c) ROI detection (d) ROI by red boundary

Fig. 12.8 An image during ROI selection

12.3.1.3 Feature Extraction

This subsection discusses important geometrical features and texture features that are used in this study. These features are considered for accurate analyses of the images. We use a back tracking process to standardize different feature ranges.

12.3.1.3.1 Geometrical Features

Geometrical features (such as radius, area, and perimeter) represent simplest attributes of breast cancer images, but these aspects are important for breast cancer analysis. Physicians usually identify the geometrical features by viewing mammogram images, which may introduce inaccuracies in the diagnosis. In this study, we perform computer-assisted analyses on the following geometrical features: radius, perimeter, and area.

Radius: Radius is the distance from the center of a circle to a point on the circle. Tumors can be circular (i.e., approximately round) or irregular in shape. For circular tumors, the number of pixels from the center of a ROI to a point on the perimeter of the ROI is considered as the radius of the ROI. For irregular-shaped tumors, the longest radius (with maximum quantity of pixels) that covers the entire ROI is considered as the radius of the ROI.

Area: An area is a measure that articulates the extent of a two-dimensional (2D) shape and helps understand the size of the shape. The area of a circular ROI is calculated using the radius (i.e., the quantity of pixels on the radius) of the ROI as shown in Eq. (1a).

$$Area(of\,a\,circular\,shape) = \pi r^2 \tag{1a}$$

Where, π (Pi) is approximately 3.14159, and r is the radius of the ROI.

The area of an irregular-shaped ROI is found by adding the number of pixels inside the ROI using Eq. (1b).

$$Area\,(of\,an\,irregular\,shape) = \sum_i \sum_j A_{i,j} \tag{1b}$$

Where, i = x_ROI[] and j = y_ROI[]. x_ROI[] is a vector containing pixels along the x-axis. Also, y_ROI[] contains pixels along the y-axis. $A_{i,j}$ is the product of the number of horizontal pixels and the number of vertical pixels for one rectangular area inside the ROI.

Perimeter: A perimeter (or boundary) is a path that surrounds a 2D shape. The perimeter of a ROI helps understand the shape of the ROI. The perimeter of a circular ROI is calculated using the radius of the ROI as presented in Eq. (2a).

$$Perimeter(of\,a\,circular\,shape) = 2\pi r \tag{2a}$$

The perimeter of an irregular-shaped ROI is the number of pixels along the horizontal and vertical edges as shown in Fig. 12.3(b). The perimeter calculation for irregular-shaped ROIs is done after edge detection of the ROIs using Eq. (2b).

$$Perimeter\ (of\ an\ irregular\ shape) = \sum_i \sum_j P_{i,j} \tag{2b}$$

Where, i = x_edge[] and j = y_edge[]. These vectors represent the coordinates of the ith and jth pixels, respectively.

12.3.1.3.2 Textural Features

The ability to realize the consistency of a surface is named as texture. In this work, we consider a number of basic statistical functions to characterize texture features such as mean value, standard deviation, entropy, and skewness. For texture feature analysis, matrices are generated for ROIs using the equivalent numeric values of the corresponding pixels as illustrated in Fig. 12.9.

Mean Value: The average amount of a set of numbers is represented as mean. As we know, a ROI consists of pixels, and pixels contain grayscale values. For texture features, we consider each ROI as a matrix of pixels. We calculate the mean (pixel) values in two steps: row-mean and column-mean. The mean feature helps to shed light on the brightness of ROIs.

Standard Deviation: The variation of a data set from its average value is known as standard deviation. If the standard deviation is high, the data set is away from the average. If the standard deviation is low, the data set is very close to the average. Breast cancers are naturally non-uniform in growth, for this reason the standard deviation of cancers is high. MATLAB function *std2* is used to return a scalar standard deviation value. *GpuArray* is used to perform operations.

Entropy of ROI: Entropy can be defined as the level of disorder or randomness of a system. Entropy is an important aspect to grasp texture features of images. Entropy

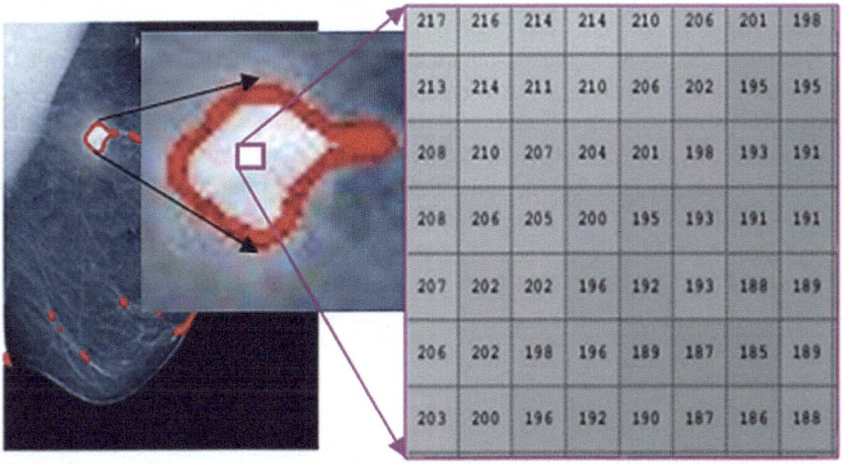

Fig. 12.9 A 7 × 8 matrix is created from 7 × 8 pixels of a ROI

of an image is zero means the image texture is dull. If the entropy value escalates, then the unevenness also escalates. Scientifically, entropy is given in Eq. (12.3).

$$Entropy = -\sum(I \times log_2(I)) \tag{12.3}$$

where, I is the histogram total of an image, and entropy is a scalar value.

Skewness of ROI: The irregularity of the probability distribution function along average value is skewness. Skewness values can be negative or positive. Negative skewness signifies that the bell-shaped probability curve spreads to the left of the mean and positive skewness signifies that the bell-shaped curve spreads to the right of the mean. Microcalcifications usually have negative skewness, but there are some exceptions.

12.3.1.4 Data Analysis

To investigate a complex disease including breast cancer, a maximum number of features should be considered. As a result, the data analysis for breast cancer becomes very complicated. In order to avoid errors during data analyses, we use two popular methods: LDA and SVM. The results of these methods are used to validate each other. The actual extracted feature pixel-values are used in the LDA method. Using the SPSS tool, we first calculate the LDA weight for each feature value, and then we determine the LDA value using the actual feature values and the calculated weights using Eq. (12.4).

$$\begin{aligned} LDA\ Value = {} & Value\text{-}1 \times Weight\text{-}1 + Value\text{-}2 \times Weight\text{-}2 + \ldots \\ & + Final\text{-}Value \times Final\text{-}Weight \end{aligned} \tag{12.4}$$

However, the SVM value is calculated by counting the normalized values. Microsoft Excel software is used to compute the normalized values of the feature values. The Microsoft Excel *Standardize* function computes a normalized value for a distribution by average and standard deviation. Using the Weka tool, we calculate the SVM weight for each feature value, and then we determine the SVM value using the normalized values and the calculated weights using Eq. (12.5).

$$\begin{aligned} SVM\ Value = {} & Normalized\text{-}Value\text{-}1 \times Weight\text{-}1 + Normalized\text{-}Value\text{-}2 \times Weight\text{-}2 + \ldots \\ & + Normalized\text{-}Final\text{-}Value \times Final\text{-}Weight \end{aligned} \tag{12.5}$$

12.3.2 Guiding Surgical Procedures

Determining which portion(s) from a cancerous breast should be taken off during surgery is very crucial. Due to poor contrast, typical mammograms are prone to lead to wrong decisions. Guidance to determine infected region(s) through image processing during surgery should be very helpful. Image guided surgical procedures for treating breast cancer require real-time responses from image processing units. Most image processing techniques take a significantly large amount of processing time. In this work, we introduce a high-performance imaging technique that is capable of identifying ROIs in real-time to help surgical procedures.

In order to guide the surgeons in real-time during the surgery, equipment used is properly interfaced to feed images to a high-performance imaging system. Each image is broken down into multiple contours to pinpoint the ROIs. Each pixel of the image is placed into one of two states depending on the intensity value of that pixel. Mammography images are grayscale and thus only have an intensity of the color for a pixel value and this value ranges from 0–255. The threshold value is set to a desired value by the operator of the program. Figure 12.10 (b) shows the threshold image of the original image shown in Fig. 12.10 (a) using a value of 150 as

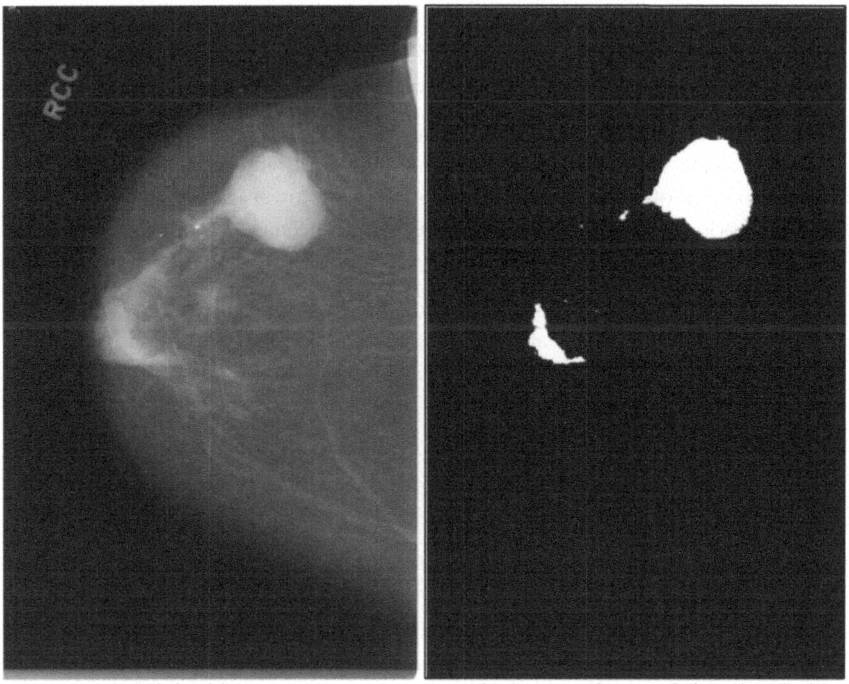

(a) Original image (b) Image after applying threshold

Fig. 12.10 Pinpointing an ROI on a malignant image for surgery

the threshold value. Anything above this threshold value is set to solid white (255), while conversely everything else is set to absolute black (0). The density value may be changed to allow different contours to be drawn on the same area. With a higher density value, the contour will be drawn closer to the centroid of the object, whereas with a lower density value the contour will be drawn further away from the centroid. There needs to be a breakpoint on what the density value is chosen to be, because if the value is set too high then the contour could possibly be drawn on the outside of the desired area of interest. Centroid of each contour is extracted using the OpenCV library. The centroids are used to calculate the rate of change in pixel value.

12.4 Evaluation

In this section, we discuss experimental details and important results to evaluate the proposed image processing methodology.

12.4.1 Assumptions

The following assumptions are made to address some limitations and to assess the proposed ML-based technique for processing images.

- We ignore the compression of images of the MATLAB software. MATLAB may compress a picture file up to 2% of the initial size, affecting accuracy.
- During image format conversion, the loss is much lower ($\leq$1%) and negligible.
- In this work, we use mammogram images only because of their importance in breast cancer treatment.

Among many geometrical and textural features, we consider some important ones (including area, perimeter, mean value, standard deviation, entropy, and skewness) to explain the proposed imaging methodology.

12.4.2 Tools and Languages Used

We use several software packages and programming languages. While selecting images, Photomania DX tool is used to view and assess them. MATLAB is used to process the images, such as generating pixel-value matrices from mammogram images. Microsoft Excel, SVM (via Weka), and LDA (via SPSS) are used to analyze the pixel-value matrices. C programming is used to post-process and verify some results collected from MATLAB programs.

12.4.3 Training and Testing

In the experiments, a total of 1500 mammogram images are used: 1380 DDSM images and 120 MIAS images. The images are split into two sets: (i) training and (ii) testing. First, the image processing (training) function is developed using 1000 cases of the training set. Then the developed method is evaluated using 500 cases of the testing set.

12.4.4 Mammogram Images Used

Mammogram images used in breast cancer study are accessible from the MIAS and DDSM databases. DDSM has a collection of more than 2600 images and MIAS has collection of about 320 digitized images. Herein, we use 1380 DDMS and 120 MIAS images; these images are 12- or 16-bit long with resolution greater than 42 microns. Table 12.1 shows important characteristics of a number of images. The images are divided into two major categories: malignant and benign. As shown in the table, a benign tumor may have false positives (i.e., an area looks malignant but it is actually benign), dense area, benign mass, speculated mass, etc. As noted in Table 12.1, there are many reasons for an image to be malignant.

12.4.5 Experimental Results

In this subsection, we discuss some experimental results. Results for geometrical features are presented first. After that results from texture features are presented. Using the proposed method, geometrical features radius, area, and perimeter are examined. As shown in Fig. 12.11, the radius values for the benign ROIs found are between 0 and 197 pixels and those for the malignant ROIs found are between 13 and 174 pixels; these values overlap.

For area and perimeter, the benign and malignant values also overlap. The minimum and maximum area values for the benign tumors are 0 and 121,922 pixels, respectively. The minimum and maximum area values for the malignant tumors are 907 and 95,114 pixels, respectively. Perimeter values due to benign tumors are between 0 and 1237 pixels, and the same due to malignant tumors are between 106 and 1093 pixels. Because the radius, area, and perimeter values due to malignant and benign ROIs overlap, we conclude that the geometrical values may introduce errors while analyzing cancer cells. Texture attributes such as average and entropy values are considered for more accurate analyses.

Essential texture attributes used in this study are average (pixel) value, entropy, standard deviation, and skewness. Experiments are performed on the pixel counts of the related ROI matrices. In Fig. 12.12, we plot the mean values from the benign

Table 12.1 Sample of Mammogram Images Considered

Image	Source	Known category	Remark
mdb148.pgm	MIAS	Malignant	Speculated mass
mdb179.pgm	MIAS	Malignant	Tumor on entire breast
mdb184.pgm	MIAS	Malignant	Tumor is clearly visible
mdb202.pgm	MIAS	Malignant	Big tumor at middle
mdb099.pgm	MIAS	Benign	Dense normal breast
mdb091.pgm	MIAS	Benign	Dense normal breast
mdb069.pgm	MIAS	Benign	Clean normal breast
mdb063.pgm	MIAS	Benign	Benign mass at middle
mdb032.pgm	MIAS	Benign	Normal breast
mdb021.pgm	MIAS	Benign	False positive at bottom
A-1075-1	DDMS	Malignant	Tumor in the middle
A-1077-1	DDMS	Malignant	False positive tumor
D-4032-1	DDMS	Malignant	Malignant mass
D-4126-1	DDMS	Malignant	Tumor – first stage
D-4141-1	DDMS	Malignant	Malignant tumor

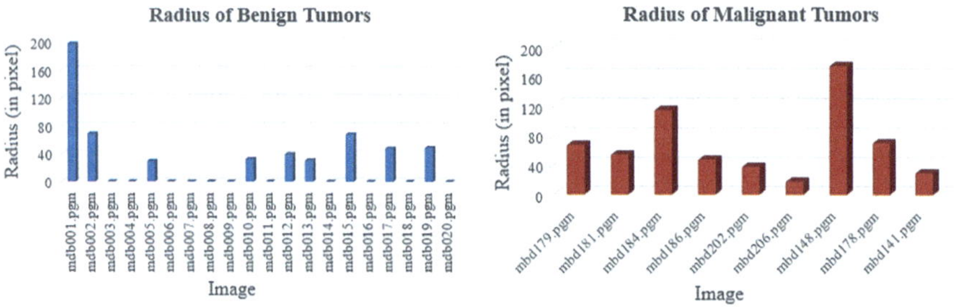

(a) Benign (b) Malignant

Fig. 12.11 Radius of (a) benign tumors and (b) malignant tumors. Some values overlap

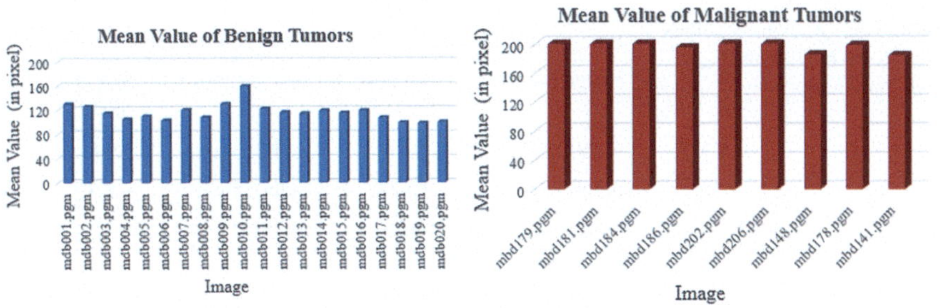

(a) Benign (b) Malignant

Fig. 12.12 Mean values of (a) benign ROIs and (b) malignant ROIs

tumors (98 to 160) in (8a) and the average values from the malignant tumors (165 to 226) in (8b). We observe that the average pixel values using malignant and benign ROIs do not overlap. Hence, the average values can be trusted to distinguish between malignant and benign tumors.

Standard deviation due to benign tumors are between 0.32 and 10.23 and the same due to malignant tumors are between 19.32 and 36.23. The standard deviations due to malignant and benign ROIs evidently fits into two distinct ranges. Therefore, the standard deviation can certainly be considered to distinguish among malignant and benign ROIs.

We examine the entropy values from malignant and benign ROIs as shown in Table 12.2. Entropy values for benign tumors are between 9.25 and −32.36 and entropy values for malignant tumors are between −85.23 and −198.23. Again, there is clear distinction between the entropy values from benign tumors and those due to malignant ROIs. Therefore, the entropy should be used to determine malignant and benign ROIs.

Table 12.2 Entropy values of selected benign and malignant tumors

Benign tumors		Malignant tumors	
Image (.pgm)	Entropy (pixel)	Image (.pgm)	Entropy (pixel)
mdb001.pgm	−5.21	mdb023.pgm	−120.38
mdb010.pgm	9.25	mdb058.pgm	−181.71
mdb013.pgm	−32.37	mdb092.pgm	−91.84
mdb069.pgm	0.98	mdb125.pgm	−175.32
mdb91.pgm	−2.25	mdb141.pgm	−85.37
mdb104.pgm	0.75	mdb170.pgm	−140.41
mdb132.pgm	2.33	mdb181.pgm	−198.23
mdb163.pgm	4.87	Mdb202.pgm	−128.74
mdb198.pgm	−1.94	mdb213.pgm	−175.63
mdb222.pgm	8.47	mdb267.pgm	−110.58

Table 12.3 Skewness values of selected benign and malignant tumors

Benign tumors		Malignant tumors	
Image (.pgm)	Skewness (pixel)	Image (.pgm)	Skewness (pixel)
mdb001.pgm	−1.23	mdb023.pgm	1.99
mdb009.pgm	−10.24	mdb058.pgm	−2.37
mdb010.pgm	51.45	mdb092.pgm	1.37
mdb015.pgm	43.73	mdb125.pgm	14.96
mdb069.pgm	−2.33	mdb141.pgm	2.02
mdb104.pgm	−1.25	mdb179.pgm	25.13
mdb132.pgm	9.81	mdb184.pgm	−6.72
mdb163.pgm	19.56	Mdb209.pgm	−2.46
mdb198.pgm	10.94	mdb265.pgm	13.53
mdb223.pgm	−1.26	mdb270.pgm	−1.64

We also explore the skewness due to malignant and benign ROIs. As shown in Table 12.3, the skewness values due to benign tumors (−10.24 to 51.45) and skewness values due to malignant tumors (−6.24 to 25.13), respectively. There are overlapping values of skewness. Therefore, the impact due to skewness is less significant to segregate among malignant and benign ROIs.

It is noticed that values of several features, such as radius and skewness, due to malignant and benign ROIs coincide. Therefore, additional analysis is required to make these values useful in breast cancer treatment. Using SVM (via Weka) and LDA (via SPSS), we generate classification functions based on the extracted features. The analysis uses the training data set to develop the functions. Then, we test the predictive accuracy of the classification functions by applying them to testing cases.

In addition to calculating the discriminant function, the LDA method is a step-by-step procedure to provide information on the respective significance of the variables in grouping the images. The given dataset is divided into a training dataset (1000 cases) and a testing dataset (500 cases). For MIAS images, the discriminant function to calculate LDA value during training is shown in Eq. (12.6). Because geometrical features are less important to differentiate benign and malignant images, they are not included in Table 12.4. The discriminant function is applied to the testing cases and we find that all images are classified as benign or malignant (as shown in Table 12.4).

$$\begin{aligned}\text{LDA Value(MIAS)} = {} & 0.026 \times \text{Mean-Value} + 0.179 \times \text{Standard-Deviation} \\ & -0.019 \times \text{Entropy} - 0.017 \times \text{Skewness} - 7.887 \qquad (12.6)\end{aligned}$$

In the SVM method, the normalized values are used. Microsoft Excel function calculates normalized values for the original texture values. The SVM sequential minimal optimization (SMO) function is calculated using Weka. For DDSM images, the discriminant function to calculate the SVM value is given in Eq. (12.7), where

Table 12.4 LDA and SVM values to separate benign and malignant images

Known Decision	Mean Value		Std. Deviation		Entropy		Skewness		LDA Value	SVM Value
	Actual	Norm.	Actual	Norm.	Actual	Norm.	Actual	Norm.		
Benign	153.00	−0.080	2.00	−1.096	−6.33	1.000	40.230	−0.039	−3.318	−3.593
Benign	89.00	−1.732	3.60	−0.994	−16.24	0.858	19.236	−0.070	−3.988	−5.307
Benign	162.00	0.153	1.30	−1.141	−10.32	0.943	10.326	−0.083	−2.613	−3.402
Benign	113.00	−1.113	2.40	−1.071	−19.24	0.814	20.370	−0.068	−3.660	−4.691
Malignant	186.00	0.772	26.00	0.443	−170.00	-1.356	−2.326	−0.101	3.405	5.952
Malignant	203.00	1.211	31.00	0.764	−100.00	-0.348	−4.236	−0.104	2.739	4.707
Malignant	198.00	1.082	20.00	0.059	−190.00	-1.644	−4.690	−0.105	3.582	6.227
Malignant	210.00	1.392	36.00	1.085	−200.00	-1.788	−5.236	−0.105	5.653	9.347

only the textural features are considered. Using the SMO function, all DDSM images are separated as benign or malignant.

$$\begin{aligned} \text{SVM Value (DDSM)} = {} & 1.3406 \times \text{Normalized-Mean-Value} + \\ & 1.7355 \times \text{Norm-Std-Deviation} - 2.6728) \times \text{Norm-Entropy} \\ & -0.1070 \times \text{Norm-Skewness} + 0.5316 \end{aligned} \tag{12.7}$$

The LDA value and SVM value are found in agreement between LDA and SVM methods on all MIAS and DDSM images of the test sets. A perfect categorization using the LDA and SVM methods verifies the validity of the results and value of the extracted features. A more than 99% taxonomy rate in the experiments on more than 500 test cases (hold out samples) provides strong evidence on the effectiveness of the discriminant functions and the predictive value of the extracted features in accurately predicting breast cancer. The proposed approach outperforms similar recently proposed approaches while identifying benign and malignant images (Ragab 2019; Shen 2019; [49]).

12.4.6 Guiding Surgical Procedures

The proposed imaging technique is capable of drawing contours of ROIs in real-time during surgical procedures to help surgeons. Figure 12.13 (a) represents a contour of the ROI shown in Fig. 12.10 (b). In Fig. 12.13 (a), the gray area (with a centroid) is the suspected region that needs to be treated. Figure 12.13 (b) is to illustrate the rate of change in pixel value. First, we consider the rate of change from the centroid to a point on the contour line; this is then done for every point on the contour line and averaged out to give the final rate of change value. Then, we calculate the second

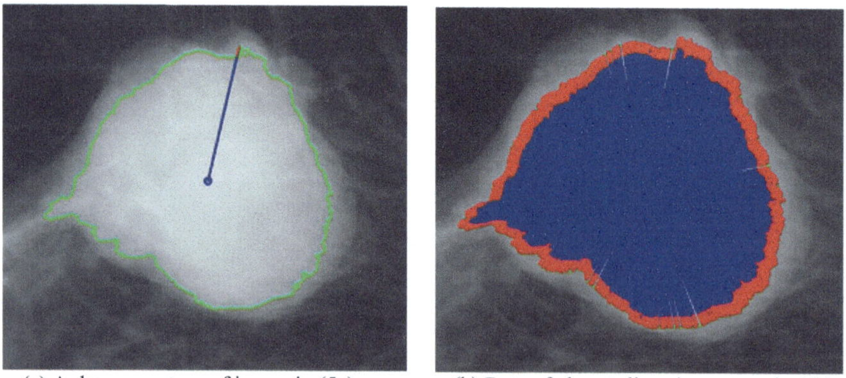

(a) A drawn contour of image in (5a) (b) Rate of change lines in contour in (5b)

Fig. 12.13 Pinpointing an ROI on a malignant image for surgery

rate of change value at the edge of the contour alone; this shows how the shape is growing on the exterior edge. The edge values are calculated over the last 30 pixels towards the edge of the contour; this is again done over every point of the contour and averaged out to give a final rate of change value. While calculating the rate of change at the edge of the contour some inconsistencies in the image itself were found and this was causing the rate of change values to become skewed about 0.05%. This is a major research area that needs more data points and analyses to develop a system for properly guiding the surgical procedures.

12.5 Conclusion and Future Scopes

There are growing demands for building effective methods to treat breast cancer patients. The image processing method can be improved by analyzing mammogram images using ML techniques to address poor contrast issues. In this study, we present a computer-assisted imaging technique to correctly categorize mammogram images as malignant or benign and identify infected areas in real-time during surgical procedures. In the proposed technique, suspicious areas on mammogram images are determined, concealed characteristics including texture and geometrical features of the detected ROIs are obtained, and the extracted ROI matrices are analyzed to categorize the images as malignant or benign. An image enhancement platform is built using MATLAB to extract and analyze the feature values. Then, pattern analysis techniques, LDA (via SPSS) method and SVM (via Weka) model, are used to develop a training function and to evaluate the developed training function using the obtained feature values. The derived formulas of LDA and SVM, which have independent weights, are used to classify images as malignant or benign.

From the DDSM and MIAS databases, 1500 images are carefully selected for experiments. Randomly selected 1000 of those images are used for training the proposed imaging method and the remaining 500 images are used for testing the training method. According to the experimental results, some features such as *area pixel-value* are less meaningful in distinguishing benign from malignant images. However, features such as *mean pixel-value, entropy, and standard deviation* are extremely significant, and therefore, effective to correctly differentiate malignant and benign images. According to the experimental results, the proposed methodology correctly assesses more than 99% of the test set images as malignant or benign. In addition, the rate of change in pixel values can be used to determine the ROIs with more than 98% accuracy. Experimental results convincingly suggest that the proposed imaging technique is effective in reducing errors due to poor contrast, classifying mammogram images as malignant or benign, and identifying cancerous infected regions to assist surgical procedures.

In view of the growing demands on 3D images analysis, we plan to extend this work to 3D medical imaging and 3D data processing.

Appendix A

This appendix briefly discusses some selected images showing the initial image, image after preprocessing, and image after Region of Interest (ROI) detection.

Table 12.5 shows Images mdb021.pgm and mdb063.pgm, which are predetermined as benign images. During the experiments, we observe that there is a false positive in Image mdb021.pgm at the lower bottom. Similarly, Image mdb063.pgm has a false positive in the middle. No ROI is determined for these images.

Table 12.6 shows two other benign images, Images mdb069.pgm and mdb080.pgm. During the experiments, we observe that these images show visible milk ducts and small ROI. However, after completing the analysis of the extracted feature values, these images are categorized as benign, as expected

As shown in Table 12.7, Images mdb150.pgm and mdb184.pgm are predetermined as malignant. After preprocessing, mdb150.pgm shows several ROIs; after analysis,

Table 12.5 Images after preprocessing and detecting ROI - no ROI, benign tumors

Original Image	After Preprocessing	After ROI Detection
mdb021.pgm		
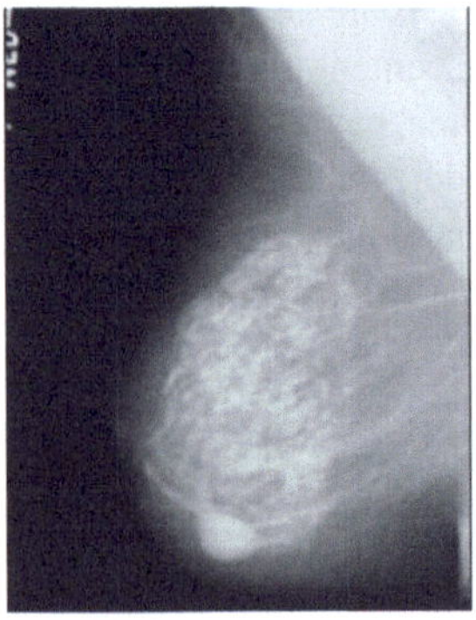	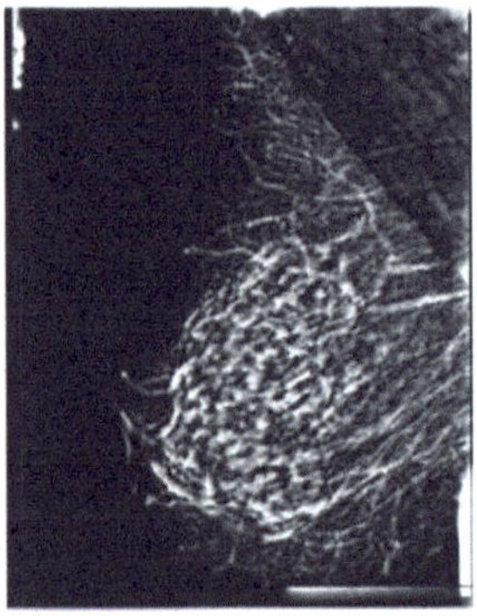	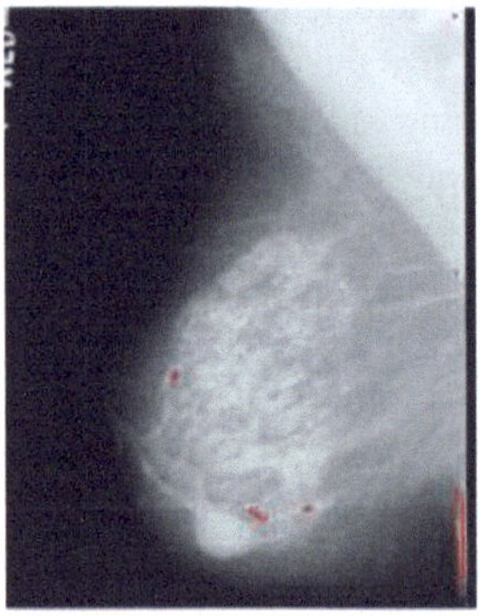
mdb063.pgm 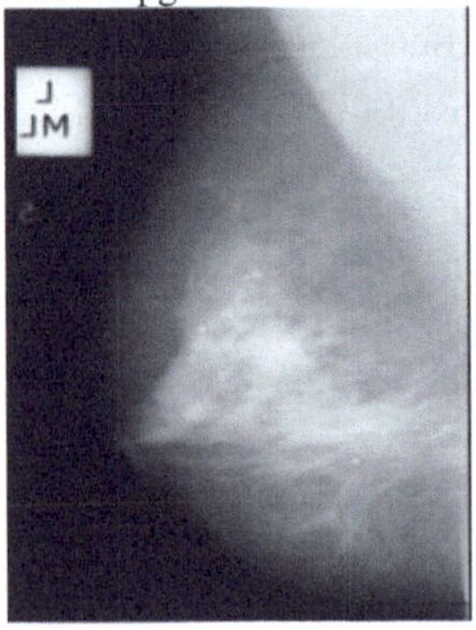	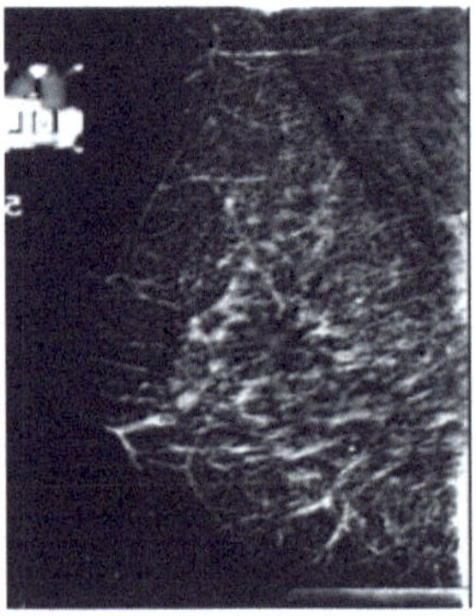	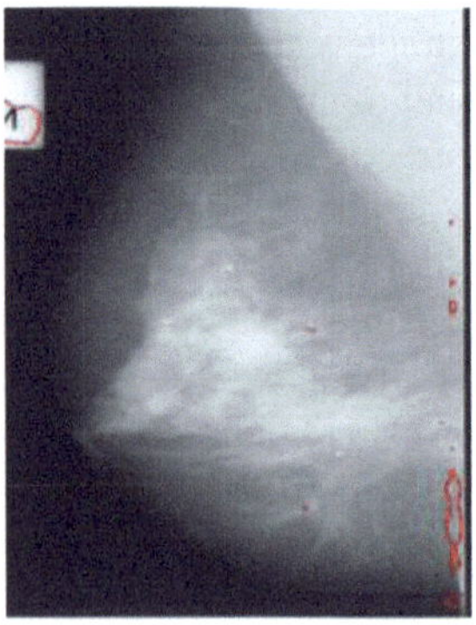

Table 12.6 Images after preprocessing and detecting ROI - ROI detected, benign tumors

Original Image	**After Preprocessing**	**After ROI Detection**
mdb069.pgm		
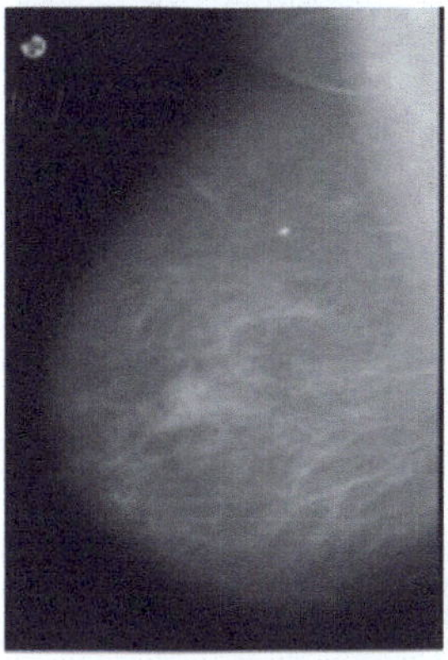	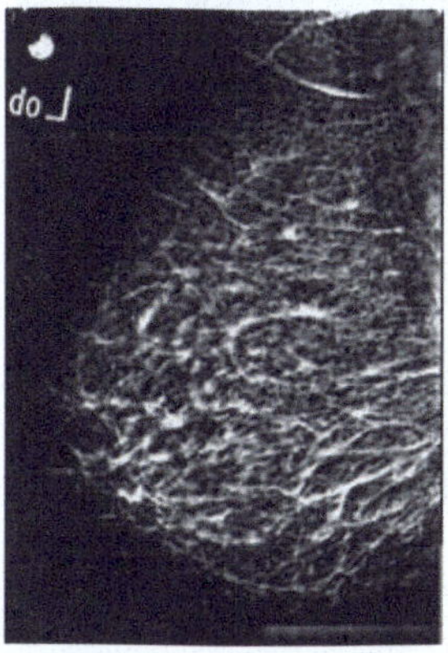	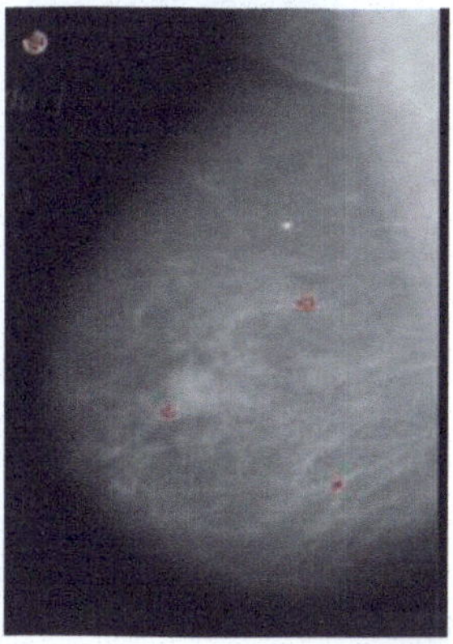
mdb080.pgm 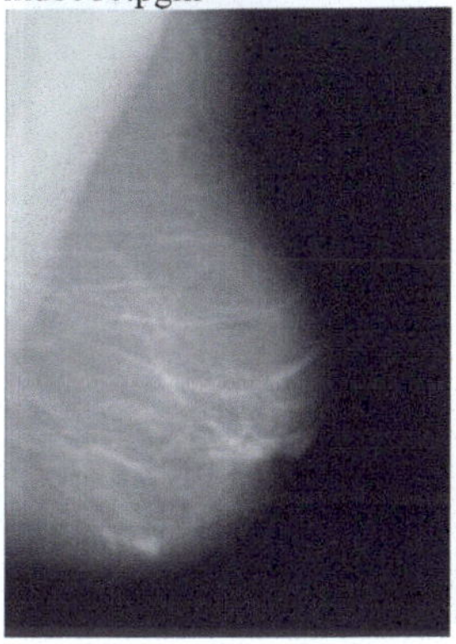	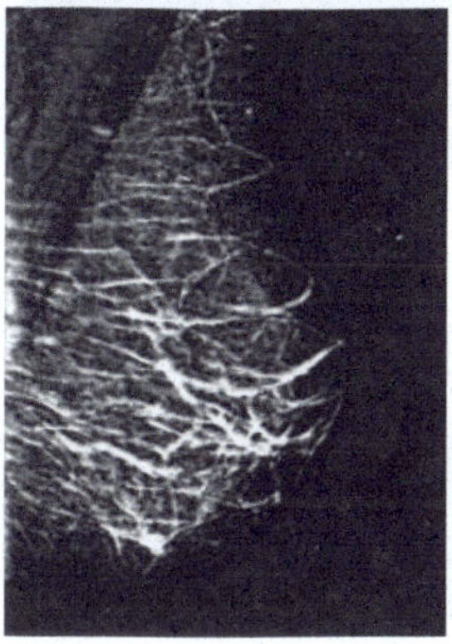	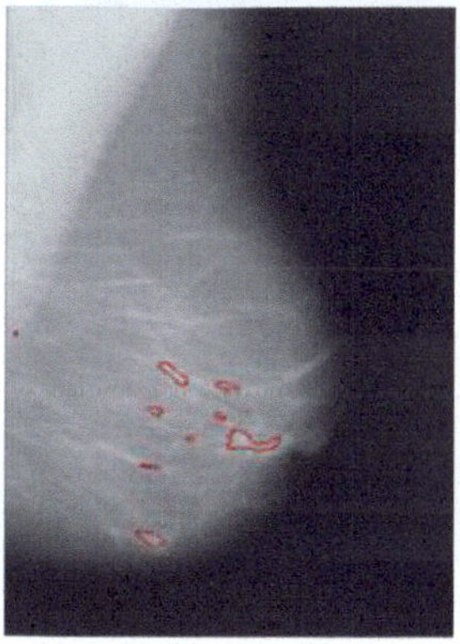

upper middle tumor is classified as malignant. After preprocessing mdb184.pgm, two dark spots in the image are detected at the upper part of breast. The spots represent two skeptical regions; one region is circular and is a malignant tumor; but the other one is skin of the chest.

Table 12.7 Images after preprocessing and detecting ROI - ROI detected, malignant tumors

Original Image	After Preprocessing	After ROI Detection
mdb150.pgm 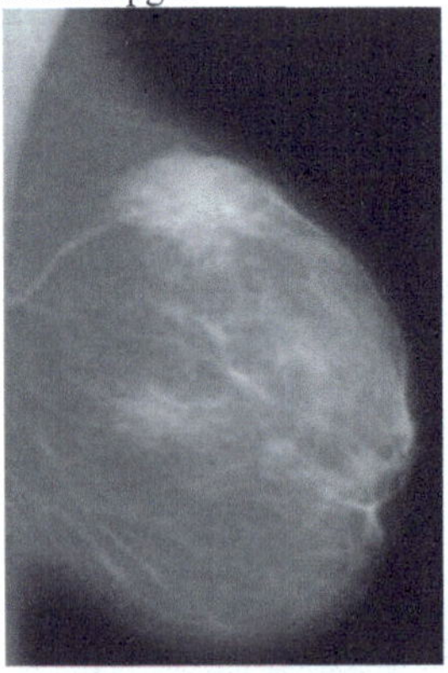	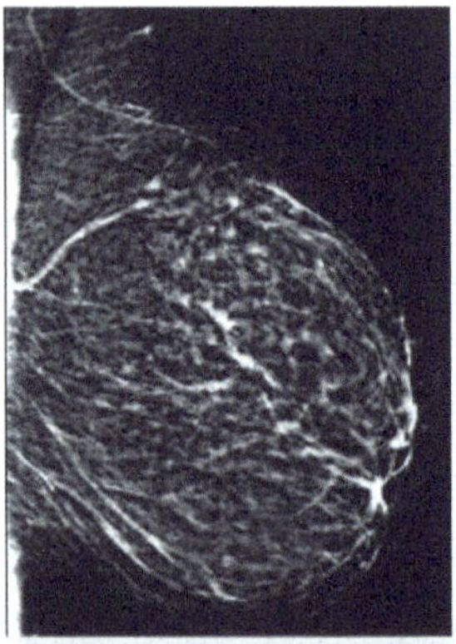	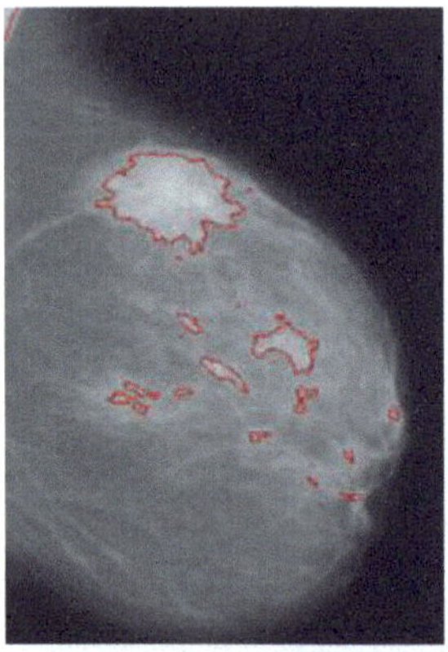
mdb184.pgm 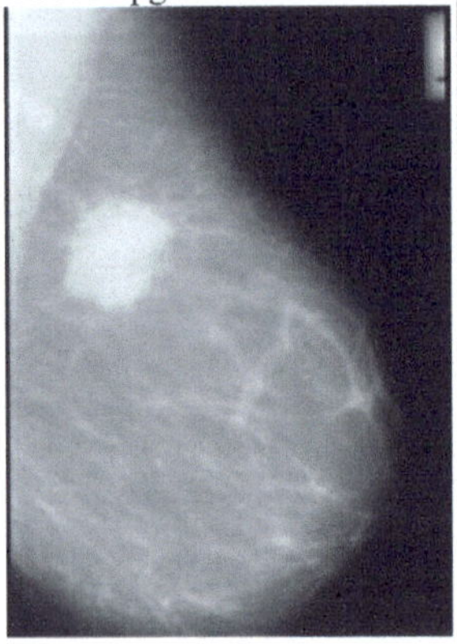	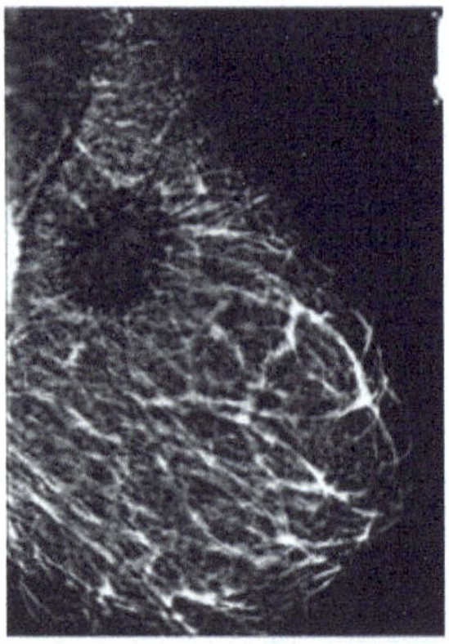	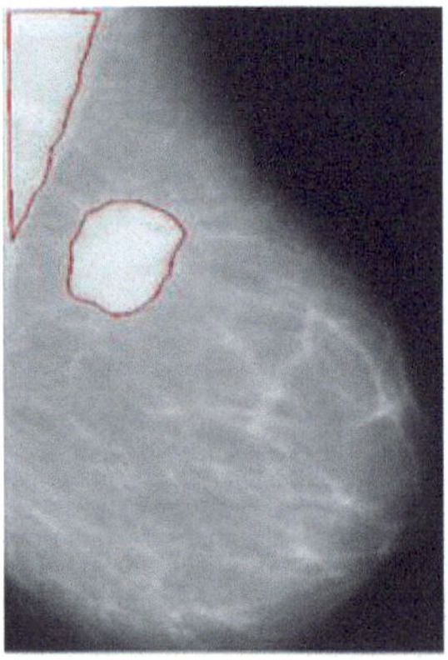

References

1. Anami, B.S., Unki, P.H.: Multilevel thresholding and fractal analysis-based approach for classification of brain MRI images into tumor and non-tumor. Int. J. Med. Eng. Inform. **8**(1), 1–13 (2016)
2. Asaduzzaman, A., Mitra, P., Chidella, K.K., Saeed, K.A., Cluff, K., Mridha, M.F.: A computer-assisted mammography technique for analyzing breast cancer. In: IEEE International Conference on Advances in Electrical Engineering (ICAEE), Bangladesh (2017)
3. Bird, R.E., Wallace, T.W., Yankaskas, B.C.: Analysis of cancers missed at screening mammography. Radiology **184**(3), 613–617 (1992)
4. Bleyer, A., Welch, H.G.: Effect of three decades of screening mammography on breast-cancer incidence. N. Engl. J. Med. **367**(21), 1998–2005 (2012)

5. Calonge, N., Petitti, D.B., DeWitt, T.G., Dietrich, A.J., Gregory, K.D., Grossman, D., Isham, G., LeFevre, M.L., Leipzig, R.M., Marion, L.N., Melnyk, B.: Screening for breast cancer. Ann. Intern. Med. **151**(10), 716–726 (2009)
6. Carney, P.A., Cook, A.J., Miglioretti, D.L., Feig, S.A., Bowles, E.A., Geller, B.M., Elmore, J.G.: Use of clinical history affects accuracy of interpretive performance of screening mammography. J. Clin. Epidemiol. **65**(2), 219–230 (2012)
7. Chagpar, A.B., McMasters, K.M.: Trends in mammography and clinicalbreast examination: a population-based study. J. Surg. Res. **140**(2), 214–219 (2007)
8. Cheng, H.D., Shan, J., Ju, W., Guo, Y., Zhang, L.: Automated breast cancer detection and classification using ultrasound images: a survey. Pattern Recogn. **43**(1), 299–317 (2010)
9. Cluff, K., Miserlis, D., Naganathan, G.K., Pipinos, I.I., Koutakis, P., Samal, A., McComb, R.D., Subbiah, J., Casale, G.P.: Morphometric analysis of gastrocnemius muscle biopsies from patients with peripheral arterial disease: objective grading of muscle degeneration. Am. J. Physiol.-Regul. Integr. Comp. Physiol. **305**(3), R291–R299 (2013)
10. Dhahri, H., Maghayreh, E.A., Mahmood, A., Elkilani, W., Nagi, M.F.: Automated breast cancer diagnosis based on machine learning algorithms. Hindawi J. Healthc. Eng. **2019** (2019). https://doi.org/10.1155/2019/4253641
11. Edwards, B.K., Brown, M.L., Wingo, P.A., Howe, H.L., Ward, E., Ries, L.A., Friedman, C.: Annual report to the nation on the status of cancer, 1975–2002, featuring population-based trends in cancer treatment. J. Natl. Cancer Inst. **97**(19), 1407–1427 (2005)
12. Fernández-Ovies, F.J., Santiago Alférez-Baquero, E., de Andrés-Galiana, E.J., Cernea, A., Fernández-Muñiz, Z., Fernández-Martínez, J.L.: Detection of breast cancer using infrared thermography and deep neural networks. In: IWBBIO-2019 Bioinformatics and Biomedical Engineering, vol. 11466, no. 1, pp. 514–523 (2019)
13. Garner, S.R.: WEKA: the waikato environment for knowledge analysis. In: Proceedings of the NZCSRSC, NZ, pp. 57–64 (1995)
14. Gowri, D.S., Amudha, T.: A review on mammogram image enhancement techniques for breast cancer detection. In: ICICA, pp. 47–51 (2014)
15. Holleczek, B., Brenner, H.: Trends of population-based breast cancer survival in Germany and the US: decreasing discrepancies, but persistent survival gap of elderly patients in Germany. BMC Cancer **12**(1), 317 (2012)
16. Houfani, D., Slatnia, S., Kazar, O., Zerhouni, N., Merizig, A., Saouli, H.: Machine learning techniques for breast cancer diagnosis: literature review. In: Ezziyyani, M. (eds.) Advanced Intelligent Systems for Sustainable Development (AI2SD-2019). Springer Journal on Advances in Intelligent Systems and Computing, vol. 1103 (2020). https://doi.org/10.1007/978-3-030-36664-3_28
17. Humphrey, L.L., Helfand, M., Chan, B.K., Woolf, S.H.: Breast cancer screening: a summary of the evidence for the US Preventive Services Task Force. Ann. Intern. Med. **137**(5), 347–360 (2002)
18. Kelly, K.M., Dean, J., Comulada, W.S., Lee, S.J.: Breast cancer detection using automated whole breast ultrasound and mammography in radio-graphically dense breasts. Eur. Radiol. **20**(3), 734–742 (2010)
19. Kerlikowske, K., Miglioretti, D.L., Buist, D.S., Walker, R., Carney, P.A.: Declines in invasive breast cancer and use of postmenopausal hormone therapy in a screening mammography population. J. Natl Cancer Inst. **99**(17), 1335–1339 (2007)
20. Krainer, M., Hoover, I., O'Neil, E., Unsal, H., Silva-Arrieto, S., Finkelstein, D.M., Beer-Romero, P., Englert, C., Sgroi, D.C.: Germline BRCA1 mutations in Jewish and non-Jewish women with early-onset breast cancer. N. Engl. J. Med. **334**(3), 143–334 (1996)
21. Kuusisto, K.M., Bebel, A., Vihinen, M., Schleutker, J., Sallinen, S.L.: Screening for BRCA1, BRCA2, CHEK2, PALB2, BRIP1, RAD50, and CDH1 mutations in high-risk Finnish BRCA1/2-founder mutation-negative breast and/or ovarian cancer individuals. Breast Cancer Res. **13**(1), R20 (2011)
22. Lambson, B.: Computer scientists take on cancer research. Berkeley Science Review, Berkeley, CA (2020). https://berkeleysciencereview.com/computer-scientists-take-on-cancer-research/. Accessed 5 Jan 2021

23. Lasztovicza, L., Pataki, B., Szekely, N., Toth, N.: Neural network based micro-calcification detection in a mammographic CAD system. In: IEEE International Workshop on Intelligent Data Acquisition and Advanced Computing Systems: Technology and Applications, pp. 319–323 (2003)
24. Mainiero, M.B., Lourenco, A., Mahoney, M.C., Newell, M.S., Bailey, L., Barke, L.D., D'Orsi, C., Harvey, J.A., Hayes, M.K., Huynh, P.T., Jokich, P.M., Lee, S.-J., Lehman, C.D., Mankoff, D.A., Nepute, J.A., Patel, S.B., Reynolds, H.E., Linda Sutherland, M., Haffty, B.G.: ACR appropriateness criteria breast cancer screening. J. Am. Coll. Radiol. **10**(1), 11–14 (2013)
25. Miller, A.B., Wall, C., Baines, C.J., Sun, P., To, T., Narod, S.A.: Twenty-five years follow-up for breast cancer incidence and mortality of the Canadian National Breast Screening Study: randomized screening trial. BMJ **348**, g366 (2014)
26. Narod, S.A., Ford, D., Devilee, P., Barkardottir, R.B., Lynch, H.T., Smith, S.A., Ponder, B.A., Weber, B.L., Garber, J.E., Birch, J.M., Cornelis, R.S.: An evaluation of genetic heterogeneity in 145 breast-ovarian cancer families. Am. J. Hum. Genet. **56**(1), 254 (1995)
27. Nelson, H.D., Tyne, K., Naik, A., Bougatsos, C., Chan, B.K., Humphrey, L.: Screening for breast cancer: an update for the US Preventive Services Task Force. Ann. Intern. Med. **151**(10), 727–737 (2009)
28. Online, bigthink.com. Can computer scientists stop cancer? (2020). https://bigthink.com/ideafeed/can-computer-scientists-stop-cancer/. Accessed 7 Jan 2021
29. Online, breastcancer.org. Breast cancer symptoms (2020). https://www.breastcancer.org/symptoms/diagnosis/invasive. Accessed 5 Jan 2021
30. Online, cancer.gov. General definition of cancer (2020). https://www.cancer.gov/about-cancer/understanding/what-is-cancer/. Accessed 5 Jan 2021
31. Online, cancer.org. Cancer facts and figures 2017, American Cancer Society. Atlanta, GA (2020). https://www.cancer.org/content/dam/cancer-org/research/cancer-facts-and-statistics/annual-cancer-facts-and-figures/2017/cancer-facts-and-figures-2017.pdf. Accessed 7 Jan 2021
32. Online, cancerstatisticscenter.cancer.org. 2020 Estimates, American Cancer Society Cancer Statistics Center (2020). https://cancerstatisticscenter.cancer.org/. Accessed 5 Jan 2021
33. Online, ch.ic.ac.uk. Normal cell division and cancer cell division image (2020). https://www.ch.ic.ac.uk/local/projects/burgoine/origins.txt.html. Accessed 5 Jan 2021
34. Online, hopkinsmedicine.org. Anatomy of the Breasts, Johns Hopkins Medicine (2021). https://www.hopkinsmedicine.org/health/wellness-and-prevention/anatomy-of-the-breasts. Accessed 7 Jan 2021
35. Online, lincolnradiology.com. Digital Mammography, Lincoln Radiology Group (2020). https://lincolnradiology.com/radiology-services/digital-mammography/. Accessed 7 Jan 2021
36. Online, mammoimage.org. Mammographic Image Analysis Homepage – Databases, University of Zagreb, Zagreb, Croatia (2020). https://www.mammoimage.org/databases/. Accessed 7 Jan 2021
37. Online, softonic.com. Photomania DX – a full featured, photo editing, viewing, and cataloguing app (2020). https://photomania.en.softonic.com/. Accessed 5 Jan 2021
38. Palm, W.J.: MATLAB for Engineering Applications. McGraw-Hill Higher Education, 1st edn. (2020). https://www.mheducation.com/highered/product/matlab-engineering-applications-palm-iii/M9781259405389.html. Accessed 5 Jan 2021
39. Patterson, D.: Do Computer Scientists Hold the Key to Treating Cancer?" the Association for Computing Machinery, the Huffington Post (2017). https://www.huffingtonpost.com/acm-the-association-for-computing-machinery/do-computer-scientists-ho_b_9111292.html. Accessed 7 Jan 2021
40. Ragab, D.A., Sharkas, M., Marshall, S., Ren, J.: Breast cancer detection using deep convolutional neural networks and support vector machines. PeerJ **7**, e6201 (2019). https://doi.org/10.7717/peerj.6201. Accessed 5 Jan 2021
41. Ravdin, P.M., Cronin, K.A., Howlader, N., Berg, C.D., Chlebowski, R.T., Feuer, E.J., Berry, D.A.: The decrease in breast-cancer incidence in 2003 in the United States. N. Engl. J. Med. **356**(16), 1670–1674 (2007)

42. Robinson, B.D., Sica, G.L., Liu, Y.F., Rohan, T.E., Gertler, F.B., Condeelis, J.S., Jones, J.G.: Tumor microenvironment of metastasis in human breast carcinoma: a potential prognostic marker linked to hematogenous dissemination. Clin. Cancer Res. **15**(7), 2433–2441 (2009)
43. Rodriguez-Martinez, S.: What Is Contrast in Photography and How to Use It Correctly (2019). https://expertphotography.com/contrast-in-photography/. Accessed 7 Jan 2021
44. Shen, L., Margolies, L.R., Rothstein, J.H., Fluder, E., McBride, R., Sieh, W.: Deep learning to improve breast cancer detection on screening mammography. Sci. Rep. **9**, 12495 (2019). https://doi.org/10.1038/s41598-019-48995-4
45. Siegel, R.L., Miller, K.D., Jemal, A.: Cancer statistics, 2020. CA Cancer J. Clin. **70**(1), 7–30 (2020). https://doi.org/10.3322/caac.21590
46. Smith, R.A., Cokkinides, V., Brooks, D., Saslow, D., Brawley, O.W.: Cancer screening in the United States, 2010: a review of current American Cancer Society guidelines and issues in cancer screening. CA Cancer J. Clin. **60**(2), 99–119 (2010)
47. SPSS: IBM SPSS Statistics (2017). https://en.wikipedia.org/wiki/SPSS. Accessed 7 Jan 2021
48. Udayangani, S.: Difference Between Cancer Cells and Normal Cells (2010). https://www.differencebetween.com/difference-between-cancer-cells-and-normal-cells/. Accessed 7 Jan 2021
49. Vázquez, F.: Detecting Breast Cancer with Deep Learning (2018). https://towardsdatascience.com/detecting-breast-cancer-with-a-deep-learning-10a20ff229e7. Accessed 7 Jan 2021
50. Yang, W., Dempsey, P.J.: Diagnostic breast ultrasound: current status and future directions. Radiol. Clin. North Am. **45**(5), 845–861 (2007)

Chapter 13
Role-Framework of Artificial Intelligence in Combating the COVID-19 Pandemic

Mohammad Shorif Uddin, Sumaita Binte Shorif, and Aditi Sarker

Abstract COVID-19 caused by the SARS-CoV-2 (Corona) virus, first detected in China in December 2019 is now a worldwide pandemic due to its rapid spreading. People all over the world are fighting against it to combat the pandemic. Maintaining social aloofness and lockdowns can prevent the infection of COVID-19 but if this situation continues then the whole world must be confronted with economic catastrophe. Technology governed by artificial intelligence (AI) is a promising logistic that confirms its effectiveness for the benefits in different sectors like spread prediction, population screening, social awareness, hospital management, healthcare logistics, vaccine and drug delivery, surveillance and tracking, continuation of education and industrial production, etc. This article describes a framework of the role of AI in combating the effects of COVID-19 pandemic in dividing into nine sectors: i) Early trace-out, detection and diagnosis, ii) Disease surveillance, control, awareness build-up, and disease prevention, iii) Monitoring the treatment and predicting the risk of developing severe cases, iv) Screening and helping patients through chatbots, v) Service management through intelligence drones and robots, vi) Management of stress and the spread of rumors through social networks, vii) Understanding the virus through analysis of protein–protein interactions, viii) Speeding up the vaccine and drug discoveries and development, ix) Continuation of education and prediction of economic loss. In addition, an overview of commercialization of the AI strategies by highlighting some success stories is presented.

Keywords COVID-19 pandemic · Artificial intelligence · Machine learning · Social distancing and lockdown

M. S. Uddin (✉) · S. B. Shorif
Jahangirnagar University, Savar, Dhaka 1342, Bangladesh

A. Sarker
Comilla University, Cumilla 3506, Bangladesh

M. A. R. Ahad and A. Inoue (eds.), *Vision, Sensing and Analytics: Integrative Approaches*, Intelligent Systems Reference Library 207, https://doi.org/10.1007/978-3-030-75490-7_13

13.1 Introduction

Started from Wuhan, China in December 2019, the COVID-19 (Coronavirus) is now a world pandemic. As of 16 January 2021, throughout the globe there are more than 94.475 million confirmed cases of COVID-19, more than 2.021 million deaths and more than 67.499 million recovered [1]. The whole world wants to see a more suitable way to control the impact as well as to overcome the pandemic situation of the novel coronavirus rather than having to shut down borders, business, and similar. We are now in the domain of intelligent system developments. Artificial Intelligence (AI) has many finest tools in coping with the current pandemic. If we can take advantage of AI tools and techniques, we can save many lives as well as solve the economic crisis.

In this chapter, we predominantly cultivate about the applications of AI in diverse sectors in combating the COVID-19 pandemic. The main intention of this article is to point out the research issues whose outcomes may serve as inputs for rapid responses in policymaking, medical treatments, and economic rolling. The cost of this Corona pandemic in terms of lives and economic damage is extremely bad and a great uncertainty surrounded us with a gloomy future. AI, one of the most promising data analytic tools, may help in reducing these uncertainties and will pave a way-out. Time demands that data scientists should come forward to take up this challenging task. AI has the opportunity to work against this virus by population screening, medical treatments, continuing education, and running economy, controlling different occurrences, research to understand the virus, mental care, and suggestions about infection control [2–8].

The remainder of this chapter sequentially contains data acquisition, AI role framework, commercialization of AI strategies, and the conclusion.

13.2 Data

The databases of PubMed, Google Scholar, and Scopus using the keyword of COVID-19 for reviewing the literature of coronavirus and the prospects of AI in handling this pandemic. The scarcity of data is hindering the development of many promising AI-based applications. Some open access data sources are Kaggle, a COVID-19 Open Research Dataset Challenge; Research papers on COVID-19 for data mining from Elsevier and Springer Nature; Data from Johns Hopkins Corona Virus Resource Center; Human Coronavirus Innovation Landscape Patent and Research Works Open Datasets to support the search for new and repurposed drugs by Lens, and COVID-19 publicly available Twitter dataset, etc. [9–11]. In addition, there are several datasets [12–22] have been developed for the development of AI-based COVID-19 diagnosis algorithms using radiological images, which are summarized in Table 13.1.

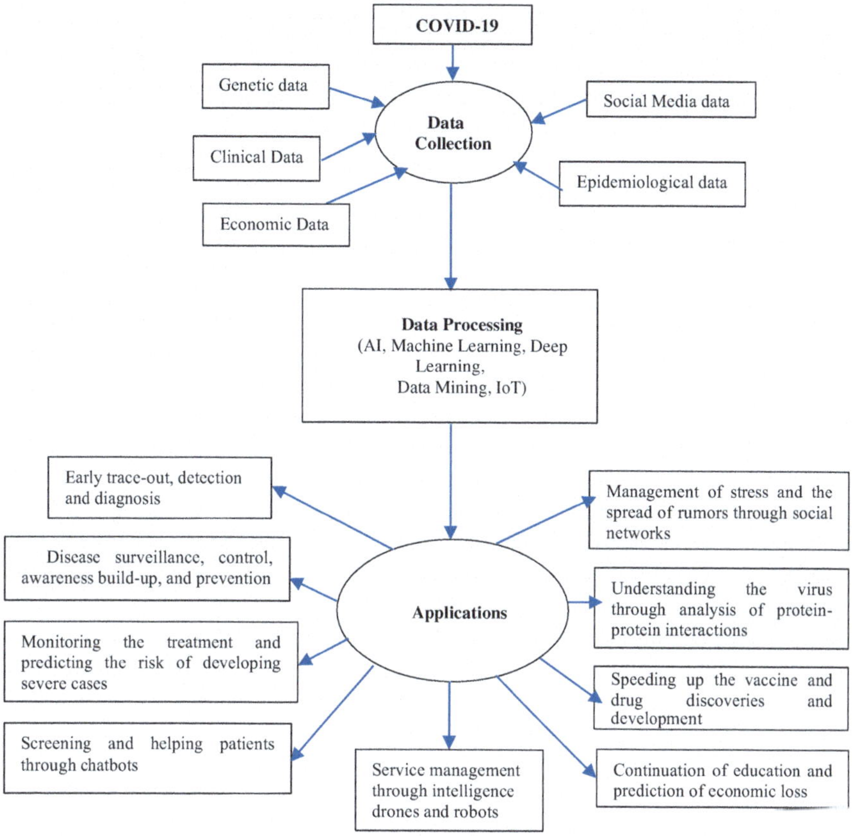

Fig. 13.1 Application framework of AI to combat the COVID-19 pandemic

Table 13.1 Summarization of important open-source datasets for COVID-19 radiological diagnosis

Dataset	Description
Zhao et al. [12]	275 CT scans for COVID-19 positive cases
Jun et al. [13, 14]	A benchmark dataset for annotated COVID-19 segmentation from CT scans
MedSeg [15]	100 axial CT images with COVID-19 from about 40 patients for infected area segmentation
SIRM [16]	Chest X-ray and CT data repository from the Italian society of medical and interventional radiology
BSTI [17]	Imaging database of COVID-19 by British Society of Thoracic Imaging
Radiopaedia [18]	Open source COVID 19 radiographic imaging repository (X-ray+CT)
Kaggle [19–21]	X-ray and CT snapshots of COVID 19 and pneumonia patients
Born et al. [22]	Ultrasound diagnosis image of COVID 19 lungs

13.3 Role Framework of Artificial Intelligence

There are several areas where AI can contribute significantly. Figure 13.1 shows a framework of the role of AI in combating COVID-19.

13.3.1 Early Trace-Out, Detection, and Diagnosis

AI technology helps in analyzing the symptoms and other health conditions and aware of the patients for the prevention of this virus [3–5]. Figure 13.2 shows a schematic diagram of the early trace-out and detection of COVID-19 infection by analyzing different smart sensors, radiological, and social media data. Mining mainstream news help to detect epidemiological pattern for early trace-out. Cameras with AI-based multisensory technology deployed in airports, shopping complexes, hospitals, nursing homes, public places, etc. can help in the detection of individuals with fever, track their movement, face recognition, and checking individuals wearing a face mask or not. Fever level detection can be done by temperature sensor [23], human fatigue detection can be done by gait analysis [24, 25].

AI learning and classification algorithms are playing a vital role in the detection and diagnosis of COVID-19 viral infection with the help of medical imaging systems like X-ray, computed tomography (CT), and magnetic resonance imaging (MRI) scan data of human body parts [26–29]. Linda Wang and Alexander Wong [30] of Canadian start-up DarwinAI designed and developed a deep convolutional neural network (named COVID-Net) for the detection of COVID-19 patients from chest radiography images. Some COVID-Net-based diagnosis apps [31] were developed that can detect infection through cough sample and CT/MRI images. In addition, some smartphone-based methods [23, 32] were developed that can use cameras, microphones, temperature, and inertial sensors for the detection and rapid diagnosis. A detailed survey on architectures, protocols, applicability, effectiveness, security and privacy issues and data management of available tracing apps is presented in [33].

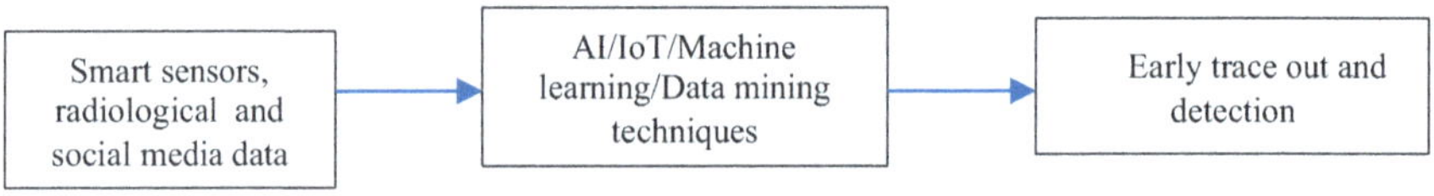

Fig. 13.2 Process of early trace out and detection of COVID-19 infection

13.3.2 Disease Surveillance, Control, Awareness Build-Up, and Prevention

AI can help to detect the risk of infection of an individual and spread in a community at an early stage by monitoring some attributes like age, awareness, general hygiene habits, frequency of human social interactions, location, climate, socioeconomic status, pre-existing health diseases, etc. Travel history and common manifestation of individuals can be collected using a smartphone to predict the infection risk [32–36]. We can use AI-based apps or smart wearable devices to (a) measure the distance between two people in the market area and public places, (b) track whether a person had contact with the confirmed patients of COVID-19, who traveled from other countries with a declared outbreak, or who were living in regions with a high infection of the COVID-19, (c) track individual needs immediate help. AI-based clustering and classification algorithms can identify the community having high spread susceptibility of infection. To assess the infectious risk of a given geographical area at community levels α-Satellite—an AI-based prototype [37] is proposed. Transmission dynamics in a specific area can be predicted by applying contact tracking [38]. Besides, the epidemiological SIR model [39, 40] is used for containment measures by the government.

13.3.3 Monitoring the Treatment and Predicting the Risk of Developing a Severe Case

AI is useful to predict the probability of a patient who will survive with certain initial or existing symptoms. On the other hand, if we can find through machine learning and data mining the outcomes of specific treatment methods, then the doctor can provide treatment more effectively. AI is also used in the prediction of a COVID-19 patient developing ARDS (acute respiratory distress syndrome) as well as the risk of death, just by looking at the initial symptoms [35, 41]. For extracting the visual features of this disease and for monitoring and proper treatment neural networks can be deployed [3, 4, 32].

13.3.4 Screening and Helping Patients Through Chatbots

A chatbot is an AI-based software that is capable of human conversations through voice or texts in natural languages by identifying and analyzing the requests of a user to extract relevant entities. Some well-known chatbots, such as Alexa of Amazon, Siri of Apple, Cortana of Microsoft, s-voice of Samsung, etc. are giving satisfactory performance [42, 43]. Chatbots show promising abilities that could help us by sharing information quickly, symptom detection, encouraging behavioral change support, and reducing the physical and mental damages caused by fear, depression, and isolation [4, 43]. Due to the COVID-19 pandemic, the World Health Organization (WHO) and the Centers for Disease Control and Prevention (CDC) have begun utilizing chatbots to share information, suggest behavior, and offer cognitive support. Though chatbots are very useful in a pandemic like COVID-19, there exist some challenges like misinformation and conflict between global and local authorities.

13.3.5 Service Management Through Intelligence Drones and Robots

Robots and drones can provide a safe contact-free alternative, as COVID-19 is a highly infectious disease. Most of the developed countries have already worked with robots. For example, Tommy [43], a robot nurse is helping doctors in Italian hospitals with the treatment of COVID-19 patients. This robot nurse can measure blood pressure and oxygen saturation level of the patients, and, also can remotely communicate audio-visually with the medical team. COVID-19 spreads not only through contact respiratory droplet transfer among people but also from the contaminated surfaces. Many companies, such as UVD Robots of Blue Ocean Robotics, Denmark, Sunay Healthcare Supply, China, Keenon Robotics, China, Xinhengjia Supply Chain, Hong Kong, etc. have already designed and manufactured many robots for disinfecting hospitals, corridors, roads, and public places. Researchers developed an automatic robotic system based on ultrasound to safely and accurately perform blood draws through guiding the needle on peripheral forearm veins for automated venepuncture [44]. Shenzhen-based company MicroMultiCopter deployed many drones in Chinese cities to efficiently patrol areas and observe crowds who are not wearing masks, and, also monitor traffic. A brief role of robots and drones in combating COVID-19 can be found in [45, 46].

13.3.6 Management of Stress and the Spread of Rumors Through Social Network

Many people are suffering from mental stress due to fear of COVID-19 infection, isolation, death news, and long-term lockdown. AI algorithms can help in monitoring stress through social media like Facebook and Twitter along with some confirmed cases, deaths, recoveries, demographics [8]. Based on social media data analysis AI technology can be used to provide health care and counseling service after extraction of people's activities and mental situations. Due to the rapid spread of coronavirus people are concern about knowing the information, safety, and support in this crisis. But some fraud people try to abuse this helpless situation by spreading fake news about the COVID-19 outbreak, vaccination fear, and uncertainty. The impacts of

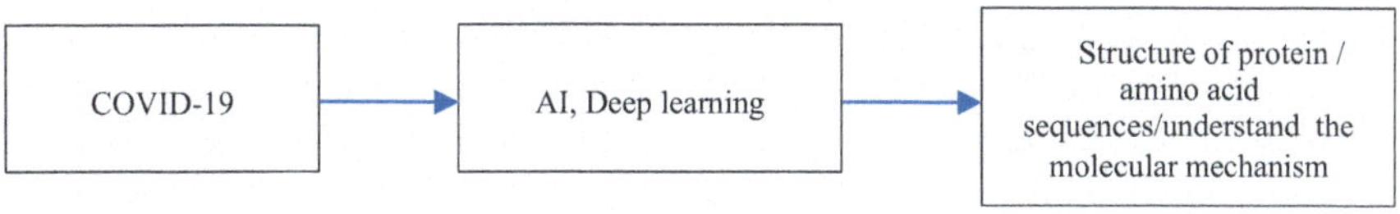

Fig. 13.3 Process of understanding the molecular structure of COVID-19 virus

this news are awful. Big technology companies, such as Google, Facebook, and YouTube have come forward to counter the conspiracy theory of rumors, fake news, misinformation, phishing, and malware [47–49].

13.3.7 Understanding the Virus Through Analysis of Protein–Protein Interactions

Protein–protein interactions between the virus and human body cells are extremely important in determining our body's reaction to pathogens that are quite helpful in the development of new treatment strategies and the discovery of new drugs. In addition, protein–protein interaction is an effective way to understand the molecular mechanism of viral infection analysis [50]. A protein structure provides a vital clue for understanding its functions. However, usual experiments to determine the function of a protein are too time-consuming, hence, the development of computational models using 'template modeling' is a good choice to predict the protein structure from the amino acid sequence. A deep neural network-based computation model named AlphaFold of DeepMind [51] has been developed to predict protein structure accurately through "free modeling." Besides, the design of novel molecules that can inhibit COVID-19 has been made using an AI-based generative chemistry approach [52]. Also, Randhawa et al. [53, 54] proposed an AI-based alignment-free method based on a genomic signature and a decision tree approach to predict the taxonomy of COVID-19. Recently, Nguyen et al. [55] developed an AI-based clustering method to search for the origin of the COVID-19 virus. A flow diagram to understand the molecular mechanism of Corona virus (COVID-19) is shown in Fig. 13.3.

13.3.8 Speeding up the Vaccine and Drug Discoveries and Development

Four steps are involved in vaccine and drug development and discoveries. AI tools and techniques are highly useful in speeding up the required processes [56–61].

a) Identifying targets for intervention: Because of the learning ability of AI algorithms, these can analyze the available protein data to identify good target proteins [56, 57].

b) Discovering drug candidates: AI and machine learning techniques are applicable in finding the molecules that have minimal side effects through filtering millions of potential molecules by analyzing suitability prediction [58, 59].
c) Speeding up clinical trials: AI and machine learning techniques can automatically identify suitable participants as well as ensuring the correct distribution for groups of trial participants. As a result, the design of clinical trials has been speeded up [60–62].
d) Finding biomarkers for diagnosing the diseases: The molecules found in human blood that provide absolute certainty as to whether a patient has a specific disease are known as biomarkers. AI can speed up the process by automating a large portion of the manual works in finding the biomarkers [63, 64].

13.3.9 Continuation of Education and Prediction of Economic Loss

In this COVID-19 global crisis, our education system is seriously hampered. AI can help in coping with this situation in the following ways [6, 65].

(a) Content making: AI technology can make smarter content by analyzing available huge content.
(b) Intelligent tutoring systems: AI can help in designing personalized electronic tutoring based on the adaptivity of students' learning styles, preferences, and comprehension.
(c) Virtual learning environment: Virtual classroom is quite useful in this situation. Google Classroom is a good example of virtual learning system.
(d) Administrative task: Different administrative tasks in an educational institution can be done through AI technology.

Lockdown and blockade are essential to control the spread of COVID-19 infection. But these result in a huge loss of economy. AI techniques are useful in predicting economic loss and this prediction helps the policymakers as well as governments to make viable decisions [7, 66]. In addition, AI can help in running the office and industry through virtual platforms.

13.4 Commercial Applications of AI

Recently, several AI strategies such as deep learning, machine learning, neural networks, evolutionary algorithms, etc. have been developed to perform diverse types of applications to efficiently combat the COVID-19. These research based applications can be categorized as clinical applications, processing COVID-19 related images, pharmaceutical studies, epidemiology, and so on, which are summarized in Table 13.2 [67, 68].

Table 13.2 An overview of AI applications for COVID-19

Applications	Commercial AI strategies	Functions
Tracking and forecasting	BlueDot (Canadian company)	Tracks and predicts the symptoms and activities of the COVID-19 disease over time and space
Diagnosis	Infervision (AI company)	Developed an AI solution to efficiently detect and monitor the coronavirus by increasing the diagnosis speed of CT scan
	Alibaba (Chinese e-commerce)	Introduced an AI based diagnosis system that is capable of diagnosing the virus with an accuracy of 96%
Screening for infected individuals	SenseTime (Chinese surveillance system)	Identifies people, might have a fever and prone to be infected by the virus using the face recognition and temperature detection software
	Smart Helmets (Sichuan official technology)	Screens individuals with fevers
	WeChat or Alipay (Chinese health code monitoring apps)	People are indicated by a color code (red, yellow, or green) using the big data technology to be confirmed that either they should be quarantined or allowed in public
Treatment	Google's DeepMind	Helps to develop new drugs by predicting the protein structure of the virus
	Benevolent AI (UK-based startup)	Recognized a drug—Baricitinib through AI technologies as a potential treatment for COVID-19 that is normally used for rheumatoid arthritis and myelofibrosis
	Gero (Singaporean firm)	Recognized a drug—Afatinib (that is used for lung-cancer treatment) by AI for using in COVID-19 treatment
Supplements delivery by robots and drones	Terra Drone	Transports necessary medical supplies in Xinchang County's disease control centres and hospitals. It is also used to patrol public spaces with minimal risk in quarantine

(continued)

Table 13.2 (continued)

Applications	Commercial AI strategies	Functions
	UVD robots from Blue Ocean Robotics	Automatically kill the corona virus by using ultraviolet
	Chinese robots deployed by Pudu Technology	Deliver food and medicines at more than 40 hospitals around the country
Mask Protection	Sonovia (Israeli startup)	Developed face masks through AI initiatives using anti-pathogen and anti-bacterial fabrics, made from metal-oxide nanoparticles
Healthcare claims processing	Ant Financial (A blockchain platform)	Enhances the claims processing by minimizing the face-to-face interaction between patients and healthcare staffs
Social Control	WeChat (Tencent chatbot to share information)	Provides free online health consultation services and keeps track of the latest information
	Baidu (Chinese search engine company)	Recognizes people, scanning from crowds whose body temperature exceeds the normal reading through thermal imaging by using infrared cameras
	USA computer vision-based startup	Uses camera images to detect when the rules of social distancing are breached for taking actions
	Russian app	Tracks infected people by using a QR system to control the movement
Vaccine Development	Companies e.g., Tencent, DiDi, and Huawei	Working on developing the vaccine for the COVID-19 virus by using cloud computing technology and supercomputers for faster processing

13.5 Conclusion

Artificial Intelligence (AI) finds wide applications in combating the pandemic effects of COVID-19. This work discusses a framework of AI applications in diverse sectors including spread prediction, population screening, tracing apps, social awareness, hospital management, healthcare logistics, vaccine and drug development as well as delivery, surveillance and tracking, continuation of education, running office and industry, impacts on the economy, and so on. The framework divided the AI applications into nine sectors and highlighted AI strategies with some success stories which will certainly help the stakeholders and researchers to focus on pragmatic steps. Besides, this chapter illustrates some commercial uses of AI technologies in COVID-19 detection, forecasting, treatment, vaccine invention, etc. AI has strong potential in combating COVID-19 as well as any viral pandemic. However, AI-based prediction, analysis, and identification need a large as well as an accurate data set, which is a challenging issue due to data scarcity. Smart and accurate data acquisition techniques might be an important area of research to focus on.

References

1. Worldometer: COVID-19 Coronavirus Pandemic. https://www.worldometers.info/coronavirus/? Accessed 16 Jan 2021. 15:48 GMT
2. Haleem, A., Javaid, M.: Vaishya Effects of COVID 19 pandemic in daily life. Curr. Med. Res. Pract. **10**(2), 78–79 (2020)
3. Hu, Z., Ge, Q., Jin, L., Xiong, M.: Artificial intelligence forecasting of COVID-19 in China. https://arxiv.org/abs/2002.07112
4. Miner, A.S., Laranjo, L., Kocaballi, A.B.: Chatbots in the fight against the COVID-19 pandemic. NPJ Digit. Med. **3**(65), 1–14 (2020)
5. Schmitt, M.: Artificial Intelligence in Medicine. https://towardsdatascience.com/artificial-intelligence-in-medicine-1fd2748a9f87
6. Erdemir, M.: Using web-based intelligent tutoring systems in teaching physics subjects at undergraduate level. Univ. J. Educ. Res. **7**(7), 1517–1525 (2019)
7. Bloom, N., Bunn, P., Chen, S., Mizen, P., Smietanka, P.: The Economic Impact of Coronavirus on UK Businesses: Early Evidence from the Decision Maker Panel. https://voxeu.org/article/economic-impact-coronavirus-uk-businesses
8. Madhuri, V.J., Mohan, M.R., Kaavya, R.: Survey: stress management using artificial intelligence. In: Third International Conference on Advances in Computing and Communications, vol. 3, pp. 54–57 (2013). IARJSET
9. Xu, B., et al.: Epidemiological data from the COVID-19 outbreak, real-time case information. Sci. Data **7**, 106 (2020). https://doi.org/10.1038/s41597-020-0448-0
10. Dong, E., Du, H., Gardner, L.: An interactive web-based dashboard to track COVID-19 in real-time. Lancet Infect. Dis. **20**(5), 533–534 (2020). https://doi.org/10.1016/S1473-3099(20)30120-1
11. Chen, E., Lerman, K., Ferrara, E.: Tracking social media discourse about the COVID-19 pandemic: development of a public coronavirus Twitter data set. JMIR Public Health Surveill. **6**(2) (2020). https://publichealth.jmir.org/2020/2/e19273/
12. Yang, X., He, X., Zhao, J., Zhang, Y., Zhang, S., Xie, P.: COVID-CT-Dataset: A CT Scan Dataset about COVID-19 (2020). https://arxiv.org/abs/2003.13865
13. Ma, J., et al.: COVID-19 CT Lung and Infection Segmentation Dataset (2020). https://doi.org/10.5281/zenodo.3757476
14. Ma, J., et al.: Towards Efficient COVID-19 CT Annotation: A Benchmark for Lung and Infection Segmentation (2020). https://gitee.com/junma11/COVID-19-CT-Seg-Benchmark
15. COVID-19 CT Segmentation Dataset (2020). https://www.medseg.ai/
16. COVID-19 Database (2020). https://www.sirm.org/en/category/articles/covid-19-database/
17. COVID-19 Imaging Database, British Society of Thoracic Imaging Database. https://www.bsti.org.uk/training-and-education/covid-19-bsti-imaging-database/
18. COVID-19, Radiopaedia. https://radiopaedia.org/articles/covid-19-4?lang=us

19. COVID-19 X Rays. https://kaggle.com/andrewmvd/convid19-x-rays
20. Chest X-Ray Images (Pneumonia). https://kaggle.com/paultimothymooney/chest-xray-pneumonia
21. COVID-19 Radiography Database. https://kaggle.com/tawsifurrahman/covid19-radiography-database
22. Born, J., Brändle, G., Cossio, M., Disdier, M., Goulet, J., Roulin, J., Wiedemann, N.: POCOVID-Net: Automatic Detection of COVID-19 From a New Lung Ultrasound Imaging Dataset (POCUS).https://arxiv.org/abs/2004.12084. https://github.com/jannisborn/covid19_pocus_ultrasound
23. Maddah, E., Beigzadeh, B.: Use of a smartphone thermometer to monitor thermal conductivity changes in diabetic foot ulcers: a pilot study. https://pubmed.ncbi.nlm.nih.gov/31930943/
24. Karvekar, S.B.: Smartphone-based human fatigue detection in an industrial environment using gait analysis. https://scholarworks.rit.edu/theses/10275/
25. Roldan, J.C., Bennett, P., Ortiz, G.A., Cuesta, V.: Fatigue detection during sit-to-stand test based on surface electromyography and acceleration: a case study. Sensors (Basel) **19**(19), 4202 (2019). https://pubmed.ncbi.nlm.nih.gov/31569776/
26. Bai, H.X., Hsieh, B., Xiong, Z., Halsey, K., Choi, J.W., Tran, T.M., Pan, I., Shi, L.B., Wang, D.C., Mei, J., Jiang, X.L.: Performance of radiologists in differentiating COVID-19 from viral pneumonia on chest CT. Radiology (2020). https://pubs.rsna.org/doi.org/10.1148/radiol.2020200823
27. Ai, T., Yang, Z., Hou, H., Zhan, C., Chen, C., Lv, W., Tao, Q., Sun, Z., Xia, L.: Correlation of chest CT and RT-PCR testing in coronavirus disease 2019 (COVID-19) in China: a report of 1014 cases. Radiology. https://pubs.rsna.org/doi/full/10.1148/radiol.2020200642
28. Zhang, J., Xie, Y., Li, Y., Shen, C., Xia, Y.: COVID-19 Screening on Chest X-ray Images Using Deep Learning based Anomaly Detection. https://arxiv.org/abs/2003.12338
29. Li, L., et al.: Artificial intelligence distinguishes COVID-19 from community acquired pneumonia on chest CT. Radiology. https://doi.org/10.1148/radiol.2020200905
30. Wang, L., Wong, A.: COVID-Net: A Tailored Deep Convolutional Neural Network Design for Detection of COVID-19 Cases from Chest X-Ray Images. https://arxiv.org/abs/2003.09871
31. Imran, A., Posokhova, I., Qureshi, H.N., Masood, U., Riaz, M.S., Ali, K., John, C.N., Hussain, I., Nabeel, M.: AI4COVID-19: AI Enabled Preliminary Diagnosis for COVID-19 from Cough Samples via an App. https://arxiv.org/abs/2004.01275
32. Rao, A.S.S., Vazquez, J.A.: Identification of COVID-19 can be quicker through artificial intelligence framework using a mobile phone-based survey in the populations when cities/towns are under quarantine. Infect. Control Hosp. Epidemiol. https://doi.org/10.1017/ice.2020.61
33. Ahmed, N., et al.: A survey of COVID-19 contact tracing apps. IEEE Access **8**, 134577–134601 (2020). https://doi.org/10.1109/ACCESS.2020.3010226
34. Giordano, G., Blanchini, F., Bruno, R., Colaneri, P., Di Filippo, A., Di Matteo, A., Colaneri, M.: Modelling the COVID-19 epidemic and implementation of population-wide interventions in Italy. Nat. Med. https://rdcu.be/b4utX
35. Yan, L., Zhang, H.-T., Xiao, Y., Wang, M., Guo, Y., Sun, C., Tang, X., Jing, L., Li, S., Zhang, M., Xiao, Y., Tang, X., Cao, H., Tan, X., Huang, N., Luo, A., Cao, B.J., Xu, Z.H., Yuan, Y.: Prediction of criticality in patients with severe Covid-19 infection using three clinical features: a machine learning-based prognostic model with clinical data in Wuhan. https://www.medrxiv.org/content/10.1101/2020.02.27.20028027v3
36. Xu, G., Yang, Y., Du, Y., Peng, F., Hu, P., Wang, R., Yin, M., Li, T., Tu, L., Sun, J., Jiang, T, Chang, C.: Clinical Pathway for Early Diagnosis of COVID-19: Updates from Experience to Evidence-Based Practice. https://rdcu.be/b4BaL
37. Binti Hamzah, F.A., et al.: CoronaTracker: world-wide Covid-19 outbreak data analysis and prediction. Bulletin of World Health Organization (2020). https://www.who.int/bulletin/online_first/20-255695.pdf
38. Ye, Y., Hou, S., Fan, Y., Qian, Y., Zhang, Y., Sun, S., Peng, Q., Laparo, K.: α-Satellite: An AI-driven System and Benchmark Datasets for Hierarchical Community-level Risk Assessment to Help Combat COVID-19. https://arxiv.org/abs/2003.12232

39. Kim, S., Seo, Y.B., Jung, E.: Prediction of COVID-19 transmission dynamics using a mathematical model considering behavior changes in Korea. Epidemiol. Health **42**, e2020026 (2020)
40. Yang, Z., et al.: Modified SEIR and AI prediction of the epidemics trend of COVID-19 in China under public health interventions. J. Thoracic Dis. **12**(3), 165–174 (2020)
41. Gong, J., Ou, J., Qiu, X., Jie, Y., Chen, Y., Yuan, L., Cao, J., Tan, M., Xu, W., Zheng, F., Shi, Y., Hu, B.: A Tool to Early Predict Severe 2019-Novel Coronavirus Pneumonia (COVID-19): A Multicenter Study using the Risk Nomogram in Wuhan and Guangdong, China. https://www.medrxiv.org/content/10.1101/2020.03.17.20037515v2
42. Nobles, A.L., et al.: Responses to addiction help-seeking from Alexa, Siri, Google Assistant, Cortana, and Bixby intelligent virtual assistants. NPJ Digit. Med. **3**(11), 1-3 (2020)
43. Tommy - the robot nurse helps Italian doctors care for COVID-19 patients. https://www.pri.org/stories/2020-04-08/tommy-robot-nurse-helps-italian-doctors-care-covid-19-patients
44. Leipheimer, J.M., Balter, M.L., Chen, A.I., Pantin, E.J., Davidovich, A.E., Labazzo, K.S., Yarmush, M.L.: First-in-human evaluation of a hand-held automated venipuncture device for rapid venous blood draws. Technology **7**, 98–107 (2019)
45. Marr, B.: Robots and drones are now used to fight COVID-19. https://www.forbes.com/sites/bernardmarr/2020/03/18/how-robots-and-drones-are-helping-to-fight-coronavirus/#86a32ed2a12e
46. Yang, G.-Z., Nelson, B.J., Murphy, R.R., Choset, H., Christensen, H., Collins, S.H., Dario, P., Goldberg, K., Ikuta, K., Jacobstein, N., Kragic, D., Taylor, R.H., McNutt, M.: Combating COVID-19—The role of robotics in managing public health and infectious diseases. https://robotics.sciencemag.org/content/5/40/eabb5589
47. Buntain, C., Golbeck, J.: Automatically identifying fake news in popular Twitter threads. In: IEEE International Conference on Smart Cloud, pp. 208–215 (2017)
48. OECD: Ensuring data privacy as we battle COVID-19 (2020). https://www.oecd.org/coronavirus/policy-responses/ensuring-data-privacy-as-we-battle-covid-19-36c2f31e/
49. Ajao, O., Bhowmik, D., Zargari, S.: Fake News Identification on Twitter with Hybrid CNN and RNN Models. https://arxiv.org/abs/1806.11316
50. Yang, S., Fu, C., Lian, X., Dong, X., Zhang, Z.: Understanding human-virus protein-protein interactions using a human protein complex-based analysis framework. mSystems (Am. Soc. Microbiol.) **4**(2) (2019)
51. Senior, A.W., Evans, R., Jumper, J., Kirkpatrick, J., Sifre, L., Green, T., Penedones, H.: Improved protein structure prediction using potentials from deep learning. Nature **577**, 706–710 (2020)
52. Zhavoronkov, A., Aladinskiy, V., Zhebrak, A., Zagribelnyy, B., Terentiev, V., Bezrukov, D.S., Polykovskiy, D., Shayakhmetov, R., Filimonov, A., Orekhov, P., Yan, Y.: Potential COVID-2019 3C-like protease inhibitors designed using generative deep learning approaches. ChemRxiv. https://doi.org/10.26434/chemrxiv.11829102
53. Randhawa, G.S., Hill, K.A., Kari, L.: MLDSP-GUI: an alignment-free standalone tool with an interactive graphical user interface for DNA sequence comparison and analysis. Bioinformatics **36**(7), 2258–2259 (2020)
54. Randhawa, G.S., Soltysiak, M.P., El Roz, H., de Souza, C.P., Hill, K.A., Kari, L.: Machine learning using intrinsic genomic signatures for rapid classification of novel pathogens: COVID-19 case study. PLoS ONE **15**(4), e0232391 (2020)
55. Nguyen, T.T., Abdelrazek, M., Nguyen, D.T., Aryal, S., Nguyen, D.T., Khatami, A.: Origin of novel coronavirus (COVID-19): a computational biology study using artificial intelligence. BioRxiv. https://doi.org/10.1101/2020.05.12.091397
56. Verdicchio, M., Kim, S.: Identifying targets for intervention by analyzing basins of attraction. In: Pacific Symposium on Biocomputing, pp. 350–361 (2011). https://doi.org/10.1142/9789814335058_0036
57. Stochl, J., Soneson, E., Wagner, A.P., Khandaker, G.M., Goodyer, I., Jones, P.B.: Identifying key targets for interventions to improve psychological wellbeing: replicable results from four UK cohorts. Psychol. Med. **49**(14), 2389–2396 (2019)

58. Faggella, D.: Machine Learning Drug Discovery Applications – Pfizer, Roche, GSK, and More. https://emerj.com/ai-sector-overviews/machine-learning-drug-discovery-applications-pfizer-roche-gsk/
59. Keserü, G., Makara, G.: The influence of lead discovery strategies on the properties of drug candidates. Nat. Rev. Drug Discov. **8**, 203–212 (2009). https://doi.org/10.1038/nrd2796
60. Woo, M.: An AI boost for clinical trials. Nature **573** (2019)
61. Le, T.T., Andreadakis, Z., Kumar, A., Román, R.G., Tollefsen, S., Saville, M., Mayhew, S.: The COVID-19 vaccine development landscape. Nat. Rev. Drug Discov. **19**, 305–306 (2020)
62. Ho, D.: Addressing COVID-19 drug development with artificial intelligence. In: Advanced Intelligent Systems, vol. 2, 2000070 (2020).https://doi.org/10.1002/aisy.20200007
63. Ko, J., Baldassano, S.N., Loh, P.-L., Kording, K., Litt, B., Issadore, D.: Machine learning to detect signatures of disease in liquid biopsies - a user's guide. Lab Chip **18**(3), 395–405 (2018)
64. Reddy, V.: Using AI to identify biomarkers that facilitate personalized medicine. https://www.proxzar.ai/blog/using-ai-to-identify-biomarkers-that-facilitate-personalized-medicine/
65. Giuffra, C.E.P., Silveira, R.A.: An agent based model for integrating intelligent tutoring system and virtual learning environments. In: Advances in Artificial Intelligence – IBERAMIA 2012. Lecture Notes in Computer Science, vol. 7637. Springer (2012)
66. Bobdey, S., Ray, S.: Going viral – COVID-19 impact assessment: a perspective beyond clinical practice. J. Marine Med. Soc. **22**(1), 9 (2020)
67. Marr, B.: Coronavirus: How Artificial Intelligence, Data Science And Technology Is Used To Fight The Pandemic. https://www.linkedin.com/pulse/coronavirus-how-artificial-intelligence-data-science-technology-marr
68. Naudé, W.: Artificial Intelligence against COVID-19: An Early Review. https://towardsdatascience.com/artificial-intelligence-against-covid-19-an-early-review-92a8360edaba

Chapter 14
Time Series Analysis for CoVID-19 Projection in Bangladesh

Kawser Ahammed and Mosabber Uddin Ahmed

Abstract The coronavirus disease-19 (CoVID-19) caused by severe acute respiratory syndrome coronavirus-2 (SARS-CoV-2) has been spreading rapidly at different divisions in Bangladesh since April 12, 2020. As CoVID-19 is highly infectious, the national impact of this disease must be analysed. Therefore, we need to project the spread of infected cases across the country. In this chapter, we discuss different epidemic models for modeling infectious disease. After that, we apply Logistic growth model and SIR (susceptible-infectious-recovered) model to the CoVID-19 time series data publicly available online for modelling CoVID-19 epidemic in Bangladesh. Also, we project the probable ending time of the epidemic. To do this, the CoVID-19 time series data from March 17, 2020 to December 31, 2020 is analysed and after that, the projection is performed.

14.1 Introduction

Coronaviruses (CoVs) ranging from 65 nm to 125 nm in diameter are enveloped, non-segmented positive sense single stranded RNA viruses [14, 39]. They hold the largest RNA genome of length 30 kilobase (kb) approximately. Apart from founding in animals, they are found in humans. Prior to the CoVID-19 outbreak, six human coronaviruses (HCoVs) were identified, including OC43, 229E, NL63, HKU1, SARS-CoV (severe acute respiratory syndrome coronavirus), and MERS-CoV (middle east respiratory syndrome coronavirus) [9, 14, 16, 19, 35, 36]. Among these six human coronaviruses, four HCoVs (HCoV-229E, HCoV-NL63, HCoV-OC43 and HCoV-HKU1) are circulated in the human population and can cause common cold infections, life threatening diseases such as pneumonia and bronchiolitis in human.

K. Ahammed (✉)
Department of Electrical and Electronic Engineering, Jatiya Kabi Kazi Nazrul Islam University, Trishal, Mymensingh, Bangladesh
e-mail: kawser@jkkniu.edu.bd

M. U. Ahmed
Department of Electrical and Electronic Engineering, University of Dhaka, Dhaka, Bangladesh
e-mail: mosabber.ahmed@du.ac.bd

M. A. R. Ahad and A. Inoue (eds.), *Vision, Sensing and Analytics: Integrative Approaches*, Intelligent Systems Reference Library 207, https://doi.org/10.1007/978-3-030-75490-7_14

In November 2002, a new coronavirus was identified in the Guangdong province, China. This new virus, named as SERS-CoV, was highly transmissible among humans and swiftly spread across 29 countries [31]. As a result, it infected more than 8000 people mainly in China and Hong Kong and caused a mortality rate of about 10% [28, 39]. The source of origin of SARS-CoV was bat and the palm civet cats were the intermediary host to transmit this virus to humans. [20, 39].

In 2012, another new coronavirus designated as MERS-CoV emerged in Saudi Arabia [29]. Although the spread of this virus was geographically limited [18], it infected more than 2000 people and caused a fatality rate of approximately 35% [10, 15]. The origin of MERS-CoV was also bat and this virus was transmitted to humans via the intermediate host of dromedary camels [12].

In December 2019, multiple cases with severe pneumonia of unknown cause occurred in some hospitals in Wuhan city, Hubei province, China [41]. The pathogen of pneumonia was identified and designated as 2019 novel coronavirus (2019-nCoV) [43]. But the 2019-nCoV was subsequently named as severe acute respiratory syndrome coronavirus-2 (SARS-CoV-2) with the International Committee on Taxonomy Viruses [17]. On 30 January 2020, the outbreak of SARS-CoV-2 was declared as public health emergency of international concern by World Health Organization (WHO) [27]. Since the SARS-CoV-2 was spreading rapidly in China, people were being infected with disease. As the disease caused by SARS-CoV-2 was epidemic in china, on 11 February 2020, the WHO declared a new name, coronavirus disease-19(CoVID-19), for this epidemic disease.

Although there are some debates about the source of origin of the SARS-CoV-2, some researchers have found that the virus has 96% homology with the genome level of bats coronavirus [33, 43]. As a result, bats are the most potential host for this virus. But this virus was transmitted from bats to humans. Though the intermediary host through which the virus was transmitted to humans is unknown, pangolins and snakes are the possible suspects. Figure 14.1 shows a model of transmission of SARS-CoV, MERS-CoV and SARS-CoV-2.

Though the mortality rate of SARS-CoV-2 is lower than SARS-CoV and MERS-CoV, the number of infected cases is higher. That means SARS-CoV-2 spreads rapidly from people to people. This infection is transmitted through large respiratory droplets produced during coughing and sneezing by symptomatic patients. The asymptomatic people can also infect healthy people before the onset of symptoms. The droplets can spread about 1–2 m. As a result, the people will be infected if they do not maintain this minimum distance while talking. Moreover, the droplets can also deposit on the different surfaces such as table, steel, and door etc. If people touch one of these surfaces and then touch nose, mouth and eyes, they will be infected. The common symptoms of CoVID-19 caused by SARS-CoV-2 are fever, cough, sore throat, fatigue, headache and breathlessness. The incubation period of CoVID-19 varies from 2 to 14 days.

As CoVID-19 was spreading rapidly around the world, the WHO declared CoVID-19 as a pandemic [7] On March 11, 2020. On March 8, 2020, the Institute of Epidemiology, Disease Control and Research (IEDCR) of Bangladesh identified the first CoVID-19 confirmed case in Bangladesh [38]. After that, the number of confirmed

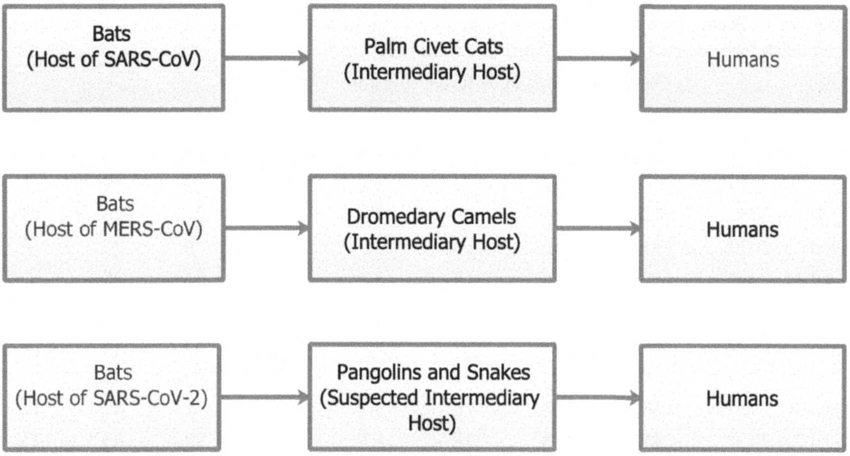

Fig. 14.1 Model of transmission of SARS-CoV, MERS-CoV and SARS-CoV-2

cases were increasing day by day. To date (December 31, 2020), the total number of confirmed cases is 513510. As the number of confirmed cases is increasing rapidly in Bangladesh, the projection of CoVID-19 is necessary so that we can take necessary initiatives for controlling the epidemic before further increasing.

To model epidemic diseases such as Ebola, AIDS and CoVID-19, the logistic growth model and SIR model can be used as an efficient projection model. The Logistic growth model was employed by Chowell et al. [11] to estimate the spread of Ebola virus. This model was also used by Pell et al. [32] to project the final epidemic size and the peak time for the infection of the 2015 Ebola virus. Also, the SIR model was employed by researchers to model Ebola [26] and AIDS [42]. Moreover, both the Logistic growth model and SIR model were used by the researchers to model the epidemic of CoVID-19 [3, 4, 13, 37]. Although various techniques including machine learning [30, 34], artificial intelligence [21], and ARIMA [8] have recently been used by the researchers for modelling the spread of CoVID-19, two epidemic models (logistic growth model and SIR model) have been used to project the cumulative number of confirmed cases and the probable ending time of CoVID-19 epidemic in Bangladesh.

14.2 Materials and Mathematical Models to Epidemic

14.2.1 Materials

The CoVID-19 data used in this study is publicly available online [22, 23]. The CoVID-19 data represents the time series of cumulative confirmed cases, cumulative death cases and cumulative recovered cases of different countries of the world. But we have only analysed the time series data of cumulative confirmed cases for

Table 14.1 Break down of CoVID-19 cumulative confirmed cases at different divisions in Bangladesh from April 12, 2020 to April 29, 2020 [22]

Date	Confirmed cases at different divisions							
	D[a]	C [b]	S[c]	R[d]	K[e]	M[f]	B[g]	Ra[h]
April 12, 2020	529	35	3	15	1	14	7	0
April 13, 2020	674	41	4	15	1	14	10	2
April 14, 2020	820	50	6	19	3	21	16	3
April 15, 2020	934	62	5	34	3	26	23	4
April 16, 2020	1181	69	7	36	3	32	25	3
April 17, 2020	1370	92	7	37	6	42	31	8
April 18, 2020	1606	97	7	44	6	59	36	8
April 19, 2020	1869	105	7	47	6	66	41	9
April 20, 2020	2280	118	8	50	9	81	47	12
April 21, 2020	2452	125	18	52	9	99	65	21
April 22, 2020	2756	143	20	60	24	133	71	25
April 23, 2020	2985	149	33	63	25	139	74	31
April 24, 2020	3502	156	49	70	38	139	86	32
April 25, 2020	3682	170	69	72	38	168	86	33
April 26, 2020	4053	185	79	76	65	181	102	37
April 27, 2020	4397	201	89	82	83	201	102	44
April 28, 2020	4857	227	102	104	101	219	111	62
April 29, 2020	5338	270	108	113	136	237	112	101

[a]D represents Dhaka [b]C represents Chattogram [c]S represents Sylhet [d]R represents Rangpur [e]K represents Khulna [f]M represents Mymensingh [g]B represents Barishal [h]Ra represents Rajshahi

Bangladesh along with most affected countries of the world. Moreover, the time series of cumulative confirmed cases for different divisions in Bangladesh has also been analysed. The cumulative number of confirmed cases for different divisions from April 12, 2020 to April 29, 2020 has been shown in Table 14.1.

14.2.2 Logistic Growth Model

The logistic growth model originated from the population dynamics can be used to model the epidemic of CoVID-19. This model was first introduced by Pierre-Francois Verhulst [40] for biological systems. According to the underlying assumption of this model, the rate of change of new cases per capita decreases linearly when the cumulative number of cases approaches to the final epidemic size. Based on this assumption, the model is defined as

$$\frac{dP}{dt} = rP(1 - \frac{P}{K}) \tag{1}$$

The Eq. (1) can also be written as

$$P' = rP(1 - \frac{P}{K}) \tag{2}$$

where P is the cumulative number of confirmed cases, P'/P or $\frac{1}{P}\frac{dP}{dt}$ is the rate of growth per capita, r is the infection rate and K is the final epidemic size. From Eq. (1), we can write

$$\frac{KdP}{P(K-P)} = rdt \tag{3}$$

Integrating both sides of Eq. (3), we get

$$\int \frac{KdP}{P(K-P)} = \int rdt$$

$$\int [\frac{1}{P} + \frac{1}{K-P}]dP = \int rdt$$

$$ln|p| - ln|K-P| + lnB = rt \tag{4}$$

$$ln[\frac{PB}{K-P}] = rt$$

$$\frac{PB}{K-P} = e^{rt} \tag{5}$$

In Eq. (4), lnB is the integrating constant. If the cumulative number of confirmed cases at time t – 0 is P_0, thc Eq. (5) will bc

$$P_0B = (K - P_0)e^{r.0}$$

$$B = \frac{K - P_0}{P_0} \tag{6}$$

Putting the value of B into Eq. (5), we get

$$[\frac{K-P_0}{P_0}]P = (K-P)e^{rt}$$

$$\frac{KP}{P_0} - P + Pe^{rt} = Ke^{rt}$$

$$KP - PP_0 + PP_0e^{rt} = KP_0e^{rt}$$

$$P(K - P_0 + P_0e^{rt}) = KP_0e^{rt}$$

$$P = \frac{KP_0e^{rt}}{K + P_0(e^{rt} - 1)}$$

$$= \frac{K}{1 + e^{-rt}KP_o^{-1} - e^{-rt}}$$

$$= \frac{K}{1 + e^{-rt}(KP_0^{-1} - 1)} \tag{7}$$

The Eq. (7) can be represented as

$$P(t) = \frac{K}{1 + Be^{-rt}} \tag{8}$$

where $B = \frac{K-P_0}{P_0}$ and P_0 is the initial cases. Differentiating Eq. (8) with respect to time, we get

$$\begin{aligned} \frac{dP}{dt} &= \frac{1 + Be^{-rt}.0 - K(-rBe^{-rt})}{(1 + Be^{-rt})^2} \\ &= \frac{rBKe^{-rt}}{(1 + Be^{-rt})^2} \end{aligned} \tag{9}$$

The maximum growth rate, $\frac{dP}{dt}$, occurs when $\frac{dP^2}{dt^2} = 0$. To find the peak time, t_p, at which the maximum growth rate occurs, differentiate Eq. (9) with respect to time,

$$\begin{aligned} \frac{dP^2}{dt^2} &= \frac{-r^2BKe^{-rt} - 2r^2B^2Ke^{-2rt} - r^2B^3Ke^{-3rt} + 2r^2B^2Ke^{-2rt} + 2r^2B^3Ke^{-3rt}}{(1 + Be^{-rt})^4} \\ &= \frac{-r^2BKe^{-rt} + r^2B^3Ke^{-3rt}}{(1 + Be^{-rt})^4} \end{aligned} \tag{10}$$

Setting the value of Eq. (10) to 0, we find

$$\begin{aligned} -r^2BKe^{-rt} + r^2B^3Ke^{-3rt} &= 0 \\ r^2BKe^{-rt} &= r^2B^3Ke^{-3rt} \\ e^{2rt} &= B^2 \end{aligned} \tag{11}$$

Taking *ln* on both sides of Eq. (11),

$$\begin{aligned} lnB^2 &= lne^{2rt} \\ 2lnB &= 2rt \\ t_p &= \frac{lnB}{r} \end{aligned} \tag{12}$$

Equation (12) demonstrates the peak time at which the maximum growth rate is observed. Putting the value of peak time into Eq. (8), we get

$$\begin{aligned} P(t_p) &= \frac{K}{1+\frac{B}{e^{\ln B}}} \\ &= \frac{K}{1+\frac{B}{B}} \\ &= \frac{K}{2} \end{aligned} \tag{13}$$

Equation (13) represents the cumulative number of cases at the peak time. Putting the value of peak time into Eq. (9), we find

$$\begin{aligned} \left(\frac{dP}{dt}\right)_{t=t_p} &= \frac{\frac{rBK}{e^{\ln B}}}{(1+\frac{B}{e^{\ln B}})^2} \\ &= \frac{rK}{4} \end{aligned} \tag{14}$$

Equation (14) shows the value of growth rate peak at the peak time. For modelling CoVID-19 epidemic using this model, fitVirus MATLAB program provided by Batista [5] has been used.

14.2.3 SIS (Susceptible-Infectious-Susceptible) Epidemic Model

The SIS model proposed by Kermack and McKendrick [25] consists of two compartments. One compartment is called the susceptible class (S) and the other compartment is defined as the infectious class (I). The SIS model is appropriate for the infectious diseases that have no immunity against reinfection. In this model, the individuals from the susceptible class pass to the infectious class and the recovered infectives who get recovered from the infectious class return to the susceptible class. The transfer diagram of the SIS model is shown in Fig. 14.2. The SIS model is defined by the following differential equations:

$$\frac{dS(t)}{dt} = -\frac{\beta S(t)I(t)}{N} + \gamma I(t) \tag{15}$$

$$\frac{dI(t)}{dt} = \frac{\beta S(t)I(t)}{N} - \gamma I(t) \tag{16}$$

Where β is the infection rate at which the disease is transmitted from infected individuals to susceptible individuals during interaction, γ is the recovery rate at which the individuals recover from the infectious groups after an infectious period so that they can return to the susceptible class and $N = S(t) + I(t)$ is the total population.

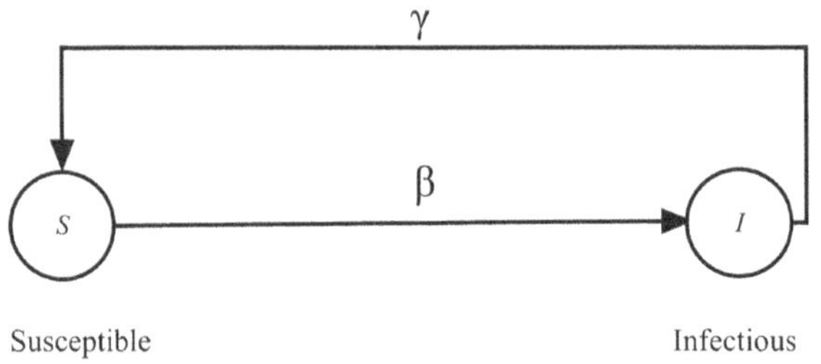

Fig. 14.2 Diagram of SIS model

14.2.4 SIR Epidemic Model

The SIR model is used to analyse how a disease spreads through a population. This model divides the entire population into three groups: susceptible individuals, infectious individuals and recovered individuals. Susceptible individuals are those who are healthy but capable of becoming infected. Infectious individuals are those who have the disease and can cause infection. Recovered individuals are defined as such kind of people who had disease but are now immune or isolated until recovered or deceased. In this model, the susceptible individuals pass to the infected class and the infected individuals pass to the recovered class. The direction of individuals transmission in SIR model is shown in Fig. 14.3. In this subsection, we discuss the SIR epidemic model [2, 24] and this model can be used to project the spread of CoVID-19 in Bangladesh. This model was developed based on some assumptions. The first assumption was that the size of the population is large and constant. The second assumption was that no natural birth or no natural deaths can occur in the population. The third assumption was that an individual recovered from infection confers life time immunity. The last assumption was that the individuals are well mixed in the population. Based on these assumptions, the SIR model was defined with the following nonlinear system of ordinary differential equations:

$$\frac{dS(t)}{dt} = -\frac{\beta S(t)I(t)}{N} \tag{17}$$

$$\frac{dI(t)}{dt} = \frac{\beta S(t)I(t)}{N} - \gamma I(t) \tag{18}$$

$$\frac{dR(t)}{dt} = \gamma I(t) \tag{19}$$

Fig. 14.3 Diagram of SIR model

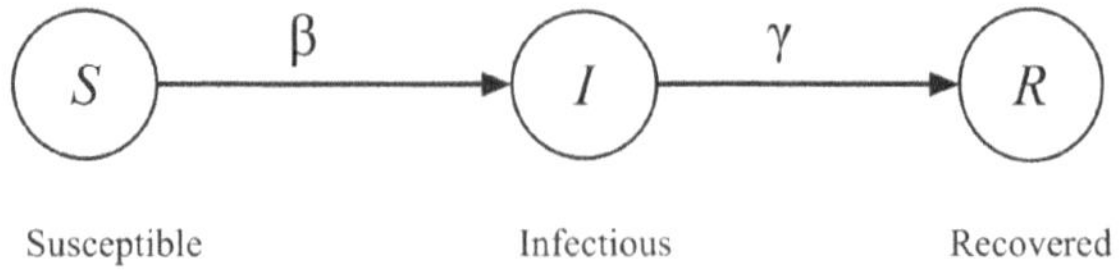

Where N is the total population and $N = S(t) + I(t) + R(t)$, $S(t)$ is the number of susceptible individuals at time t, $I(t)$ is the number of infected individuals at time t, $R(t)$ is the number of recovered individuals at time t, β is the infection rate, γ is the recovering rate and $\frac{1}{\gamma}$ means the average infectious period. The ratio of β to γ, $\Re_0 = \frac{\beta}{\gamma}$, is called the basic reproduction number. It is defined as the average number of secondary cases generated by a primary infected individual in the susceptible population. From the value of the basic reproduction number, one can assume whether an epidemic will occur or not. The value of $\Re_0$ that is less than <1 means that on average an infected individual produces less than one secondary infected cases. Since the infection can not grow due to $\Re_0 <1$, the epidemic will not occur. The condition $\Re_0 >1$ indicates that on average more than one secondary infected cases is produced by an infected primary individual. As the infection can spread due to $\Re_0 >1$, an epidemic will occur. The infection rate β indicates that each person has an average β contacts per unit time. The recovering rate γ means that each person in the infected state get recovered or dies at the average rate γ. From Eq. (19), we get

$$I = \frac{1}{\gamma}\frac{dR}{dt} \tag{20}$$

Putting this value of I into Eq. (17), we can get

$$\frac{dS}{dt} = -\frac{\beta S}{\gamma N}\frac{dR}{dt} \tag{21}$$

Integrating both sides of Eq. (21),

$$\int \frac{dS}{S} = -\int \frac{\beta}{\gamma N} dR$$
$$lnS = \frac{-\beta R}{\gamma N} + c \tag{22}$$

If the number of susceptible individuals and the number of recovered individuals at time $t = 0$ are S_0 and R_0 respectively, we can write Eq. (22) as

$$lnS_0 = -\frac{\beta R_0}{\gamma N} + c$$
$$c = lnS_0 + \frac{\beta R_0}{\gamma N}$$

where c is an integrating constant. Putting the value of integrating constant into Eq. (22), we find

$$
\begin{aligned}
lnS &= \frac{-\beta R}{\gamma N} + lnS_0 + \frac{\beta R_0}{\gamma N} \\
ln(\frac{S}{S_0}) &= -\frac{\beta}{\gamma N}(R - R_0) \\
\frac{S}{S_0} &= e^{-\frac{\beta}{\gamma N}(R-R_0)} \\
S &= S_0 e^{-\frac{\beta}{\gamma N}(R-R_0)}
\end{aligned} \tag{23}
$$

When $t \rightarrow \infty$, the final number of susceptible individuals and the final number of recovered individuals are represented as S_∞ and R_∞ respectively. Therefore, at $t \rightarrow \infty$, the Eq. (23) becomes

$$
S_\infty = S_0 e^{-\frac{\beta}{\gamma N}(R_\infty - R_0)} \tag{24}
$$

Since the final number of infected individuals is assumed to be 0 at $t \rightarrow \infty$, the total number of population will be $N = S_\infty + R_\infty$. From this and using Eq. (24), we can write

$$
R_\infty = N - S_0 e^{-\frac{\beta}{\gamma N}(R_\infty - R_0)} \tag{25}
$$

To project the CoVID-19 epidemic using this model, we have used fitVirus CoVID-19 program provided by Batista [6] using MATLAB.

14.2.5 SEIR (Susceptible-Exposed-Infectious-Recovered) Epidemic Model

The SEIR model proposed in [1] divides the entire population into four groups: Susceptible(S), Exposed(E), Infectious (I) and Recovered (R). In susceptible group, the individuals are not yet infected but can be infected. In exposed category, the individuals have been infected but are not yet capable of transmuting pathogens to others. To become infectious, they experience a latent period (the time of infection to time of being infectious) during which the pathogens attack them severely so that the infected individuals become symptomatic and transmit pathogens to others. After the latent period, the infected individuals from the exposed class become infectious. The individuals who are not only infected but also infectious are included in the infectious group. The people who recover from the infectious class remain in the recovered class. The flow chart of the SEIR model in which how individuals transmit from one group to other is shown in Fig. 14.4. In SEIR model, the transmission of the disease is defined by the following ordinary differential equations:

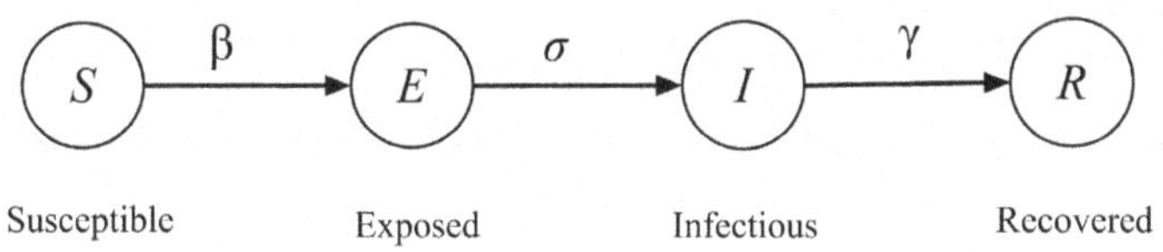

Fig. 14.4 Diagram of SEIR model

$$\frac{dS(t)}{dt} = -\frac{\beta S(t)I(t)}{N} \tag{26}$$

$$\frac{dE(t)}{dt} = \frac{\beta S(t)I(t)}{N} - \sigma E(t) \tag{27}$$

$$\frac{dI(t)}{dt} = \sigma E(t) - \gamma I(t) \tag{28}$$

$$\frac{dR(t)}{dt} = \gamma I(t) \tag{29}$$

Where β is the transmission rate, σ is the progression rate at which an infected individual of exposed group become infectious per unit time, γ is the recovery rate, $\frac{1}{\sigma}$ is the average latent period, $\frac{1}{\gamma}$ is the mean infectious period, $S(t)$ is the number of susceptible individuals at time t, $E(t)$ is the number of exposed individuals at time t, $I(t)$ is the number of infectious individuals at time t, $R(t)$ is the number of recovered individuals at time t and $N = S(t) + E(t) + I(t) + R(t)$ is the total population.

14.3 Analysis of CoVID-19 Cases

14.3.1 Analysis of Confirmed Cases, Death Cases and Recovered Cases in Bangladesh

The blue line (first subplot) of Fig. 14.5 demonstrates the daily cumulative number of confirmed cases. The red line (second subplot) of Fig. 14.5 reflects the daily cumulative number of death cases which is 1.47% of the cumulative number of confirmed cases till now (December 31, 2020) (Table 14.2). The magenta color of (third subplot) Fig. 14.5 represents the daily cumulative number of recovered cases up to December 31, 2020. The recovery rate is 89.08% of the cumulative confirmed cases as of December 31, 2020 (Table 14.2). It is also clear from Fig. 14.5 (fourth subplot) that the daily cumulative number of confirmed cases was extremely higher at Dhaka division compared to other divisions upto April 29, 2020. To date (December 31, 2020), the cumulative number of confirmed cases is higher at Dhaka division compared to other divisions (Fig. 14.6). So, it can be inferred that if the necessary initiatives are not taken to reduce the confirmed cases, the Dhaka division could likely be the most dangerous place for people to be infected rapidly compared to

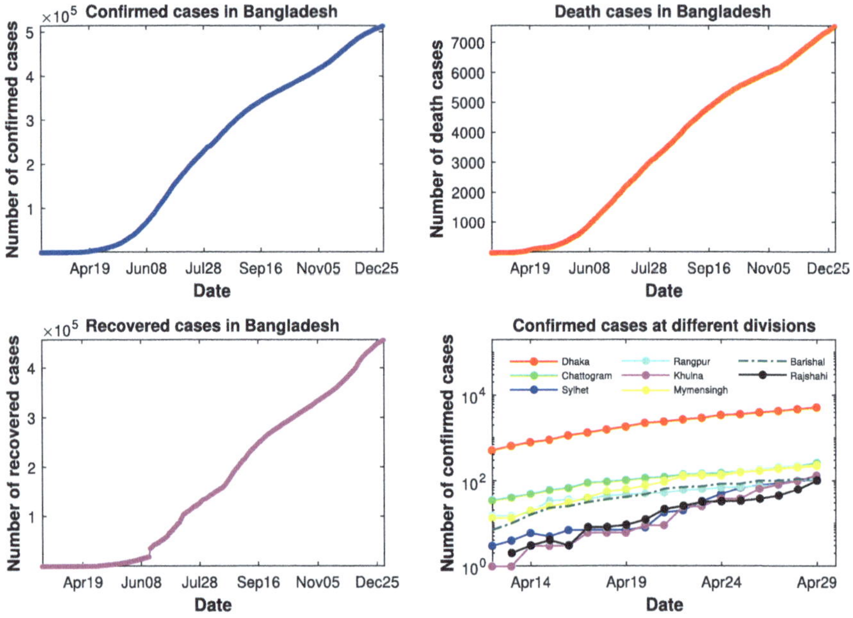

Fig. 14.5 Daily cumulative cases of CoVID-19 from day 1 (March 8, 2020) to day 299 (December 31, 2020). The vertical axis of subfigure that represents the daily cumulative confirmed cases at different divisions is in log scale and the daily cumulative number of confirmed cases from April 12, 2020 to April 29, 2020 is used for different divisions

Table 14.2 Status of CoVID-19 in Bangladesh from March 8, 2020 to December 31, 2020 [22, 23]

Confirmed Cases	Death Cases	Recovered Cases	Mortality (Percentage of confirmed cases)	Recovery (Percentage of confirmed cases)
513510	7559	457459	1.47%	89.08%

other divisions. As a result, the mortality rate could be extremely higher compared to other divisions. The lowest daily cumulative number of confirmed cases was observed at Khulna division up to April 23, 2020. After that, it started to become higher compared to Rajshahi division. However, the lowest cumulative number of confirmed cases is now (December 15, 2020) observed at Mymensingh division (Fig. 14.6).

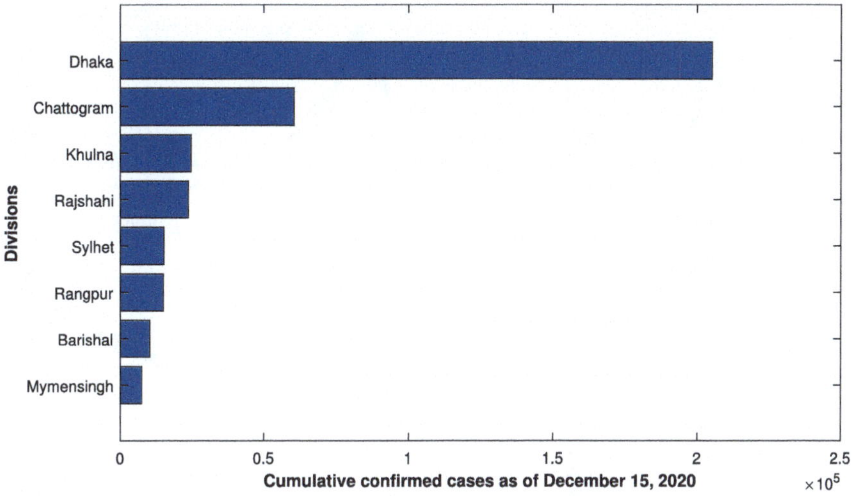

Fig. 14.6 Comparison of cumulative confirmed cases at different divisions in Bangladesh

14.3.2 Analysis of Daily New Cases and Cumulative Confirmed Cases

Like most affected countries of the world, Bangladesh is still fighting against CoVID-19. So, we need to compare the daily new cases of Bangladesh with the most affected countries of South Asia and South East Asia region. Also, the daily new cases of most affected countries of the world need to be analysed so that we can compare Bangladesh with these most affected countries. By comparing the daily new cases, we will be able to identify how fast the number of daily new cases in Bangladesh is increasing compared to most affected countries of the world. From Fig. 14.7, it is clear that the number of daily new cases is approaching very quickly to zero in Afghanistan, Nepal and Singapore. It can be inferred that these three countries are controlling the spread of CoVID-19 successfully. However, the number of daily new cases is approaching very slowly to zero in India, Pakistan and Bangladesh. Although the number of daily new cases of Bangladesh is approaching slowly to zero, it seems that there is a probability of ending CoVID-19 epidemic very quickly in Bangladesh compared to India and Pakistan. On the other hand, the daily new cases is increasing in Indonesia, Malaysia and Srilanka. Figure 14.8 demonstrates that the number of daily new cases of Bangladesh could approach quickly to zero compared to most affected countries of the world. Therefore, it can be assumed that the epidemic could end very soon in Bangladesh compared to these countries. From Fig. 14.9, it is obvious that the number of cumulative confirmed cases is highest in India and lowest

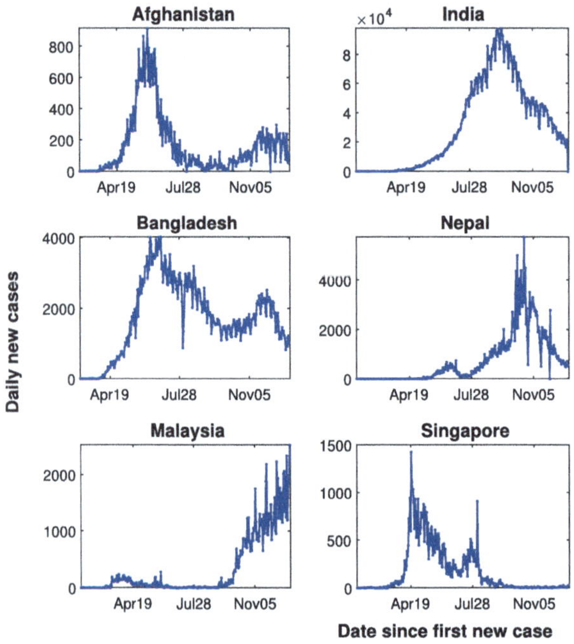

Fig. 14.7 Daily new cases of eight most affected countries along with Bangladesh in South Asia and South East Asia region as of December 31, 2020

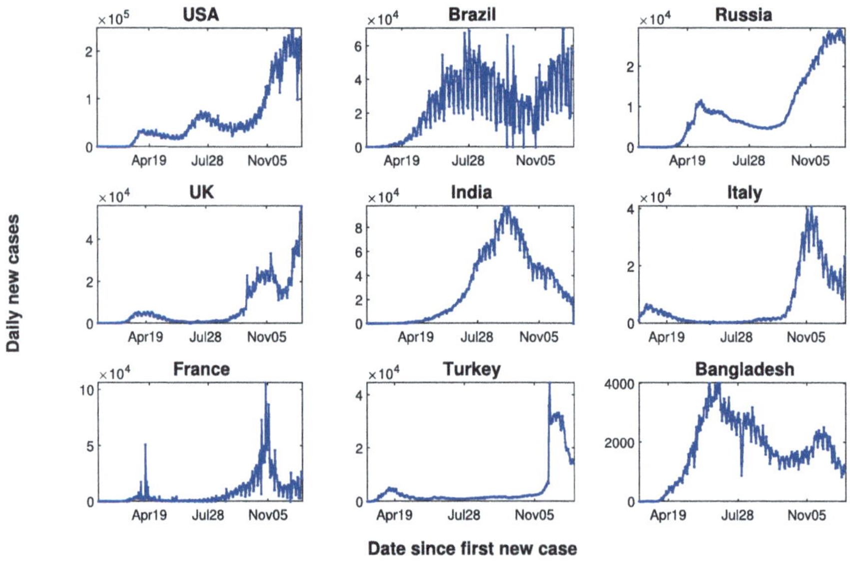

Fig. 14.8 Daily new cases of eight most affected countries along with Bangladesh in the world (Data as of December 31, 2020)

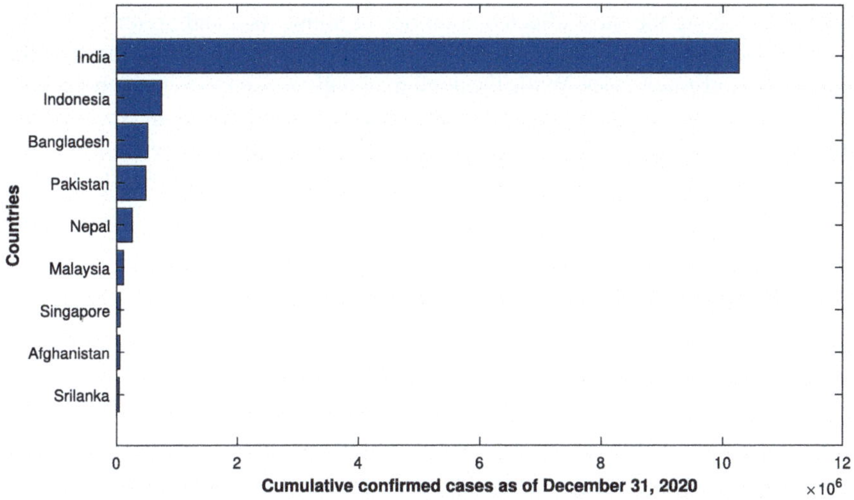

Fig. 14.9 Comparison of cumulative confirmed cases among most affected countries in South Asia and South East Asia as of December 31, 2020

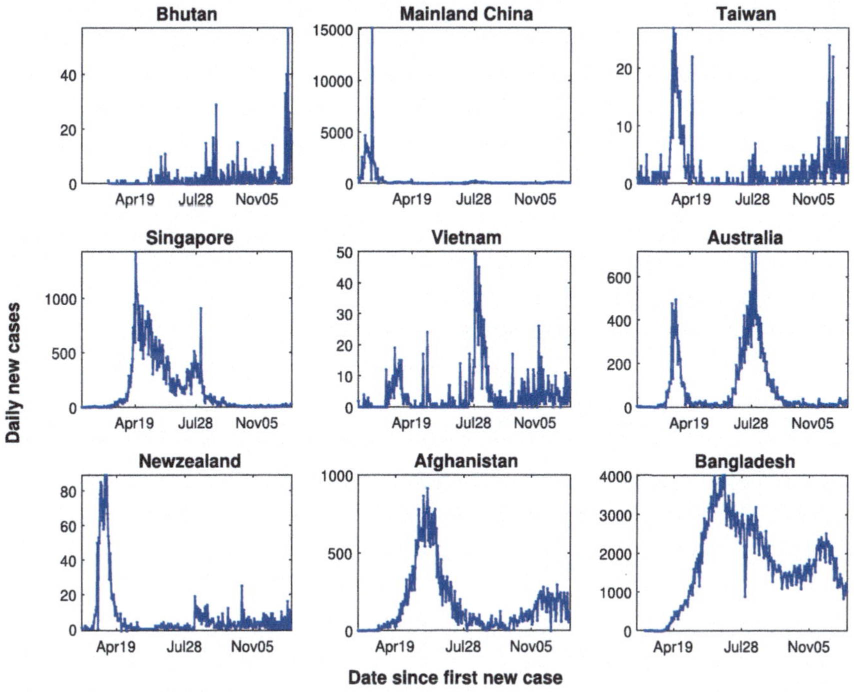

Fig. 14.10 Comparison of Bangladesh with some successful countries of the world where the spread of CoVID-19 has been controlled (Data as of December 31, 2020)

in Srilanka among the most affected countries in South Asia and South East Asia. As of December 31, 2020, Bangladesh lies in the highest third position based on the cumulative confirmed cases. While the number of daily new cases is not approaching quickly to zero in Bangladesh, some of the countries of the world are successfully controlling the spread of CoVID-19 demonstrating the noticeable reduction in the number of daily new cases that is approaching very quickly to zero (Fig. 14.10).

14.4 Projection of CoVID-19 in Bangladesh

14.4.1 Application of Logistic Growth Model

In the first subplot of Fig. 14.11, the magenta color represents the actual cumulative infected cases and the black line indicates the projection of cumulative infected cases. In the second subplot, the blue color represents the actual new cases per day and the red line demonstrates the projection of new cases per day. From Table 14.3, it can be projected with logistic growth model that the final epidemic size of CoVID-19 in Bangladesh could reach 481705. From Table 14.4, the p-value (0) and the value of R^2 (close to 1) of the model indicate the statistical significance of the model results. Table 14.5 shows the short term projection of CoVID-19 cases (cumulative cases and daily cases) in Bangladesh. We can see that the maximum error between the actual cumulative cases and predicted cumulative cases is 8.35%. Also, the maximum accuracy between actual cases and predicted cases is 92.33%. While the significant error is observed in the predicted daily cases, the promising accuracy is observed in the predicted cumulative cases. According to Table 14.6 and second subplot of

Table 14.3 Estimated logistic model coefficients for Bangladesh

	Estimate	SE (Standard Error)	tStat	p-Value
K (Epidemic Size)	**481705**	4055.1	118.79	2.0088e−245
r (Infection rate)	0.02591	0.00057696	44.908	1.4532e−131
$B = \frac{K-P_o}{P_o}$, P_o = initial cases	40.969	2.8425	14.413	8.9115e−36

Table 14.4 Statistical Parameters of the logistic growth model for Bangladesh

Number of Observations	Degrees of Freedom	Root Mean Squared Error	R^2 (Coefficient of determination)	Adjusted R-Squared	F-Statistic vs. Zero model	p-value
289	286	2.08e+04	0.986	0.986	1.94e+04	0

Table 14.5 Short-term projection for Bangladesh using logistic growth model

Day	Date	Cumulative cases (Actual)	Cumulative cases (Predicted)	Error (%)	Accuracy (%)	Daily cases (Actual)	Daily cases (Predicted)	Error (%)
284	25-Dec-2020	508099	469139	7.67	92.33	834	321	61.51
285	26-Dec-2020	509148	469452	7.80	92.2	1049	313	70.16
286	27-Dec-2020	510080	469758	7.91	92.09	932	306	67.17
287	28-Dec-2020	511261	470056	8.06	91.94	1181	298	74.77
288	29-Dec-2020	512496	470347	8.22	91.78	1235	291	76.44
289	30-Dec-2020	513510	470631	8.35	91.65	1014	284	71.99

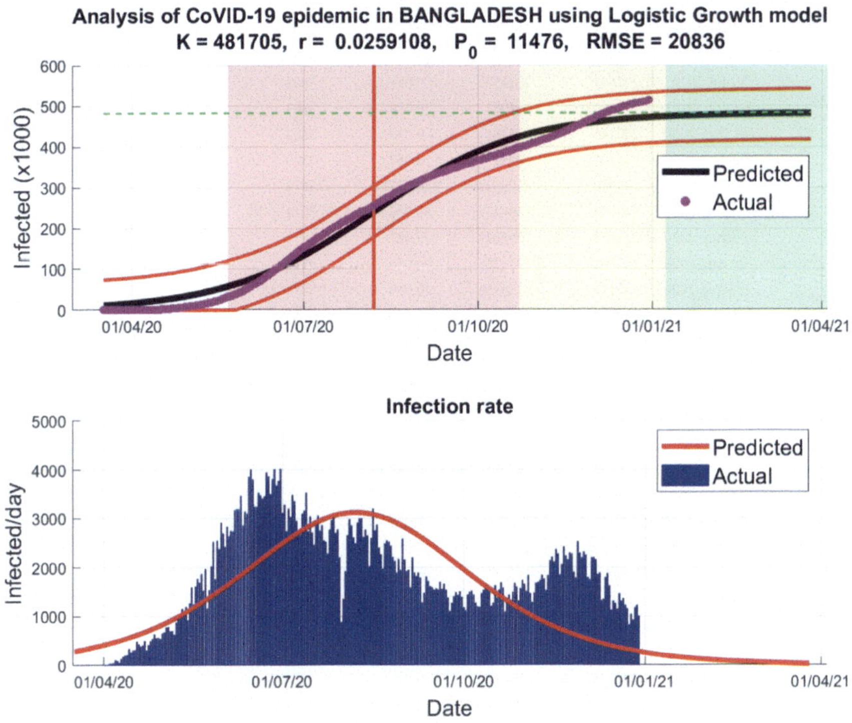

Fig. 14.11 Projection of infected cases and daily new cases. Here, K is the estimated final epidemic size, r is the estimated infection rate, P_0 is the estimated initial cases and RMSE means root mean squared error

Table 14.6 Estimated logistic growth model parameters for Bangladesh

Day	Date	P (cases)	K (cases)	r (1/day)	dP/dt (cases/day)	$t_{peak}(day)$	$Date_{Peak}$
116	10-Jul-2020	181129	236869	0.062	3696	96	21-Jun-2020
117	11-Jul-2020	183795	237445	0.062	3700	96	21-Jun-2020
118	12-Jul-2020	186894	238188	0.062	3705	96	21-Jun-2020
119	13-Jul-2020	190057	239123	0.062	3711	96	21-Jun-2020
120	14-Jul-2020	193590	240413	0.062	3719	96	21-Jun-2020
121	15-Jul-2020	196323	241615	0.062	3725	97	22-Jun-2020
122	16-Jul-2020	199357	242904	0.061	3733	97	??-Jun-2020
123	17-Jul-2020	202066	244141	0.061	3739	97	22-Jun-2020
124	18-Jul-2020	204525	245253	0.061	3745	97	22-Jun-2020
125	19-Jul-2020	207453	246465	0.061	3750	97	22-Jun-2020
126	20-Jul-2020	210510	247818	0.061	3756	98	23-Jun-2020
127	21-Jul-2020	213254	249189	0.060	3762	98	23-Jun-2020
128	22-Jul-2020	216110	250622	0.060	3768	98	23-Jun-2020
129	23-Jul-2020	218658	252017	0.060	3773	98	23-Jun-2020
130	24-Jul-2020	221178	253384	0.060	3778	98	23-Jun-2020
131	25-Jul-2020	223453	254665	0.059	3782	99	24-Jun-2020
132	26-Jul-2020	226225	256026	0.059	3786	99	24-Jun-2020
133	27-Jul-2020	229185	257512	0.059	3790	99	24-Jun-2020
134	28-Jul-2020	232194	259123	0.059	3794	99	24-Jun-2020
135	29-Jul-2020	234889	260765	0.058	3798	100	25-Jun-2020
136	30-Jul-2020	237661	262457	0.058	3802	100	25-Jun-2020
137	31-Jul-2020	239860	264063	0.058	3805	100	25-Jun-2020
138	01-Aug-2020	240746	265301	0.057	3807	100	25-Jun-2020
139	02-Aug-2020	242102	266339	0.057	3809	100	25-Jun-2020
140	03-Aug-2020	244020	267333	0.057	3810	101	26-Jun-2020
141	04-Aug-2020	246674	268445	0.057	3811	101	26-Jun-2020
142	05-Aug-2020	249651	269724	0.057	3812	101	26-Jun-2020
143	06-Aug-2020	252502	271126	0.056	3813	101	26-Jun-2020
144	07-Aug-2020	255113	272595	0.056	3814	101	26-Jun-2020
145	08-Aug-2020	257600	274105	0.056	3815	102	27-Jun-2020
146	09-Aug-2020	260507	275724	0.055	3815	102	27-Jun-2020
147	10-Aug-2020	263503	277458	0.055	3815	102	27-Jun-2020
148	11-Aug-2020	266498	279295	0.055	3815	103	28-Jun-2020
149	12-Aug-2020	269115	281166	0.054	3814	103	28-Jun-2020
150	13-Aug-2020	271881	283093	0.054	3813	103	28-Jun-2020
151	14-Aug-2020	274525	285052	0.054	3813	104	29-Jun-2020
152	15-Aug-2020	276549	286950	0.053	3811	104	29-Jun-2020
153	16-Aug-2020	279144	288876	0.053	3810	104	29-Jun-2020
154	17-Aug-2020	282344	290920	0.052	3808	104	29-Jun-2020
155	18-Aug-2020	285091	293003	0.052	3807	105	30-Jun-2020
156	19-Aug-2020	287959	295145	0.052	3805	105	30-Jun-2020
157	20-Aug-2020	290360	297275	0.051	3802	106	01-Jul-2020
158	21-Aug-2020	292625	299376	0.051	3800	106	01-Jul-2020
159	22-Aug-2020	294598	301413	0.050	3798	106	01-Jul-2020
160	23-Aug-2020	297083	303461	0.050	3795	107	02-Jul-2020
161	24-Aug-2020	299628	305524	0.050	3793	107	02-Jul-2020
162	25-Aug-2020	302147	307600	0.049	3790	107	02-Jul-2020
163	26-Aug-2020	304583	309679	0.049	3787	108	03-Jul-2020
164	27-Aug-2020	306794	311733	0.049	3784	108	03-Jul-2020
165	28-Aug-2020	308925	313755	0.048	3781	108	03-Jul-2020
166	29-Aug-2020	310822	315721	0.048	3778	109	04-Jul-2020

(continued)

Table 14.6 (continued)

Day	Date	P (cases)	K (cases)	r (1/day)	dP/dt (cases/day)	$t_{peak}(day)$	$Date_{Peak}$
167	30-Aug-2020	312996	317667	0.048	3775	109	04-Jul-2020
168	31-Aug-2020	314946	319569	0.047	3772	109	04-Jul-2020
169	01-Sep-2020	317528	321502	0.047	3769	110	05-Jul-2020
170	02-Sep-2020	319686	323421	0.047	3765	110	05-Jul-2020
171	03-Sep-2020	321615	325301	0.046	3762	110	05-Jul-2020
172	04-Sep-2020	323565	327142	0.046	3758	111	06-Jul-2020
173	05-Sep-2020	325157	328919	0.046	3755	111	06-Jul-2020
174	06-Sep-2020	327359	330695	0.045	3752	111	06-Jul-2020
175	07-Sep-2020	329251	332436	0.045	3748	112	07-Jul-2020
176	08-Sep-2020	331078	334147	0.045	3745	112	07-Jul-2020
177	09-Sep-2020	332970	335831	0.045	3741	112	07-Jul-2020
178	10-Sep-2020	334762	337484	0.044	3738	113	08-Jul-2020
179	11-Sep-2020	336044	339057	0.044	3734	113	08-Jul-2020
180	12-Sep-2020	337520	340577	0.044	3731	113	08-Jul-2020
181	13-Sep-2020	339332	342078	0.044	3727	113	08-Jul-2020
182	14-Sep-2020	341056	343556	0.043	3724	114	09-Jul-2020
183	15-Sep-2020	342671	345001	0.043	3720	114	09-Jul-2020
184	16-Sep-2020	344264	346414	0.043	3717	114	09-Jul-2020
185	17-Sep-2020	345805	347799	0.043	3713	114	09-Jul-2020
186	18-Sep-2020	347372	349157	0.043	3710	115	10-Jul-2020
187	19-Sep-2020	348918	350487	0.042	3706	115	10-Jul-2020
188	20-Sep-2020	350621	351807	0.042	3703	115	10-Jul-2020
189	21-Sep-2020	352178	353106	0.042	3699	115	10-Jul-2020
190	22-Sep-2020	353844	354393	0.042	3696	116	11-Jul-2020
191	23-Sep-2020	355384	355661	0.042	3692	116	11-Jul-2020
192	24-Sep-2020	356767	356901	0.041	3688	116	11-Jul-2020
193	25-Sep-2020	357873	358092	0.041	3685	116	11-Jul-2020
194	26-Sep-2020	359148	359254	0.041	3681	116	11-Jul-2020
195	27-Sep-2020	360555	360395	0.041	3678	117	12-Jul-2020
196	28-Sep-2020	362043	361524	0.041	3674	117	12-Jul-2020
197	29-Sep-2020	363479	362640	0.040	3670	117	12-Jul-2020
198	30-Sep-2020	364987	363751	0.040	3667	117	12-Jul-2020
199	01-Oct-2020	366383	364845	0.040	3663	118	13-Jul-2020
200	02-Oct-2020	367565	365911	0.040	3659	118	13-Jul-2020
201	03-Oct-2020	368690	366951	0.040	3656	118	13-Jul-2020
202	04-Oct-2020	370132	367984	0.040	3652	118	13-Jul-2020
203	05-Oct-2020	371631	369017	0.040	3649	118	13-Jul-2020
204	06-Oct-2020	373151	370048	0.039	3645	119	14-Jul-2020
205	07-Oct-2020	374592	371078	0.039	3641	119	14-Jul-2020
206	08-Oct-2020	375870	372091	0.039	3637	119	14-Jul-2020
207	09-Oct-2020	377073	373089	0.039	3633	119	14-Jul-2020
208	10-Oct-2020	378266	374073	0.039	3630	119	14-Jul-2020
209	11-Oct-2020	379738	375054	0.039	3626	119	14-Jul-2020
210	12-Oct-2020	381275	376044	0.039	3622	120	15-Jul-2020
211	13-Oct-2020	382959	377046	0.038	3618	120	15-Jul-2020
212	14-Oct-2020	384559	378056	0.038	3614	120	15-Jul-2020
213	15-Oct-2020	386086	379070	0.038	3609	120	15-Jul-2020
214	16-Oct-2020	387295	380071	0.038	3605	120	15-Jul-2020
215	17-Oct-2020	388569	381064	0.038	3601	121	16-Jul-2020
216	18-Oct-2020	390206	382068	0.038	3597	121	16-Jul-2020
217	19-Oct-2020	391586	383066	0.038	3592	121	16-Jul-2020

(continued)

Table 14.6 (continued)

Day	Date	P (cases)	K (cases)	r (1/day)	dP/dt (cases/day)	$t_{peak}(day)$	Date$_{Peak}$
218	20-Oct-2020	393131	384076	0.037	3588	121	16-Jul-2020
219	21-Oct-2020	394827	385097	0.037	3583	121	16-Jul-2020
220	22-Oct-2020	396413	386125	0.037	3579	122	17-Jul-2020
221	23-Oct-2020	397507	387137	0.037	3574	122	17-Jul-2020
222	24-Oct-2020	398815	388146	0.037	3570	122	17-Jul-2020
223	25-Oct-2020	400251	389155	0.037	3565	122	17-Jul-2020
224	26-Oct-2020	401586	390160	0.037	3560	122	17-Jul-2020
225	27-Oct-2020	403079	391171	0.036	3556	123	18-Jul-2020
226	28-Oct-2020	404760	392194	0.036	3551	123	18-Jul-2020
227	29-Oct-2020	406364	393225	0.036	3546	123	18-Jul-2020
228	30-Oct-2020	407684	394254	0.036	3541	123	18-Jul-2020
229	31-Oct-2020	409252	395288	0.036	3536	124	19-Jul-2020
230	01-Nov-2020	410988	396340	0.036	3531	124	19-Jul-2020
231	02-Nov-2020	412647	397398	0.035	3526	124	19-Jul-2020
232	03-Nov-2020	414164	398464	0.035	3521	124	19-Jul-2020
233	04-Nov-2020	416006	399549	0.035	3515	124	19-Jul-2020
234	05-Nov-2020	417475	400634	0.035	3510	125	20-Jul-2020
235	06-Nov-2020	418764	401715	0.035	3505	125	20-Jul-2020
236	07-Nov-2020	420238	402797	0.035	3499	125	20-Jul-2020
237	08-Nov-2020	421921	403889	0.035	3494	125	20-Jul-2020
238	09-Nov-2020	423620	404992	0.034	3488	126	21-Jul-2020
239	10-Nov-2020	425353	406112	0.034	3483	126	21-Jul-2020
240	11-Nov-2020	427198	407245	0.034	3477	126	21-Jul-2020
241	12-Nov-2020	428965	408388	0.034	3471	126	21-Jul-2020
242	13-Nov-2020	430496	409540	0.034	3465	126	21-Jul-2020
243	14-Nov-2020	432333	410704	0.034	3459	127	22-Jul-2020
244	15-Nov-2020	434472	411896	0.034	3453	127	22-Jul-2020
245	16-Nov-2020	436684	413116	0.033	3447	127	22-Jul-2020
246	17-Nov-2020	438795	414359	0.033	3441	128	23-Jul-2020
247	18-Nov-2020	441159	415632	0.033	3434	128	23-Jul-2020
248	19-Nov-2020	443434	416937	0.033	3427	128	23-Jul-2020
249	20-Nov-2020	445281	418253	0.033	3420	128	23-Jul-2020
250	21-Nov-2020	447341	419590	0.033	3414	129	24-Jul-2020
251	22-Nov-2020	449760	420960	0.032	3407	129	24-Jul-2020
252	23-Nov-2020	451990	422356	0.032	3399	129	24-Jul-2020
253	24-Nov-2020	454146	423773	0.032	3392	130	25-Jul-2020
254	25-Nov-2020	456438	425213	0.032	3385	130	25-Jul-2020
255	26-Nov-2020	458711	426685	0.032	3377	130	25-Jul-2020
256	27-Nov-2020	460619	428164	0.031	3370	131	26-Jul-2020
257	28-Nov-2020	462407	429654	0.031	3362	131	26-Jul-2020
258	29-Nov-2020	464932	431177	0.031	3354	131	26-Jul-2020
259	30-Nov-2020	467225	432725	0.031	3346	132	27-Jul-2020
260	01-Dec-2020	469423	434292	0.031	3339	132	27-Jul-2020
261	02-Dec-2020	471739	435888	0.031	3330	132	27-Jul-2020
262	03-Dec-2020	473991	437507	0.030	3322	133	28-Jul-2020
263	04-Dec-2020	475789	439134	0.030	3314	133	28-Jul-2020
264	05-Dec-2020	477545	440765	0.030	3306	133	28-Jul-2020
265	06-Dec-2020	479743	442418	0.030	3298	134	29-Jul-2020
266	07-Dec-2020	481945	444088	0.030	3290	134	29-Jul-2020
267	08-Dec-2020	484104	445783	0.029	3282	134	29-Jul-2020
268	09-Dec-2020	485965	447476	0.029	3274	135	30-Jul-2020

(continued)

Table 14.6 (continued)

Day	Date	P (cases)	K (cases)	r (1/day)	dP/dt (cases/day)	$t_{peak}(day)$	Date $_{Peak}$
269	10-Dec-2020	487849	449192	0.029	3265	135	30-Jul-2020
270	11-Dec-2020	489178	450893	0.029	3257	136	31-Jul-2020
271	12-Dec-2020	490533	452582	0.029	3249	136	31-Jul-2020
272	13-Dec-2020	492332	454276	0.029	3241	136	31-Jul-2020
273	14-Dec-2020	494209	455985	0.028	3233	137	01-Aug-2020
274	15-Dec-2020	495841	457694	0.028	3225	137	01-Aug-2020
275	16-Dec-2020	496975	459379	0.028	3218	138	02-Aug-2020
276	17-Dec-2020	498293	461058	0.028	3210	138	02-Aug-2020
277	18-Dec-2020	499560	462731	0.028	3202	138	02-Aug-2020
278	19-Dec-2020	500713	464379	0.028	3195	139	03-Aug-2020
279	20-Dec-2020	502183	466025	0.027	3188	139	03-Aug-2020
280	21-Dec-2020	503501	467662	0.027	3180	139	03-Aug-2020
281	22-Dec-2020	504868	469299	0.027	3173	140	04-Aug-2020
282	23-Dec-2020	506102	470917	0.027	3166	140	04-Aug-2020
283	24-Dec-2020	507265	472509	0.027	3159	141	05-Aug-2020
284	25-Dec-2020	508099	474094	0.027	3152	141	05-Aug-2020
285	26-Dec-2020	509148	475648	0.026	3145	141	05-Aug-2020
286	27-Dec-2020	510080	477183	0.026	3139	142	06-Aug-2020
287	28-Dec-2020	511261	478703	0.026	3132	142	06-Aug-2020
288	29-Dec-2020	512496	480214	0.026	3126	142	06-Aug-2020
289	30-Dec-2020	513510	481705	0.026	3120	143	07-Aug-2020

Fig. 14.11, it can be predicted using the data from March 17, 2020 to December 31, 2020 that the estimated epidemic peak was on August 7, 2020. The estimated potential ending time of the epidemic could be at the end of March, 2021 (Fig. 14.11).

14.4.2 Application of SIR Model

Table 14.7 represents the estimated model parameters of the SIR model and Table 14.8 shows the statistical parameters of the SIR model. From Table 14.8, we see that the p-value of the model is close to 0 and the value of R^2 (coefficient of determination) of the model is close to 1 for Bangladesh. This indicates the statistical significance of the model results. In the first subplot of Fig. 14.12, the magenta color represents the actual cumulative infected cases and the blue curve indicates the projection of cumulative infected cases. The estimated outbreak of CoVID-19 started on March 17, 2020 and started to increase slightly. The estimated acceleration phase started on May 8, 2020 and lasted 78 days. After acceleration phase, the epidemic curve reached to its turning point on July 25, 2020. Since the deceleration phase started

Table 14.7 Estimated SIR model parameters for Bangladesh

Parameters	Value
Contact frequency (β)	0.048 (1/day)
Removal frequency (γ)	0.024 (1/day)
Population size (N)	655983
Initial number of cases (P_0)	9323
Basic reproduction number (R_0)	2.048
Reproduction number (R)	0.48
Time between contacts	20.7 (day)
Infectious period	42.3 (day)
Final Epidemic Size (C_{end})	**533890**
Final Susceptible Individuals (S_{end})	122092

Table 14.8 Statistical Parameters of the SIR model for Bangladesh

Number of Observations	Degrees of Freedom	Root Mean Squared Error	R^2	Adjusted R-Squared	F-Statistic vs. Zero model	p-value
289	285	30542.3	0.97	0.97	3209.53	1.3354e-185

from the turning point, the projection curve shows that this phase would continue upto October 30, 2020. After that, the cumulative infected cases could grow steadily up to January 26, 2021. The estimated ending phase of the curve would start on January 27, 2021 and could continue upto August 17, 2022. The total epidemic duration could be of 994 days since the outbreak. The projection curve also shows that the final epidemic size would reach $C_{end} = 533890$ (Fig. 14.12) and the final number of susceptible individuals left at the end of the epidemic could be 122092 (Fig. 14.12). In the second subplot, the green line represents the projection of the new cases per day. From this subplot, it is expected that the epidemic could end on August 18, 2022 showing only 5 infected cases and would end on December 6, 2022 reflecting only 1 infected cases. In the third subplot, the magenta line refers to the actual daily cases (%)) of cumulative confirmed cases of that day and the blue line demonstrates the projected daily cases (%) of cumulative confirmed cases of that day.

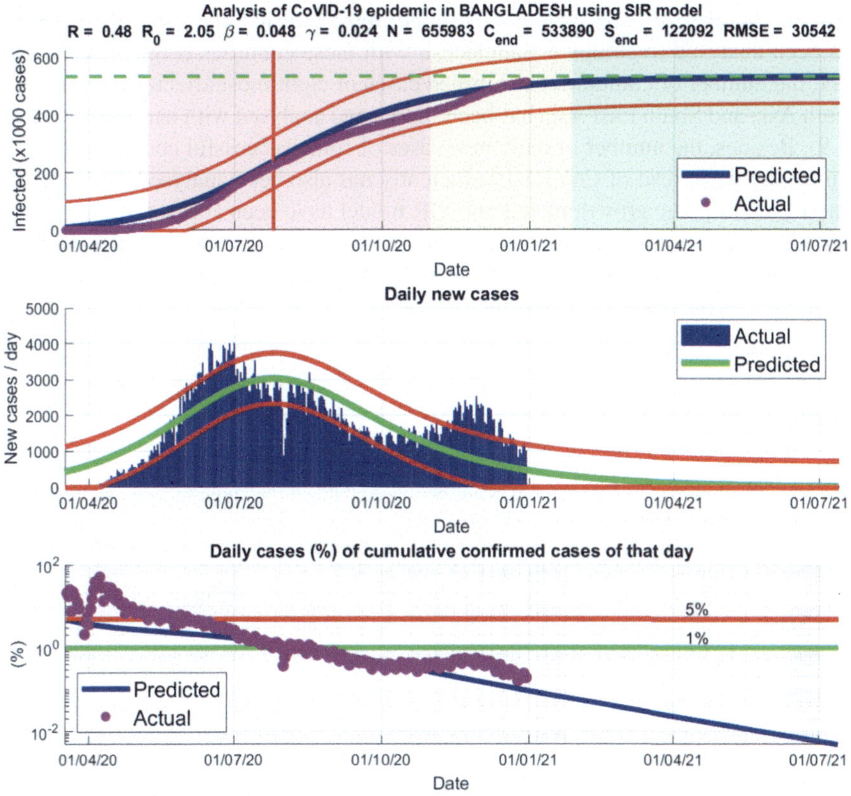

Fig. 14.12 Projection for daily cases (%) of cumulative confirmed cases of that day, daily new cases and infected cases. Here, R_0 is basic reproduction number, R is reproduction number, β is contact rate, γ is recovering rate, N is total population, C_{end} is total infected cases, S_{end} is total susceptible individuals and RMSE is root mean squared error

14.5 Conclusion

CoVID-19 identified on March 8, 2020 in Bangladesh is spreading though out this country until now. For this reason, containment activities must be taken before increasing the confirmed cases extremely. Moreover, we should maintain the social distancing at workplace to mitigate the spread of CoVID-19. In this chapter, we have discussed some epidemic models such as Logistic growth, SIS (Susceptible-Infectious-Susceptible), SIR (Susceptible-Infectious-Recovered) and SEIR (Susceptible-Exposed-Infectious-Recovered) in detail with mathematical equations. We have analysed pictorially the cumulative confirmed cases, cumulative death cases, cumulative recovered cases and cumulative confirmed cases at different divisions in Bangladesh (Fig. 14.5). We have also analysed pictorially the daily new cases of eight most affected countries of South Asia and South East Asia region (Fig. 14.7).

Also, the number of daily new cases of eight most affected countries of the world has been analysed to compare Bangladesh with these countries (Fig. 14.8). Moreover, the number of cumulative confirmed cases of eight most affected countries of South Asia and South East Asia has been shown and analysed with bar diagram (Fig. 14.9). Besides, the number of daily new cases of some successful countries that are controlling the spread of CoVID-19 efficiently has also been analysed (Fig. 14.10). Finally, the Logistic growth model and SIR model have been applied to time series data of CoVID-19 for the projection of cumulative confirmed cases in Bangladesh. Although no projection is certain, it can be assumed that the future might continue to follow the past pattern of confirmed cases.

References

1. Aron, J.L., Schwartz, I.B.: Seasonality and period-doubling bifurcations in an epidemic model. J. Theoret. Biol. **110**(4), 665–679 (1984)
2. Bailey, N.T., et al.: The Mathematical Theory of Infectious Diseases and Its Applications. Charles Griffin & Company Ltd., High Wycombe (1975)
3. Batista, M.: Estimation of the final size of the covid-19 epidemic. MedRxiv. **10**, 16–20023606 (2020)
4. Batista, M.: Estimation of the final size of the second phase of coronavirus epidemic by the logistic model. MedRxiv. [Internet] (2020)
5. Batista, M.: fitvirus. MATLAB Central File Exchange (2020). https://www.mathworks.com/matlabcentral/fileexchange/74411-fitvirus. Accessed 14 Nov 2020
6. Batista, M.: fitviruscovid19. MATLAB Central File Exchange (2020). https://de.mathworks.com/matlabcentral/fileexchange/74658-fitviruscovid19. Accessed 14 Nov 2020
7. Bedford, J., Enria, D., Giesecke, J., Heymann, D.L., Ihekweazu, C., Kobinger, G., Lane, H.C., Memish, Z., Oh, M.D., Schuchat, A., et al.: Covid-19: towards controlling of a pandemic. Lancet (2020). https://doi.org/10.1016/S0140-6736(20)30673-5
8. Benvenuto, D., Giovanetti, M., Vassallo, L., Angeletti, S., Ciccozzi, M.: Application of the arima model on the covid-2019 epidemic dataset. Data in brief, p. 105340 (2020)
9. Berry, M., Gamieldien, J., Fielding, B.C.: Identification of new respiratory viruses in the new millennium. Viruses **7**(3), 996–1019 (2015). https://doi.org/10.3390/v7030996
10. Chafekar, A., Fielding, B.C.: Mers-cov: understanding the latest human coronavirus threat. Viruses **10**(2), 93 (2018). https://doi.org/10.3390/v10020093
11. Chowell, G., Simonsen, L., Viboud, C., Kuang, Y.: Is west Africa approaching a catastrophic phase or is the 2014 ebola epidemic slowing down? Different models yield different answers for Liberia. PLoS Curr. **6**, 1–11 (2014)
12. Corman, V.M., Ithete, N.L., Richards, L.R., Schoeman, M.C., Preiser, W., Drosten, C., Drexler, J.F.: Rooting the phylogenetic tree of middle east respiratory syndrome coronavirus by characterization of a conspecific virus from an African bat. J. Virol. **88**(19), 11297–11303 (2014). https://doi.org/10.1128/JVI.01498-14
13. Dhanwant, J.N., Ramanathan, V.: Forecasting covid 19 growth in India using susceptible-infected-recovered (sir) model. arXiv preprint arXiv:2004.00696 (2020)
14. Fehr, A.R., Perlman, S.: Coronaviruses: an overview of their replication and pathogenesis. In: Coronaviruses, pp. 1–23. Springer (2015). https://doi.org/10.1007/978-1-4939-2438-7_1
15. Fung, T.S., Liu, D.X.: Human coronavirus: host-pathogen interaction. Ann. Rev. Microbiol. **73**, 529–557 (2019). https://doi.org/10.1146/annurev-micro-020518-115759
16. Gerna, G., Campanini, G., Rovida, F., Percivalle, E., Sarasini, A., Marchi, A., Baldanti, F.: Genetic variability of human coronavirus oc43-, 229e-, and nl63-like strains and their associ-

ation with lower respiratory tract infections of hospitalized infants and immunocompromised patients. J. Med. Virol. **78**(7), 938–949 (2006). https://doi.org/10.1002/jmv.20645
17. Gorbalenya, A.E., Baker, S.C., Baric, R.S., de Groot, R., Drosten, C., Gulyaeva, A., et al.: Severe acute respiratory syndrome-related coronavirus: the species and its viruses-a statement of the coronavirus study group. BioRxiv (2020). https://doi.org/10.1101/2020.02.07.937862
18. Graham, R.L., Donaldson, E.F., Baric, R.S.: A decade after SARS: strategies for controlling emerging coronaviruses. Nat. Rev. Microbiol. **11**(12), 836–848 (2013). https://doi.org/10.1038/nrmicro3143
19. Van der Hoek, L., Pyrc, K., Jebbink, M.F., Vermeulen-Oost, W., Berkhout, R.J., Wolthers, K.C., Wertheim-van Dillen, P.M., Kaandorp, J., Spaargaren, J., Berkhout, B.: Identification of a new human coronavirus. Nat. Med. **10**(4), 368–373 (2004). https://doi.org/10.1038/nm1024
20. Hu, B., Ge, X., Wang, L.F., Shi, Z.: Bat origin of human coronaviruses. Virol. J. **12**(1), 221 (2015). https://doi.org/10.1186/s12985-015-0422-1
21. Hu, Z., Ge, Q., Jin, L., Xiong, M.: Artificial intelligence forecasting of covid-19 in China. arXiv preprint arXiv:2002.07112 (2020)
22. IEDCR: Distribution of confirmed cases in Bangladesh (2020). https://iedcr.gov.bd/
23. JHU: 2019 novel coronavirus covid-19 (2019-ncov) data repository by Johns Hopkins center for systems science and engineering (JHU CSSE) (2019). https://github.com/CSSEGISandData/COVID-19
24. Kermack, W.O., McKendrick, A.G.: A contribution to the mathematical theory of epidemics. Proc. R. Soc. Lond. Ser. A Containing Papers Math. Phys. Charact. **115**(772), 700–721 (1927)
25. Kermack, W.O., McKendrick, A.G.: Contributions to the mathematical theory of epidemics. ii.–the problem of endemicity. Proc. R. Soc. Lond Ser. A Containing Papers Math. Phys. Charact. **138**(834), 55–83 (1932)
26. Khaleque, A., Sen, P.: An empirical analysis of the Ebola outbreak in West Africa. Sci. Rep. **7**, 42594 (2017)
27. Li, X., Wang, W., Zhao, X., Zai, J., Zhao, Q., Li, Y., Chaillon, A.: Transmission dynamics and evolutionary history of 2019-nCoV. J. Med. Virol. (2020). https://doi.org/10.1002/jmv.25701
28. Lim, Y.X., Ng, Y.L., Tam, J.P., Liu, D.X.: Human coronaviruses: a review of virus-host interactions. Diseases **4**(3), 26 (2016). https://doi.org/10.3390/diseases4030026
29. Mackay, I.M., Arden, K.E.: MERS coronavirus: diagnostics, epidemiology and transmission. Virol. J. **12**(1), 222 (2015). https://doi.org/10.1186/s12985-015-0439-5
30. Ndiaye, B.M., Tendeng, L., Seck, D.: Analysis of the covid-19 pandemic by sir model and machine learning technics for forecasting. arXiv preprint arXiv:2004.01574 (2020)
31. Peiris, J., Guan, Y., Yuen, K.: Severe acute respiratory syndrome. Nat. Med. **10**(12), S88–S97 (2004). https://doi.org/10.1038/nm1143
32. Pell, B., Kuang, Y., Viboud, C., Chowell, G.: Using phenomenological models for forecasting the 2015 Ebola challenge. Epidemics **22**, 62–70 (2018)
33. Perlman, S.: Another decade, another coronavirus (2020). https://doi.org/10.1056/NEJMe2001126
34. Pinter, G., Felde, I., Mosavi, A., Ghamisi, P., Gloaguen, R.: Covid-19 pandemic prediction for Hungary; a hybrid machine learning approach. Mathematics **8**(6), 890 (2020)
35. Pyrc, K., Berkhout, B., van der Hoek, L.: The novel human coronaviruses NL63 and HKU1. J. Virol. **81**(7), 3051–3057 (2007). https://doi.org/10.1128/JVI.01466-06
36. Pyrc, K., Berkhout, B., Van Der Hoek, L.: Identification of new human coronaviruses. Expert Rev. Anti-infect. Ther. **5**(2), 245–253 (2007). https://doi.org/10.1586/14787210.5.2.245
37. Ranjan, R.: Predictions for covid-19 outbreak in India using epidemiological models. medRxiv (2020)
38. Reuters: Bangladesh confirms its first three cases of coronavirus (2020). https://www.reuters.com/article/us-252health-coronavirus-bangladesh-idUSKBN20V0FS. Accessed 9 May 2020
39. Singhal, T.: A review of coronavirus disease-2019 (Covid-19). Indian J. Pediatr. 1–6 (2020). https://doi.org/10.1007/s12098-020-03263-6
40. Verhulst, P.F.C.C.O.: Mathematical research on the population growth law. New Mem. R. Acad. Sci. **18**, 1–41 (1845)

41. Wang, D., Hu, B., Hu, C., Zhu, F., Liu, X., Zhang, J., Wang, B., Xiang, H., Cheng, Z., Xiong, Y., et al.: Clinical characteristics of 138 hospitalized patients with 2019 novel coronavirus-infected pneumonia in Wuhan, China. Jama (2020). https://doi.org/10.1001/jama.2020.1585
42. Zakary, O., Larrache, A., Rachik, M., Elmouki, I.: Effect of awareness programs and travel-blocking operations in the control of HIV/AIDS outbreaks: a multi-domains SIR model. Adv. Differ. Equ. **2016**(1), 169 (2016)
43. Zhou, P., Yang, X.L., Wang, X.G., Hu, B., Zhang, L., Zhang, W., Si, H.R., Zhu, Y., Li, B., Huang, C.L., et al.: A pneumonia outbreak associated with a new coronavirus of probable bat origin. Nature **579**(7798), 270–273 (2020). https://doi.org/10.1038/s41586-020-2012-7

Chapter 15
Challenges Ahead in Healthcare Applications for Vision and Sensors

Manan Binth Taj Noor, Nusrat Zerin Zenia, and M. Shamim Kaiser

Abstract The trend of todays healthcare demands prognostic, precautionary, customized and participatory care spawning an increase in stipulation for healthcare resources and services because of growing world population with rapid increase of people with special needs, such as the elderly population. Besides, the World Health Organization (WHO) conducted a survey back in 2013 which highlighted the fact that, global health workforce shortage to reach 12.9 million incoming decades; with more chronic diseases to be observed with an increasing rate of 10% each year like current pandemic situation of COVID-19. Owing to such factors, the researchers and healthcare professionals should seamlessly consolidate, coordinate and contrive new technologies to facilitate patients services; moderating healthcare transformation costs and risks at the same time. Following that, a widespread adoption of sensor technology and computer vision have been witnessed due to their superior performance for a variety of healthcare applications comprising not only remote monitoring and tracking of diseases with computer vision based diagnostic health examination methods but also early detection and prediction of stages of various chronics as well and many more. However, there remain several challenges which include the interoperability and expandability issues of technological devices and sensors including the familiarity of users with the usage of those technologies. Another prime obstacle remains the domain specific datasets, which are required for training user oriented models in terms of data driven approaches. This chapter demonstrates study of advances in modern computer vision techniques along with the development of faster and more accurate sensors for healthcare applications highlighting their challenges, open issues, and performance considerations in healthcare research. *abstract* environment.

Keywords Computer vision · Sensor · Healthcare application · Challenges

---

M. B. T. Noor · N. Z. Zenia · M. S. Kaiser (✉)
Institute of Information Technology, Jahangirnagar University, Savar-1342, Dhaka, Bangladesh
e-mail: mskaiser@juniv.edu

M. B. T. Noor
e-mail: manan.noor@juniv.edu

N. Z. Zenia
e-mail: nusratzenia@juniv.edu

M. A. R. Ahad and A. Inoue (eds.), *Vision, Sensing and Analytics: Integrative Approaches*, Intelligent Systems Reference Library 207, https://doi.org/10.1007/978-3-030-75490-7_15

15.1 Introduction

Digital technology has always played an essential role in healthcare be it for treatment, monitoring or prediction of health conditions. Although these technological facilities have traditionally been available in hospitals and acute care only, with the advent of Technological enabled Care (TEC) the focus is wide spread towards peoples own homes, health centres, supported housing and care homes as well. As a result; in the coming days; personalized healthcare would not only enable remote monitoring with tracking but also would be able to conduct diagnostics and early detection of diseases at the same time. These development works are supported by the global technological giants like Google, Apple, Amazon and active involvement of several pharmaceutical companies which are the most active health app publishers. This collaboration enhances the possibility of attaining much improved health research towards new healthcare provider models transforming the patient experience [1]. For such healthcare diegesis, solution based application of computer vision and sensor technology are playing the most prominent and vital role.

Computer vision calibrates the ability to replicate human sight with the understanding of the object in front of it by training computers with algorithms for processing images. Making a faster and more accurate diagnosis is the main goal of computer vision in healthcare [2]. Rapid application of computer vision is viable in medical imaging for patient tracking and monitoring.

The main function of electronics based medical devices equipped with sensors is conversion of various forms of stimuli into electrical signals to retrieve important health information through analysis [3]. Intelligence of medical equipment is enhanced by sensors and also enables remote monitoring of health through reading of vital signs. This remote healthcare monitoring based on wireless sensor communication comes up with such efficient solutions that let people maintain health standards while being monitored and protected in spite of being comfortably at home.

However, there are always some continual challenges that keep arising with different points in time in the application of computer vision and sensors in implementing technology enabled care in spite of conducting continuous research effort and achieving corresponding success. Because with the advancement of time; different modification requirements get cropped up; new technologies appear; network infrastructure gets modified etc. which affect the ongoing version of the current device in action. From the perspective of vision and sensor based medical equipment; some common challenging technological aspects in todays healthcare include lack of quality in terms of reliability, data overload, privacy and security with enhanced cost factor etc. Todays innovation in digital technology will impact tomorrows healthcare. The scope of this chapter is to discuss the challenging impact of those innovations due to application of vision and sensor technology in healthcare.

15.2 Application and Challenges of Computer Vision in Healthcare

At present, computer vision is being employed to a greater extent in different areas of healthcare, leveraging healthcare professionals to better diagnose and monitor their patients as well ascertaining the evolution of diseases which also includes medical image analysis, different predictive analysis ultimately resulting in prescribing the right treatments, among others. This section addresses the applications and challenges of this current and imminent cutting-edge technology in healthcare in real time monitoring and tracking of health.

Computer vision comes into context for rapid evolution of health monitoring systems which possess the prospective to change the way health care is currently delivered. According to existing medical surveys, computer vision has been arrogated to monitor the patients with cardiac diseases, diabetes, hypertension, hypothermia and hyperthermia [4–7]. Continuous and long term monitoring is required to control the threats caused by such chronic diseases. At the same time, with the elevation in global life expectancy and a rapidly increasing aging population results in challenges in health and social care as there is also a scarcity of healthcare professionals to provide required care to the senior citizens constantly. Thus real time monitoring and tracking of health could be utilized to subjugate these challenges by identifying the recurrences in health condition ultimately delegating early mediation.

Lots of research and development works in the healthcare industry are already underway. Now recent developments exploiting computer vision based real time monitoring are stated below:

15.2.1 Facial Expression Detection

One of the fundamental objectives within any medical service comprises the propensity of emotion detection and taking up of required measures accordingly. Automated emotion detection from facial expression analysis becomes a priority in those medical environments in which patients are unable to express their uneasiness verbally owing to severe medical situations such as paralysis, autism, communication disabilities, arthritis, patients under sedation or babies etc. Therefore, a useful aid would be the utilization of computer vision based facial recognition techniques for evaluating facial intimations of the patients. This technique would ultimately result in the interpretation of the emotional state of patients such as joy, sadness, surprise, anger, fear, disgust or contempt by availing appropriate assistance. Figure 15.1 depicts the basic architecture of facial expression recognition.

Several computer vision APIs are available which detect emotive states of an individual by exploitation of facial detection, eye tracking, and specific facial position intimations by applying face and emotion recognition algorithms based on curated databases. Some of those APIs include Emotient, Affectiva, EmoVu, Nviso etc.

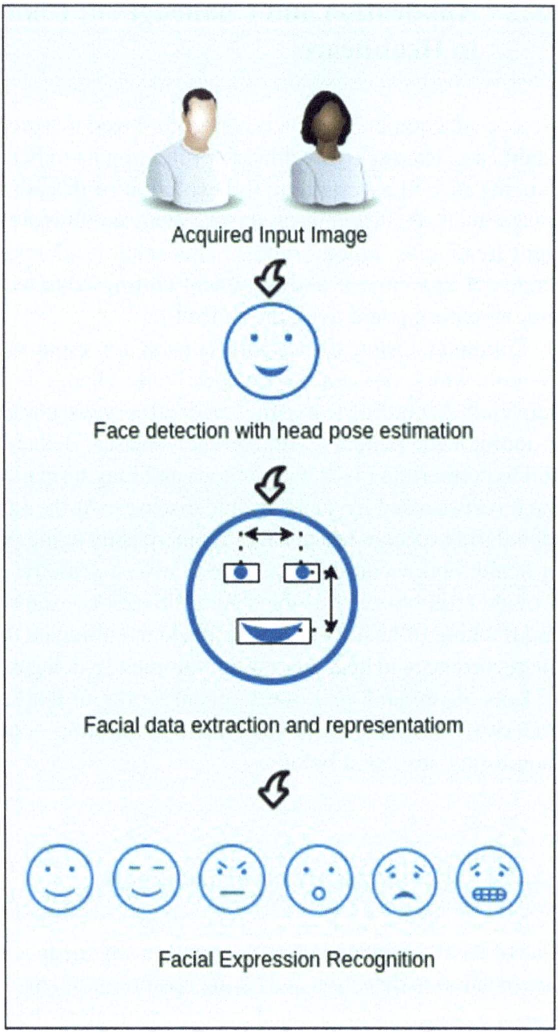

Fig. 15.1 General architecture of facial expression detection/recognition

offering Desktop SDK, Mobile SDK for developers for application development [8]. Consecutively, some of those also provide precise visual analytics to track expressions over time. SHORE (Sophisticated High-speed Object Recognition Engine) is an example of emotion detection application. This application detects emotion through built in cameras by using google glasses [9, 10].

Main challenge of facial expression analysis lies in analysis of determining the relationship between points on the face from a photo/video as it is subjugated to confusion in rendering the correct emotions owing to similarities in expression. One of the major challenging aspects also include the recognition problem of low-level emotion where the facial expressions are captured with low peak frequencies.

Most of the works conducted in this area focused on depiction of emotion being at the peak level of intensity forsaking emotion depiction at lower intensity levels owing to the difficulty level of recognition due to paucity of distinctive features in available databases. Nevertheless, having a complex and noisy target environment with unstable movements of the foreground people also pose an incredible challenge in precise emotion detection [11].

However, variation in illumination which may affect the accuracy of facial feature extraction process, changes in head pose as most of the datasets comprise frontal view of the faces which is a contradict to real world environment and subject dependency also pose greater challenges in facial expression analysis in terms of emotion detection. Subject dependency refers to the problem of recognizing pre-trained human faces only failing to perform otherwise challenging detections.

Finally, emotion analytics postulates the combination of psychology and technology in terms of emotion intensity variations in spatial and temporal domains. The upcoming future of perfection and reliability of subtle emotion detection where human emotions are made understandable through intensity of facial expression is yet to be achieved since much work and improvements remains to be done.

15.2.2 Sign Language Translation

A significant healthcare problem arises from language barrier not being inclined to second language problems only. Deaf people also come into this context as general people remain unaware about the sign language; most commonly known as American Sign Language (ASL), used by them for communication. It results in divestment of critical and timely health information with qualified healthcare for deaf and hard hearing community. Computer vision based Sign Language Recognition (SLR) comes into aiding this problem which is the technique of identifying a sequence of developed signs with subsequent translation of those signs into text or a speech providing proper meaning. SLR also requires a combination of pattern recognition, Natural Language processing and linguistics in cooperation with computer vision. The first automated sign language translation solution SignAll, based on computer vision and natural language processing. It aided communication between hearing enabled English spoken people and deaf or hard of hearing individuals who use ASL [12]. Another computer vision based application LABVIEW software Vision Assistant has been reported as well which mainly processes captured images for recognizing unique sign features producing audio output through smartphone loudspeaker after sign recognition. Some other computer vision based real time sign language translator solutions have also been reported in [13, 14].

Scope of further research and improvement is abundant in sign language translation. One of the most inhibiting factors in this case is sign language dataset which contains certain shortcomings starting with having small size compared to the demand, having native signers where novices are allowed to contribute lacking signer provenance and skill which ultimately results in insufficient variety in signs as well.

One of the most prominent challenges in achieving accuracy of SLR stands in handling simultaneous events while producing sign language where both the hands move changing the handshapes frequently. Multi channel articulation of sign language through the hands, arms, head, shoulders, torso, and parts of the face make the translation process very difficult. In real-world scenarios natural conversations take place with complete sentences and prolonged remarks in terms of sign language. Devising basic building blocks from such situations poses the biggest challenging factor as a significant number of existing sign language datasets are composed of individual signs only.

15.2.3 Detection of Safety Critical Incidents

One of the most prevalent safety critical incidents faced especially by the elderly people is falling. According to a report published in WHO, almost 28 to 35% elderly people aged over 65 face fall related accidents at least once a year which even escalates to a range of 42% for the aged over 70 [15]. Currently WHO estimates that 37.3 million falls have been severe enough requiring medical attention each year. And, the most alarming part comprises that, an estimated 6,46,000 individuals embrace death from falls globally [15]. These alarming reports make this safety critical incident a prominent issue towards healthcare to be taken care of. Vision based approaches for fall detection come into this context which could play a prominent role in future healthcare and assistance. The purely vision-based approaches focus on the real-time execution of the algorithm to be applied on the frames of videos to detect falls exploiting standard computing platforms and cameras rather than wearing any wearable devices which is a common issue as elderly people show reluctance sometime towards it. Some challenges are even faced while implementing it as cameras have to be placed in every place of actuation to cover the area of interest. Still, there are chances of improvement due to variance of viewports of cameras, handling partial or full occlusions with night and low light conditions for achieving accuracy. And the major challenge in implementing such systems aimed for elderly people specially, is that the training and testing of these systems mostly take place in controlled environments having comparatively young people as volunteers to create databases which results in reduced detection rate if applied in real life [16]. Since, it is not practically possible to subject elderly people to simulate falls, the evaluation rates deteriorate gradually.

15.2.4 Visual Prosthesis

According to WHO, there are at lease 2.2 billion people worldwide who suffer from visual impairment and blindness. It has also been reported that, of those at least 1 billion of them go through vision impairment that could have been intercepted.

Visual prosthesis comes into this backdrop which refers to the ability of restoring functional vision to those suffering from partial or total blindness is one of the most ardent acts of healing medicine that can be achieved. Visual prosthesis which is often termed as bionic eye as well is an experimental device which requires multi disciplinary knowledge consisting of computer vision, microprocessors, receivers, radio transmitters and retinal chips. This device is composed of a computer chip which is placed in the back of the eye of the affected being linked with a mini video camera built into glasses which is put on by them. When an image is captured by the camera; it is focused into the chip converting into electronic signals that can be intercepted by the brain [17]. This technology is still growing and requires massive collaboration among basic scientists, engineers and clinicians. Although great technical progress is visible in this field, certain technological challenges need to be met before visual prosthesis could be considered a convictive clinical therapy. A major challenge constitutes the presence of electrodes in the device which are too large to target individual neuron types causing distinctive change in this artificial vision compared to the normal vision. Thus the images produced lack stability, cause blurriness which possibly even dispense rudimentary control over colour.

15.2.5 Surgical Assistance Technology

Incorporation of computer vision to attain improvement fidelity in surgery help surgeons making rapid, timely and effective decisions during complicated surgeries. Following that, a computer vision based tool has been reported by RSP vision which could guide the surgeons with surgical movements during orthopedic procedures by improving visualization of input images through calibration, orientation, and navigation of those images [18]. A software called Triton has been offered by Gauss Surgical, which would help surgeons to monitor surgical blood loss using computer vision on an iPad during and after surgeries [19]. Then, there comes in the scope of robotic surgery as well which is mainly based on vision processing algorithms. In tele-surgery, surgeons operate on patients who are distantly located by directing the movements of robots in an operation environment which is provided by networking and real-time computer vision [20]. Efficiency and correctness are must in surgical assistance technologies as the lives of people are endangered and made reliant on the decisions taken at those point of time. In spite of visible improvements challenges are yet to be addressed to bring out perfection. Some of the challenging factors yet to be addressed in this field include; ultra responsive connectivity, ultra high reliability as the system must always be online, high cost in terms continuous research conductance etc.

15.2.6 Rehabilitation Aids

Computer vision aided rehabilitation in motion analysis lets people having an injury to exercise intense movement without taking assistance from a therapist daily. Rehabilitation process is a combination of pharmacological treatment and psychotherapy. Generally, rehabilitation treatment is provided to the elderly people and those who are suffering from paralysis owing to injuries. Current advancements in this field require camera based networks for continuous monitoring of physical activities [21]. Challenges ahead in this arena include lack of availability of realistic dataset as the dataset is created with comparative healthy participants; lack of diversity in the available datasets; lack of proper human activity recognition methods resulting in overlapping among the classes in dataset; training is conducted in extreme restricted lab conditions whereas produces comparative less accuracy while applied in reality based on daylight with outdoor, indoor aspects as well; camera positions with occluded viewports; and finally limited storage space with high power consumption rate impeding performance of vision based rehabilitation systems.

15.2.7 Medical Imaging

In todays time, the healthcare industry is strongly reliant on meticulous diagnostics pledged by medical imaging. Different processes and imaging methods are used to delineate internal images of the human body in medical imaging for diagnostic and treatment purposes. According to [22], a New York-based hospital has applied a computer vision based application to analyze CT scans which subsequently ascertained the presence of neurological illnesses in an average time limit of 1.2 s only which is 150 times lesser than the manual evaluation process. Almost all basic and advanced modalities of images have already been processed in computer vision. From basic modalities of general, mammographic X-ray, ultrasound to advanced modalities such as Computed Tomography (CT), Magnetic Resonance Images (MRI), Single-Photon Emission Computed Tomography (SPECT), Positron Emission Tomography (PET) etc. have been explored for numerous zonal of health issues detection and diagnostics [23]. Of those numerous domains some of them include all sorts of oncology, cardiology, neuroscience, ophthalmology, lab test automation, endocrinology (especially diabetes) and many more. Some of the major applications appear to be detection of cancer metastases from biopsy images, classification of acute leukemia, MRI-based age estimation, automatic segmentation of kidneys, red Blood based disease screening and many more.

In spite of having a huge amount of effort and research in this field, corresponding challenges of it are yet to be addressed. Major challenging factor in medical imaging for computer vision mainly consists of handling continually increasing massive amounts of complex imaging data which strongly needs proper data compression standards, mass storage space and coherent lookup/access streamlining. At the same time data gathering, training and testing processes tend to be very expensive due to hardware dependency which come into being very costly resulting in high development cost to conduct serious research. Even after the application is ready, it takes a long time to validate that application which subsequently affects the cost to a greater extent. Application of deep learning structures need to be further exploited. At the same time, data network models need to be fined with proper optimization techniques.

15.3 Application and Challenges of Sensor Technology in Healthcare

The last two decades have witnessed healthcare evolving rapidly going through an exponential growth and substantial developments by the impact of some of the technology-driven changes. Emergence of health sensors manifest to be that generation of technology paving a dedicated mechanism for healthcare monitoring applications. A significant role can be played by this technology during the time of natural disaster and pandemics in mass casualties where all the necessary information about current and past medical records, corresponding identification data etc. can be stored. Ability of providing reliability with enhanced mobility constitute the aspect of biggest advantage in exploiting wireless Sensor based technology in healthcare. In this chapter, we will focus on the recent developments and challenges ahead in portable wearable, epidermal and implantable sensor technologies in healthcare.

Wearable or implantable sensors are used for sensing biological details from either outside or inside the human body [24]. The information acquired from the human body is disseminated to the control device which is put on the body or propped in an accessible location by these sensors. Then, the data accumulated from the control device are transmitted through wireless body area networks integrated by wireless sensor networks with long-range transmission capabilities to desired locations for medical diagnosis or therapeutic purposes. Figure 15.2 elaborates this scenario of a remote healthcare monitoring application in terms of emergency situations such as heart failure, falls etc. Here, any unwanted situation is detected through processing the data availed from the wireless sensor network system by sending required message to the responsible authorities for providing immediate medical assistance. This mode of service could also be interpreted in other situations as well such as notifying patients to take medication, rest, glucose intake based on their processed physiological data received from the system. At the same time medical professionals could also remotely monitor the patient's health status if could not be present physically.

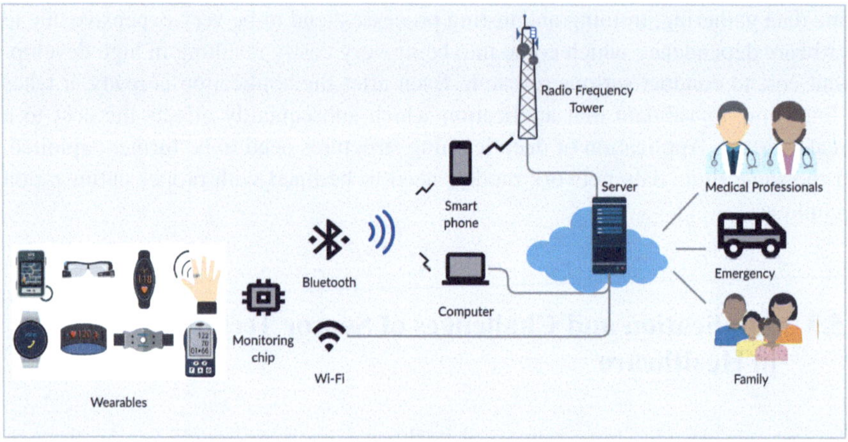

Fig. 15.2 General architecture of a wearable technology framework

15.3.1 Wearable Sensor

Wearable sensor technology is applied in healthcare by achieving huge advancements with miniature circuits, microcontroller functions, front-end amplification and wireless data transmission. Wearable sensors are integrated into devices which can be put on human body such as wristwatches, headphones, smartphones or various accessories based on electronic textiles incorporated into fabric such as garments, hats, wristbands, socks, shoes, eyeglasses etc. After integration into the healthcare devices, the wearables capture real time data from the patients and automatically turns into an alert system to communicate with doctors and other medical staff through their handheld devices, if any life endangered changes take place within the body of the corresponding patient [25]. Some common but most effective application areas of wearable sensors include wearable fitness trackers through wristbands which keeps track of the physical activities and heart rate of users; smart health watches for tracking various physical activities such as number of steps taken, distance traveled, calories burned through walking, running, swimming, and biking; wearable ECG monitors with the ability of measuring electrocardiograms; wearable blood pressure monitors to appraise blood pressure; biosensors which would allow the users to transpose while collecting data on their corresponding movement, respiratory rate, heart rate and temperature etc.; blood sugar monitoring sensors; Pulse oximeter and many more. On the other hand, textile based devices are devised into fabrics which also includes patients attire, quilts, cushion etc. as well. Some examples include WEALTHY, MY HEART which are used to measure respiratory activities, electromyograms (EMG), electrocardiograms (ECG) and body posture. These applications are embedded into cotton shirts being funded by the EU.

Wearable health monitoring systems must address certain ergonomic and medical requirements in order to ensure long-term monitoring mileage. The components of the system must be small in proportion and flexible with the prime requirement of being chemically inert and non-toxic ascertaining an overall comfortable system. In that context, certain challenging factors still need to be overcome in the application of wearable sensors in healthcare. Starting with the proper assessment problem of the device. The proper usage of these devices greatly depend upon the machine-based analysis and amalgamation of personal data with a personal calibration of the device whereas most of the time the devices are used standalone without incorporating the personal records producing poor tracking and monitoring results [26]. Misalignment issues also come into context in this case which adversely influence class and accuracy of the measurements. A crucial component of wearable devices is battery which requires more robustness as it plays a very important role in GPS tracking which requires a significant amount of battery power consumption.

One of the major technical challenges that must be overcome in the coming age is the hindrances during the feature extraction phase from signals which are induced by the motion or respiration of the user. The usage report of most of the body worn applications indicate that the systems accuracy is disrupted by the presence of those noises which are caused by several additional factors as well. These noises can not be filtered out completely owing to the limitation of the devices processing capabilities citing another limitation in this case.

Then, one of the most common issues with wearable systems is the delay in providing results and generating alerts due to data loss, buffering, network communication, monitoring or processing which is caused most of the time by hardware and limiting computation resource for the wearable on-body central node of a multi-sensor Body Sensor Network (BSN); thus there are still requirements for robust and efficient algorithm for optimizing performance of (BSN).

However, for textile based sensors, the key design challenge still comprises ensuring high signal accuracy preserving the sensitivity, SNR and stability within a textile based platform with the proper selection of sensing material as well. At the same time scopes of improvements exist in keeping up the signal integrity with time durability. Additionally, during the fabrication of smart textiles; washing cycles must also be improved for long term health monitoring. A prime fabrication issue in smart textiles comprises the exhibition of different physical properties by contrasting parts of the fibre. For example in case a woolen smart T-shirt; air humidity effects resistance of the fibre. On the contrary, electrical resistance of a metal fibre is changed with the variation of temperature. In this situation, in spite of providing sufficient strength and flexibility by the substrate; sensing applications might lose an important property of electronic functionality which would ultimately result in desecration of the conducting fibres [27]. Subsequently, considering ease of movement of users; positioning of sensors for disparate body shapes with sizes, cost-effectiveness and energy consumption by overcoming the complexity of integrating electronics into textiles for batch production comprehends the technical challenges which require more research and development.

Figure 15.3 provides a brief view to available portable wearable devices; data monitored by these devices and also addresses corresponding challenging issues which still need to be copped up with for successful application of the devices in

Fig. 15.3 Overview of different portable wearable devices worn on different parts of body, data measured by these devices with corresponding technical and physical challenges in healthcare applications for these devices

healthcare. Overview of the measured data collected from different sensors have been collected from [27] and [28].

15.3.2 *Epidermal Sensor*

Epidermal sensors in terms of skin patches are the outcome of evolution in material science being merged into electronic systems within the wearables medical field. Epidermal electronic devices which are soft and flexible in nature with stretching capabilities are placed directly on the skin. These stretchable electronic devices provides a comparatively new platform for robotic feedback and control, regenerative analysis of medicine with continuous healthcare as well. Epidermal sensors are deemed ideal wearables since these sensors can be concealed from being visible and can also record more accurate data without being deranged by physical movements. Attachment process of epidermal devices to the skin consists of mounting the corresponding device onto a svelte elastomeric supporting the substrate or directly over the skin. Epidermal sensors have already been applied for temperature, sweat [29], strain and cardiovascular sensing. A highly transparent graphene-based epidermal sensor system (GESS) can be reported as an example here which can be stowed on the patients body similarly a tattoo is put on [30].

In the context of epidermal sensors, there are yet scopes to be attempted to make the devices transparent to the users. Additionally, there are quite rigid scopes of research in realizing the full diagnostic potential of epidermal sensor based systems which must focus on the validity of the extended use. There needs to be put on more effort to technically correlate the sensor response to analyze blood concentration concurrently with coherent and controlled sampling of the target biofluids. Further attention is also needed to exploit alternative sampling routes to enhance the impact of epidermal devices in healthcare. Besides, the miniaturization of the electronic devices with proper integration process still needs to be addressed. Subsequently absence of re-calibration, data processing and wireless transmission of data constitute the ultimate challenges in healthcare applications.

15.3.3 Contact Lens Biosensor

Application of smart contact lense have paved another way to extract ophthalmological information from the eyes by monitoring the physiological traits of the eyes and of any tears non-invasively. An optical sensor embedded wearable contact lens has been reported in [31] to continuously measure glucose in physiological conditions.

However, the biggest challenge in wearable contact lens biosensors technology remains the complexity in fabricating photonic crystal sensors at a lower cost ultimately forming a user-friendly and mechanically stable device. And, a significant challenge has turned out to be the hurdles in obtaining the visual quantitative readouts. Furthermore, the limitation of curtailing the electrons into contact lenses, deterrence in power transmission and data reception from contact lenses remains a barrier [32].

15.3.4 Implantable Sensor

The most complicated sensors are implantable sensors which have recently appeared to be an emerging measurement of wireless medical healthcare. Integration of rapid advances in biology, electrical, chemistry and mechanical technology with Micro-Electromechanical Systems (MEMS) technology made it possible to implant biodevices for conducting continuous medical observation. In biomedicine, these implantable devices are going to be dominant appliances owing to the their capabilities of providing clearer and genuine depiction of the events taking place inside the body over a definite period of time which would subsequently help to monitor chronic diseases and tracking the progress post a surgery or treatment [33]. Implantable devices are more advantageous to use because these biosensors could monitor some inside biological factors which other monitoring devices would not be able to [34]. For example:

- Biological metabolites.
- Detection of electric signals.
- Nerve electrical stimulation.
- Body function restoration based activities.

Starting from cardiac pacemakers, the number of patients being treated with cardiovascular implantable electronic devices, implantable deep brain stimulators have been increasing only till now [35]. Efforts are being made as well to continuously monitor hypertension exploiting miniaturized and implantable blood pressure biosensors to conduct efficient treatment.

Implantable sensors are generally small and lightweight in nature which have been built by making them compatible with body mass and require very less power for operation. One of the prime challenging aspects is the development process of a fully implantable operational biosensor itself with heterogeneous elements. A brief description of the elements are as follows:

- Incorporating electrodes for sensing the target vital signals.
- A circuit which would be capable of measuring and transmitting the processed data.
- Power source.
- Finally the device must be biocompatible so that toxicity and chronic inflammation could be avoided being tolerable by the host as well.

Implantable biosensors are quite costly requiring specialist surgeons to implant them. Another major challenge of implantable sensors is the power requirement which is yet to be addressed. Additionally issues have been raised about communication strategy and network difficulties [36]. Then, avoidance of a complicated surgery in case of a certain home-based monitoring is still underway since a readily implantation and explantation of the implantable device is highly expected. This aspect also demands that the devices should be extremely small to get adjusted in the implementation spot through unprecedented miniaturization of disparate functional components. Furthermore, after a period of treatment, implantable sensor devices remain no longer managed which subsequently results in surgical retrieval procedure most often which causes biological, physical and economical bundles on patients. Technical efforts are required to address this issue of managing the devices efficiently. Furthermore, biocompatible materials must be chosen carefully with the proper fabrication methods since Foreign Body Response (FBR) [37] needs to be considered in this aspect implying the facet that biocompatibility and lifetime are prime challenging factors in healthcare applications other than power constraints. In the light of ensuring abundant power supply inside body area network; substitute of battery based systems are being investigated to power up the implantable biosensors from body environment itself [38].

Now, some common challenges in terms of data analytics and communication along wireless sensor network are as follows:

- Preserving data integrity is a major challenge; as the data stored in the database consist of thousands of records of different patients and medications are proposed to the concerned patients by analysing those data; any changes in the data must be maintained and protected. Another issue is privacy and confidentiality of sensitive health information; as the data are transported through different wireless connections and via the Internet; it is prone to malicious attacks. Thus, preserving the privacy of patients sensitive health information with efficient encryption techniques impose a huge challenge.
- Ensuring data reliability in terms of data acquisition, transmission and processing is an everlasting open challenge since decades in wireless sensor technology. The system must be able to handle any unwanted erroneous acquired data; path fading, interference and instability of multi-hop routing paths of transmission medium due to user mobility. Because, any erroneous result might lead to adversarial consequences providing inaccurate diagnosis which might end up to death as well. In this context supporting user mobility constitutes another significant challenge.
- Furthermore, multimodal data fusion impose a greater challenge in this wireless sensor network, where any new addition or deletion in the network infrastructure must be handled and integrated spontaneously without disturbing normal flow of information. Conflicting output is major concern in this aspect owing to repetitive false alarms because of variant interpretation of similar activities from unrelated sensor platforms. Then, subsequent increase of computational time owing to the overhead achieved during data processing in case of wearable and context-aware based data fusion owing to unnecessary complexity in the implementation of fusion algorithm.

15.4 Conclusion

Comprehensive health information of an individual can be retrieved through continuous real time monitoring over a period of time. The contribution of computer vision and sensors in this sector in healthcare is enormous. This chapter addressed current advancements with corresponding challenges ahead of exploiting computer vision and sensor technology in terms of real time monitoring in healthcare. The enormous development of the technologies in the past few decades have contributed to prospecting different aspects of healthcare. With the advancement of technologies and changes in nature of health problems with time; new challenges always need to be faced in terms of alleviating those problems. Irrespective of the difference in applicability and nature; both the technologies need to come to terms with device anomalies and architectural constraints owing to environmental constraints; reliability, compatibility, fault tolerance and security issues of health devices; diverse variance in user requirements; issues regarding ease of use, privacy and safety concerns of users. With a view to providing remarkable innovation and extenuating the continuous challenges coming with those developments; vision and sensor technologies are expected to contribute significantly in healthcare practice and research.

References

1. Ahad, M.A.R., Kobashi, S, Tavares, J.M.R.: Advancements of image processing and vision in healthcare (2018)
2. Sathyanarayana, S., Satzoda, R.K., Sathyanarayana, S., Thambipillai, S.: Vision-based patient monitoring: a comprehensive review of algorithms and technologies. J. Ambient Intell. Humaniz. Comput. **9**(2), 225–251 (2018)
3. Huynh, T.-P., Haick, H.: Autonomous flexible sensors for health monitoring. Adv. Mater. **30**(50), 1802337 (2018)
4. Gani, A., Gribok, A.V., Lu, Y., Ward, W.K., Vigersky, R.A., Reifman, J.: Universal glucose models for predicting subcutaneous glucose concentration in humans. IEEE Trans. Inf Technol. Biomed. **14**(1), 157–165 (2009)
5. Bobrie, G., Postel-Vinay, N., Delonca, J., Corvol, P.: Self-measurement and self-titration in hypertension: a pilot telemedicine study. Am. J. Hypertens. **20**(12), 1314–1320 (2007)
6. Pervez, M.A., Silva, G., Masrur, S., Betensky, R.A., Furie, K.L., Hidalgo, R., Lima, F., Rosenthal, E.S., Rost, N., Viswanathan, A., et al.: Remote supervision of IV-tPA for acute ischemic stroke by telemedicine or telephone before transfer to a regional stroke center is feasible and safe. Stroke **41**(1), e18–e24 (2010)
7. Petr, N.: Smartphones for in-home diagnostics in telemedicine. Int. J. Biomed. Biol. Eng. **6**(1), 9–13 (2012)
8. 20+ Emotion Recognition APIs That Will Leave You Impressed, and Concerned | Nordic APIs |, December 2015, section: blog. https://nordicapis.com/20-emotion-recognition-apis-that-will-leave-you-impressed-and-concerned/
9. Emotion Recognition Software SHORE: Fast, Reliable and Real-time Capable. https://www.iis.fraunhofer.de/en/ff/sse/imaging-and-analysis/ils/tech/shore-facedetection.html
10. Google Glass Helps Kids With Autism Navigate Emotions of Others, August 2018, section: Consumer News. https://consumer.healthday.com/cognitive-health-information-26/autism-news-51/google-glass-helps-kids-with-autism-navigate-emotions-of-others-736413.html
11. Tian, Y.-I., Kanade, T., Cohn, J.F.: Recognizing action units for facial expression analysis. IEEE Trans. Pattern Anal. Mach. Intell. **23**(2), 97–115 (2001)
12. Home. https://www.signall.us/
13. Uni is the world's 1st tech to convert sign language into spoken language in real time!, July 2018. https://newzhook.com/story/19304/
14. Tsuboi, K.: Sign-language translator uses gesture-sensing technology. https://www.cnet.com/news/sign-language-translator-uses-gesture-sensing-technology/
15. Falls. https://www.who.int/news-room/fact-sheets/detail/falls
16. Noury, N., Fleury, A., Rumeau, P., Bourke, A.K., Laighin, G., Rialle, V., Lundy, J.: Fall detection-principles and methods. In: 29th Annual International Conference of the IEEE Engineering in Medicine and Biology Society, pp. 1663–1666. IEEE (2007)
17. The Gold Standard for the Bionic Eye. https://www.hopkinsmedicine.org/news/articles/gold-standard-for-bionic-eye
18. Get Effective Computer Vision Consulting and R&D from RSIP Vision. https://www.rsipvision.com/
19. Home. https://www.gausssurgical.com/
20. Choi, P.J., Oskouian, R.J., Tubbs, R.S.: Telesurgery: past, present, and future. Cureus 10(5) (2018)
21. Leo, M., Medioni, G., Trivedi, M., Kanade, T., Farinella, G.M.: Computer vision for assistive technologies. Comput. Vis. Image Underst. **154**, 1–15 (2017)
22. Artificial Intelligence Platform Screens for Acute Neurological Illnesses at Mount Sinai | Mount Sinai - New York. https://www.mountsinai.org/about/newsroom/2018/artificial-intelligence-platform-screens-for-acute-neurological-illnesses-at-mount-sinai
23. Nakata, N.: Recent technical development of artificial intelligence for diagnostic medical imaging. Jpn. J. Radiol. **37**(2), 103–108 (2019)

24. Kim, J., Campbell, A.S., de Ávila, B.E.-F., Wang, J.: Wearable biosensors for healthcare monitoring. Nat. Biotechnol. **37**(4), 389–406 (2019)
25. Evangeline, C.S., Lenin, A.: Human health monitoring using wearable sensor. Sens. Rev. **39**(3), 364–376 (2019)
26. Angelov, G.V., Nikolakov, D.P., Ruskova, I.N., Gieva, E. E., Spasova, M.L.: Healthcare sensing and monitoring. In: Enhanced Living Environments, pp. 226–262. Springer (2019)
27. Koydemir, H.C., Ozcan, A.: Wearable and implantable sensors for biomedical applications. Ann. Rev. Anal. Chem. **11**, 127–146 (2018)
28. Guk, K., Han, G., Lim, J., Jeong, K., Kang, T., Lim, E.-K., Jung, J.: Evolution of wearable devices with real-time disease monitoring for personalized healthcare. Nanomaterials **9**(6), 813 (2019)
29. Alizadeh, A., Burns, A., Lenigk, R., Gettings, R., Ashe, J., Porter, A., McCaul, M., Barrett, R., Diamond, D., White, P., et al.: A wearable patch for continuous monitoring of sweat electrolytes during exertion. Lab Chip **18**(17), 2632–2641 (2018)
30. Huang, H., Su, S., Wu, N., Wan, H., Wan, S., Bi, H., Sun, L.: Graphene-based sensors for human health monitoring. Front. Chem. **7**, 399 (2019)
31. Elsherif, M., Hassan, M.U., Yetisen, A.K., Butt, H.: Wearable contact lens biosensors for continuous glucose monitoring using smartphones. ACS Nano **12**(6), 5452–5462 (2018)
32. Chen, C., Wang, J.: Optical biosensors: an exhaustive and comprehensive review. Analyst **145**(5), 1605–1628 (2020)
33. Gray, M., Meehan, J., Ward, C., Langdon, S.P., Kunkler, I.H., Murray, A., Argyle, D.: Implantable biosensors and their contribution to the future of precision medicine. Vet. J. **239**, 21–29 (2018)
34. Clausen, I., Glott, T.: Development of clinically relevant implantable pressure sensors: perspectives and challenges. Sensors **14**(9), 17 686–17 702 (2014)
35. De Santis, M., Cacciotti, I.: Wireless implantable and biodegradable sensors for postsurgery monitoring: current status and future perspectives. Nanotechnology **31**(25), 252001 (2020)
36. Dahiya, A.S., Thireau, J., Boudaden, J., Lal, S., Gulzar, U., Zhang, Y., Gil, T., Azemard, N., Ramm, P., Kiessling, T., et al.: Energy autonomous wearable sensors for smart healthcare: a review. J. Electrochem. Soc. **167**(3), 037516 (2019)
37. Means, A.K., Dong, P., Clubb, F.J., Friedemann, M.C., Colvin, L.E., Shrode, C.A., Coté, G.L., Grunlan, M.A.: A self-cleaning, mechanically robust membrane for minimizing the foreign body reaction: towards extending the lifetime of sub-q glucose biosensors. J. Mater. Sci.—Mater. Med. **30**(7), 79 (2019)
38. Wu, T., Redouté, J.-M., Yuce, M.R.: A wireless implantable sensor design with subcutaneous energy harvesting for long-term IoT healthcare applications. IEEE Access **6**, 35 801–35 808 (2018)

The manufacturer's authorised representative in the EU is Springer Nature Customer Service Centre GmbH, Europaplatz 3, 69115 Heidelberg, Germany. If you have any concerns regarding our products, please contact ProductSafety@springernature.com

Printed and bound by CPI Group (UK) Ltd, Croydon, CR0 4YY
30/09/2026
02225687-0008